# FISH PHYSIOLOGY

*Volume VI*

*Environmental Relations*
*and Behavior*

# CONTRIBUTORS

GERARD P. BAERENDS

ARTHUR L. DeVRIES

F. E. J. FRY

HENRY GLEITMAN

ARTHUR D. HASLER

W. S. HOAR

P. W. HOCHACHKA

D. J. RANDALL

PAUL ROZIN

HORST O. SCHWASSMANN

G. N. SOMERO

# FISH PHYSIOLOGY

*Edited by*

**W. S. HOAR**

DEPARTMENT OF ZOOLOGY
UNIVERSITY OF BRITISH COLUMBIA
VANCOUVER, CANADA

*and*

**D. J. RANDALL**

DEPARTMENT OF ZOOLOGY
UNIVERSITY OF BRITISH COLUMBIA
VANCOUVER, CANADA

*Volume VI*

## Environmental Relations and Behavior

ACADEMIC PRESS
New York   San Francisco   London   1971
A Subsidiary of Harcourt Brace Jovanovich, Publishers

ACADEMIC PRESS, INC.
111 Fifth Avenue, New York, New York 10003


United Kingdom Edition published by
ACADEMIC PRESS, INC. (LONDON) LTD.
24/28 Oval Road, London NW1 7DD


LIBRARY OF CONGRESS CATALOG CARD NUMBER: 76-84233


PRINTED IN THE UNITED STATES OF AMERICA
82    9 8 7 6 5 4

# CONTENTS

# LIST OF CONTRIBUTORS

Numbers in parentheses indicate the pages on which the authors' contributions begin.

GERARD P. BAERENDS (279), *Zoological Institute, University of Groningen, Haren, Holland*

ARTHUR L. DEVRIES (157), *University of California, San Diego, Scripps Institute of Oceanography, La Jolla, California*

F. E. J. FRY (1), *Department of Zoology, University of Toronto, Toronto, Canada*

HENRY GLEITMAN (191), *Department of Psychology, University of Pennsylvania, Philadelphia, Pennsylvania*

ARTHUR D. HASLER (429), *Department of Zoology, University of Wisconsin, Madison, Wisconsin*

W. S. HOAR (511), *Department of Zoology, University of British Columbia, Vancouver, Canada*

P. W. HOCHACHKA (99), *Department of Zoology, University of British Columbia, Vancouver, Canada*

D. J. RANDALL (511), *Department of Zoology, University of British Columbia, Vancouver, Canada*

PAUL ROZIN (191), *Department of Psychology, University of Pennsylvania, Philadelphia, Pennsylvania*

HORST O. SCHWASSMANN (371), *Department of Psychology, Dalhousie University, Halifax, Nova Scotia, Canada*

G. N. SOMERO (99), *Department of Zoology, University of British Columbia, Vancouver, Canada*

# PREFACE

Volume VI of this treatise is concerned with the physiological and behavioral responses of fish to a variety of environmental situations. The approach is not a detailed analysis of parts of the animal, as in previous volumes, but treats the fish as an integrated unit interacting with its environment. Chapters with a similar approach have appeared in previous volumes (e.g., M. S. Gordon, Hydrostatic Pressure, Volume IV, pp. 445–464); many contributors to Volumes I–V have discussed the significance of their observations in terms of the whole animal or even animal populations. In general, however, previous volumes have been directed toward an understanding of the physiology of systems within the animal. Volumes I–V and Volume VI are therefore complementary; the first five volumes are primarily concerned with an analysis of the parts while Volume VI is an overview of the whole animal in a changing and complex environment.

No special consideration of the responses of fish to polluted environments is included. A detailed analysis of this subject was considered beyond the terms of reference of this treatise.

The first three chapters of this volume examine physiological and biochemical adaptations of fish to a variety of environments. In some respects these chapters also reflect somewhat different approaches toward an understanding of how animals adapt to their environments. The next three chapters discuss the extensive literature on behavioral studies of fishes. Chapter 7 reviews the fascinating problem of fish migration and orientation. The final chapter is an appendix to all six volumes and presents what we consider useful information to those interested in experimenting with fishes.

In conclusion we reiterate our hope that the six volumes of this treatise on "Fish Physiology" will prove a ready and useful source of information for those interested in this diverse group of animals.

W. S. HOAR
D. J. RANDALL

# CONTENTS OF OTHER VOLUMES

# THE EFFECT OF ENVIRONMENTAL FACTORS ON THE PHYSIOLOGY OF FISH

*F. E. J. FRY*

## I. INTRODUCTION

The study of animal function is organized more or less under three heads which in everyday language are, as applied to a machine, what it can do, how it works, and what makes it go. Insofar as fields of study can be classified in biology these divisions of the subject are ordinarily considered to be autecology, physiology, and biochemistry, with a great deal of individual taste governing the label any particular worker may choose for himself. The subject of this chapter is what fish can do in relation to their environment and therefore largely autecology.

The organism can be taken to be an open system (von Bertalanffy, 1950), suitably walled off from its milieu, through which energy flows by appropriate entrances and exits. The organism uses this energy to maintain and extend its being. The energy comes from the environment, and further the environment sets to a large degree the conditions under which the organism uses the energy it has assimilated, but all organisms have regulatory powers and bargain with the environment in regard to the extent they make use of the energy they have gained. Such bargaining involves the use of some energy for regulation against the environment to free the rest for the organism's other activities.

Thus the prime subject of this chapter will be the action of the environment on metabolism and the effects of this action on the activity of the organism.

## A. Metabolism and Activity

A careful distinction will be made here in the usage of the terms metabolism and activity. Metabolism as used here is catabolism as ordinarily understood, that is, the sum of the reactions which yield the energy the organism utilizes. Activities are what the organism does with the energy derived from metabolism. Thus activities are such processes as running or fighting or other manifestations of the energy released by metabolism. These manifestations are not all movements; growth is activity and so is excretion. By this definition anabolism is an activity.

While the influence of the environment is on metabolism, the effect of that influence is displayed through the activity of the organism whose metabolism has been so affected.

The purpose of belaboring the distinction between metabolism and activity here is not to introduce a novel thought, for these generalities are what we all recognize, but rather to provide a consistent treatment of the whole organism in relation to its biochemical basis. Activity is fundamentally the result of transformation of energy from one form to another and the application of that energy to a given performance. Two generalizations arise from these circumstances. First, all the energy released will not likely be applied to the final outcome which is the object of its release. The organism will take its levy for its maintenance as a system, and there will be the ancillary costs of supply and disposal of the metabolites which pass through the system. Second, performance is qualitatively different from the power which produces it and there need not be any simple proportionality between the measures taken of the two.

These circumstances will be recognized here by considering the difference between resting and active metabolic rates, which will be termed "scope for activity," as being the power available for activity, and, where appropriate, relations will be sought between activity and scope. These concepts are, of course, regularly applied to homoiotherms by those interested in animal production (e.g., Brody, 1945) where the costs and consequences of thermoregulation are so prominent and have been simply transferred to poikilotherms over the past quarter century.

## B. Measurement of the Metabolic Rate

The metabolic rate of fish has almost universally been measured by determining oxygen consumption. The fundamental method of measuring heat production has been applied (e.g., Davies, 1966) but probably never will be suitable for measurements required for environmental physiology.

It cannot be assumed that all fish are obligate aerobes and that a measure of oxygen consumption is always a measure of the metabolic rate. Coulter (1967) reported that extensive catches of fish are regularly taken in oxygen-poor water in Lake Tanganyika under circumstances which suggest they are resident there. The goldfish (Kutty, 1968a) can live for months with a respiratory quotient of 2, and there are the dramatic reports of Blažka (1958) and Mathur (1967) on extensive survival of fish under completely anaerobic conditions.

The newer methods of easy determination of carbon dioxide in water should soon be rapidly applied to the determination of the respiratory quotient (e.g., R. W. Morris, 1967) although, as yet, the margin of error in them requires to be narrowed. At present the error inherent in the new methods is of the order of twice that for determinations with the Van Slyke apparatus or by distillation (e.g., Maros *et al.*, 1961).

Three levels of metabolism will be distinguished here. Following the usage now current among a number of fisheries workers, these will be termed "standard," "routine," and "active" levels of metabolism. Standard metabolism is an approximation of the minimum rate for the intact organism. It is preferably determined as the value found at zero activity by relating metabolic rate to random physical activity in fish in the post-absorptive state (e.g., Beamish and Mookherjii, 1964; Spoor, 1946). The fish should be able to swim freely in the respiration chamber while protected from outside disturbance and should have been in the chamber long enough to recover from the effects of transfer to it. It may also be important that the chamber is supplied with water from the aquarium in which the fish was living. Foreign water may provide disturbing chemical stimuli or perhaps more importantly may lack the familiar chemical milieu of the home tank. Standard metabolism can also be determined by extrapolation to zero activity from determinations at various levels of forced activity (e.g., Brett, 1964). The routine rate of metabolism is the mean rate observed in fish whose metabolic rate is influenced by random activity under experimental conditions in which movements are presumably somewhat restricted and the fish protected from outside stimuli. The value has usually been given only for the normal working hours of the experimenter (e.g., Beamish, 1964a). Active metabolism is the maximum sustained rate for a fish swimming steadily.

Standard and active metabolism are determined to permit calculation of scope for activity. Routine metabolism is largely to be considered as a measure of the degree of random activity and is discussed in Section VI.

Various types of apparatus used in such determinations are discussed below, together with comments on experimental precautions. Most measures of metabolism have been measurements of routine metabolism. Standard and active metabolism have not yet often been measured, and the limits of these have been still less often well worked out.

Figure 1 shows determinations of the metabolic rate of the goldfish, *Carassius auratus*, at 20°C by various workers in the same laboratory at various times over a number of years. Figure 1A shows oxygen consumption and Fig. 1B $CO_2$ output. For the sake of clarity, Kutty's points for oxygen consumption are not shown but the number of his readings under forced activity can be inferred from the number of points in Fig. 1B since he made determinations of the respiratory quotient. His curve for routine metabolism is based on 35 points. Smit's data (1965) are illustrative material based on a single fish. There are three salient points to be considered in Fig. 1A:

(1) The dots which represent oxygen consumption during routine activity show how high the metabolic rate can go when a fish is randomly active within the confines of a respiration chamber. The routine respiration rate as shown by the mean of these values approaches half the active respiration rate.

(2) An extrapolation from either forced or random activity to zero

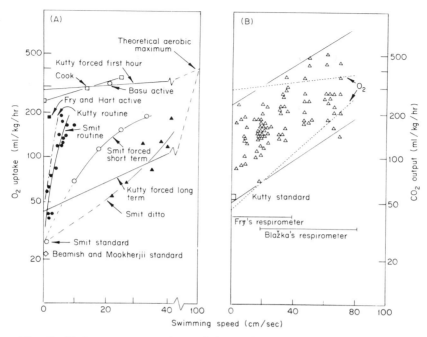

**Fig. 1.** Various measures of metabolism of the goldfish, *Carassius auratus,* under conditions of random and forced activity at 20°C. (A) Oxygen consumption; (B) carbon dioxide production. Data of Basu (1959), Beamish and Mookherjii (1964), C. N. Cook (personal communication in Fry, 1967), Kutty (1968a), and Smit (1965). Kutty's data for oxygen consumption under forced activity are omitted from A but are based on the same number of observations as are shown by the points in B. For further explanation see text.

activity gives a similar value for standard metabolism if the fish are not disturbed.

(3) A major problem in the measurement of metabolism of fish is the wide range of values that may be obtained for a given fish in the same state of overt physical activity. A fish resting quietly may be consuming oxygen at a given rate from its standard to well over half its maximum level. Ordinarily a fish need only be moved from its bath to the respiration chamber to elicit almost its active metabolic rate. Cook's data illustrate this phenomenon in Fig. 1. Her data are for the first 15 min after the fish were transferred from their acclimation tank to the respiration chamber. The effect of swimming is to somewhat increase oxygen uptake in an excited fish, possibly by facilitating venous return, but more than half the extrapolated maximum rate of oxygen consumption associated with the most vigorous sustained activity can be displayed by the goldfish with

no overt movement at all. Kutty (1968a) found that under forced activity the metabolic rate of goldfish fell after about 3 hr from these initially high values to a steady state which more truly reflected the degree of swimming effort, but his data were for well-trained fish. Smit (1965), who did not train his fish, found the rate declining sometimes over at least 8 hr.

The active rate most frequently reported in the literature (e.g., Basu, 1959; shown in Fig. 1) has been an acute measurement at a moderate rate of swimming speed. Only the more recent authors (Brett, 1964; Kutty, 1968a; Smit, 1965) have pushed their measurements to the point where the maximum sustained rate could be estimated. Similarly, most earlier workers concerned with standard metabolism (e.g., Fry and Hart, 1948) approximated that value by taking the lowest point in the daily cycle. Beamish (1964a) and Fry and Hochachka (1970) gave comparisons of many of the remaining published values for the metabolic rate of the goldfish.

Routine metabolism has, of course, been the level of metabolism most frequently reported, usually with the implication that the fish were in a quiescent state when the measurements were made. Most fish show a daily cycle of activity, so that the degree of quiescence depends to a considerable extent on the time of day when the measurements are made in spite of all the usual precautions to confine the fish within an appropriately limited space and to protect it from disturbance (e.g., Kausch, 1968). The significance of measuring routine metabolism is that it is a reflection of random activity, the degree of which reflects response to the directive effects of the environment. Routine metabolism is unsuitable as an approximation of standard metabolism because of the high metabolic rate which may be achieved by a fish within a restricted space. In Fig. 1 the peak routine rate for the goldfish is approximately six times the standard rate. These rates reflect presumably the cost of continual small accelerations as the fish starts to swim and then checks its progress.

The emphasis above has been on physical activity as a variable influencing the metabolic rate, together with what has been termed "excitement." There are of course other well-known influences, the two major ones being the cost of assimilation and of ion-osmoregulation. The latter ugly term seems necessary because we have not distinguished the cost of transfer of ions from that of the transfer of water. The cost of assimilation has been eliminated by a fast of some 48 hr (Beamish, 1964d). In the species Beamish investigated assimilation accounted for about 50% of the inactive metabolism. In general, major costs of assimilation have been removed from most measurements since it is usual to fast the animals, if only to avoid feces in the respiration chamber. In fresh water or salt water the cost of ion-osmoregulation in the rainbow trout, as determined by subtracting the minimum rate in an isosmotic dilution of seawater, is 20–30% of the metabolic rate (Rao, 1968). Such a cost, however, is a proper fraction of standard metabolism.

In addition to such costs of regulation and activity as may be included in the resting metabolism, there may be changes in the residual (standard) metabolism with season, a subject which has been little explored, although Beamish (1964e) showed an approximate doubling of the standard metabolic rate of the eastern brook trout from its low to its high in the annual cycle.

Figure 1B shows the metabolism of the goldfish as carbon dioxide output. The chief purpose of the comparison is to show to what extent the metabolic rate of this species under these circumstances can be expressed by oxygen consumption alone. The points shown for carbon dioxide output were all obtained under forced activity. Goldfish which are randomly active in water high in oxygen have a respiratory quotient ($RQ$) of approximately unity (Kutty, 1968a). Under forced activity at air saturation the $RQ$ is again unity or lower for the long term and also excitement alone can be satisfied by aerobic respiration. However, as the swimming speed is increased an increasingly large segment of the upper symbols (excitement plus activity) falls above the boundary of the area encompassed by the values for oxygen consumption. Thus, metabolism at some acute values in the goldfish cannot be estimated by oxygen consumption although all the long-term values at air saturation can be.

## C. The Relation of Metabolism to Size and Physical Activity

Most discussions of metabolism here will be divorced from a consideration of the size of the fish involved by expression as rate per unit weight.

In general the relation of metabolism to body weight has been described by the equation $y = ax^b$ where $y$ is the rate of metabolism and $x$ is the body weight (often the formula is used in the form $y/x = ax^{1-b}$). The exponent $b$ has usually been found to be of the order of 0.8 (Paloheimo and Dickie, 1966a; Winberg, 1956), and most examples have been for routine metabolism. There have been some notable exceptions to this rule. The cichlids, in particular (Ruhland, 1965; R. W. Morris, 1967), have shown some exponents of the order of 0.5 as has also been found for other species on occasion (e.g., Wells, 1935; Barlow, 1961). An exponent of unity, which indicates the metabolic rate is weight proportional, has also been found from time to time. Beamish (1964a) found the standard metabolic rate of brook trout, *Salvelinus fontinalis*, to be weight proportional. Both Brett (1965) and Rao (1968) found active metabolism to be essentially weight proportional in the two salmonids they investigated. Job (1955), however, found a decrease in the exponent for active metabolism with increasing temperature. Brett and Rao worked at or below the temperature optimum for their species and probably also stimulated their fish to greater activity. In cases where routine or standard metabolism have been compared at different temperatures the weight exponent has been temperature independent.

While the relative amount of energy the animal can produce or may require in relation to its size is of great importance in determining its relation to the environment there are no established explanations for the various differences found in the magnitude of the exponent $b$, and indeed the reality of many of these differences is in question. Glass (1969) questioned whether the best values for the exponent have been calculated in most instances. It has been the practice to fit a straight line to a logarithmic transformation of the data, like the treatment shown in Fig. 32. Glass demonstrated in a series of examples that a better fit to the points can be obtained by using the equation in its arithmetic form. The treatment here will be to use the exponent given by the author for any weight corrections, but such corrections will be largely ignored. It seems important however that the notion, now rather thoroughly fixed in the literature, that the general value for $b$ is approximately 0.8 should not yet be allowed to become a dogma. This point is emphasized in Section V where a special case of a change in the exponent in relation to osmoregulation is dealt with (Fig. 32).

There has been only one thorough investigation into the cost of swimming, that of Brett (1965) for the sockeye salmon. His data can be interpreted (Fig. 2) as showing that the metabolic cost of swimming increases approximately as the square of the swimming speed, a relation suggested earlier by fragmentary data (Fry, 1957). The restricted data of Kutty and Rao also support Brett (Fig. 2), as do the data for the haddock, *Melanogrammus aeglefinus* (Tytler, 1969).

Brett's data (1964) for fingerling sockeye indicate that the cost of swimming is essentially independent of temperature, a conclusion which differs from that of Rao (1968) who indicated a somewhat greater cost at a higher temperature, so the matter is still in question. The relation of the metabolic rate to speed of swimming needs further examination before we can generalize. While the main example in Fig. 2 is statistically strong and there are subsidiary data of a similar nature in the literature, there is at least one contradictory series. From the power relation between scope and swimming speed in Fig. 2 and the relation of scope to

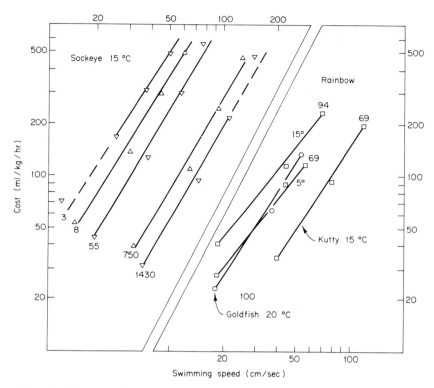

**Fig. 2.** The cost of swimming in the sockeye salmon, *Oncorhynchus nerka*, rainbow trout, *Salmo gairdneri*, and the goldfish. Data of Brett (1965), Kutty (1968a), and Rao (1968). The cost shown is scope for activity, i.e., metabolism at activity stated minus standard metabolism. The numbers associated with the curves are weights in grams unless indicated as degrees. Note Kutty's data show the same slope with a different intercept from Rao's.

size given in Brett (1965), it follows that the maximum speed of swimming at any size is related to approximately $L^{0.5}$, where $L$ is the total length, as Brett showed empirically. However, Pavlov *et al.* (1968), working with small minnows, found that the maximum swimming speed was a simple multiple of the length of the fish over lengths of 8–35 mm, which infers either a completely different relation of scope for activity to size in these small fish or a different relation of scope to swimming. speed. These workers used a gravity head of water to provide their water flow rather than the more turbulent recirculating systems of Brett and Kutty and others whose data follow the power relation. Perhaps smooth-skinned fish swimming in still water have an economy of effort not to be found in rapid streams or fine turbulence activity apparatus, or perhaps the relation for small fish differs from that for larger ones.

## D. Apparatus for the Determination of Metabolic Rate

It seems unlikely that any useful determination of the metabolic rate of fish can now be made which is not accompanied by a measure of physical activity. There are two broad approaches to this problem, one is to record the random activity of an individual free from outside disturbance while measuring its metabolic rate. The other approach is to force the fish to swim at certain constant speeds while making the measurement. Wohlschlag (1957) has somewhat combined the two approaches by driving a rotating chamber so as to counter the random swimming speed of the fish.

In the last decade several methods of measurement, or at least registration, of random activity have been reported following the pioneering work of Spoor (1946) and other workers cited in Fry (1958). The present most quantitative practical method, which under proper design should measure the energy output, is to measure the degree of turbulence of the water in the respiration chamber as affected by movements of the fish. Such measurements can be readily made by a heat loss flowmeter (Beamish and Mookherjii, 1964; Dandy, 1970; Heusner and Enright, 1966; Kausch, 1968). A convenient but less quantitative method is to equip the respiration chamber with photocells to measure the number of circuits the fish may make around it (e.g., Smit, 1965; Peterson and Anderson, 1969a). Finally, the entry of fish into an echo-sounder beam (Muir et al., 1965) offers the possibility of integrating accelerations if proper geometry of the chamber and suitably refined recording apparatus can be combined. The method of de Groot and Schuyf (1967, which see also for other references) in which the fish carry a magnet offers similar possibilities.

Undoubtedly, the best form of apparatus in which to induce the active respiration of fish is a tunnel through which the water is recirculated. Probably the best-engineered example of such an apparatus is that described by Brett (1964, see also Mar, 1959) and illustrated by Phillips, Volume I, this treatise. Rotating chambers have been used extensively in the author's laboratory because of their great simplicity and convenience, but they are suitable only for approximating the active metabolic rate. A most interesting circumstance is that it is apparently impossible to induce a fish to swim as fast in a closed rotating chamber as in a tunnel or a flume. Presumably the discrepancy is a result of the

Taylor effect (Taylor, 1923, and earlier). A body in rotating water is subject not only to form and surface drag as it is in a tunnel but carries with it a cell in which water circulates vertically from the body to the water surface. It does seem likely though that the Taylor effect can be eliminated by allowing a free meniscus at the inner and outer margins as indicated in Fig. 3.

The problem in the design of chambers for active respiration is to have a uniform cross-sectional flow through the section where the fish is held; thus, a measure of the speed of the water provides a measure of the counterspeed of the fish. A secondary problem is to keep the hydrodynamic efficiency high for quietness, ease of speed control, and reduction of heat input while keeping the circulating volume low to minimize lag in measurement. Blažka's (Blažka et al., 1960) apparatus, with appropriate modifications to improve the flow, promises to be a convenient form of apparatus for the study of active metabolism but has not yet been given appropriate engineering treatment.

Chambers for the study of routine and standard metabolism have taken a variety of forms, one even with an appropriate posterior upturn to accommodate the heterocercal tail of sturgeon (Pavlovskii, 1962), but the majority have been a horizontal cylinder (e.g., Halsband and Halsband, 1968). The author favors a chamber in which the bottom is flat so that a fish resting on the bottom has the least lateral restriction (e.g., Bullivant, 1961). Chambers which are circular in cross section do not have this feature. The top of the chamber is preferably arched to permit the easy removal of bubbles. If activity is to be measured as well as oxygen consumption, the geometry of the chamber should take that factor into consideration. Various chambers are shown in Fig. 3.

The various electrochemical methods now in vogue for measuring oxygen in water have greatly simplified the measurement of metabolism although, as mentioned above, the measurement of production of carbon dioxide cannot yet be carried out as conveniently with an equivalent accuracy. The Van Slyke apparatus still seems to be the basic tool for the measurement of carbon dioxide. The distillation method of Maros et al. (1961) has about the same precision as the Van Slyke and can probably be readily mechanized to become at least semiautomatic. No doubt gas chromatography and infrared analysis are on the verge of giving results precise enough so that the usual small difference between two relatively large values for total carbon dioxide in water will be determined to the same accuracy as can the comparable difference between two determinations of oxygen content. The use of decarbonated water adds to the accuracy when determining the respiratory quotient (Kutty, 1968a).

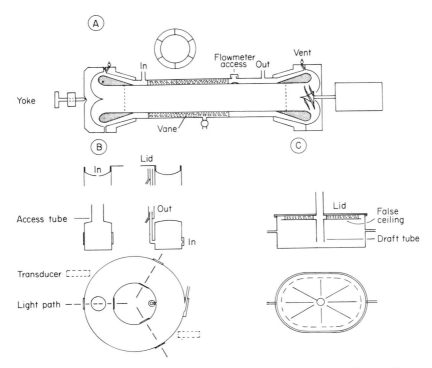

Fig. 3. Three types of apparatus for the measurement of standard and active metabolism. (A) Modified Blažka chamber for the determination of active metabolism (Blažka et al., 1960) essentially as described by Smith and Newcomb (1970). The drive indicated is a variable speed, solid-state controlled motor. (B) Annular chamber for standard or active metabolism. In the configuration shown, standard metabolism is determined by monitoring activity by lights and photocells (e.g., Smit, 1965; Peterson and Anderson, 1969a). Muir et al. (1965) immersed their annular chamber in a second water bath in which echo sounder heads were placed as indicated, whereby activity was monitored by the Doppler effect. An annular chamber can be rotated for forced activity. In that case a lid allowing a small open water surface at the inner and outer peripheries (as indicated in the upper section) promotes the best reaction from the fish. (C) Chamber for use with heat loss flowmeter. Currents induced by the fish are in general parallel to the walls of the chamber and constrained by it. The resulting constraint to a curved path induces a current down the draft tube via the false ceiling (cf. Beamish and Mookherjii, 1964; Mathur and Shrivastava, 1970; Dandy, 1970). The draft tube is not essential (Kausch, 1968). Brett's apparatus for the measurement of active metabolism is illustrated in Fig. 1, p. 420, Vol. I, this treatise.

One major nuisance in respiration chambers is the formation of gas bubbles which can provide a substantial reservoir for oxygen. It is well, therefore, to slightly undersaturate the water supply to a respiration chamber. If sufficient headroom is available in the laboratory, undersaturation can be easily achieved by passing the water down over Raschig rings in an exchange column subjected to a vacuum, say, of 50 cm $H_2O$. Otherwise the water supply can be heated 2° or 3° while it passes through an aerator and then be brought back to the desired temperature. Mount (1964) described a convenient apparatus for degassing water by vacuum which also can be used to control initial oxygen concentration.

Whenever the relation between activity and metabolism is the concern, it is necessary to pay attention to the lag in measurement resulting from the volume of the chamber. This problem can be met by using a closed system whereby activity and oxygen consumption are integrated over a fixed period, which in general is considerably longer than the lag between activity and oxygen consumption within the organism. However, there are limitations to the use of closed systems for the measurement of oxygen consumption, the greatest of which is probably that something like half the time for experiment must be wasted in flushing the chamber. There is the further problem that the action of the valves for opening and closing the chamber may be a source of disturbance to the fish. The objection sometimes put forward that products of metabolism accumulate in closed systems has no real validity since an open system is in effect merely an enlarged closed one.

Lag may be accounted for in an open system by the use of a factor to correct for changes in oxygen content in the water in the chamber during the period of measurement. The formula for calculating the oxygen consumption in a constant flow system for a period of time $t$ is

$$F(y_1 - \bar{y}_2) + V(y_{2,0} - y_{2,t})$$

where $F$ is the volume of flow through the chamber, $V$ is the volume of the chamber, $y_1$ is the inlet oxygen concentration taken to be constant throughout the period, and $y_2$ is the outlet concentration. The subscripts 0, $t$, indicate readings at the beginning and end of the period.

The correction factor $V(y_{2,0} - y_{2,t})$ in this formula has unfortunately been overlooked by most authors (e.g., Kausch, 1968).

The problem of lag under the conventional method of maintaining a constant rate of flow and monitoring change in oxygen, while amenable to correction as indicated above, still requires time-consuming tabulation or an expensive digital output. With present methods of electronic control it appears feasible to reverse the approach so that oxygen is maintained constant in the respiration chamber by a control which monitors the level there and regulates the supply of water, analogous to the method so often employed to measure the respiration of air breathers. Oxygen consumption would then be recorded in terms of volume of water pumped and chamber lag would be eliminated. Such a chamber, of course, probably could not be provided with a turbulence meter, but activity could be monitored by one of the other methods available.

Systems such as those of Scholander *et al.* (1943) and Ruhland and Heusner (1959), where oxygen pressure is kept constant in a gas phase over water, still have at least the same lag as is found in the constant flow system but offer the advantage of a simple determination of oxygen added to the system. The correction term can be found by inserting an auxiliary oxygen electrode into the water, as indeed Ruhland (1967) has already done for another purpose.

### E. Acclimation

It is well accepted that an organism is not the same organism, even from day to day, but that its physiological state is continuously being modified by its environmental history. These effects of the environment during the individual's life will be taken into account as far as possible—which can still, however, only be done in a somewhat rudimentary fashion. The two terms, "acclimation" and "acclimatization" will be applied to the conditioning of the individual by its experience. Acclimation will be used to designate the process of bringing the animal to a given steady state by setting one or more of the conditions to which it is exposed for an appropriate time before a given test. Such conditions may be fixed or cycled depending on the circumstances. A common practice is to maintain fish at a given constant temperature for such a purpose. The animal is then said to be acclimated to that particular temperature. Tests, say of its ability to swim at that temperature, will show constancy over some days or weeks after acclimation, whereas during acclimation there may have been considerable change. However, while there may be such constancy within a season, fish acclimated to the same temperature in summer may be constant at a different level from those acclimated to the same temperature in winter (Wohlschlag *et al.*, 1968). Thus there may be a major difference in the physiological state of an organism acclimated to a low temperature and one acclimatized to winter conditions, the latter term being reserved here for an organism whose history has been exposure to the total environmental complex throughout its life up to the time of test.

The most significant aspect of acclimatization as opposed to acclimation is that acclimatization allows the organism to acquire an adjustment, say to higher temperatures, in advance of the event if that event is appropriate to the seasonal cycle. Thus, acclimatization provides for anticipatory adjustment as well as reactive adjustment.

In drawing the distinctions above between acclimation and acclimatization, only reversible effects were considered. The modifications of most physiological responses by environmental history that have been investigated have been essentially reversible given sufficient time, but the possibility of irreversible changes remains, especially for influences at points of development where a given growth stanza may be prolonged or curtailed (e.g., Martin, 1949). Thus the rearing temperature has been found to have an influence on the lethal temperature of the guppy, *Poecilia reticulata* (Gibson, 1954), which cannot be eliminated by extensive thermal acclimation at a later date. To the ecologist (V. E. Shelford, personal communication), acclimatization may have also a phylogenetic implication on the subspecific level. In each locality, with its unique environment, a species is subject to different selection pressures as well as to any different ontogenetic influence which bears on the successful individuals.

The aim in the laboratory should be to duplicate the significant ontogenetic influences of acclimatization by suitable acclimations. The hope would be that residual differences then observed among populations would be the phylogenetic aspects of acclimatization.

A still unsolved problem in acclimation is how to condition animals to long-term cyclic changes such as the annual cycle of day length. Responses to such cycles appear to have an inertia which cannot be easily overcome. Moreover, the interactions between such cycles and, say, a constant temperature have not been adequately explored. On the whole it appears better at present to maintain an organism on its normal light cycle and state the season at which the work was done. Workers (e.g., Jankowsky, 1968) are now beginning to add the latter important information to their papers.

## F. A Classification of the Environment

There are two fundamental bases for a classification of the environment, that is, either by its elements according to their identities such as light, heat, and oxygen or by the manner in which the identities may influence organisms. The second method has been chosen here following Fry (1947). The term "factor," apparently introduced by Blackman (1905), from whose powerful exposition it certainly gained its widespread currency, has been commonly employed to designate such a category of effect. The effects of the environment on organisms may be grouped into five categories. These will be designated lethal, controlling, limiting, masking, and directive factors. The first factor restricts the range of the environment in which the organism can exist; beyond this range metabolism is destroyed. The second and third factors govern metabolic rate. The remaining two are exploited by the organism to achieve and maintain its being through organic regulation. These categories are defined and discussed in an introductory fashion below.

While these categories are stated to be categories of effect it must be recognized, as is the case of most classifications, that they are also categories of convenience and imperfect to the degree that this is so. Thus the category of lethal factors, while undoubtedly dealing with lethal effects, is basically set up here to deal with the statistical aspects of mortality without reference to any specific cause of death. Death often ensues because of interaction between controlling and limiting factors and when this is so it is artificial to stop dealing with these factors at the verge of annihilation and to switch to another category. Again in the discussion of controlling factors the organism is considered as having no powers of organic regulation, which again is an evasion of reality. An organism cannot be without regulation.

### 1. Lethal Factors

An environmental identity acts as a lethal factor when its effect is to destroy the integration of the organism. Properly speaking such destruction should be independent of the metabolic rate to be the result of a lethal factor.

The lethal effect of any identity may be separated into two components: (a) the *incipient lethal level*, that level of the identity concerned beyond which the organism can no longer live for an indefinite period of time, and (b) the *effective time*, the period of time required to bring about a lethal effect at a given level of the identity beyond the incipient lethal level.

### 2. Controlling Factors

Controlling factors comprise one of two categories which govern the metabolic rate. What are considered here as controlling factors are what Blackman (1905) termed "tonic effects." Controlling factors govern the metabolic rate by their influence on the state of molecular activation of the components of the metabolic chain.

Those not familiar with the general notion of normal and activated states of molecules in relation to rates of chemical reaction will find a recent general treatment of the subject in the introductory chapter in Johnson *et al.* (1954). Temperature is the most outstanding of the controlling factors.

Controlling factors place bounds to two levels of metabolism. They permit a certain maximum in the absence of a limiting factor through their influence on the rates of chemical reactions. The controlling factors also demand a certain minimum metabolic rate which, it is taken, is necessary to release the energy required for the repair reactions needed to keep the organism in being.

## 3. LIMITING FACTORS

Limiting factors make up the second category of identities that govern the metabolic rate. They are Blackman's "factors of supply" in his original treatment of "limiting factors" and the category to which Liebig's "law of the minimum" applies. Both the term and the concept of the limiting factor have been widely used in this connection. The usage here is simply to restrict the definition to what was the major burden of Blackman's exposition (1905).

Limiting factors operate by restricting the supply or removal of the materials in the metabolic chain. Thus a reduction in the supply of oxygen below a certain level can reduce the metabolic rate, and below that level it can be said that the oxygen supply is limiting.

The effect of a limiting factor is to throttle the maximum metabolic rate permitted by the existing level of controlling factors. Concentrations outside the limiting levels are to be considered as being neutral unless toxic levels are reached.

## 4. MASKING FACTORS

A masking factor is an identity which modifies the operation of a second identity on the organism. An organism achieves all its physiological regulation by the exploitation of masking factors through the channeling of energy by some anatomical device.

For example, deep-sea fishes with swim bladders have pressures of gas in these bladders far in excess of the pressure that could be generated by releasing all the atmospheric gases held in the blood. To make this gas available at the higher pressure the fish exploits a second physical law, namely, the property of dissolved gases to diffuse down a pressure gradient. The rete mirabile that connects the swim bladder gas gland with the general circulation provides a countercurrent path for such diffusion and the circuit, arteriole → gas gland → venule, forms a regenerative loop which accumulates the gas in solution at the gas gland until the final chemical release overcomes the hydrostatic pressure (e.g., Steen, 1963). In this example the essential anatomy is the loop which brings the blood leaving the gas gland in close association with the blood about to enter it. Here the physical arrangement of the blood vessels permits a result which no chemical activity in the gas gland can achieve alone. The energy to drive the fraction of the system which permits the masking factor to operate is provided by the heart. (For details see chapter by Randall, Volume IV, this treatise.)

## 5. DIRECTIVE FACTORS

These allow or require a response on the part of the organism directed in some relation to a gradient of the factor in space or in time.

The directive factors elicit the well-known forced movements (Loeb, 1913, 1918; Fraenkel and Gunn, 1961). They also provide for the animal's guidance in moving about in the environment in relation to physical obstacles and for its interactions with other organisms. The directive factors also trigger physiological responses without the mediation of the senses, as in the effect of photoperiod on the pituitary.

Directive factors operate by the impingement of energy on some appropriate target. The energy absorbed initiates a signal which appropriately channels metabolism into the appropriate response.

## II. LETHAL FACTORS

Lethal factors as used here do not fall in a pure category of effect but constitute rather a section heading under which various common aspects of bioassay can be grouped. In dealing with lethal factors the primary approach will be one of description. Such a "blinkered" consideration is compulsory when dealing with lethal temperatures, which are taken as the main example here, because of our still profound ignorance of the nature of thermal death. In any event description should logically precede and lead to analysis. Description itself, of course, can be analytical and should be so; at least things should be described to the extent that questions can be raised as to the mechanisms underlying the phenomena observed. The causes of environmental death can, for example, be divided immediately into two fundamental categories by observing the rate of dying in relation to the metabolic rate of the organism. The rate of dying at a lethal temperature, for example, is signally independent of the metabolic rate while in lethal oxygen the metabolic rate is almost paramount. The former case where the lethal effect is independent of the metabolic rate is the pure lethal factor, and lethal temperature will be considered here as such a factor. Where the metabolic rate influences the rate of dying, death is usually brought about by the interaction of limiting and controlling factors. Whenever such a case can be recognized, such as the effects of decreased oxygen, an analysis of the interaction is infinitely more valuable than a determination of the lethal level (see Section IV, D). The importance of the distinction above lies in the so-called sublethal effects which might be seen in better perspective if they were termed "prelethal."

If there is no relation of the lethal effect to the metabolic rate, then the division of the effects into zones of resistance and tolerance defined below is meaningful. The incipient lethal, the boundary between the zones of resistance and tolerance, is then a real threshold, and the factor concerned can be taken to no longer exert any direct harmful effect. Indeed, in the case of temperature, the homoiotherms have found the successful evolutionary path to be the one which has led them to live within a few degrees of their upper lethal temperature.

The range of intensity of a given identity, which at some levels has a lethal effect, can be divided into a zone of resistance—over which it will operate to kill the organism in a determinate period of time—and a zone of tolerance—over which the life span of the organism is not influenced by the direct lethal effect of the identity concerned. Thus while the life span of a poikilotherm becomes progressively shorter as temperature increases, since increasing temperature speeds metabolism, temperature is not a lethal factor until the threshold is reached above which there is a drastic change in the length of life. Below that threshold the organism will be said to be in the zone of tolerance, above it to be in the zone of resistance (see Fig. 7, goldfish). The boundary between the two will be taken as being sharp and will be designated the "incipient lethal level," which is ordinarily expressed as the median lethal dose ($LD_{50}$). Dealt with according to the scheme above, events in the zone of resistance are measured according to the principles of time mortality (Bliss, 1935) while the incipient lethal level is expressed through the determination of dosage mortality (Bliss, 1937).

In most assays of lethality where the concept of a zone of resistance is not applied, all estimates are dealt with by dosage mortality. In Fig. 7A, the various crosses are determinations by dosage mortality, i.e., the median lethal temperature for exposure for the indicated period of time. The circles are estimates by time mortality, i.e., the median time to death at the temperature indicated. Time mortality and dosage mortality yield equivalent numerical values where the two determinations can be compared.

The importance of the distinction between the zones of resistance and tolerance comes largely because the two are not necessarily correlated. In the work of V. M. Brown *et al.* (1967) (Fig. 4) at a concentration of

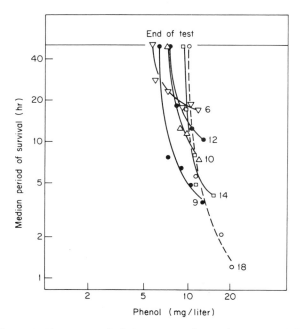

Fig. 4. Concentration vs. survival-time curves for rainbow trout, *Salmo gairdneri*, at different temperatures in phenol solutions. Slightly modified from Brown *et al.* (1967). The numbers associated with the curves indicate the test temperatures.

phenol in the mutual zone of resistance, the resistance time is longer at a lower temperature than at a higher one. However, the fish tolerate more phenol on a long-term basis at a higher temperature than a lower one, as can be seen from the way the assay lines cross each other in the figure.

While analysis into zones of resistance and tolerance as treated here gives a satisfying sense of completeness to the data, it must always be realized that there is no finality to the incipient lethal temperature short of maintaining a test throughout the whole life of the organism. The incipient lethal level should be looked on as the boundary of the imme-diate direct lethal effects, "immediate" being taken as a matter of days or weeks and "direct" as the operation of the identity directly on a site of metabolism so as to destroy it more rapidly than the organism can keep it in repair. Allen and Strawn (1968) take this point of view when they accept heat death as being complete by 20,000 min although the fish were apparently not able to live indefinitely beyond that period since their food intake could not meet their maintenance requirements.

## A. Determination of Lethal Effects

With the exception of certain determinations of lethal temperature, and to some extent the lethal effects of unsuitable concentrations of the respiratory gases, tests for lethality have been carried on by acute exposure of samples to various levels of the identity concerned until death or

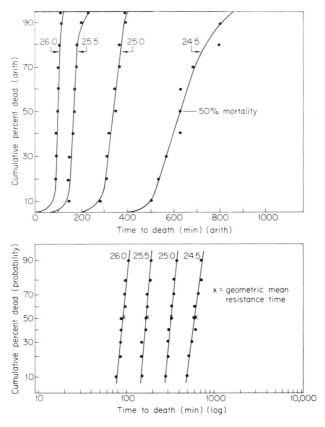

**Fig. 5.** Time mortality curves for death of fish exposed directly to various constant temperatures. Chinook salmon, *Oncorhynchus tshawytscha*, acclimated to 10°C. From Brett (1952).

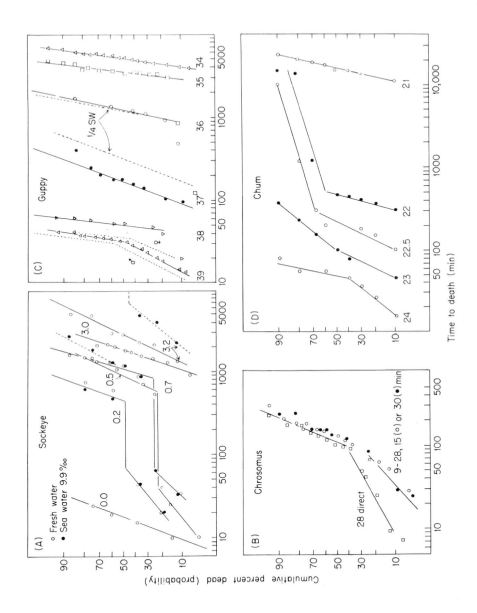

for a given period of time. There has been a tendency to extend the period as more experience has been gained; thus, now a 96-hr test period is probably the most widely approved, although 48 hr is probably most often used. These times have largely been set by rule of thumb and are essentially aimed at a period which would be long enough so that the $LD_{50}$ would represent the tolerance level as defined here. With regard to the toxicity of phenol to rainbow trout (Fig. 4) it appears probable that a 48-hr test would barely provide incipient lethal levels but that the 96-hr test would give an adequate margin.

The lethal effects of temperature will be treated by time mortality here in the zone of resistance. In the case of temperature, much work also has been devoted to the determination of lethal temperatures by placing the organism at some intermediate temperature and then heating at some convenient rate, usually a Celsius degree every few minutes. The temperature at which the animal dies or becomes visibly incapacitated, often termed the "critical thermal maximum" (CTM), is taken as a measure of its lethal temperature. This particular technique will be treated below as a special case.

### 1. MEASUREMENT OF THERMAL RESISTANCE

Time mortality curves for fish are usually surprisingly regular when the mortality in probits is plotted against the logarithm of time as Fig. 5 indicates. Here probit lines are simple, straight and parallel, indicating a statistical homogeneity both in the population and in the locus in the organism which breaks down under the influence of excessive temperature.

However, in many instances, particularly in determinations in the lower zone of thermal resistance, the regularity breaks down. Figure 6 shows four examples of such statistical heterogeneity. Figure 6A shows a case at the lower lethal which clearly displays the phenomena designated by Doudoroff (1945) as primary and secondary chill coma. All deaths at 0.0°C and the early deaths up to 0.7°C are the result of primary chill coma; the remaining deaths are the result of secondary chill coma.

Fig. 6. Statistical heterogeneity in time mortality at various lethal temperatures. (A) Cold death in *Oncorhynchus nerka*, acclimation temperature 20°C [from Brett (1952)]. (B) Heat death in *Poecilia reticulata* acclimated to 25°C [from Arai *et al.* (1963 and unpublished observations)]. (C) Heat death in the minnow, *Chrosomus eos*, acclimated to 9°C [from Tyler (1966)]. (D) Heat death in *Oncorhynchus keta* acclimated to 5°C (Brett, 1952). The numbers associated with the curves indicate the test temperatures. All results are from direct transfer from the acclimation temperature except as indicated in B.

Pitkow (1960) considered primary chill coma to be the result of failure of the respiratory center while Doudoroff (1945), who showed that some of the lethal effect at the secondary coma point could be removed by using an isosmotic solution, considered the lethal action of temperature at that point was to suppress the ion-osmoregulatory mechanism. Brett's data illustrate that point also. Exposure in dilute seawater (nearly isosmotic) prolonged life in the secondary phase, slightly at 0.7°C and significantly at 3.2°C. It is probably better to speak of a breakdown in ion-osmoregulatory regulation. Wikgren (1953) showed that carp have an excessive loss of ions at low temperatures and R. Morris (1960) made the same observation for the lamprey.

The remaining parts in Fig. 6 show statistical discontinuities at high lethal temperatures. Figure 6B shows a sudden "shock" effect which has been intuitively feared by practical fish culturists and has led to the practice of tempering the transfer of fish from one temperature to a different one by equalizing the two over a period of a fraction of an hour. In the case shown in panel B such equalization over 15 or 30 min does largely remove the first lethal effect on the section of population sensitive to it. However, at least in the laboratory, the shock effect is not a prominent feature.

Figure 6C shows two heterogeneities in the response of the guppy to high temperature. There is a sex difference in response at 39°C, while the major feature of the panel is the jump in the time–temperature sequence between 37° and 36°C where the guppy goes from a stage where exposure in 25% seawater lengthens life to one where there is no effect. The guppies used in these experiments were genetically homogeneous so that the shift in response represents a change in the locus of breakdown with a change in the intensity of the lethal factor. While not shown here (but see Fry, 1967) unselected stocks of guppies show statistical heterogeneity below 36°C. The pure line tested by Arai *et al.* (1963) were all sensitive to the first locus for temperature death at 35° and 34°C.

While the analysis of such discontinuities, as are mentioned above, is a most fruitful field for further research, the matter will be dropped at this point to take up the relation between survival at different temperatures in the zone of resistance. Figure 7 shows two typical series of determinations of thermal resistance for the upper lethal zone. The two species in the figure illustrate contrasting types of response. The goldfish has a very short zone of resistance and a very high incipient lethal temperature. The bullhead while having almost as high an incipient lethal temperature as the goldfish has a more normal zone of resistance. The salmonids (e.g., Brett, 1952) have low tolerance but high resistance.

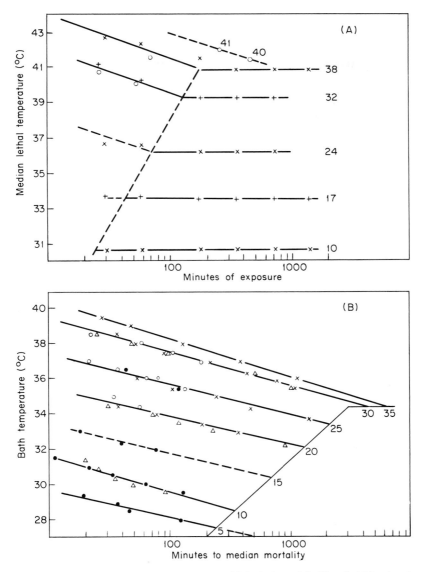

Fig. 7. Thermal resistance times for goldfish (A) and bullhead (B), *Ameiurus nebulosus*. From Fry *et al.* (1946) and Hart (1952). Circles in A represent resistance times at a given lethal temperature and thus correspond to the points in B. The various symbols for the bullhead represent samples from different localities over the range of the species. The numbers associated with the curves indicate the acclimation temperatures. The extension of the resistance line for bullheads acclimated to 5°C is discussed in Section II, A, 1.

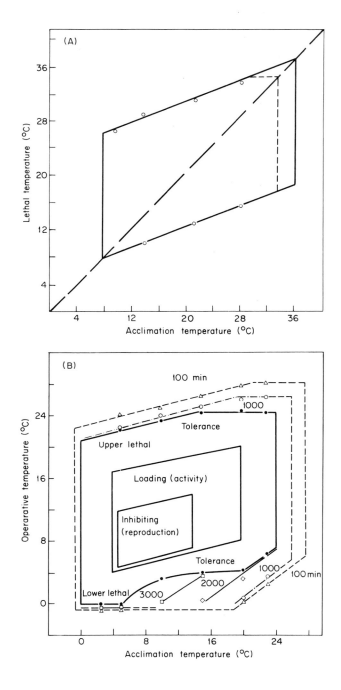

Since the goldfish has such a short zone of thermal resistance, the data for the bullhead give a more typical picture (Fig. 7B). Both species, however, are typical in their response to thermal acclimation where both the height and the extent of the resistance curves increase with increasing acclimation temperature. It is typical too for thermal resistance to continue to increase with thermal acclimation beyond the point where there is no further change in thermal tolerance, a feature also illustrated in Fig. 8.

Dosage mortality will not be dealt with extensively since this is the customary method of bioassay of lethality (e.g., Bliss, 1952). The only point to be made is that the period of exposure for the determination of tolerance cannot be arbitrary but must be based on a knowledge of the extent of the zone of resistance. If in the case of the bullhead (Fig. 7B) samples of fish acclimated to 5°C were exposed to various constant temperatures from 26.5° to 28.5°C and observed for 1000 min, mortality would have been complete at 28.5°C while no mortality would have been observed at 27°C as indicated by the dotted line. By extrapolation of the relation between time and mortality at the higher temperatures, 50% mortality would have been expected at 500 min at 27°C. Since the latter did not take place, the total mortalities in such a series of baths would then be plotted against temperature to give the incipient lethal temperature since it could be expected that mortality was complete in the lowest temperature although only a fraction of the sample had died in the test.

For other species, or the bullhead at another acclimation temperature, another period of observation would be more suitable for the determination of dosage mortality so that an arbitrary period of determination is not practical. A good method of determining the period of exposure is thus to expose the samples in which mortalities are not complete for the period of time indicated by extrapolation from events at higher lethal temperatures in which 50% of the sample in the lowest test temperature might have died. There are still dangers to such extrapolation. Figure 6D illustrates the difficulty of being sure that an assay proceeds to the tolerance level when there is a discontinuity in the time–temperature response. Here, suppose the assay had been carried out over a 5-day week, a most likely practical operation, which, when one considers the time at which the working week would start and stop, allows for a period of

---

Fig. 8. Typical thermal tolerance diagrams. (A) Puffer, *Spheroides maculatus* [from Hoff and Westman (1966)]; (B) sockeye salmon, *Oncorhynchus nerka* [from Brett (1952, 1958) and Brett and Alderdice (1958)]. See text for definitions.

about 6200 min. In that case the mortality at 21°C would have been entirely overlooked.

## 2. THERMAL TOLERANCE

The incipient lethal temperatures can be plotted as shown in Fig. 8. Here the various typical responses for a thermal tolerance diagram are to be seen. Typically the upper incipient lethal temperature changes approximately 1° for a 3° change in acclimation temperature. The lower incipient lethal shows a somewhat greater response, usually shifting 1° for about 2° change in acclimation temperature. The response of the lower incipient lethal temperature to acclimation temperature is not always linear. In the sockeye salmon (Fig. 8B) Brett found a sigmoid response in the lower incipient lethal to the acclimation temperature, the flex no doubt coming where the cause of cold death passes from primary to secondary chill coma.

The example Fig. 8A was chosen because it displays almost completely the simplest relation to be expected in the response of the incipient lethal to thermal history, a regular linear change in lethal temperature to acclimation temperature so that a trapezoidal figure bounds the zone of thermal tolerance. The lowest and highest incipient lethal temperatures which an organism can attain by extreme acclimation have been termed the "ultimate incipient lethal temperatures" (Fry et al., 1942). In Fig. 8A these ultimate lethals have the ideal values where the acclimation temperature equals the lethal temperature. Figure 8B, on the other hand, shows the various modifications that have been encountered in the tolerance diagram: a plateau in the upper incipient lethal at high acclimation temperatures and a floor to the lower lethal at low acclimation temperatures together with a flexure in the course of the lower incipient lethal temperature to acclimation temperature, as mentioned above. The precise level of this latter floor has been little explored. In many freshwater species the ultimate lower lethal is indeterminate since the fish can still be active at the freezing point of water. Most marine species apparently will freeze in seawater before the latter itself freezes, and the floor there may be set by the freezing point of blood. In the case of the sockeye in Fig. 8, however, the floor comes a little above the freezing point of the blood and has no direct explanation. Death is obviously not a result of the formation of ice crystals in this case.

Some marine species have the ability to supercool, as is discussed in Chapter 3 by DeVries. Such species then may also have ultimate incipient lower lethal temperatures which are indeterminate in their normal

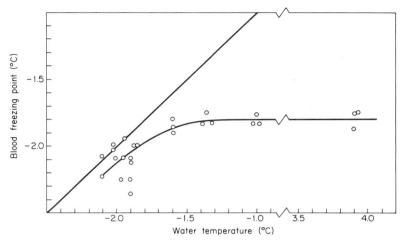

**Fig. 9.** Relation between water temperature and blood freezing point in *Trematomus bernachii.* Modified from Potts and Morris (1968). The diagonal passes through the points where blood and water would have identical freezing points.

habitat. There is some question as to whether the increases in salt content, noted in the bloods of some marine species (e.g., Fig. 9), are adjustments to low temperature. To some extent at least, they may be symptoms of lack of acclimative capacity in the ion-osmoregulatory system and represent approaches of secondary chill coma. Thus, Woodhead (1964) shows a decided upward drift in the serum sodium concentration at about 3°C in the sole, *Solea vulgaris,* a species which he considers from field evidence to have its ultimate lower incipient lethal temperature at about 2°C, well above the freezing point of its blood.

The Antarctic species, for which data are shown in Fig. 9, apparently restricts the blood flow through its gills at −2°C, possibly to restrict loss of water or influx of salt.

The thermal tolerances of fish vary greatly. Antarctic species, for which unfortunately we do not have any tolerance diagrams, die at 5°C or a little above (Wohlschlag, 1964). Low temperate species like the goldfish have ultimate lethal temperatures of 0°C and in the vicinity of 40°C and tolerance diagrams that bound an area of some 1200°C². High temperate species have ultimate upper lethals ranging from 20°C to approximately 35°C. Tropical species appear to be distinguished by high, low lethal temperatures and as investigated have not shown higher upper lethal temperatures than some temperate species (e.g., Allanson and Noble, 1964). Indeed, the guppy cannot be carried through its life cycle above 32°C.

Brett (1958) has extended the concept of the tolerance diagram by considering that the lethal temperature is the ultimate response to thermal stress and recognized loading and inhibiting stresses, which terms he uses to designate limits within the zone of thermal tolerance where growth and activity and reproduction, respectively, are suppressed His suggested boundaries for these levels are shown for the sockeye. However, these concepts may be modified, particularly in view of his recent work on growth (Brett *et al.*, 1969).

It has long been recognized that there can be differences in lethal temperature among fish acclimated to the same temperature at different seasons of the year. In this respect most attention has been paid to the upper lethal temperature. An example of the magnitude of such differences is given in Fig. 10. Some modification of the lethal temperature in the appropriate directions has been achieved by manipulation of the photoperiod, a long day bringing an increase in thermal resistance and in the incipient lethal temperature (Hoar and Robinson, 1959; Tyler, 1966), but neither the seasonal effect nor the extent to which manipulation of the photoperiod can modify the lethal temperature has yet been

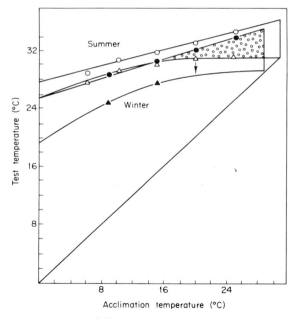

**Fig. 10.** Summer–winter differences in lethal temperature in the minnow *Chrosomus eos.* The upper boundary to each area is the 30-minute resistance level (circles), the lower, the incipient lethal level (triangles). Modified from Tyler (1966). Solid symbols are winter values; the arrow represents an incomplete winter determination. Stippling indicates the area of overlap.

the subject of any exhaustive analysis. Recently (Johansen, 1967), it has been shown that the pituitary must be intact if the goldfish is to acclimate to a higher temperature, which is the first clear indication of endocrine involvement in the response to lethal temperature.

Ushakov (e.g., 1968) and his school have shown that the thermal resistance of muscle is relatively unaffected by acclimation temperature although some slight seasonal effects are to be seen. Interestingly the upper incipient lethal temperature of excised muscle (or tissue cultures) approximates the ultimate upper incipient lethal temperature in four species (Fry, 1967; Fry and Hochachka, 1970). However, the whole problem is still obscured by our ignorance of the specific sites of breakdown in thermal death and, indeed, by lack of unequivocal comparisons between the whole animal and its tissues.

## 3. Rates of Thermal Acclimation

Brett (1944, 1946) early showed that the lethal temperature of various freshwater species adjusted so rapidly to changes in water temperature that changes in the weather were reflected as well as the seasonal cycle (Fig. 11). He and various workers, in particular Doudoroff (1942) and Cocking (1959), have addressed themselves to the measurement of the rate at which fish adjust their lethal temperature in relation to a change in acclimation temperature. Such changes can be very rapid when the temperature is adjusted upward while downward changes are much slower. The main feature of the observations of Brett and Doudoroff on rates of adjustment of the upper and the lower lethal temperature to changes in acclimation temperature are shown in Fig. 12. It would appear from these data that the change of heat resistance follows a different course from change in cold resistance indicating that the two responses do not operate on the same site. Change in resistance to high temperature follows a sigmoid course and can show a long latent period which is about 1 week at 12°C for goldfish when moved from 4°C, while the curves for change in resistance to low temperature are simply convex or concave. Heinicke and Houston (1965) found a distortion in the plasma sodium:chloride ratio in goldfish transferred abruptly from 20° to 30°C which reached its extreme in the first 4 days of exposure. The ratio returned to normal by about 10 days. Another major difference made clear in the figure is that, as the data for *Girella* show, while adjustments in cold resistance are about equally rapid whether a given step is up or down over a given acclimation range, heat resistance is gained much more rapidly than it is lost. The upper lethal temperature of goldfish is adjusted within a day when they are shifted from 20° to 28°C,

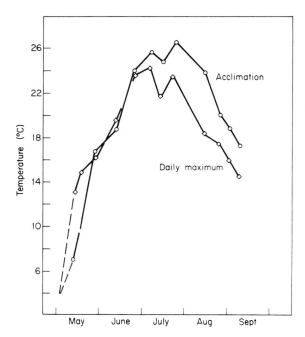

**Fig. 11.** Acclimation temperature of the brown bullhead, *Ameiurus nebulosus*, as determined from the lethal temperature, in relation to mean daily maximum water temperature. From Brett (1946).

while the shift from 24° to 16°C requires over 2 weeks for the reciprocal adjustment in *Pimephales*. It is not likely that these differences are the result of different species being used in the two experiments. In consequence of this differential in rate, acclimation of the upper lethal temperature tends to follow the daily maximum with increasing temperatures, as Fig. 11 indicates, where the acclimation temperature rapidly catches up with the daily maximum temperature in approximately the first 3 weeks after breakup. Heath (1963) found that in the cutthroat trout, *Salmo clarki*, a 24-hr period of fluctuating temperature resulted in the highest acclimation temperature as compared to the mean temperature throughout the whole period. There is no information on the effect of fluctuating acclimation temperatures on the lower lethal temperature.

Rates of adjustment to acclimation temperature (neglecting the $Q_{10}$ effect), vary from approximately 1°C/day in the goldfish and the roach (Cocking, 1959) to the same change in an hour or so in *Girella*, the bullhead, and various salmonids.

There are experimental difficulties in the determination of rates of acclimation which have not yet been thoroughly explored. Brett (1946)

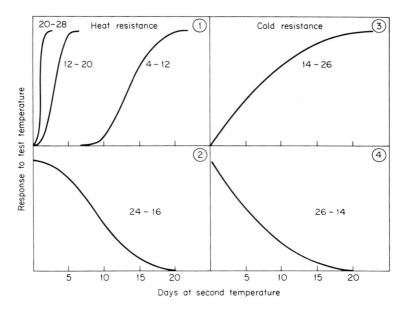

**Fig. 12.** The four basic courses of acclimation to lethal temperature. Heat resistance determined as average survival time; cold resistance as 24-hr median tolerance limit. (1) Goldfish (Brett, 1946). (2) The minnow, *Pimephales promelas* (Brett, 1944). (3 and 4) *Girella nigricans* (Doudoroff, 1942). Numbers associated with the curves indicate the magnitude and direction of shift of acclimation temperature.

noted that the brown bullhead would not respond to a change in acclimation temperature if the oxygen in the acclimation bath were reduced to approximately 10% air saturation. Hart (1952) was not able to acclimate yellow perch, *Perca flavescens,* to high temperature in winter and noted that he was unable to get them to feed. Later, in our laboratory, perch which were feeding did acclimate. On the other hand, Brett found bullheads starved for a total of 40 days from June 16 acclimated precisely as did fish which were feeding up to a day or so of test.

### 4. DEATH IN CHANGING TEMPERATURES

Lethal temperatures in the environment are ordinarily temporary changes brought about by unusual seasons or extreme weather. Thus, time is a major element in lethality as well as is temperature, which indeed is the reason for segregating thermal resistance from thermal tolerance. It is of importance therefore to be able to integrate the temperature experience. The progress to death at any one lethal temperature can be taken as linear ( e.g., Jacobs, 1919; Olson and Stevens, 1939). Thus fractions of the respective resistance times spent at each of a series of lethal temperatures can be summed to indicate total lethal experiences as Table I indicates. If the temperature changes continuously then the summation can be made by relating the time-temperature curve to the temperature-resistance time curve. This may be done graphically as in Fig. 13 where temperatures are plotted as equivalent to the reciprocals of resistance times appropriate to them, or by calculation based on the same principle, as is done by engineers interested in heat sterilization (Olson and Jackson, 1942).

To simplify the problem in the examples given above the experiment began with a sudden transfer from the acclimation temperature to the zone of resistance in order to eliminate the effects of acclimation during the heating period before the incipient lethal temperature was passed. There is acclimation in the zone of resistance, too, as can be demonstrated by subjecting a fish to a sublethal exposure at temperatures above the incipient lethal and then returning it to its original acclimation temperature for a period of recovery, after which its resistance to a given lethal temperature is tested again (Fry *et al.*, 1946). Cocking (1959)

**Table I**

Summation of Lethal Experience in Various Temperatures in the
Eastern Brook Trout, *Salvelinus fontinalis*[a,b]

| Acclimation temp. (°C) | Thermal experience | Observed time to death (min) | Theoretical time to death (min) | Summed lethal dose |
|---|---|---|---|---|
| 11 | 27.1, 33 min (0.50) to 26.3 till death (0.45) | $102 \pm 18$ | 107 | 0.95 |
| 11 | 26.3, 60 min (0.39) to 27.1 till death (0.80) | $112 \pm 10$ | 100 | 1.19 |
| 11 | 25.8, 90 min (0.34) to 26.1, 40 min (0.21) to 26.5 till death (0.48) | $190 \pm 22$ | 177 | 1.03 |
| 20 | 26.5, 230 min (0.48) to 28.0 till death (0.64) | $291 \pm 16$ | 281 | 1.10 |
| 20 | 27.0, 125 min (0.44) to 28.0 till death (0.60) | $183 \pm 20$ | 179 | 1.04 |
| 20 | 27.5, 75 min (0.46) to 28.0 till death (0.55) | $128 \pm 10$ | 122 | 1.01 |

[a] Error given is $2\sigma$.

[b] Based on Fry *et al.* (1946), their Tables 4 and 6. Numbers in parentheses indicate calculated fraction of lethal experience corresponding to the period of exposure at the particular temperature.

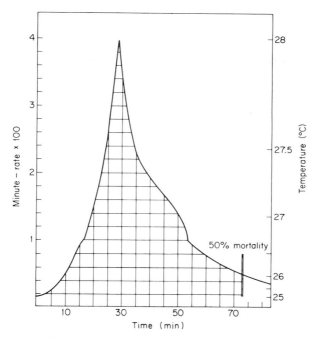

**Fig. 13.** Accumulative effects of exposure to changing lethal temperature in eastern brook trout, *Salvelinus fontinalis*, acclimated to 11°C. From Fry *et al.* (1946). The "minute-rate" is the fraction of the resistance time at any lethal temperature represented by an exposure time of one minute. The temperature scale on the right is based on the minute-rate scale. Thus, for example, the resistance time at 27° was 72.5 min so that the minute-rate × 100 (1/72.5 × 100) is 1.37. The median mortality point indicated was the observed one. Each square under the curve indicates 1% of the total theoretical lethal exposure. There are approximately 101 squares under the curve up to the median mortality point.

slowly heated from the acclimation temperature and continued to the death of the fish to get a measure of the rate of thermal acclimation.

As mentioned earlier, lethal temperatures (CTM) are often determined by steadily increasing the temperature a degree every few minutes and recording the temperature at which the sample dies or is incapacitated. Incapacity is taken as the equivalent of death for two reasons: first, the indicator must provide an unambiguous point under such rapidly changing conditions; second, it is assumed that if the animal becomes incapacitated it will not be able to escape from further stress and will be trapped in a lethal situation. From the narrow ecological point of view the CTM is useful and is above all extremely economical of material. It can even be used to determine the ultimate upper incipient lethal temperature if rate of heating is slow enough, say, 0.5°C/day (Cocking, 1959; Spaas, 1959).

As a means of physiological analysis however the CTM has many failings. The determination confounds time and temperature, but in particular discontinuities of response such as are illustrated in Fig. 6 are lost. In practice when the rates of heating are of the order of minutes per degree only the most acute cause of death will be displayed. There is no time for the slower breakdowns. Moreover, the plateau often to be found in thermal tolerance diagrams is not likely to be displayed in the response of the CTM to acclimation temperature, as is well shown for the guppy, *Poecilia reticulata* (Fry, 1967, Fig. 3) in which the incipient lethal temperature is approximately 32°C at all acclimation temperatures above 20°C while the CTM increases steadily about 1° for every 3° change in acclimation temperature throughout the whole range of observations. Thus, whereas the ultimate incipient lethal temperature of this species is below 33°C, the maximum CTM is somewhat over 40°C.

Finally, it should be remarked that when a fish is in water warmed at the rate of a degree every few minutes its internal temperature will lag somewhat behind the ambient temperature. It can be calculated from Harvey's data (1964) that the lag for a 50-g sockeye, *Oncorhynchus nerka,* would be 0.4°C if the heating range were 1°C/5 min (Fry, 1967). A tenfold change in weight would bring about a twofold change in lag on the basis of present data on the size-thermal conductivity relation.

## B. Toxicity Studies

This section is written almost with reluctance. It may be said that pollution biologists have backed into bioassay. It is proper for those interested in chemical control of pests to assay their biocides by determining the lethal dose with care and precision, for their purpose is to load the environment with the minimum proper dose. They wish to be sure they have killed their target at a minimum of cost and further damage to the organic community. The pollution biologist, on the other hand, wishes to protect the organism of his concern and to see it prosper—again with a minimum of interference with man's other interests. Thus, his aim is to protect the organism from damage, not from death alone. To deal then, even learnedly, with lethal levels of a pollutant is somewhat to serve orthodoxy at the expense of progress. Lethal levels are to be considered only as the boundaries of the zone within which the real work goes on.

Statistical response to toxic materials is similar to response to lethal temperatures. For example, Fig. 14 shows that the same statistical discontinuity can be found in stress from a poison as is found with the physical identity, temperature. With regard to acclimation—a prominent feature in temperature death, and presumably also for setting the lethal levels of toxic substances—there is little that can yet be said. Most assays are made with animals which have not been previously exposed to the lethal agent under test; thus, they present the most severe case.

One important difference between the harmful effects of toxic agents and the lethal action of temperature as discussed above is that the effects of incomplete exposure to a lethal level are by no means necessarily reversible. Sublethal exposure can lead to permanent destruction of critical tissue such as gill (Scheier and Cairns, 1966) or sensory epithelium (Bardach *et al.*, 1965).

An extensive review on toxic substances as they affect aquatic organisms with an exhaustive bibliography has been prepared by McKee and Wolf (1963). Here only a few remarks on the interactions of multiple materials in the environment will be offered.

An effluent often contains a mixture of various toxic materials

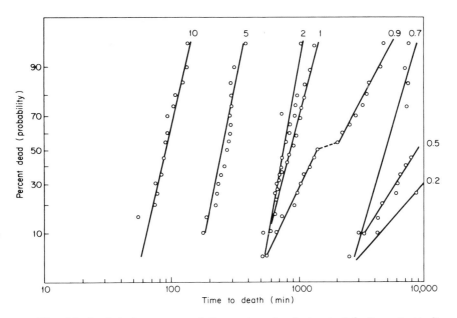

**Fig. 14.** Statistical response of the eastern brook trout, *Salvelinus fontinalis*, to various dosages of 20% dinitro-*o*-cyclohexylphenol, dicyclohexylamine salt, at approximately 9°C. Data of D. F. Alderdice (personal communication). Numbers associated with the probit lines indicate the dosage in milligrams per liter.

and it is necessary to consider whether each of these acts independently, in which case all that is necessary is to be certain that each is kept below its threshold level, or whether the effects are additive or even synergistic. Lloyd (1961c) demonstrated that fractions of the incipient lethal doses were additive in the case of copper and zinc. This rule also held for resistance times at higher doses in hard water, but mixtures of the two metals were relatively more toxic than either alone at high doses in soft water. Sprague and Ramsay (1965) reported similar findings. Such additivity has been reported for various other combinations (e.g., Herbert and Vandyke, 1964). Much of this work is reviewed compactly in Herbert (1965). Again, it has long been known that the toxicity of metals, for instance, is highly variable in natural waters when the concentration of the toxic agent is expressed as the total of that element present in solution. It has also long been well-recognized that such differences are in large part the result of differences in pH, but the principles of chemical dissociation appear to have first been applied extensively to the effects of toxic substances on fish by Wuhrmann and Woker (1948) who demonstrated that the amount of un-ionized ammonia in a given solution was the significant measure to take to determine its toxicity. Cyanide, which complexes with metals, offers another example (Fig. 15).

Determination of the concentration of the toxic form in the environment may not entirely resolve the question of toxicity. The dose the fish absorbs may be further modified by the same agent which influences the state of the toxic agent. Thus, Lloyd and Herbert (1960) calculated that the interaction of various ambient concentrations of free $CO_2$ with the respiratory exchange so modified conditions within the interlamellar spaces that the same concentration of un-ionized ammonia (0.40 mg/ liter) was presented to the gill under circumstances where the outside concentrations varied from 0.84 to 0.49 mg/liter.

An agent which changes respiratory flow will also modify the dosage received from a given ambient concentration, again through the agency of the countercurrent exchange system in the gill (Lloyd, 1961a,b; Herbert and Shurben, 1963).


## III. CONTROLLING FACTORS


The operation of controlling factors will be illustrated exclusively by the effects of temperature. The effects of pressure, which must have a major controlling effect in the oceanic depths, have been reviewed in the chapter by Gordon, Volume IV, this treatise.

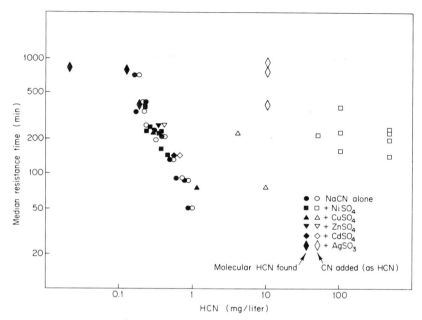

**Fig. 15.** Median resistance time of bluegill, *Lepomis macrochirus*, in relation to free molecular HCN and total cyanide concentrations (expressed as HCN equivalents) in various simple and complex cyanide solutions. Modified from Doudoroff *et al.* (1966).

The thesis taken here, which seems to be the one generally implied, is that, other things being equal, the metabolic rate is a function of molecular activity while otherwise under the regulation of agents which operate in ways not yet much understood (but see Chapter 2 by Hochachka and Somero). On these premises, the controlling factor, setting a limit to molecular activity, sets an upper limit to the metabolic rate. By conferring a given level of molecular activity the controlling factor also imposes a given level of instability on the living system, which must be counteracted by repair through energy-yielding reactions. Thus a given level of controlling factor is taken to permit an upper limit to the metabolic rate and to require a lower one. As defined in the Introduction, the upper limit is active metabolism, the lower one standard metabolism. Activity, a transformation of the energy released by metabolism and a function of some fraction of the difference between these two levels, may or may not bear the same relation to temperature as either of them.

Controlling factors operate at the cellular level, the site of the metabolism yielding the energy. The potential for energy yield in the cells may not always be permitted full expression, even in the normal environment, because of restrictions imposed by the nature of the whole organism, as will be discussed in Section IV. There is also the complication that random activity varies with temperature. Thus, the relation of routine metabolism to temperature cannot be expected to fit directly any of the curves described by the temperature formulas. The relation of routine metabolism to temperature will be considered in Section VI.

## A. Formulas Relating Temperature to Metabolism and Activity

Figure 16 gives a crude genealogy of the major formulas which have been applied to the effects of temperature on living processes. Briefly, the earliest formulation appears to have been the rule of thermal sums proposed by Réaumur and applied to the effect of temperature on the date of the appearance of such phenological events as the ripening of crops or the development of larvae in relation to whether the season is advanced or retarded in different climates. Much later Berthelot, studying the rate of fermentation, found the effect of temperature could be described as a process which increased geometrically as temperature increased arithmetically. From this point of view effects of temperature

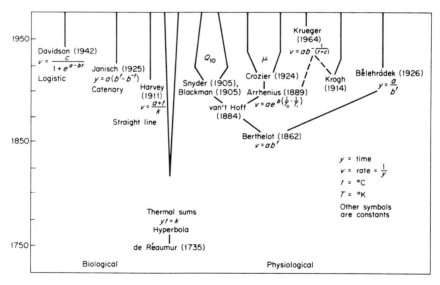

Fig. 16. Historical and algebraic relations between the various formulas applied to the effect of temperature on the rate of processes in organisms. The dates given indicate original or early references to each formula.

could be characterized by a coefficient comparing rates over a stated interval. This coefficient is the well-known $Q_{10}$ and the rates of biological processes usually double or treble over an interval of 10°C ($Q_{10} = 2$ to 3). Through van't Hoff to Arrhenius the effect of temperature on rates of chemical reaction was related to molecular theory and much later the coefficient $\mu$ (which represents $e$, the energy of molecular activation) came into widespread use in biology in America, largely through the work of Crozier and his colleagues. Bělehrádek proposed a third coefficient $b$, derived from the slope of the relation between log temperature and log time, not being satisfied that the $\mu$ as found in whole organisms necessarily strictly reflected the $e$ of the gaseous reactions of the physical chemists. Bělehrádek considered that the rates of reaction in organisms were more likely to be limited by diffusion than to be set by the level of molecular activation. The final item given on the right hand side of Fig. 16 is Krogh's curve for standard metabolism, which was not formulated but simply a descriptive curve of proportional change with temperature that fitted various observations that Krogh and his colleagues had gathered, among them the metabolic rate of the goldfish. Krueger later proposed a method to fit a formula to the curve.

The first formula given to the left of the rule of thermal sums is Harvey's fit of the straight line to various rates such as heartbeat, which is mathematically identical to the rule of thermal sums. The remaining two formulas were proposed by workers concerned with the rate of development of insect eggs to meet the long-recognized inadequacy of the rule of thermal sums as a fit over the whole range of temperature at which development can take place, by formulating curvilinear fits. Janisch fitted a catenary to the time curve to permit inclusion of the cases at high temperature where time increases over that required for development at a somewhat lower temperature. Davidson fitted the logistic formula to the rate curve to provide for the deviation from linearity at low temperatures where the rates are higher than would be predicted from extrapolation of the straight line which fits the central range. He ignored the response to the extreme upper temperatures which were the special concern of Janisch, as Janisch ignored the response to the extreme lower ones.

A good review through which to enter the earlier literature on this subject is that of Bělehrádek (1930). The essentially chemical approach is well presented in Precht et al. (1955) and Johnson et al. (1954). In particular, the latter authors state the general biochemical case for the controlling factors in their treatment of the interaction between temperature and pressure, which is not dealt with in the summary above.

The two series of formulas are segregated in the figure under the terms "biological" and "physiological." Andrewartha and Birch (1954) recognized the same segregations but preferred the terms "empirical" and "theoretical"; but while the terminology of these authors is correct concerning the origins of the formulas (except for Berthelot's), it misses the point of their application. Their terminology seems to imply that the "theoretical" series will ultimately prevail over the "empirical" one. However, the distinction is really in kind not in worth. The biological series relate directly to activity in relation to temperature. The physiological formulas relate to the chemical basis which yields the energy for activity.

The difference between the biological and physiological needs for formulation are well expressed by Booij and Wolvekamp (1944, p. 212) as follows:

> Whilst the chemist takes care that the form of the reaction vessels and the properties of the substances of which they are manufactured do not interfere with the processes under investigation, the engineer on the contrary will design special structures and make use of their physical properties in order to obtain a harmonious cooperation of physico-chemical processes. . . .
>
> It is to the segregated parts of the more complex processes taking place within the organism (just as those taking place within an engine) only that the fundamental laws of physical chemistry may be applied.

In the discussion below, concerned as it is with the whole organism, there will be little direct application of the physiological series of formulas, only the $Q_{10}$ being used in general terms. The physiological relation looked for will be the effect of temperature on the relation of active and standard metabolism and on the relation of the difference between these values to activity.

## B. Active and Standard Metabolism in Relation to Temperature

The most thorough determinations of active and standard metabolism have been made by Brett (1964) for his stock A. His data are summarized in Table II and were obtained in his apparatus which is illustrated by Phillips in Volume I of this treatise. Brett's determinations were the product of experiments each several hours long in which the fish were stimulated to swim faster and faster by small steps. He took the maximum rate so achieved as his measure of active metabolism while he extrapolated the rate determined at various speeds to zero speed as his measure of standard metabolism. Thus not only did he obtain a close approximation to the maximum continuous rate of oxygen consumption but also probably largely eliminated the early effects of

**Table II**

Swimming Speed and Active and Standard Metabolism of the Sockeye
Salmon in Relation to Acclimation Temperature[a]

| Temp. (°C) | $O_2$ (mg/liter) | Metabolism (ml/kg/hr) | | | | Swimming speed (length/sec) | Ratio Active/ Std. |
|---|---|---|---|---|---|---|---|
| | | Active | Standard | Scope | Scope$^{0.55}$ | | |
| A | | | | | | | |
| 5 | 12.8 | 364 | 29 | 335 | 24.5 | 3.26 | 12.5 |
| 10 | 11.2 | 445 | 42 | 403 | 27.1 | 3.65 | 10.6 |
| 15 | 10.1 | 635 | 50 | 585 | 33.3 | 4.12 | 12.7 |
| 20 | 14.0 | 921 | 85 | 836 | 40.5 | 4.27 | 10.8 |
| B | | | | | | | |
| 20 | 9.1 | 604 | 85 | 519 | 31.2 | 3.94 | — |
| 24 | 8.5 | 601 | 139 | 462 | 29.2 | 3.75 | — |

[a] Data of Brett (1964) for stock A, from his Tables 1 and 6 and Fig. 16. "Scope" is the difference between standard and active metabolism. Fish weight, approximately 50 g.

excitement from his data and thus obtained as well a close approximation for the minimum resting level.

Brett determined only oxygen consumption, but it probably is safe to infer from Kutty's work (1968a) on the respiratory quotients of goldfish and rainbow trout in similar experiments that the respiration of the salmon was essentially aerobic in these long-term determinations.

In general, Brett worked with the dissolved oxygen concentration in the water at approximately air saturation. At 20°C, however, a special experiment was performed in which the water was considerably enriched to 14 mg/liter $O_2$. Table II is divided into two parts: part A shows experiments to 15°C plus the one at increased oxygen at 20°C, and part B, shows experiments at air saturation at 20° and 24°C. Part B will be dealt with again in Section IV. All Brett's measurements were made with the fish acclimated to the test temperature.

Looking at the data in part A it is apparent, as would be expected, that increasing temperature accelerates both active and standard metabolism. The point of major interest here though is that within the limits of error, standard metabolism is the same fraction of active metabolism at the four temperatures concerned. Thus, standard metabolism reflects the possibilities for active metabolism.

The various data in the table are plotted on a semilogarithmic grid in Fig. 17. Here the data of part B are plotted as well as those of part A of Table II, again for further consideration below.

The curve for active metabolism with supplementary oxygen and that for standard metabolism are parallel and almost straight on the

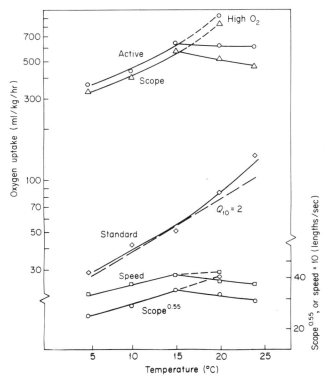

**Fig. 17.** The effect of temperature on metabolism and activity in the sockeye salmon, *Oncorhynchus nerka*. From Brett (1964), his stock A. For further explanation see text.

semilogarithmic plot and have a $Q_{10}$ of approximately 2. The points for standard metabolism at 20° and 24°C with oxygen at air saturation continue the trend of the points at the lower temperatures. The corresponding points for active metabolism, of course, do not, because of the limiting effect of oxygen in air-saturated water. The curve for standard metabolism can therefore be taken as reflecting the potential for active metabolism even though the latter may not be attained. However, such a conclusion can only be highly tentative. In particular, the measurement of standard metabolism is only yet in its infancy as is, of course, also the measurement of active metabolism, most especially at high temperatures without oxygen being limiting.

## 1. The $Q_{10}$ of Standard Metabolism

Few general statements can yet be made of the relation of standard metabolism to temperature except that different species are adapted to different temperature ranges [see Wohlschlag's (1964) summary diagram], a point the fish make well themselves without recourse to a respiration chamber. Another general point is that over the biokinetic range of a species the metabolic rate is approximately 75 ml/$O_2$/hr for a 100-g individual at the mid range. The relation of standard metabolism to temperature is often described by Krogh's "standard curve" (Ege and Krogh, 1914) which has been formulated by Krueger (1964) and which Winberg (1956) applied so effectively in his generalizations. However, Krogh's curve as an empirical formula applies much more generally to curves of routine metabolism than to standard metabolism, at least as the latter has been determined by extrapolation to zero physical activity.

The relations of the extrapolated values for standard metabolism to temperature (Beamish, 1964a; Beamish and Mookherjii, 1964; Brett, 1964; Rao, 1968) have been various, ranging from a response similar to the Krogh curve with constantly decreasing $Q_{10}$ through a case of constant $Q_{10} = 2$ in the goldfish—which was the fish species on which the Krogh curve was determined—to the case of the sockeye (Fig. 17) where $Q_{10}$ shows a slight constant increase with increasing temperature (as the dotted line indicates). The convex course of Krogh's curve with decreasing $Q_{10}$ with increasing temperature will be considered in Section VI.

## 2. THE RELATION OF ACTIVITY TO TEMPERATURE

The remaining curves in Fig. 17 are concerned with the effect of temperature on activity, directly or indirectly. The uppermost of these curves, labeled "scope" is the difference between active and standard metabolism and is shown both for the case where oxygen was not limiting and for the case of air saturation over all temperatures. The curve for scope where oxygen is not limiting is parallel to the corresponding curves for active and standard metabolism and therefore can be interpreted by the same temperature coefficient. Thus, under circumstances where the activity concerned has a linear relation to the metabolic scope available for that activity and no limiting factor intervenes, then the $Q_{10}$ for the activity will be the same as the $Q_{10}$ for standard metabolism. However, no specific example of such a relation is at hand.

The series of curves grouped within the central box on the figure represent the relation of swimming speed to temperature and the function of scope related to that activity derived from Fig. 2. The $Q_{10}$ for the maximum sustained swimming speed up to 15°C is much less than 2 and approximates $\sqrt{2}$. Scope$^{0.55}$, the power relation derived from Fig. 2, has a similar temperature response, as would be expected since Brett found that the increase in metabolism brought about by a given increase in swimming speed was the same at all temperatures. The $Q_{10}$ therefore for the activity, maximum continuous swimming speed, when no limiting factor intervenes is approximately 1.45 but it is based on a $Q_{10}$ of 2 for metabolism. In the example oxygen is limiting above 15°C and the swimming speed curve drops. The course however is still explained by the same function applied to the metabolic scope still available. Various other similar curves for the effect of temperature on swimming speed are collected in Fry (1967).

Larimore and Duever (1968) give a curve for the swimming speed of smallmouth bass fry, *Micropterus dolomieui*, where the $Q_{10}$ approximates 2 from 5° to 25°C. These fish, about 20 mm long, may not show the same relation between speed and metabolism. Pavlov *et al.* (1968) and Houde (1969) showed that maximum sustained swimming speed increases directly with length in small smooth-skinned fish rather than as length$^{0.5}$, which was the case for the sockeye salmon (Brett, 1965). However, such evidence is not conclusive since the divergence may be in the capabilities for metabolism, which have not been measured.

In the example given the $Q_{10}$ of active and standard metabolism is approximately 2, the commonly found relation for biochemical reactions. One example, cruising speed, is worked out to show a quantitative relation between temperature and metabolism. The example is perhaps deceptively simple and probably oversimplified (for example, ancilliary costs are ignored) so that only the general principle of activity being related to some function of some fraction of total metabolism (in the present case aerobic, but not necessarily so) should be retained after considering these paragraphs. In fact, while the relation of muscular activity to temperature can be justified in terms of metabolic cost in the straightforward fashion indicated above the response of other activities cannot yet be so easily analyzed. The rate of embryonic development (e.g., Krogh, 1914; Garside, 1966; Kinne and Kinne, 1962) has a mean $Q_{10}$ ranging up to 6. Perhaps these high $Q_{10}$'s are a reflection of the multiplicative nature of growth which accelerates the oxygen supply as growth is faster.

## 3. THE RULE OF THERMAL SUMS

The rule of thermal sums, which states that time × temperature is a constant for a given developmental or phenological event, has widespread use in practical fish culture. Normally the "thermal unit" (cf.

Embody, 1934) is employed which is the Fahrenheit expression ( °F — 32 × days). As a practical tool it is highly useful but some biologists, particularly those concerned with the morphometric consequences, have endowed the rule with an undue constancy and an unproven physiological significance (e.g., Tåning, 1952). In such cases the confidence in the rule is often misplaced, and it is doubtful whether it should be used. Shelford (1929) is a useful reference for anyone to consult who wishes to apply this rule. Basically the relation between temperature and rate of development, as ordinarily observed, is sigmoid so that the central section about the point of flexure, which is rather gradual, is well approximated by a straight line. If the linear section is extrapolated to zero rate the intercept $T_0$ provides a correction so that $(T_1 - T_0)$ provides a corrected temperature, which when multiplied by the *time* for development gives a constant. The rule, with the correction when $T_0$ is not 0°C, therefore is good for a median range of temperatures within the total range over which a given species can develop. At extreme temperatures the linear approximation of rate breaks down and the thermal sum is no longer constant. Hence, while the rule is useful in phenology and in hatchery practice, where the normal annual fluctuations are not likely to have a mean far from the median range, in physiological investigations, where controlled temperatures are used over the total range for development, the rule will break down and should not be used. The formulas of Janisch (1925) and Davidson (1944) are empirical fits to the development curve. Since there is evidence the inflection in the curve is the result of the limiting effect of oxygen, as is discussed below, these formulations are likely to apply to only the specific case of air saturation and have no general application to aquatic organisms which often develop under various degrees of oxygen deficiency.

## C. Acclimation to Controlling Factors

The relation of cruising speed to temperature given in Fig. 17 is for the fish acclimated to each test temperature before test. The general case for response at a given test temperature over all acclimation temperatures is given for the cruising speed of goldfish in Fig. 18 as a response surface (see Alderdice, 1971, for a general review).

The figure is to be viewed as a hillock rising to a peak at the point "S" with contours as shown by the curved lines. The contours are elliptical with the major axis inclined at approximately 45° to the temperature axes. The major axes are joined by a broken line which represents the so-called ridge line, the path of most gradual ascent

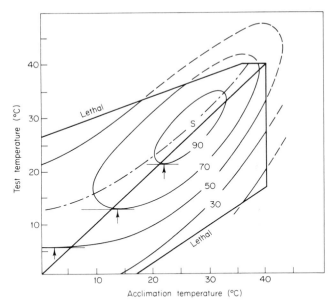

**Fig. 18.** The relation between acclimation temperature and test temperature as these affect the sustained swimming speed of 5-g goldfish within the zone of thermal tolerance. From Lindsey *et al.* (1970), their model B, and Fry *et al.* (1942). Numbers on isopleths indicate speed in feet per minute. For further explanation see text.

up the surface. In terms of adjustment to temperature the ridge line represents the combinations of test temperatures and acclimation temperatures which gives the least variation in swimming speed over the tolerance domain. The ridge line is above the 45° line (dotted) at low acclimation temperatures, indicating that performance is better at a temperature somewhat higher than the acclimation temperature when the acclimation temperature is low, perhaps an adaptation to favor activity in the spring warming period. The conditions for peak performance "S" come at a point where the ridge line, the acclimation, and the test temperature coincide, which may or may not have any significance.

The arrows impinging on the horizontal tangents extending to the test temperature axis indicate the conditions where performance is maximized for each test temperature concerned. These points come where test temperature is also the acclimation temperature. All in all, therefore, the process of thermal acclimation, at least as exemplified by this analysis of the response of the cruising speed of the goldfish, fits the organism to best meet the problem it faces at a given temperature. Similar studies of cruising speed on four other species (McCrimmon, 1949; Roots and Prosser, 1962; unpublished data of Ferguson, in Fry, 1964; Fry, 1967) indicate similar responses.

Accordingly, while it is most likely that the present skimpy data are overinterpreted in the paragraph above, it does seem clear that there are major and meaningful adjustments to temperature to make the organism able to perform more effectively, as indeed is the current general opinion.

However, it is not possible here to give a clear-cut analysis of the changes in the level of metabolism which bring about the change in activity, in spite of the number of contributions to the subject (e.g., see the review of Precht, 1968). The problem is that most shifts in metabolism of the whole organism observed with changing temperature have undoubtedly resulted from changes in random activity which have neither been observed nor controlled (see analysis in Fry and Hochachka, 1970). Except for the work of Kanungo and Prosser (1959), changes in active metabolism do not appear to have been followed in the course of thermal acclimation. Kanungo and Prosser, while finding values only about one-fourth of those reported by Basu (1959) and Kutty (1968a), got a shift such that up to 25°C active metabolism was higher for fish acclimated to 10°C than for those acclimated to 30°C. Above 25°C the positions were reversed. Both curves reached the same peak, about 200 ml/kg/hr, the curve for 10°C acclimation at 25°C, the other at 30°C. The two curves appear to be essentially parallel with a lateral shift when plotted on semilogarithmic paper. The authors themselves, by plotting logarithms of the rates on a logarithmic scale, suggested that the curves rotate, but they did not give the mathematical justification for their procedure.

It seems impossible to measure standard metabolism in fish at any but the acclimation temperature because of the stimulus brought about by temperature change which may not be reflected in overt movement and thus escape detection by current methods of accounting for departure from the standard state. Even changes in active metabolism will have to be viewed with the reservation that excitement metabolism may also enter in, and, in addition, that the cost of the various regulatory functions may, and probably does, vary with any departure from the acclimation temperature. Perhaps the problem will be seen more clearly after the consideration of the costs of regulation in Section V.

The rate and degree of adjustment to change of temperature as reflected in tissue metabolism and adjustments in the capacities of organ systems have not often been investigated but all the work done indicates appropriate changes in capacity to compensate for change in temperature. Thus, Prosser and his associates (Prosser and Fahri, 1965; Roots and Prosser, 1962) have found the activity of the nervous system of the goldfish to show almost complete compensation, i.e., to change almost to the same degree as the change in acclimation temperature. Four degrees' change in acclimation temperature brought about three degrees' change in cold block temperature for nervous activity. Compensatory changes in the function of the circulatory system are suggested by the work of Hart (1957), Das and Prosser (1967), and Jankowsky (1968). Smit (1967) showed temperature compensation in the digestive system, in the rate of secretion of pepsin and acid.

Because of the difficulties of interpretation (see, e.g., Peterson and Anderson, 1969b), the effect of acclimation temperature on the metabolic rate of various tissue slices, minces and breis will not be discussed. The question of cellular restructuring in relation to temperature adjustment will be dealt with by Hochachka and Somero in Chapter 2.

## IV. LIMITING FACTORS

The limiting factors are first of all the metabolites, food, water, and the respiratory gases. Other identities operate as secondary limiting factors when they influence the rate of exchange of the metabolites between organism and environment.

The discussion of limiting factors will be largely confined to consideration of the effects of varying the concentration of the respiratory gases, oxygen and carbon dioxide. Basically a decrease in oxygen or an increase in carbon dioxide, over the ranges of concern here, operate in the same way. The supply of oxygen to the tissues is restricted.

The effect of a limiting factor is to restrict activity. Two examples of such restrictions are shown in Fig. 19. A feature common to both these examples, and indeed ordinarily to be found, is that the limiting factor (here oxygen) becomes operative at a relatively high value. Thus, in Fig. 19A, the ability to swim is first affected by oxygen concentration at about 6 mg/liter $O_2$ at 10°C and 10 mg/liter at 20°C. The coho, a salmon, would of course be expected to be sensitive to relatively slight decreases in oxygen concentration, but the goldfish, which in contrast is expected to withstand low oxygen, surpasses another salmonid, the rainbow trout, only below 2.5 mg/liter (Fig. 22) and in this range probably does so because it can operate to a certain degree

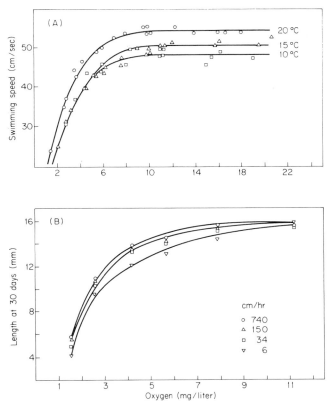

Fig. 19. (A) Swimming speed of underyearling coho salmon, *Oncorhynchus kisutch*, in relation to oxygen concentration. From Davis *et al.* (1963). (B) Effect of oxygen concentration and rate of percolation on the growth of embryos of steelhead trout, *Salmo gairdneri*, at 9.5°C. From Silver *et al.* (1963).

anaerobically (Kutty, 1968a) not because it can take up a great deal more oxygen (Basu, 1959).

Panel B not only shows an example of the operation of a limiting factor on another type of activity, namely, development, but emphasizes the problem of "supply." When the fish is in the egg, oxygen supply depends on three circumstances, one chemical—concentration, one physical—diffusion pressure, and one mechanical—rate of flow. A complete unit of oxygen supply has not yet been proposed. In the present section, as a compromise, the unit of concentration will be used since the countercurrent system by which the fish in general gains its oxygen when out of the egg seems to be more dependent on the mass of oxygen presented to the respiratory surface than on the partial pressure.

The response of swimming to the limiting effect of low oxygen is under the control of the central nervous system as Fig. 20 shows. In this example the fish was stimulated to swim steadily at a moderate speed (about one-half its maximum capacity for steady swimming, cf. Kutty, 1968a) and the oxygen allowed to fall gradually. After about 2½ hr the fish abruptly stopped swimming and fell back against the screen when the oxygen content fell a little below 2 mg/liter. The current in the chamber was maintained and the oxygen allowed to subside a little further for another half-hour. Meanwhile the fish remained on the screen. Then the oxygen content was raised again. Within about 5 min when the oxygen content had risen above the level at which swimming had stopped the fish was again breasting the current steadily and continued to do so for some 3 hr until the oxygen became critically low once more. Again the fish abruptly stopped swimming, and again it resumed swimming promptly when the oxygen content was slightly increased.

The indication from such an experiment as Kutty's above is that an organism may adjust to a limiting factor and restrict activity during the period when it is imposed. However, there can be circumstances where the organism becomes committed to a given resource when a given identity is not limiting and then a fluctuation produces limiting conditions. Daily fluctuations in the oxygen content of well-vegetated waters is a familiar case of this sort. Brook trout will not grow in an aquarium where the oxygen is restricted only for part of the daily cycle (Fig. 21).

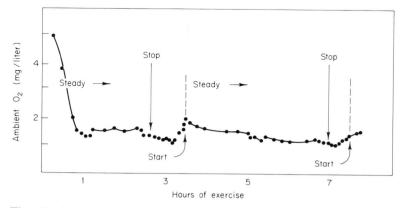

Fig. 20. Swimming response of an 18-cm goldfish to changing oxygen concentration while exposed to a water current of 60 cm/sec at 20°C. Arrows indicate times at which the fish stopped swimming and started again. From Kutty (1968b).

The effect of small inert bodies, such as silt or pulp fines, which apparently operate by displacing an equivalent amount of water and the oxygen supply it contains and otherwise interfering with the respiratory flow, is a case of operation of a limiting factor of special interest to pollution biologists.

Some effects of limiting factors may be extremely obscure. For example, Kinne and Kinne (1962) observed a reduction in the rate of development of the desert pupfish, *Cyprinodon macularius*, in relation to increased salt content in the water in which they were incubated. These authors were able to correlate the change with the change in the saturation value of oxygen as influenced by the presence of the salt. There was apparently no other substantial effect of the widely differing salt content in the water in which the various samples were hatched.

## A. Acclimation to Low Oxygen

A clear shift in the lethal level of oxygen in relation to acclimation level was shown by Shepard (1955) for the eastern brook trout, *Salvelinus fontinalis,* and for three warm-water species by Moss and Scott

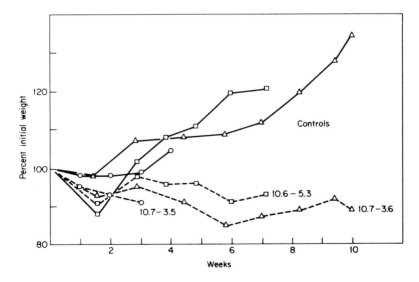

**Fig. 21.** Growth of yearling eastern brook trout, *Salvelinus fontinalis,* at constant high (solid lines) and various daily fluctuating levels of oxygen (broken lines). Numbers indicate upper and lower levels of oxygen in milligrams per liter. From Whitworth (1968).

(1961). To explain the change, Shepard found an increase in the ability to extract oxygen from water when the concentration was low, as did Prosser *et al.* (1957) also, for goldfish. Similarly, MacLeod and Smith (1966) found that the hematocrit of the fathead minnow, *Pimephales promelas,* changed in response to lowered oxygen or increased content of pulp fiber. Thus changes have been found in the supply system, as is well known for mammals.

In the brook trout the relation between lethal and acclimation levels of oxygen was linear and can be expressed by the formula $y = 0.88 + 0.08x$, where $y$ is the lethal level and $x$ the acclimation level of oxygen, both being expressed in milligrams per liter. The lower limit of the formula is when $y = x$; the upper limit of Shepard's observations was air saturation. MacLeod and Smith also found that the response of the hematocrit was linear over the whole range of oxygen concentration they investigated and for concentrations of pulp fiber from zero to 800 mg/liter.

All adjustments, however, do not seem to be so simple. There appears to be a good deal of accommodation. Kutty (1968b) found no difference between goldfish acclimated to low oxygen and those acclimated to air saturation with respect to the speed at which they could be induced to

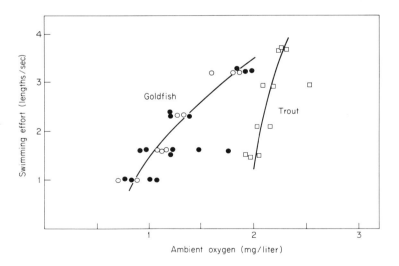

Fig. 22. Swimming effort of 18-cm goldfish at 20°C and 20-cm rainbow trout, *Salmo gairdneri,* at 15°C in relation to ambient oxygen. Open symbols denote fish acclimated to oxygen at air saturation, closed (goldfish only) acclimated to 15% air saturation. Both species acclimated to their respective test temperatures. From Kutty (1968b).

swim at limiting levels of oxygen (Fig. 22), although Prosser *et al.* (1957) showed the potential for increased oxygen uptake. Furthermore, Kutty demonstrated (Fig. 23) that at a given swimming speed the oxygen consumption was reduced greatly in goldfish acclimated to low oxygen as compared with those acclimated to high oxygen. A similar reduction in the overt cost of running was found by Segrem and Hart (1967) for the white-footed mouse. Thus, assuming all these observations to be valid, it seems that acclimation to low oxygen involves both an enhancement of supply and a restraint in utilization. With regard to the reduction in consumption at a given speed, comparison of Figs. 1 and 23 suggests, since the oxygen consumption of the fish acclimated to low oxygen lies on the lower boundary of the triangle relating oxygen consumption to speed, that when acclimated to low oxygen the fish is applying its energy only to the business at hand. Such a behavioral modification is also indicated in the response of fish to restricted food supply. Paloheimo and Dickie (1966b) showed the efficiency of food conversion was inversely

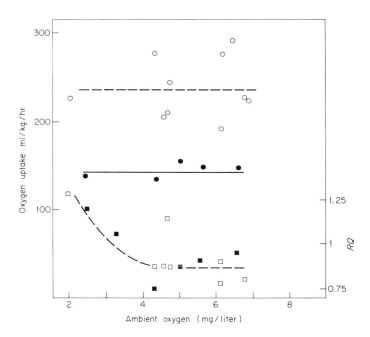

**Fig. 23.** Oxygen consumption (circles) and RQ determinations (squares) of goldfish acclimated to oxygen at air saturation (open symbols) and 15% air saturation (closed symbols). Fish approximately 18 cm long swimming continuously at 45 cm/sec. Determinations after the first two hours of swimming. From Kutty (1968a).

related to the size of the ration, but it must be noted as Brett *et al.* (1969) state that the examples available to those authors did not have data for cases of severe restriction.

There are few data on the rate of acclimation to low oxygen. Shepard (1955) demonstrated that acclimation of the eastern brook trout, *Salvelinus fontinalis,* to a change of oxygen concentration was 95% complete in 100–200 hr at 10°C, being somewhat slower if the fish were in the dark and presumably thus not so active. The data of Moss and Scott (1961) support Shepard's findings.

It seems likely that a moderate imposition of a limiting factor does not involve the imposition of stress, although there will be restriction in activity. It is normal of course, as shown in Fig. 18, for air saturation to limit active metabolism at higher temperatures which otherwise are well within the normal range. Dahlberg *et al.* (1968) showed that while the growth rate of the largemouth bass was restricted in their experiments from an oxygen concentration of 8 mg/liter down, the food conversion ratio remained stable at least down to 4 mg/liter and perhaps 3 mg/liter. In this case, food consumption was progressively restricted throughout the total range of oxygen in question, presumably through some central control as in the case of swimming speed (Fig. 20). Their experiments were carried out at constant levels of oxygen in contrast to the fluctuating levels employed by Whitworth (1968) (Fig. 21).

## B. Oxygen Concentration and Metabolic Rate

There is an extensive early literature on this subject which can now be considered to be of only historic interest and which can largely be found through the reviews of Tang (1933) and von Ledebur (1939). The confusion in the earlier literature lies in a lack of distinction between the various levels of metabolism. The point of view expressed in the present discussion has its origins in the work of Van Dam (1938) on resting metabolism and of Lindroth (1940), who appreciated that decreased oxygen limited active metabolism. Lindroth (1942) stated the concept designated below as the "level of no excess activity." Figure 24 shows the typical response of the standard and active metabolic rates to oxygen concentration. Standard metabolism in the brook trout is relatively unaffected by oxygen concentration until the level of oxygen drops to approximately 50% air saturation. Below that point there is first an increase, as was well-demonstrated by Van Dam (1938), which it is presumed takes care of the increased needs of ventilation. The need for increased ventilation seems to vary with the species and with the tem-

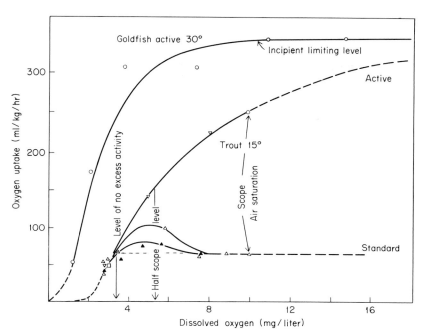

**Fig. 24.** Active metabolism of the goldfish, and active and standard metabolism of the eastern brook trout, *Salvelinus fontinalis*, in relation to oxygen concentration. Closed symbols, fish acclimated to test level of oxygen; open, acclimated to air saturation. From Basu (1959) (○), Beamish (1964c) (△,▲), Graham (1949) (▽), and Job (1955) (□).

perature. In the goldfish (Beamish, 1964b) such cost of respiration can hardly be detected at 10°C when the standard oxygen consumption is only 15 ml/kg/hr for a fish of approximately 100 g, while there is an approximate doubling of the standard rate at approximately 40% air saturation if the goldfish are at 20°C. As the oxygen concentration drops still further the fish can finally no longer increase its oxygen consumption, even with heavy breathing, and the total rate of oxygen consumption then falls progressively with further lowering of the oxygen concentration. This is not to say that ventilation may not still increase, but the fish then has to resort to anaerobic support. In his study of respiratory efficiency in relation to respiratory flow, Saunders (1962) took advantage of the increased ventilation induced by lowering the ambient oxygen to stimulate the higher rates of ventilation. Acclimation to low oxygen possibly reduces the cost of ventilation, as is indicated by the solid triangles in Fig. 24.

In contrast to the course of standard metabolism in relation to oxygen concentration, which is relatively unaffected, active metabolism may be strongly influenced by oxygen concentration at all levels up to air saturation and even higher. The experimenter often must artificially increase the oxygen content of the water if he wishes to obtain the full extent of active metabolism (e.g., Fig. 17). In Fig. 24 the curve for the brook trout was carried only as far as air saturation but that for the goldfish was continued to higher levels of oxygen concentration. Both are for fish acclimated to air-saturated water. As mentioned above, acclimation to lower oxygen would have displaced the curves to the left (e.g., Prosser et al., 1957; Shepard, 1955). An earlier speculation of Fry (1947) that the maximum would be reduced still has not been tested, in spite of the apparent confirmation shown by Prosser et al., since it is probable that they did not increase the swimming speed to the limit.

The terminology chosen here to describe the relations of metabolism to a limiting factor is indicated in Fig. 24. As was the usage in considering controlling factors, the scope for activity is taken as the difference between active and standard metabolism. Two restrictions of scope are designated: the "half-scope concentration"—that concentration of oxygen where the active metabolism is reduced to the point where the scope is one-half that at air saturation—and the "level of no excess activity"—the point where the active metabolism is reduced to the standard level. The half-scope level (Basu, 1959) is simply a convenient arbitrary point to take in discussing the restrictive effects of a limiting factor. The level of no excess activity approximates the lethal level. It will be noted in the diagram that both these points have been estimated from an extrapolation of the line for standard metabolism determined at concentrations of oxygen higher than those at which the cost of respiration increases. In the case of the half-scope value, it can be taken that the cost of respiration is often absorbed in the general cost of physical activity, at least within the limits of accuracy of the index, since swimming fish frequently, and perhaps ordinarily, passively irrigate their gills by their forward movement, except of course in start and stop activity. Swimming with the mouth open will, of course, contribute to drag so that some cost is still there, but the efficiency of irrigation is greatly increased since there is no longer the need to accelerate each mouthful of water to take it in nor to accelerate it again to expel it from the epibranchial cavity on exhalation. Only the friction to pass water steadily through the branchial sieve remains to be overcome in the work of respiration (see C. E. Brown and Muir, 1970, for a quantitative analysis). Similarly, it can be expected that the activity of the swimming muscles will promote the circulation and reduce the work of the heart. The position of the level of no excess activity is decidedly more arbitrary, for at that point the fish is breathing heavily and the cost of respiration must be maximal. Extrapolation in this case really relies on an overestimate in the determination of standard metabolism to compensate for a change in the cost of respiration and on the possibility of some anaerobic support for activity.

The final term indicated in Fig. 24 is the "incipient limiting level" shown for the goldfish, that is, the point where a further reduction in oxygen begins to restrict the active metabolic rate. The point is of little ecological interest as a datum since it does not necessarily appear under natural conditions, being often above air saturation. Moreover, it is not a precisely defined point.

## C. Combinations of Oxygen and Carbon Dioxide

The effects of low oxygen and high carbon dioxide show the typical consecutive interaction of limiting factors (Fig. 25A). Here, to take the dashed line for the effect of 18 mg/liter $CO_2$ at 20 hr of acclimation, oxygen below 4 mg/liter is limiting and no effect is seen from the additional carbon dioxide present, while the effect of carbon dioxide as indicated by the horizontal position of the curve is complete at about 10 mg/liter $O_2$. At 61 mg/liter $CO_2$, carbon dioxide exerts a graduated limiting effect over the whole region of the data up to 15 mg/liter $O_2$. Thus, there is a wide transition phase between the operation of oxygen as the preponderant limiting factor to that of carbon dioxide. The data are imperfect here, and the effect is better shown in Fig. 28. There is, of course, a substantial early literature on the precise nature of the operation of limiting factors which is to a large degree nowadays irrelevant but which is admirably discussed in Booij and Wolvekamp (1944). There seems to be no point in dealing with the sharpness of the transition phase, which was the subject of most of the early work, nor indeed in the complete preponderance of one limiting factor over another. The best present approach appears to be empirical description of the effects of limiting factors on active metabolism.

The major point of concern to the fisheries biologist with respect to the respiratory gases is that under natural conditions oxygen lack is a much more likely limiting factor than carbon dioxide excess, particularly since it is only under anaerobic conditions that free carbon dioxide can ordinarily reach major levels. It is only under rather special conditions that carbon dioxide becomes a limiting factor. The commonest such condition is when fish are transported (e.g., see review of Fry and Norris, 1962).

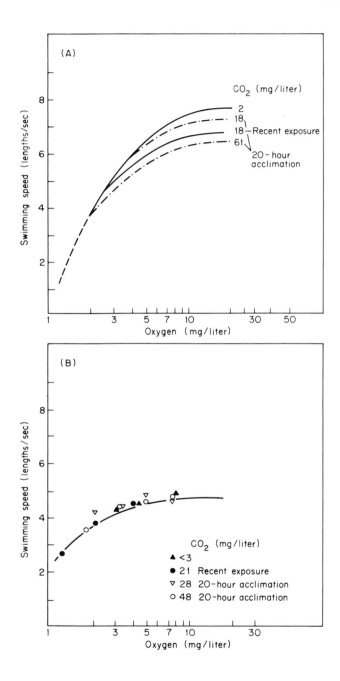

Different species display different sensitivities to carbon dioxide. The largemouth bass (Fig. 25B) showed no effect on exposure to 48 mg/liter free $CO_2$.

A final point to be noted in Fig. 25A is that there may be extensive acclimation to carbon dioxide in a comparatively short time, as is shown by the difference between the curve for acute exposure to 18 mg/liter free $CO_2$ and for 20-hr acclimation to that level. Similarly, Saunders (1962) found that the efficiency with which oxygen is taken up at the gills, which is reduced by the presence of increased carbon dioxide, is recovered in a few hours of continuous exposure to moderate levels of $CO_2$. Again, the data of Lloyd and White (1967) suggest that the change in blood bicarbonate (Lloyd and Jordan, 1964) in response to increased $CO_2$ is largely complete in 24 hr in rainbow trout at 12°–16°C.

Beamish (1964c) and Basu (1959) have carried out the most extensive researches to date on the interaction of various levels of oxygen and carbon dioxide on standard and active metabolic rates. Beamish (Fig. 26) showed that standard metabolism was essentially uninfluenced by the level of free carbon dioxide until the total uptake of oxygen was reduced below the requirements for standard metabolism. On the other hand, as Basu found, the active metabolic rate at any given level of oxygen concentration is progressively reduced by increase in carbon dioxide concentration (Figs. 26 and 27). There is, however, a great limitation in Basu's data in that his results are for acute exposure of animals acclimated to low $CO_2$. No one has yet, apparently, measured the active metabolism of fish acclimated to high $CO_2$.

The essential feature of Basu's results is that the response to increasing carbon dioxide is an exponential decrease in active oxygen consumption. The same proportionate effect was found at all oxygen concentrations down to a low value, which in the case of the carp shown is 12.5% air saturation; below this the rate of reduction of oxygen consumption with increasing carbon dioxide sharply increased. The increased slope shown by the response to $CO_2$ at low oxygen in the carp was found also in the bullhead, *Ameiurus nebulosus*, and in the goldfish, *Carassius auratus*. Basu had no explanation for this phenomenon but did show the change in response was statistically significant, whereas there was no significant difference in slopes at the various higher concentrations of oxygen.

A special feature in the data for the eastern brook trout is shown by

---

**Fig. 25.** Swimming speed of (A) coho salmon, *Oncorhynchus kisutch*, at 20°C and (B) largemouth bass, *Micropterus salmoides*, at 25°C. From Dahlberg *et al.* (1968). Fish approximately 8 cm long.

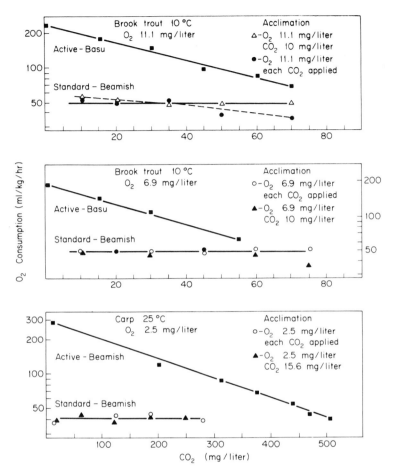

**Fig. 26.** Effect of various concentrations of oxygen and carbon dioxide on active and standard metabolism. From Basu (1959) and Beamish (1964c). Measurements with active metabolism made only with fish acclimated to air-saturated water.

the points enclosed in the dashed ellipse. Basu was unable to reduce the oxygen consumption of this species below approximately 70 mg/kg/hr $O_2$ under the conditions of the experiment and concluded that the species was able to transport that much oxygen by serum transport alone at the oxygen level indicated.

Figures 27C and D show the response to temperature. The characteristic shown here, namely, that the effect of a given concentration of carbon dioxide was least at the highest temperature investigated for a given species, is true of goldfish and the bullhead also (Basu, 1959).

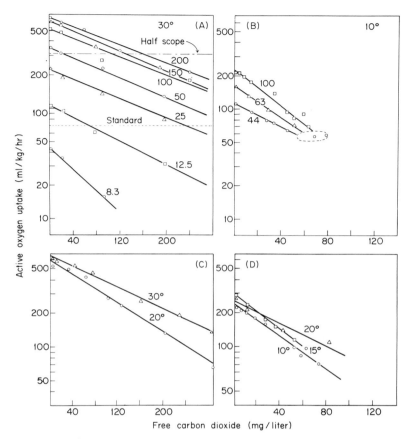

**Fig. 27.** Effect of various concentrations of oxygen and carbon dioxide on active metabolism of brook trout, *Salvelinus fontinalis* (B,D) and carp, *Cyprinus carpio* (A,C). From Basu (1959) and Beamish (1964a).

Basu was able, on the assumption there would be no effect of carbon dioxide on standard metabolism, which Beamish (1964c) later demonstrated, to show that the curves for respiratory sensitivity as found, for example, by E. C. Black *et al.* (1954) could be calculated from the data on oxygen consumption with a high degree of approximation. Curves based on his calculations of the effects of the interaction between oxygen and carbon dioxide on various levels of activity are shown in Fig. 28.

The various isopleths in Fig. 28 are constructed by taking points at which the rate of oxygen consumption had been reduced to the same

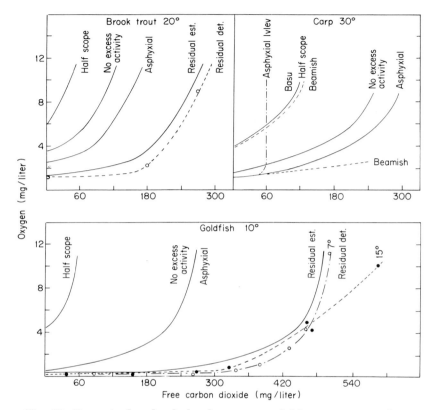

**Fig. 28.** Determined and calculated sensitivity of fish to various combinations of oxygen and carbon dioxide. From Basu (1959).

given level from such curves as are in Figs. 27A and B. Values for respiration at the half-scope level and the level of no excess activity based on Beamish's (1964a) more recent determinations of standard metabolism are included in Fig. 27A. Points determined from the intersection of these lines are also plotted in Fig. 28 as indicated.

The analysis presented above glosses over a number of points. First, as already mentioned, the active rates are for fish not previously acclimated to the test levels of carbon dioxide, except for the lowest. Second, Basu's calculations did not take into account any change in the cost of irrigation of the gills (Beamish, 1964c) or transport of oxygen in the blood. Finally, Basu's work was carried out with water of 270 mg/liter $CaCO_3$ hardness. Ivlev (1938), using water with a bicarbonate alkalinity of about 40 mg/liter, found the asphyxial level of $CO_2$ to be independent of oxygen concentration above 60 mg/liter $CO_2$, whereas Basu's respiratory data indicate a continuing interaction between oxygen and carbon dioxide up at least to air saturation. It is probable that the difference is due to the difference in water hardness between the two tests. Lloyd and Jordan (1964) reported an interaction between high $CO_2$ and low pH on the time of survival of rainbow trout, *Salmo gairdneri,* such that, with oxygen at air saturation, approximately 20 mg/liter free $CO_2$ was fatal to rainbow trout at approximately pH 5.5. In neutral water, trout could be expected to survive continuous exposure to at least twice that concentration of free $CO_2$. A free carbon dioxide concentration of 60 mg/liter may possibly therefore represent a lethal pH for carp in soft water.

## D. Interaction of Limiting and Controlling Factors

Figure 18, as pointed out at the time it was introduced, could not be completely discussed from the point of view of controlling factors alone. In air-saturated water, the normal environment for the sockeye, the limiting effect of oxygen intervened above 15°C to suppress the full potentiality for active metabolism permitted by temperature. Thus the effect is for the sockeye to have an optimum for activity at 15°C, much below its lethal temperature. Oxygen at air saturation is often limiting for fish as the example shows. Such a temperature optimum is frequently called a "conditioned" optimum, e.g., an optimum temperature conditioned by the existing level of oxygen.

Figure 29 shows the generalized response of the eastern brook trout to various combinations of oxygen and temperature. Unfortunately, there is not sufficient information to provide a strictly quantitative picture. In particular, there are no data for active metabolism in relation to acclimation to oxygen except those of Shepard (1955). Figure 29A shows the typical effect of temperature on the rate of oxygen uptake when oxygen is limiting. At a higher temperature more oxygen can be taken up at a given concentration of oxygen over the whole range, whether oxygen is limiting or not, although in the brook trout that effect is not seen above 13°C. In part the difference is the result of the imperfect expression of supply, as mentioned earlier, but also there is a change in regulation. There may be more irrigation and flow will be higher for a given pressure head at a higher water temperature, since the gill is essentially a capillary sieve and flow through it at a given pressure depends on viscosity. Utilization may also be more complete with increas-

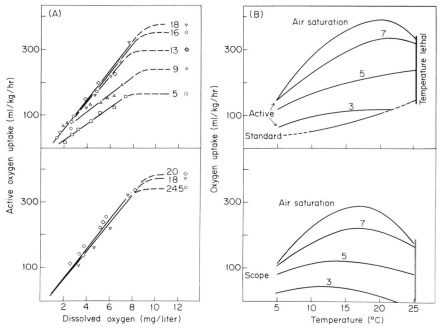

**Fig. 29.** (A) Limiting effect of oxygen on active respiration at various temperatures. (B) The interaction between oxygen concentration and temperature on the metabolic rate of the eastern brook trout, *Salvelinus fontinalis*. Based on the data of Basu (1959), Beamish and Mookherjii (1964), and Graham (1949). [Note that ppm in Fig. 9, Graham (1949) should have been ml/liter.] Numbers in A indicate temperature; in B, mg/liter $O_2$.

ing temperature because of an increase in blood flow, more rapid diffusion, and more rapid progress toward equilibration in the hemoglobin.

An important consequence of the interaction of oxygen concentration and temperature is that the optimum temperature in terms of scope for activity cannot be predicted from the lethal temperature. The relation of maximum scope to lethal temperature depends on whether and where the normal oxygen content begins to exert a limiting effect. Species with similar lethal temperatures may have quite different optima as conditioned by oxygen at air saturation. While the data must now be considered only semiquantitative, Fry (1957) showed that various salmonids have similar lethal temperatures but while two species of trout (*Salmo*) show increasing active oxygen consumption right up to the lethal temperature, two chars (*Salvelinus*) have their active oxygen consumption limited by air saturation above about 15°C.

Consideration of the previous section indicates how a second limiting factor, say, carbon dioxide, can be added to the interaction with temperature. Basu (1959) gave an example of such a plot in his Fig. 9.

## V. MASKING FACTORS

The masking factors and the directive factors, discussed below, deal with the channeling of the energy available to the organism, which in the broad sense is all applied to organic regulation.

As an organized segment of the universe it goes without saying that all the independence an organism achieves comes through the interaction of the various identities in an appropriate matrix, itself fashioned from the environment or a pattern in it. Life is governed by all the laws of nature, but like a good corporation flourishes by playing one against another to its own advantage. Also, like the corporation it must pay its lawyer.

Organic regulation can be broadly classified as mechanical, physiological, and behavioral, the latter term being taken to include all the manifestations of the central nervous system, some of which as we know them ourselves may be internal but which in fish, if solely so, would not be accessible to us. Structure apparently always enters into regulation and the least costly regulation is achieved by some structural isolation which, after the investment has been made in ancestry and individual development, calls for next to no cost for operation and maintenance.

The present section will be brief since physiological regulation is the main subject of most of the treatise and will be confined to a summary of some recent work on the cost of regulation of the body fluids in the rainbow trout and an outline of our knowledge of the regulation of the body temperature in the tunas and the lamnid sharks. The latter is taken as an example of a regulation which is brought about almost entirely by the appropriate development of form.

## A. Cost of Ion-Osmoregulation

The physiological details of water economy and ion exchange are now becoming clear (see Chapters 1–3, Volume I) but we still have only fragmentary information concerning the metabolic cost of such regulations. Various workers (e.g., Keys, 1931; Leiner, 1938; Veselov, cited by V. S. Black, 1951; Hickman, 1959; Job, 1959) have noted differences in the routine metabolic rate of fish in waters of various salinities. Recent preliminary studies by Rao (1968) and Farmer and Beamish

(1969) of the relation between activity, metabolic rate, and the salinity of the medium appear to be the only ones to offer a quantitative estimate of the cost of regulation of the body fluids in fish. Rao's work is the basis of the present section.

Rao measured the metabolic rate of the rainbow trout, *Salmo gairdneri*, at rest and at various swimming speeds over a modest size range (approximately 40–120 g) in freshwater and various dilutions of seawater. He measured standard metabolism in Fry's apparatus and active metabolism in Blažka's chamber (Fig. 3). He acclimated his subjects to the test conditions of temperature and salinity, taking cognizance of the observation of Conte and Wagner (1965) that this species has a seasonal cycle in its tolerance to seawater. Thus, he worked with high salinities in late summer and autumn. However, he maintained his fish under a constant 12-hr light period.

Rao's data calculated for a 100-g fish are given in Table III. The boldface numbers are his observations, those in italics are the differences between the metabolic rate in the respective medium and that in 7.5‰ salinity, which is approximately isosmotic with fish blood. Rao obtained the result to be expected from the work of his predecessors (e.g., Job, 1959). The metabolic rate for a given level of activity was least in an

**Table III**

Cost of Ion-Osmoregulation in 100-g Rainbow Trout, *Salmo gairdneri*, in Relation to Swimming Speed and Salinity[a,b]

| Speed (cm/sec) | Oxygen uptake (ml/kg/hr) | | | |
|---|---|---|---|---|
| | Uptake at 7.5‰ | Excess over uptake at 7.5‰ | | |
| | | FW | 15‰ | 30‰ |
| | | | 5° | |
| 0 | **38** | *2* | *4* | *14* |
| 18.5 | **62** | *11* | *12* | *20* |
| 45.1 | **94** | *29* | *32* | *43* |
| 57.5 | **123** | *36* | *50* | *65* |
| Max. speed | **186** | *62* | *66* | *92* |
| | | | 15° | |
| 0 | **66** | *13* | *11* | *23* |
| 18.5 | **92** | *26* | *28* | *41* |
| 45.1 | **159** | *36* | *29* | *62* |
| 72.7 | **246** | *59* | *62* | *87* |
| Max. speed | **340** | *69* | *78* | *97* |

[a] As indicated by excess in oxygen consumption at a given salinity over oxygen consumption at 7.5‰ salinity (approximately isosmotic).

[b] Based on Rao (1968), his Table 2.

isosmotic dilution of seawater. Further, he found that the cost of ion-osmoregulation was proportional to the metabolic rate and hence presumably a function of the respiratory flow (Fig. 30). Thus, isolation plays a large part in osmotic regulation, the general body surface being largely impermeable to water, as was already considered to be the case. Another finding of considerable ecological significance was that the cost of regulation is little affected by temperature. Indeed, if the solubility of oxygen in water is taken into account and with the assumption that the efficiency of extraction is constant, then the cost of regulation bears the same relation to respiratory flow at both temperatures investigated. Further, the increase in cost of regulation in seawater is not proportional to the increase in osmotic gradient. The metabolic data confirm the findings of Houston (1959) and Gordon (1963) that there is a decrease in the permeability of the exposed membranes as an adjustive response to increased salinity.

The second major point of ecological interest in Rao's work is that under the conditions of his tests the systems for uptake and transport of oxygen could handle the cost of ion-osmoregulation in addition to the cost of physical activity; thus, no penalty with regard to the ability to physically compete was imposed at air saturation by the increased regulatory load. Under his circumstances oxygen was not limiting, and each

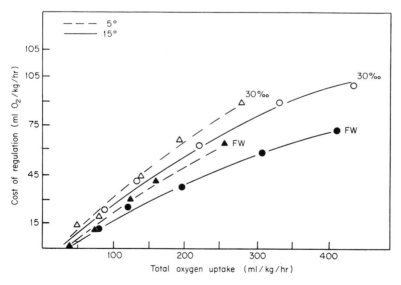

Fig. 30. Relation between total metabolism and the cost of ion-osmoregulation in fresh water and 30‰ salinity for 100-g rainbow trout. From Rao (1968).

organ system had full scope to carry on its appropriate activity as required and as permitted by the controlling factors.

We can only speculate concerning the result if oxygen were limiting, but such speculation at least leads to an interesting hypothesis. Figure 31 suggests that as oxygen becomes limiting the cost of internal regulation must compete with the scope for activity. In consequence, with increasing limiting conditions the scope for activity can be expected to be reduced more quickly than the metabolic rate. Two courses for such a decrease are suggested in Fig. 31 labeled "absolute" and "modulated," respectively. The absolute curve postulates that all regulatory processes will have precedence over external activity; the modulated curve postulates that activity will compete with regulation so that regulation is not perfect and the internal condition can be allowed to drift within limits toward the external one. In the rainbow trout (Rao, 1969) there appears to be such modulation in osmotic concentration, but there is no profit at present in carrying the speculation further.

Since Rao has shown that the cost of ion-osmoregulation can be added to the scope for activity it can be presumed that other costs may be similarly added, at least up to some limit of the oxygen supply system we have not yet determined. Such an important cost is the cost of assimilation, and Fry's suggestion (1957) that the rate of oxygen consumption might be the limit to growth, which has been questioned by Swift (1964), needs examination. Measurements of active metabolism have been made till now with fasted fish so that the cost of assimilation has been largely removed from them. Again experiment needs to catch up with speculation.

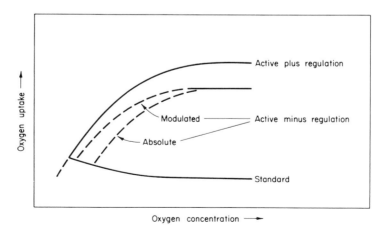

**Fig. 31.** Hypothetical effects of cost of physiological regulation on scope for activity in the presence of a limiting factor. See text for explanation.

Small rainbow trout (approximately 50 g) are difficult to maintain in seawater and probably do not fully adjust. Rao (1969) suggested that when the capacity to regulate is overreached then metabolism (irrigation) may be restricted. A selection of his data are plotted in Fig. 32. In the lowermost series of curves where the fish are swimming slowly (about 1 length per second for a 100-g fish), the cost of regulation is proportional to the metabolic rate up to a salinity of 15‰ (about half-strength seawater) as shown by the parallel course of the log weight–log metabolism lines. At the highest salinities (30‰) the cost increases for

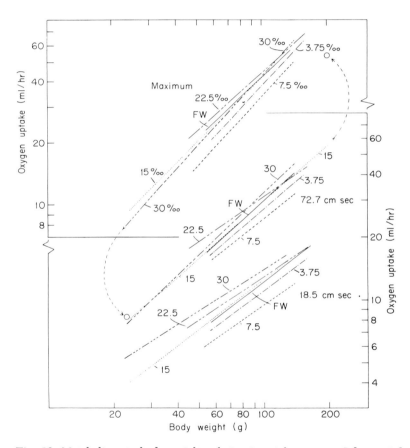

**Fig. 32.** Metabolic rate–body weight relation in rainbow trout, *Salmo gairdneri,* as related to swimming speed and salinity at 15°C. From Rao (1967, 1968). Note that the scale for the ordinate is displaced upward for curves relating to maximum swimming speed. The curved arrows give the positions of the indicated ends of the 30‰ and 15‰ curves on the respective scales. Numbers indicate salinity in parts per thousand; FW is fresh water.

smaller fish. The convergence of the lines at the upper ends may be taken as an artifact. It is probable that if a range of large fish had been examined the whole line for 30‰ salinity would have been a concave curve, with the slope at the upper weights being parallel to the slopes found at lower salinities.

In the next series of curves the fish were forced to swim about four times as fast. All sizes still seem capable of regulation up to 15‰ salinity, and the same relative increased cost of regulation for small fish is shown at 22.5‰ as was found at the lower speed. However, there is a decided change at 30‰ so that now the metabolism of the smaller fish is depressed. The same picture is found in the metabolism at maximum speed where there is also a suggestion that metabolism is beginning to be depressed in the smaller fish at 22.5‰.

Rao presented further data for standard metabolism and at an intermediate swimming speed which are concordant with the series shown. At 5°C he found little if any evidence of inability to regulate in the small fish within a similar size range but did find a relatively higher metabolic rate for standard metabolism in small fish at 30‰ salinity.

Unfortunately, Rao had to stop his investigations at this point so that an analysis of the response is not available. In particular, the osmotic pressure of the blood in the small fish is not known. Rao (1969) found a significant increase in the osmotic pressure of swimming fish weighing 100 g, but he did not examine either smaller or larger specimens. There may then be a reduction in cost of regulation because the fish allows the osmotic gradient to be reduced, although it would take a major departure in serum values from those found in 100-g fish for such an effect to account for the changes in metabolic rate. Another possibility is that the fish reduces its irrigation or perhaps increases mucus secretion at the gills. These various possibilities, however, are simply speculation.

An important consideration is that rainbow trout have little capacity for anaerobic metabolism (Kutty, 1968a). However, if the oxygen consumption data of Rao (1968) and Kutty (1968a) for this species are compared it will be found that Rao's data for fresh water are somewhat higher than Kutty's. Since he did not extend his readings at any one speed over as long a time as did Kutty, it is probable that he did not ordinarily find the minimum metabolic rate at a given swimming speed. Accordingly, part of the result of increasing ion-osmotic stress could also be an adjustment in behavior to conserve energy while swimming at a given speed, as appeared to be the case in the goldfish acclimated to reduced oxygen (Fig. 23). In any event, it appears likely that activity is accommodated to the limits of regulation in this activity as it may be in others.

## B. Thermoregulation in Fish

In general the body temperature of fish is slightly above the ambient temperature, the difference being of the order of 0.5°C (Nicholls, 1931). Such a difference can be explained by the use of gills for respiration. The countercurrent system of exchange in the gill assures that the temperature of the blood will be reduced almost to that of the water on every circuit. The potential heat capacity of chemical transport is of the order of 0.5 cal/ml; thus, an excess temperature of the order of 0.5°C can be gained each circuit. Consequently, if the heat is also dissipated each circuit, the excess temperature at the site of metabolism will be of the order of 0.5°C. Lindsey (1968) has found excess deep muscle temperatures up to 2.6°C in very large fish (Fig. 34A). Since arteries and veins tend to run parallel in tissues, there is the probability that there may be some incidental local conservation of heat by countercurrent exchange, which can account for the excess temperatures these authors reported. The highest excess temperature found by Lindsey was in white muscle of fish which had recently struggled. Probably the anaerobic activity is not immediately balanced by increased circulation to drain away the heat. There do not appear to be any body temperature data for predominantly air-breathing fish.

Two groups of fish, the tunas (various authors to Carey and Teal, 1969b) and the lamnid sharks (Carey and Teal, 1969a), have deep muscle temperatures which may be greatly in excess of the water temperature (Fig. 33). These fish have retia mirabilia which in particular supply the red lateral muscles from the lateral artery and drain them through the lateral vein. These retia conserve the metabolic heat of the muscles they serve, which are those continuously active. The remainder of the muscle mass is presumably heated by conduction from the red muscle (Fig. 34B), although in some species there are retia associated with the dorsal and visceral as well as lateral blood vessels. The retia were figured by Kishinouye (1923) who also surmised their function. Aside from a diagram given by Carey and Teal (1966) there appears to be no recent description or specific detail with respect to the distribution from any one rete to a given muscle.

The data in Fig. 33 are for very large fish, and in these the muscle temperature can be extremely stable with respect to ambient temperature. However, when even these large fish become active their body

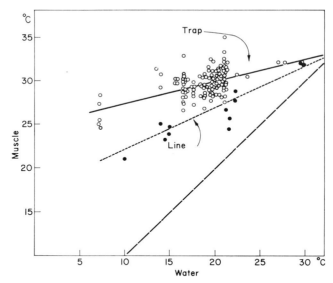

**Fig. 33.** Body temperature regulation in bluefin tuna, *Thunnus thynnus.* Data of Carey and Teal (1969b). From Fry and Hochachka (1970). Trap-caught fish were killed without a struggle. Line-caught fish were played for some time before landing.

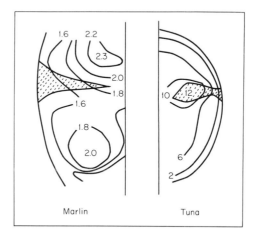

**Fig. 34.** Excess temperature profiles (°C) in muscles of·newly landed bluefin tuna, *Thunnus thynnus* (Carey and Teal, 1969b) and marlin, *Makaira mitsukurii* (Lindsey, 1968). The marlin had been played for some time, the bluefin was shot in a trap. Hatched areas are red muscle. Note centers of high temperature are in the white muscle in the marlin and in the red in the tuna.

temperature drops, presumably because the main body mass is supplied by blood from the dorsal aorta. Accordingly, the thermoregulation of these fish appears to be achieved by thermoconservation rather than by thermogenesis and thus does not call for the expenditure of extra metabolism. There is, of course, regulation, but the regulation is best expressed by the lowering of the differential in warm water, presumably by a relaxation of the countercurrent conservation system. In smaller fish, regulation is not as efficient as in the large tuna and the excess muscle temperature declines as the ambient temperature decreases although a substantial differential may still be maintained, e.g., 8°C at 20°C in the skipjack, *Katsuwonis pelamis* (Barrett and Hester, 1964).

A point of great interest on which there is at present virtually no information is whether the brains of these fishes are maintained at the temperature found in their muscles. Stevens and Fry (1971) found an excess temperature of about 4.5°C in the brain of skipjack in a sample of 20 fish in which the excess muscle temperature was 9°.

If it should transpire, as appears probable, that the tunas do not regulate the brain temperature to the degree they regulate the muscle temperature, the question then arises as to what is the utility of regulating the muscle temperature. Looking back to Fig. 17, which presumably shows the limits for acclimation of muscle, the utility of a high muscle temperature is clear. While the ratio of active to standard metabolism is constant, the difference between the two increases greatly with increasing temperature so that scope is much greater in that species at 15°C than at 5°C. Thus there are limits to temperature compensation with respect to total metabolism, which must be taken to be in large degree muscle metabolism. Brain metabolism however may compensate much more perfectly (Baslow, 1967; Baslow and Nigrelli, 1964) so that a cold brain may still be able to govern a warm muscle.

## VI. DIRECTIVE FACTORS

A directive factor is an environmental identity which exerts its effect on the organism by stimulating some transductive response. The examples of the elaborate sense organs such as the eye or the ear are of course self-demonstrative from common knowledge. Other sense organs such as those which sense water temperature (see Murray, Volume V, this treatise) have been less obvious. Transduction may not necessarily lead to sensation. Signals from the environment initiate other important events, as, for example, the effect of day length on the pituitary (see chapter by Liley, Volume I, this treatise). It is assumed that the directive factors are the basis for all behavioral and psychical regulation and for anticipatory adjustments in physiological regulation. By anticipatory adjustment is meant, for example, a hormonal change elicited by the annual photoperiod cycle which prepares the organism for future seasonal events of temperature change. An example is given in Fig. 10 for lethal temperature. The subject of directive factors is therefore a large one, but comment here will be restricted to a brief consideration of the reactions of fish to dissolved substances and temperature gradients.

Such reactions are undirected movements which are ordinarily called kineses (Fraenkel and Gunn, 1961). The definition of a kinesis implies that such activity is purely random. However, Sullivan (1954; see also in Fry, 1964), dealing with temperature, and Hemmings (1966), concerned with an odor gradient, have both pointed out that the degree of movement at any one time may depend on the immediate past experience; thus, while the direction of movement may be random the degree is directed. It seems likely too that the distinction Fraenkel and Gunn made between the undirected movements (kineses) and the taxes, the directed movements, while highly convenient, does not point to a fundamental division. A kinesis is orientation by consecutive sampling of a gradient in an opaque environment. A taxis is alignment to the source of stimulus in a transparent environment. A nice analogy that has been used is that the relation between taxes and kineses is the same as between melody and harmony.

As in physiological regulation there is a great deal of acclimation and acclimatization in behavioral regulation. There may also be a large element of what can be called transferred adjustment which has not yet been well analyzed. Transferred adjustment is typified by the well-known conditioned reflex and can serve for a taxis where the environment permits only a kinesis as the primary response to the actual gradient. The surfacing of fish when oxygen is low, and particularly their rapid gathering around a hole cut in the ice of a snow-covered pond, is probably an example of such a transferred adjustment. Here the animal may be triggered to respond to the light gradient by the stress of low oxygen, but this possibility does not yet appear to have been put to an experimental test with fish.

Experiments with the directive factors, like all experiments in behavior, require more than mechanical excellence of the apparatus to assure good results, and they depend as much on interpretation of circumstances as on the calculation of the results. A good example is the work of Ozaki (1951) on the orientation of young fish to various colors of light. He observed that a single fish alone in his apparatus could not respond and that orientation was progressively more precise as he used 2 or 3 fish; but in particular he noted that orientation was most associated with those moments when two fish were aligned as in a school. One is tempted to say that these fish were not free to respond to the subtleties of their surroundings until their primary requirement of orientation in a school had been satisfied. Again Verheijen (1958) makes a good case for the point of view that a positive phototaxis to a point source of light is really a strong disorientation occasioned by glare. Thus, while good results may be achieved by fishing with lights, such light experiments may say little about the responses of fish to natural light in its various forms. Fry (1958) has assembled a number of other similar examples. Harden Jones (1968) has an interesting chapter on the reactions of fish to stimuli.

Up to now there has been little standardization of apparatus or refinement in approach in the study of these mechanical aspects of behavior. Fry (1958) compiled a bibliography which contained references to most types of gradient apparatus used up to that time. Kleerekoper (1967) has probably produced the most elaborate means of monitoring the paths taken by fish in responding to various stimuli. His largest tank to date had an area of 25 meter² with a grid of 2500 photocells linked to a computer.

## A. Reactions to Dissolved Substances

### 1. GRADIENT EXPERIMENTS

Figure 35 shows the results of two recent workers and relates the gradient experiments to the lethal levels. In each case the fish react to avoid a level far below the incipient lethal. There are not sufficient comparisons of this sort yet to generalize, but the reactions in the two examples have high statistical reliability. At present, however, such reactions are not recognized by a standard method of bioassay (McKee and Wolf, 1963; American Public Health Association, 1965). It would seem desirable that they should be since they afford a rapidly measured prelethal test. However, Sprague (1968) pointed out that in contrast to the sharp avoidance of low concentrations of metals (e.g., Fig. 35A) rainbow trout do not avoid phenol at 10 mg/liter, a near-lethal concentration, and apparently could not discriminate between a lethal concentration and clean water in his gradient although such a situation provoked high activity. A similar confusion resulted when an alkyl

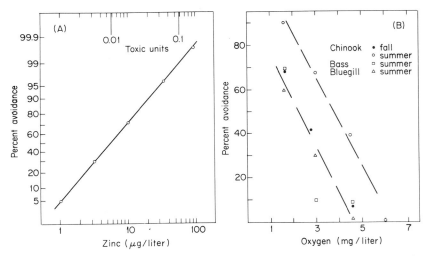

Fig. 35. (A) Avoidance of zinc sulfate solution by rainbow trout, *Salmo gairdneri*, at 9.5°C, water hardness 15 mg CaCO₃. From Sprague (1968). The avoidance is essentially complete at 0.1 toxic unit. One toxic unit is the incipient lethal concentration (Sprague and Ramsay, 1965). (B) Avoidance of low oxygen by chinook salmon, *Oncorhynchus tshawytscha*, largemouth bass, *Micropterus salmoides*, and the bluegill, *Lepomis macrochirus*, at existing river temperatures. Data from Whitmore *et al.* (1960), their Table 1, "periodic count." Avoidance of low oxygen is apparently not complete until the incipient lethal level is approached.

benzene sulfonate (ABS) detergent was presented at 10 mg/liter, which Sprague suggested may be the result of damage to the olfactory receptors (Bardach, 1956). When the fish were presented with a slowly lethal concentration of chlorine (0.1 mg/liter), they showed a net preference for it although they avoided higher or lower concentrations (0.01 and 1.0 mg/liter). The threshold for avoidance of Kraft pulp mill effluent was also approximately the incipient lethal level. Thus, Sprague concluded that it cannot be arbitrarily assumed that any given pollutant will automatically repel fish.

There do not appear to be any reaction experiments in which there has been acclimation to the test substances so that the ecological meaning of such tests is not clear.

## 2. Changes in Activity

There have been other tests in which the reaction to dissolved substances has been monitored by changes in random activity (e.g., Dandy, 1967, 1970), opercular movements (e.g., Halsband and Halsband, 1968), or routine metabolic rate (e.g., Kutty, 1968a). In continuous exposure to 100 μg/liter Cu in Toronto tapwater (survival time $> 7$ days) the increase in activity shown on the introduction of the metal subsided to the pre-introduction level in about 6 hr and continued at that level thereafter (Dandy, 1967). Similar behavior was shown in the response to $H_2S$ down to the threshold found at 100 $\mu$g/liter, at which level no initial increase in activity was found. It appears from Dandy's data that such clear-cut selection as shown in the examples in Fig. 35 may only appear in acute experiments.

The subsidence of response described by Dandy may be habituation of the sense organ rather than acclimation in terms of the whole animal's increase in ability to resist the influence of the toxicant, but further work is needed, particularly with fish acclimated to a given level and then tested over the whole range of reaction.

## B. Temperature Selection

A typical example of the response of fish to a temperature gradient is given in Fig. 36. Such responses usually have the sort of statistical precision shown in the figure, but the whole pattern of all responses for a given species still shows variations that have not all been explained. Figure 37 shows the observations available for the rainbow trout. While

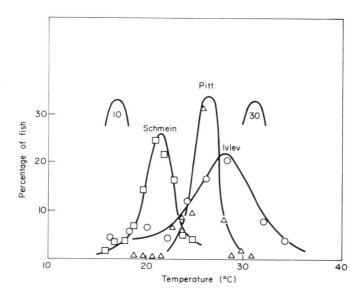

Fig. 36. Distribution of carp, *Cyprinus carpio*, in temperature gradients. Data of Ivlev (1960), Pitt *et al.* (1956), and Schmein-Engberding (1953). All fish acclimated to approximately 20°C prior to test. The numbers 10 and 30 indicate modes for fish acclimated to those temperatures. From Fry and Hochachka (1970).

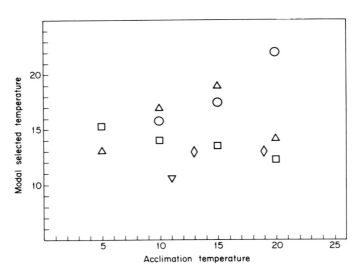

**Fig. 37.** Various modal selected temperatures in relation to thermal history for the rainbow trout, *Salmo gairdneri.* Data of W. J. Christie (personal communication) ( △ ), Garside and Tait ( 1958) ( □ ), Javaid and Anderson ( 1967) ( ○ ), Mantelman ( 1958) ( ◇ ), and Schmein-Engberding ( 1953) ( ▽ ).

rainbow trout may not be genetically homogeneous, particularly with respect to the various domestic stocks in different parts of the world, three of the groups whose work is illustrated in Fig. 37 worked within a few hundred miles of each other. There is as yet no complete explanation for the differences in behavior these various workers have reported. Three different types of apparatus were used, but the differences found were not necessarily related to differences between apparatus.

Among the major sources of variability in temperature selection is time of year, as was first pointed out by Sullivan and Fisher ( 1953). Unfortunately, there are still no complete annual series of observations on the relation of the preferred temperature to acclimation temperature. The most complete are those of Zahn ( 1963) ( Fig. 38) who, however, dampened the annual light cycle somewhat by superimposing a minimum day length in the case of the bitterling, while with the plaice he used a constant day length for the various relatively short periods of time that he maintained these fish after capture and before experiment. Despite this, the influence of season is profound in both cases in Fig. 38 and leads to the conclusion that any discussion of the relation between acclimation temperature and temperature preferendum without reference to season should be in the most general terms.

Both Ferguson ( 1958) and Zahn ( 1962) have noted that the response

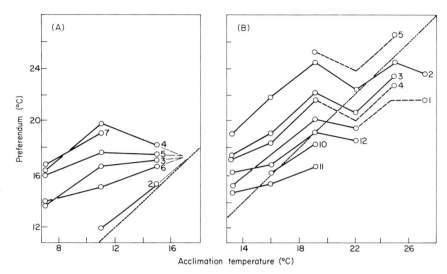

Fig. 38. Effect of season and acclimation temperature on temperature selection in the (A) plaice, *Pleuronectes platessa,* and (B) bitterling, *Rhodeus sericeus.* After Zahn (1963), from Fry and Hochachka (1970). Numbers indicate month of test.

of the temperature preferendum to acclimation temperature has been most diverse, ranging from a slightly negative one such as shown by the data of Garside and Tait in Fig. 37 to cases like the bitterling (Fig. 38B, any single curve), where the temperature preferendum increases almost in step with increase in acclimation temperature. In the plaice, Zahn found almost the whole range of response in the one species at different times of the year. There is clear need for some patient description of the temperature preferendum in relation to season and latitude under various conditions of acclimation.

The response of the fish to the gradient apparatus itself also requires further analysis. There are three fundamentally different methods of presenting a temperature gradient to a fish. The first and most widely used method has been the longitudinal horizontal gradient (e.g., Norris, 1963; Schmein-Engberding, 1953; Alabaster and Downing, 1966) whereby the fish are placed in a tube or trough in which the water changes in temperature from one end to the other so that the gradient and the swimming path of the fish are constrained into the same plane and are parallel. The second method has been to establish a vertical thermal gradient (e.g., Brett, 1952) in a tank large enough to allow the fish some freedom in a horizontal path and in which temperature selection is effected by the fish swimming higher or lower. Thus, the fish has two degrees of freedom in its swimming path and the temperature gradient is at right angles to the longitudinal axis of the fish in its normal orientation. In this chamber the fish may react to depth as well as temperature (Javaid and Anderson, 1967). The third method has been to place the fish in a central chamber into which all choice chambers open individually. In its simplest form such apparatus is a divided trough (e.g., Collins, 1952). A more elaborate form is the rosette (e.g., Kleerekoper, 1969) in which the central chamber is surrounded by a number of choice chambers. In the third method both the direction and the magnitude of the change experienced by the animal in passing from the central condition to a given test condition is randomized. In the first and second methods the change is gradual as the animal passes up or down the gradient. The direction is random in the horizontal gradient but has a fixed association with gravity in the vertical gradient. In general, although no extensive comparisons have been made, these three types of apparatus yield similar results, or at least they are not consistently associated with any given result. A thorough statistical comparison of the three types in a single laboratory is highly desirable.

## 1. RANDOM ACTIVITY IN RELATION TO TEMPERATURE CHANGE

Sullivan's distinction (1954; see also in Fry, 1964) cleared up a great deal of confusion with regard to temperature selection. In effect, she pointed out that certain activity is stimulated by recent temperature change and thus is a response to temperature as a directive factor, while activity at a given constant temperature is rather a facilitation of response by temperature as a controlling factor. The latter response has been widely but mostly unwittingly reported in the literature as a homeostatic response in the metabolic rate. Its general effect is to produce a central horizontal section or even a dip in the curve relating routine metabolism to temperature, or at least make that curve decidedly convex on a semilogarithmic plot. The clearest statement of such influence of random activity on the course of the temperature metabolism curve is probably that of Schmein-Engberding (1953) whose data are illustrated in Fig. 39 and who demonstrated that the anomaly could be removed by anesthesia. That author also pointed out there was a correspondence between the anomaly in the curve and the temperature preferendum.

Unfortunately, the works of Schmein-Engberding and Sullivan have been largely overlooked in the physiological literature, where discussion has followed the lines of workers such as Meuwis and Heuts (1957) whose data are summarized in Fig. 39 and who concluded there could be a "broad homeostatic zone of independence of the breathing frequency on temperatures" (Meuwis and Heuts, 1957, p. 107, 1.12). Such a notion

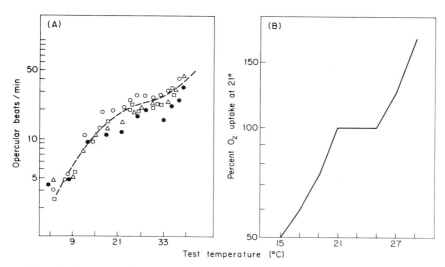

Fig. 39. Measures of routine metabolism of the carp, *Cyprinus carpio,* in relation to temperature. (A) From Meuwis and Heuts (1957) and (B) from Schmein-Engberding (1953). The various symbols in A represent different individuals.

of metabolic regulation is still widely held although Beamish (1964a) and Roberts (1966), for example, refer clearly to the effect of random activity on the level of metabolism as ordinarily measured (routine metabolism). The apparent homeostasis in the routine metabolism curve is brought about by a peak of random activity associated with the temperature that is the thermal preferendum, presumably for the state of thermal acclimation of the fish in question. Scope for activity appears to be ordinarily greatest at this temperature, and the animal presumably reacts most vigorously here to any stray stimuli. As the temperature increases beyond the preferendum, then increase in standard metabolism counteracts lessening random activity; thus, the routine metabolism curve rises again at higher temperatures. The interaction of these two effects is responsible for the so-called homeostasis. Figure 40 shows the relation of random movement to temperature when fish are equilibrated to successive temperature levels in turn.

The movements associated with thermoregulation are of two degrees. There is a gross response of activity (the response temperature of Rubin, 1935) whereby a fish resting quietly will show activity in warming water as its lethal temperature is reached. Fisher and Sullivan (1958) also showed the response temperature in the brook trout, together with the controlling effect. It is the second peak in Fig. 40.

There is also a tendency, much harder to demonstrate, for fish which are randomly active during a period of temperature change to be at first

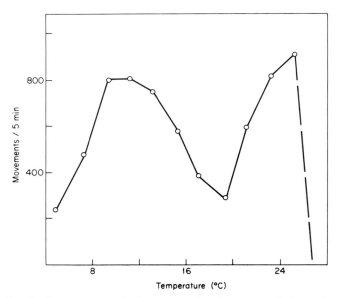

**Fig. 40.** Random activity of the eastern brook trout, *Salvelinus fontinalis,* in relation to temperature after approximately 30 min exposure to a given temperature. From Fisher and Sullivan (1958).

progressively more active as the temperature departs from their thermal preferendum. Figure 41 shows the latter phenomenon. Line A is the immediate response to temperature change which can be taken to be the response to temperature as a directive factor. The final preferendum of the Atlantic salmon is close to 18°C (Javaid and Anderson, 1967) so that any departure from the preferred temperature evokes an immediate increase in activity of the fish.

Line B approximates the response to temperature as a controlling factor. It seems probable, both from the data of Peterson and Anderson and from Fisher and Sullivan (1958), who intentionally allowed time for thermal equilibrium, that what is "stabilized" to give the difference between curves A and B in Fig. 41 is the body temperature of the fish.

## VII. RECAPITULATION

Figure 42 is presented as a summary of the various relations of the organism to its environment insofar as these can be pictured in one plane by quantitative expressions of metabolism.

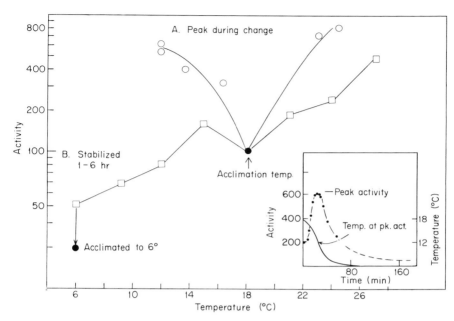

**Fig. 41.** Random activity of Atlantic salmon, *Salmo salar,* in relation to changing and stabilized temperatures. From Peterson and Anderson (1969a). The inset gives an example of the details of the short-term effects of temperature change on random activity.

The pervading environmental factor is the controlling factor which sets the upper and lower limit to the metabolic rate. Thus the potential range for activity is set by the controlling factors and by the capacity of the organism to satisfy the requirements for metabolism that the controlling factors require or permit. The potential range can be envisioned as extending from a level where the state of activation, and with it the metabolic rate, is so low that the organism cannot respond at all, to a level where molecules are so activated that the standard metabolic rate absorbs the organism's total metabolic capacity (Fig. 42A).

The potential range over which the controlling factors can operate may be restricted by a controlling factor becoming lethal without any effect on scope for activity within the zone of tolerance (Fig. 42B).

The scope for activity within the limits set by the controlling factors is restricted by the operation of any limiting factor. Limiting factors operate to suppress the active metabolic rate and therefore to reduce scope for activity. The limiting factor will have its greatest effect on scope where the controlling factors require the highest level of standard

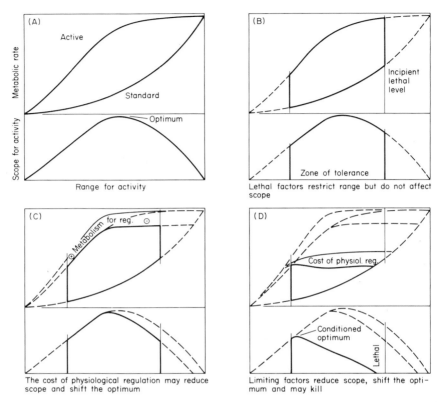

**Fig. 42.** Summary of relations of metabolism to environmental factors modified from Fry (1947). The solid lines represent the boundaries to scope for activity as various factors interact from A to D. For further explanation see text.

metabolism; if severe enough, it will have a lethal effect down to the level of controlling factors where the limiting factor permits sufficient metabolism to satisfy the standard requirement. Thus, a limiting factor also reduces the range of controlling factors available for activity when the restriction to the active metabolic rate is sufficiently severe (Fig. 42D).

The organism regulates to maintain its own integrity and the continuation of the species. Such regulation involves mechanical barriers, physiological and biochemical activity, and behavioral responses. Regulation is achieved by channeling energy through appropriate form and is accomplished through the masking and directive factors. Standard metabolism probably represents the regulatory energy required by the quiescent animal and is related to the level of controlling factors which impinge on the organism. Beyond that the cost regulation is some function of activity (Fig. 42C).

Regulation mitigates the effects of the controlling factors and facilitates the uptake, discharge, and transport of metabolites. In the broad sense all physical activity is also regulation, but attention is usually focused on ancilliary activities which support growth or behavior. The masking factors provide the direct machinery of regulation by providing form and energy. The directive factors provide signals which permit the organism to respond both behaviorally and physiologically to its environment. The physiological response to a directive factor allows regulation to meet some future event by anticipatory adjustment linked to the precursor which acts as the directive factor in question.

Where no limiting factor operates, the immediate cost of the ancillary activity may be added to the cost of the behavioral activity or growth and so be shown as an increment to the curve for active metabolism. Under limiting conditions such costs become competitive to a greater and greater degree, but the interaction between these costs is not known. Certainly, as in Figs. 20 and 32, the behavioral activity is restricted to permit regulation to continue, either to lessen the requirement for regulation or to meet the lessened possibilities.

Organisms are continually adjusting to the fluctuating environment so that their reaction always depends on their history. Many aspects of their history may be stabilized by exposure to various constant conditions for various periods (acclimation). However, in a cyclic environment the organism may have evolved endogenous cyclic adjustments that cannot be entirely dampened by acclimation; moreover, structural characteristics determined during various critical periods of development may not be reversible at a later date. Hence, the ideal description in terms of reproducible response which can be evoked at any time by a stabilized environment is impossible to attain. Descriptions of metabolism require statements of history such as latitude and time of year that are not usually provided. Figure 42 is a compromise in which certain gross aspects of stability have been required.

## REFERENCES

Alabaster, J. S., and Downing, A. L. (1966). A field and laboratory investigation of the effect of heated effluents on fish. *Min. Agr. Fish Food, U. K., Fish Invest.,* Ser. I. 6, No. 4, 1–42.

Alderdice, D. F. (1971). Factor combinations. *In* "Marine Ecology" (O. Kinne, ed.). Wiley, New York (in press).

Allanson, B. R., and Noble, R. G. (1964). The tolerance of *Tilapia mossambica* (Peters) to high temperature. *Trans. Am. Fisheries Soc.* 93, 323–332.

Allen, K. O., and Strawn, K. (1968). Heat tolerance of channel catfish *Ictalurus punctatus. Proc. 21st Ann. Conf., Southeast. Assoc. Game Fish Comm., 1967* pp. 399–411.

American Public Health Association. (1965). "Standard Methods for the Examination of Water and Wastewater," 12th ed. Am. Public Health Assoc., New York.

Andrewartha, H. G., and Birch, L. C. (1954). "The Distribution and Abundance of Animals." Univ. of Chicago Press, Chicago, Illinois.

Arai, M. N., Cox, E. T., and Fry, F. E. J. (1963). An effect of dilutions of seawater on the lethal temperature of the guppy. *Can. J. Zool.* 41, 1011–1015.

Arrhenius, S. (1889). Über die Reaktionsgeschwindigkeit bei der Inversion von Rohrzucker durch Säuren. *Z. Physik. Chem.* 4, 226–248.

Bardach, J. E. (1956). The sensitivity of the goldfish (*Carassius auratus* L.) to point heat stimulation. *Am. Naturalist* 90, 309–317.

Bardach, J. E., Fujiya, M., and Holl, A. (1965). Detergents: Effects on the chemical senses of the fish *Ictalurus natalis* (leSueur). *Science* 148, 1605–1607.

Barlow, G. W. (1961). Intra- and interspecific differences in rate of oxygen consumption in gobiid fishes of the genus *Gillichthys. Biol. Bull.* 121, 209–229.

Barrett, I., and Hester, F. (1964). Body temperature of yellowfin and skipjack tunas in relation to sea surface temperatures. *Nature* 203, 96–97.

Baslow, M. H. (1967). Temperature adaptation and the central nervous system of fish. In "Molecular Mechanisms of Temperature Adaptation," Publ. No. 84, Am. Assoc. Advance. Sci., Washington, D. C.

Baslow, M. H., and Nigrelli, R. F. (1964). The effect of thermal acclimation on brain cholinesterase activity of the killifish, *Fundulus heroclitus. Zoologica* 49, 41–51.

Basu, S. P. (1959). Active respiration of fish in relation to ambient concentrations of oxygen and carbon dioxide. *J. Fisheries Res. Board Can.* 16, 175–212.

Beamish, F. W. H. (1964a). Respiration of fishes with special emphasis on standard oxygen consumption. II. Influence of weight and temperature on respiration of several species. *Can. J. Zool.* 42, 176–188.

Beamish, F. W. H. (1964b). III. Influence of oxygen. *Can. J. Zool.* 42, 355–366.

Beamish, F. W. H. (1964c). IV. Influence of carbon dioxide and oxygen. *Can. J. Zool.* 42, 847–856.

Beamish, F. W. H. (1964d). Influence of starvation on standard and routine oxygen consumption. *Trans. Am. Fisheries Soc.* 93, 103–107.

Beamish, F. W. H. (1964e). Seasonal changes in the standard rate of oxygen consumption of fishes. *Can. J. Zool.* 42, 189–194.

Beamish, F. W. H., and Mookherjii, P. S. (1964). Respiration of fishes with special emphasis on standard oxygen consumption. I. Influence of weight and temperature on respiration of goldfish, *Carassius auratus* L. *Can. J. Zool.* 42, 161–175.

Bělehrádek, J. (1926). Sur la formule générale exprimant l'action de la tempèrature sur les processus biologiques. *C. R. Soc. Biol.* 95, 1449–1452.

Bělehrádek, J. (1930). Temperature coefficients in biology. *Biol. Rev.* 5, 30–58.

Berthelot, M. (1862). Essai d'une theorie sur la formation des ethers. *Ann. Chim. Phys. 3ᵉ serie* 66, 110–128.

Black, E. C., Fry, F. E. J., and Black, V. S. (1954). The influence of carbon dioxide on the utilization of oxygen by some freshwater fish. *Can. J. Zool.* 32, 408–420.

Black, V. S. (1951). Osmotic regulation in teleost fishes. *Univ. Toronto Biol. Ser.* 59, 53–89.

Blackman, F. F. (1905). Optima and limiting factors. *Ann. Botany (London)* **19**, 282–295.

Blažka, P. (1958). The anaerobic metabolism of fish. *Physiol. Zool.* **31**, 117–128.

Blažka, P., Volf, M., and Čepela, M. (1960). A new type of respirometer for the determination of the metabolism of fish in an active state. *Physiol. Bohemoslov.* **9**, 553–558.

Bliss, C. I. (1935). The calculation of the dosage mortality curve. *Ann. Appl. Biol.* **22**, 134–167.

Bliss, C. I. (1937). The calculation of the time-mortality curve. *Ann. Appl. Biol.* **24**, 815–852.

Bliss, C. I. (1952). "The Statistics of Bioassay," pp. 445–628. Academic Press, New York.

Booij, H. L., and Wolvekamp, H. P. (1944). Catenary processes, master reactions and limiting factors. *Bibliotheca Biotheor.* **D1**, 145–224.

Brett, J. R. (1944). Some lethal temperature relations of Algonquin Park fishes. *Univ. Toronto Studies Biol. Ser.* **52**, 1–49.

Brett, J. R. (1946). Rate of gain of heat-tolerance in goldfish (*Carassius auratus*). *Univ. Toronto Studies Biol. Ser.* **53**, 1–28.

Brett, J. R. (1952). Temperature tolerance in young Pacific salmon genus *Oncorhynchus*. *J. Fisheries Res. Board Can.* **9**, 265–323.

Brett, J. R. (1958). Implications and assessments of environmental stress. In "Investigations of Fish-power Problems" (P. A. Larkin, ed.), pp. 69–83. H. R. MacMillan Lectures in Fisheries, Univ. of Brit. Columbia.

Brett, J. R. (1964). The respiratory metabolism and swimming performance of young sockeye salmon. *J. Fisheries Res. Board Can.* **21**, 1183–1226.

Brett, J. R. (1965). The relation of size to rate of oxygen consumption and sustained swimming speed of sockeye salmon (*Oncorhynchus nerka*). *J. Fisheries Res. Board Can.* **22**, 1491–1501.

Brett, J. R., and Alderdice, D. F. (1958). The resistance of cultured young chum and sockeye salmon to temperatures below 0°C. *J. Fisheries Res. Board Can.* **15**, 805–813.

Brett, J. R., Shelbourn, J. E., and Shoop, C. T. (1969). Growth rate and body composition of fingerling sockeye salmon, *Oncorhynchus nerka*, in relation to temperature and ration size. *J. Fisheries Res. Board Can.* **26**, 2363–2394.

Brody, S. (1945). "Bioenergetics and Growth." Reinhold, New York.

Brown, C. E., and Muir, B. S. (1970). Analysis of ram ventilation of fish gills with application to skipjack tuna (*Katsuwonus pelamis*). *J. Fisheries Res. Board Can.* **27**, 1637–1652.

Brown, V. M., Jordan, D. H. M., and Tiller, B. A. (1967). The effect of temperature on the acute toxicity of phenol to rainbow trout in hard water. *Water Res.* **1**, 587–594.

Bullivant, J. S. (1961). The influence of salinity on the rate of oxygen consumption of young Quinnat salmon, *Oncorhynchus tschawytscha*. *New Zealand J. Sci.* **4**, 381–391.

Carey, F. G., and Teal, J. M. (1966). Heat conservation in tuna fish muscle. *Proc. Natl. Acad. Sci. U. S.* **56**, 1464–1469.

Carey, F. G., and Teal, J. M. (1969a). Mako and porbeagle: Warm-bodied sharks. *Comp. Biochem. Physiol.* **28**, 199–204.

Carey, F. G., and Teal, J. M. (1969b). Regulation of body temperature by the bluefin tuna. *Comp. Biochem. Physiol.* **28**, 205–213.

Cocking, A. W. (1959). The effects of high temperatures on roach (*Rutilus rutilus*). II. The effects of temperature increasing at a known constant rate. *J. Exptl. Biol.* **36**, 217–226.

Collins, G. B. (1952). Factors influencing the orientation of migrating anadromous fishes. *U. S. Fish Wildlife Serv., Fishery Bull.* **73**, 52, 375–396.

Conte, F. P., and Wagner, H. H. (1965). Development of osmotic and ionic regulation in juvenile steelhead trout *Salmo gairdneri*. *Comp. Biochem. Physiol.* **14**, 603–620.

Coulter, G. W. (1967). Low apparent oxygen requirements of deep-water fishes in Lake Tanganyika. *Nature* **215**, 317–318.

Crozier, W. J. (1924). On the critical thermal increment for the locomotion of a diplopod. *J. Gen. Physiol.* **7**, 123–136.

Dahlberg, M. L., Shumway, D. L., and Doudoroff, P. (1968). Influence of dissolved oxygen and carbon dioxide on swimming performance of largemouth bass and coho salmon. *J. Fisheries Res. Board Can.* **25**, 49–70.

Dandy, J. W. T. (1967). The effects of chemical characteristics of the environment on the activity of an aquatic organism. Ph.D. Thesis, University of Toronto (National Library of Canada, Canadian theses on microfilm, No. CM. 68-616).

Dandy, J. W. T. (1970). Activity response to oxygen in the brook trout, *Salvelinus fontinalis* (Mitchill). *Can. J. Zool.* **48**, 1067–1072.

Das, A. B., and Prosser, C. L. (1967). Biochemical changes in tissues of goldfish acclimated to high and low temperatures. I. Protein synthesis. *Comp. Biochem. Physiol.* **21**, 449–467.

Davidson, J. (1942). On the speed of development of insect eggs at constant temperatures. *Aust. J. Exp. Biol. Med. Sci.* **20**, 233–239.

Davidson, J. (1944). On the relationship between temperature and rate of development of insects at constant temperatures. *J. Animal Ecol.* **13**, 26–38.

Davies, P. M. C. (1966). The energy relations of *Carassius auratus* L. II. The effect of food, crowding and darkness on heat production. *Comp. Biochem. Physiol.* **17**, 983–995.

Davis, G. E., Foster, J., Warren, C. E., and Doudoroff, P. (1963). The influence of oxygen concentration on the swimming performance of juvenile Pacific salmon at various temperatures. *Trans. Am. Fisheries Soc.* **92**, 111–124.

DeGroot, S. J., and Schuyf, A. (1967). A new method for recording the swimming activity of flatfishes. *Experientia* **23**, 1–6.

de Réaumur, R. A. F. (1735). *Mém. Acad. Roy. Sci. Paris* [cited in Shelford (1929)].

Doudoroff, P. (1942). The resistance and acclimatization of marine fishes to temperature changes. I. Experiments with *Girella nigricans* (Ayres). *Biol. Bull.* **83**, 219–244.

Doudoroff, P. (1945). II. Experiments with fundulus and Atherinops. *Biol. Bull.* **88**, 194–206.

Doudoroff, P., Leduc, G., and Schneider, C. R. (1966). Acute toxicity to fish of solutions containing complex metal cyanides, in relation to concentrations of molecular hydrocyanic acid. *Trans. Am. Fisheries Soc.* **95**, 6–22.

Ege, R., and Krogh, A. (1914). On the relation between temperature and the respiratory exchange in fishes. *Intern. Rev. Ges. Hydrobiol. Hydrog.* **7**, 48–55.

Embody, G. C. (1934). Relations of temperature to the incubation periods of eggs of four species of trout. *Trans. Am. Fisheries Soc.* **64**, 281–292.

Farmer, G. J., and Beamish, F. W. H. (1969). Oxygen consumption of *Tilapia nilotica* in relation to swimming speed and salinity. *J. Fisheries Res. Board Can.* **26**, 2807–2821.

Ferguson, R. G. (1958). The preferred temperature of fish and their midsummer distribution in temperate lakes and streams. *J. Fisheries Res. Board Can.* **15**, 607–624.

Fisher, K. C., and Sullivan, C. M. (1958). The effect of temperature on the spontaneous activity of speckled trout before and after various lesions of the brain. *Can. J. Zool.* **36**, 49–63.

Fraenkel, G. S., and Gunn, D. L. (1961). "The Orientation of Animals," 2nd ed. Dover, New York.

Fry, F. E. J. (1947). Effects of the environment on animal activity. *Univ. Toronto Studies Biol. Ser.* **55**, 1–62.

Fry, F. E. J. (1957). The aquatic respiration of fish. *In* "The Physiology of Fishes" (M. E. Brown, ed.), Vol. 1, pp. 1–63. Academic Press, New York.

Fry, F. E. J. (1958). Laboratory and aquarium research. II. The experimental study of behaviour in fish. *Proc. Indo-Pacific Fishery Council* pp. 37–42.

Fry, F. E. J. (1964). Animals in aquatic environments: Fishes. *In* "Handbook of Physiology" (Am. Physiol. Soc., J. Field, ed.), Sect. 4, pp. 715–728. Williams & Wilkins, Baltimore, Maryland.

Fry, F. E. J. (1967). Responses of vertebrate poikilotherms to temperature. *In* "Thermobiology" (A. H. Rose, ed.), pp. 375–409. Academic Press, New York.

Fry, F. E. J., and Hart, J. S. (1948). The relation of temperature to oxygen consumption in the goldfish. *Biol. Bull.* **94**, 66–77.

Fry, F. E. J., and Hochachka, P. W. (1970). Fish. *In* "Comparative Physiology of Thermoregulation" (G. C. Whittow, ed.), pp. 79–134. Academic Press, New York.

Fry, F. E. J., and Norris, K. S. (1962). The transportation of live fish. *In* "Fish as Food" (G. Borgstrom, ed.), Vol. 2, pp. 595–608. Academic Press, New York.

Fry, F. E. J., Brett, J. R., and Clawson, G. H. (1942). Lethal limits of temperature for young goldfish. *Rev. Can. Biol.* **1**, 50–56.

Fry, F. E. J., Hart, J. S., and Walker, K. F. (1946). Lethal temperature relations for a sample of young speckled trout, *Salvelinus fontinalis. Univ. Toronto Studies Biol. Ser.* **54**, 1–47.

Garside, E. T. (1966). Effects of oxygen in relation to temperature on the development of embryos of brook trout and rainbow trout. *J. Fisheries Res. Board Can.* **23**, 1121–1134.

Garside, E. T., and Tait, J. S. (1958). Preferred temperature of rainbow trout (*Salmo gairdneri* Richardson) and its unusual relationship to acclimation temperature. *Can. J. Zool.* **36**, 563–567.

Gibson, M. B. (1954). Upper lethal temperature relations of the guppy, *Lebistes reticulatus. Can. J. Zool.* **32**, 393–407.

Glass, N. R. (1969). Discussion of calculation of power function with special reference to respiratory metabolism in fish. *J. Fisheries Res. Board Can.* **26**, 2643–2650.

Gordon, M. S. (1963). Chloride changes in rainbow trout (*Salmo gairdneri*) adapted to different salinities. *Biol. Bull.* **124**, 45–54.

Graham, J. M. (1949). Some effects of temperature and oxygen pressure on the metabolism and activity of the speckled trout, *Salvelinus fontinalis. Can. J. Res.* **D27**, 270–288.

Halsband, E., and Halsband, I. (1968). Eine Apparatur zur Messung der Stoffwechselintensität von Fischen und Fischnährtieren. *Arch. Fischereiwiss.* **19**, 78–82.

Harden Jones, F. R. (1968). "Fish Migration." St. Martin's Press, New York.

Hart, J. S. (1952). Geographic variations of some physiological and morphological characters in certain freshwater fish. *Univ. Toronto Biol. Ser.* **60**, 1–79.

Hart, J. S. (1957). Seasonal changes in $CO_2$ sensitivity and blood circulation in certain freshwater fishes. *Can. J. Zool.* **35**, 195–200.

Harvey, E. N. (1911). Effect of different temperatures on the medusa *Cassiopea*, with special reference to the rate of conduction of the nerve impulse. *Carnegie Inst. Wash. Publ., Pap. Tortugas Lab.* **3**, 27–39.

Harvey, H. H. (1964). Dissolved nitrogen as a tracer of fish movements. *Verhandl. Intern. Ver. Limnol.* **15**, 947–951.

Heath, W. G. (1963). Thermoperiodism in sea-run cutthroat trout (*Salmo clarki clarki*). *Science* **142**, 486–488.

Heinicke, E. A., and Houston, A. H. (1965). Effect of thermal acclimation and sublethal heat shock upon ionic regulation in the goldfish, *Carassius auratus* L. *J. Fisheries Res. Board Can.* **22**, 1455–1476.

Hemmings, C. C. (1966). The mechanism of orientation of roach, *Rutilus rutilus* L. in an odor gradient. *J. Exptl. Biol.* **45**, 465–474.

Herbert, D. W. M. (1965). Pollution and fisheries. Ecology and the industrial society. *Ecol. Ind. Soc., Symp., 1964* pp. 173–195.

Herbert, D. W. M., and Shurben, D. S. (1963). A preliminary study of the effect of physical activity on the resistance of rainbow trout (*Salmo gairdneri* Richardson) to two poisons. *Ann. Appl. Biol.* **52**, 321–326.

Herbert, D. W. M., and Vandyke, J. M. (1964). The toxicity to fish of mixtures of poisons. II. Copper-ammonia and zinc-phenol mixtures. *Ann. Appl. Biol.* **53**, 415–421.

Heusner, A., and Enright, J. T. (1966). Long-term activity in small aquatic animals. *Science* **154**, 532–533.

Hickman, C. P., Jr. (1959). The osmoregulatory role of the thyroid gland in the starry flounder *Platichthys stellatus*. *Can. J. Zool.* **37**, 997–1060.

Hoar, W. S., and Robinson, G. B. (1959). Temperature resistance of goldfish maintained under controlled photoperiods. *Can. J. Zool.* **37**, 419–428.

Hoff, J. G., and Westman, J. R. (1966). The temperature tolerances of three species of marine fishes. *J. Marine Res.* (*Sears Found. Marine Res.*) **24**, 131–140.

Houde, E. D. (1969). Sustained swimming ability of larvae of walleye (*Stizostedion vitreum*) and yellow perch (*Perca flavescens*). *J. Fisheries Res. Board Can.* **26**, 1647–1659.

Houston, A. H. (1959). Osmoregulatory adaptation of steelhead trout (*Salmo gairdnerii* Richardson) to sea water. *Can. J. Zool.* **37**, 729–748.

Ivlev, V. S. (1938). The effect of temperature on the respiration of fish. *Zool. Zh.* **17**, 645–660. (Engl. transl. by E. Jermolajev.)

Ivlev, V. S. (1960). Analiz mekhanizma raspredelniia ryb v usloviiakh temperaturnovo gradienta. *Zool. Zh.* **39**, 494–499 [for translation, see *Fisheries Res. Board Can., Transl. Ser. 364*, (1961)].

Jacobs, M. H. (1919). Acclimatization as a factor affecting the upper thermal death points of organisms. *J. Exptl. Zool.* **27**, 427–442.

Janisch, E. (1925). Über die Temperaturabhängigkeit biologischer Vorgänge und ihre kurvenmässige Analyse. *Arch. Ges. Physiol.* **209**, 414–436.

Jankowsky, H.-D. (1968). Versuche zur Adaptation der Fische im normalen Temperaturbereich. *Helgolaender Wiss. Meeresuntersuch.* 18, 317–362.

Javaid, M. Y., and Anderson, J. M. (1967). Thermal acclimation and temperature selection in Atlantic salmon, *Salmo salar,* and rainbow trout, S. *gairdneri. J. Fisheries Res. Board Can.* 24, 1507–1513.

Job, S. V. (1955). The oxygen consumption of *Salvelinus fontinalis. Univ. Toronto Biol. Ser.* 61, 1–39.

Job, S. V. (1959). The metabolism of *Plotosus anguillaris* (Bloch) in various concentrations of salt and oxygen in the medium. *Proc. Indian Acad. Sci.* B50, 267–288.

Johansen, P. H. (1967). The role of the pituitary in the resistance of the goldfish (*Carassius auratus* L.) to a high temperature. *Can. J. Zool.* 45, 329–345.

Johnson, F. H., Eyring, H., and Polissar, M. J. (1954). "The Kinetic Basis of Molecular Biology." Wiley, New York.

Kanungo, M. S., and Prosser, C. L. (1959). Physiological and biochemical adaptation of goldfish to cold and warm temperatures. I. Standard and active oxygen consumptions of cold- and warm-acclimated goldfish at various temperatures. *J. Cellular Comp. Physiol.* 54, 259–263.

Kausch, H. (1968). Der Einfluß der Spontanaktivität auf die Stoffwechselrate junger Karpfen (*Cyprinus carpio* L.) im Hunger und bei Futterung. *Arch. Hydrobiol.* Suppl. 33, No. 3, 263–330.

Keys, A. B. (1931). A study of the selective action of decreased salinity and of asphyxiation on the Pacific killifish, *Fundulus parvipinnis. Bull. Scripps Inst. Oceanog. Univ. Calif., Tech. Ser.* 2, 417–490.

Kinne, O., and Kinne, E. M. (1962). Rates of development in embryos of a cyprinodont fish exposed to different temperature-salinity-oxygen combinations. *Can. J. Zool.* 40, 231–253.

Kishinouye, K. (1923). Contributions to the comparative study of the so-called scombroid fishes. *J. Coll. Agr., Tokyo Imp. Univ.* 8, 293–470.

Kleerekoper, H. (1967). Some aspects of olfaction in fishes, with special reference to orientation. *Am. Zool.* 7, 385–395.

Kleerekoper, H. (1969). "Olfaction in Fishes." Indiana Univ. Press, Bloomington, Indiana.

Krogh, A. (1914). On the influence of the temperature on the rate of embryonic development. *Z. Allgem. Physiol.* 16, 163–177.

Krueger, F. (1964). Neuere mathematisch Formulierung der biologischen Temperaturfunktion und des Wachstums. *Helgolaender Wiss. Meeresuntersuch.* 9, 108–124.

Kutty, M. N. (1968a). Respiratory quotients in goldfish and rainbow trout. *J. Fisheries Res. Board Can.* 25, 1689–1728.

Kutty, M. N. (1968b). Influence of ambient oxygen on the swimming performance of goldfish and rainbow trout. *Can. J. Zool.* 46, 647–653.

Larimore, R. W., and Duever, M. J. (1968). Effects of temperature acclimation on the swimming ability of smallmouth bass fry. *Trans. Am. Fisheries Soc.* 97, 175–184.

Leiner, M. (1938). "Die Physiologie der Fischatmung." Akad. Verlagsges., Leipzig.

Lindroth, A. (1940). Sauerstoffverbrauch der Fische bei verschiedenem Sauerstoffdruck und verschiedenem Sauerstoffbedarf. *Z. Vergleich. Physiol.* 28, 142–152.

Lindroth, A. (1942). Sauerstoffverbrauch der Fische. II. Verschiedene entwicklungs- und altersstadien vom Lachs und Hecht. *Z. Vergleich. Physiol.* 29, 583–594.

Lindsey, C. C. (1968). Temperatures of red and white muscle in recently caught marlin and other large tropical fish. *J. Fisheries Res. Board Can.* **25**, 1987–1992.

Lindsey, J. K., Alderdice, D. F., and Pienaar, L. V. (1970). Analysis of nonlinear models—the nonlinear response surface. *J. Fisheries Res. Board Can.* **27**, 765–791.

Lloyd, R. (1961a). The toxicity of ammonia to rainbow trout (*Salmo gairdneri* Richardson). *Water Waste Treat. J.* **8**, 278–279.

Lloyd, R. (1961b). Effect of dissolved oxygen concentrations on the toxicity of several poisons to rainbow trout (*Salmo gairdneri* Richardson). *J. Exptl. Biol.* **38**, 447–455.

Lloyd, R. (1961c). The toxicity of mixtures of zinc and copper sulphates to rainbow trout (*Salmo gairdneri* Richardson). *Ann. Appl. Biol.* **49**, 535–538.

Lloyd, R., and Herbert, D. W. M. (1960). The influence of carbon dioxide on the toxicity of un-ionized ammonia to rainbow trout (*Salmo gairdneri* Richardson). *Ann. Appl. Biol.* **40**, 399–404.

Lloyd, R., and Jordan, D. H. M. (1964). Some factors affecting the resistance of rainbow trout (*Salmo gairdneri* Richardson) to acid waters. *Air Water Pollution* **8**, 393–403.

Lloyd, R., and White, W. R. (1967). Effect of high concentration of carbon dioxide on the ionic composition of rainbow trout blood. *Nature* **216**, 1341–1342.

Loeb, J. (1913). Die Tropismen. *In* "Handbuch der Vergleichenden Physiologie" (H. Winterstein, ed.), Vol. 4, pp. 451–519. Fischer, Jena.

Loeb, J. (1918). "Forced Movements, Tropisms and Animal Conduct." Lippincott, Philadelphia, Pennsylvania.

McCrimmon, H. R. (1949). The survival of planted salmon (*Salmo salar*) in streams. Ph.D. Thesis, University of Toronto.

McKee, J. E., and Wolf, H. W. (1963). "Water Quality Criteria," 2nd ed., Publ. 3A. Res. Agency Calif. State Water Qual. Control Board, Sacramento, California.

MacLeod, J. C., and Smith, L. L., Jr. (1966). Effect of pulpwood fiber on oxygen consumption and swimming endurance of the fathead minnow, *Pimephales promelas. Trans. Am. Fisheries Soc.* **95**, 71–84.

Mantelman, I. I. (1958). O raspredelenii molodi nekotorykh vidov ryb v termogradientnykh usloviiakh. *Izv. Vses. Nauch-Issled. Inst. Ozer. Rechn. Ryb Khoz.* **47**(1), 1–63 [for translation, see *Fisheries Res. Board Can., Transl. Ser.* **257** (1960)].

Mar, J. (1959). "A Proposed Tunnel Design for a Fish Respirometer," Tech. Memo. 59–3. Pacific Naval Lab., D. R. B. Esquimalt, B. C.

Maros, L., Schulek, E., Molnar-Perl, I., and Pinter-Szakacs, M. (1961). Einfaches destillationsverfahren zur titrimetrischen bestimmung von Kohlendioxyd. *Anal. Chim. Acta* **25**, 390–399 [for translation, see *Fisheries Res. Board Can., Transl. Ser.* **596** (1965)].

Martin, W. R. (1949). The mechanics of environmental control of body form in fishes. *Univ. Toronto Biol. Ser.* **58**, 1–91.

Mathur, G. B. (1967). Anaerobic respiration in a cyprinoid fish *Rasbora daniconius* (Ham). *Nature* **214**, 318–319.

Mathur, G. B., and Shrivastava, B. D. (1970). An improved activity meter for the determination of standard metabolism in fish. *Trans. Am. Fisheries Soc.* **99**, 602–603.

Meuwis, A. L., and Heuts, M. J. (1957). Temperature dependence of breathing rate in carp. *Biol. Bull.* **112**, 97–107.

Morris, R. (1960). General problems of osmoregulation with special reference to cyclostomes. *Symp. Zool. Soc. London* **1**, 1–16.

Morris, R. W. (1967). High respiratory quotients of two species of bony fishes. *Physiol. Zool.* **40**, 409–423.

Moss, D. D., and Scott, D. C. (1961). Dissolved-oxygen requirements of three species of fish. *Trans. Am. Fisheries Soc.* **90**, 377–393.

Mount, D. I. (1964). Additional information on a system for controlling the dissolved oxygen content of water. *Trans. Am. Fisheries Soc.* **92**, 100–103.

Muir, B. S., Nelson, G. J., and Bridges, K. W. (1965). A method for measuring swimming speed in oxygen consumption studies on the aholehole *Kuhlia sandvicensis. Trans. Am. Fisheries Soc.* **94**, 378–382.

Nicholls, J. V. V. (1931). The influence of temperature on digestion in *Fundulus heteroclitus. Contrib. Can. Biol. Fisheries* **7**, 47–55.

Norris, K. S. (1963). The functions of temperature in the ecology of the percoid fish *Girella nigricans* (Ayres). *Ecol. Monographs* **33**, 23–62.

Olson, F. C. W., and Jackson, J. M. (1942). Heating curves: Theory and practical application. *Ind. Eng. Chem.* **34**, 334–341.

Olson, F. C. W., and Stevens, H. P. (1939). Thermal processing of canned foods in tin containers. II. Nomograms for graphic calculation of thermal processes for non-acid canned foods exhibiting straight-line semi-logarithmic heating curves. *Food Res.* **4**, 1–20.

Ozaki, H. (1951). On the relation between the phototaxis and the aggregation of young marine fish. *Rept. Fac. Fisheries Prefect. Univ. Mie* **1**, 55–66.

Paloheimo, J. E., and Dickie, L. M. (1966a). Food and growth of fishes. II. Effects of food and temperature on the relation between metabolism and body weight. *J. Fisheries Res. Board Can.* **23**, 869–908.

Paloheimo, J. E., and Dickie, L. M. (1966b). Food and growth of fishes. III. Relations among food, body size and growth efficiency. *J. Fisheries Res. Board Can.* **23**, 1209–1248.

Pavlov, D. S., Sbikin, Yu. N., and Mochek, A. D. (1968). The effect of illumination in running water on the speed of fishes in relation to features of their orientation. *Vopr. Ikhtiol.* **8**, 250–255 (for translation, see "Problems of Ichthyology." Am. Fisheries Soc., Washington, D. C., 1968).

Pavlovskii, E. N., ed. (1962). "Techniques for the Investigation of Fish Physiology." Izd. Akad. Nauk S.S.S.R. (Transl. No. OTS 64-11001. Off. Tech. Serv., U. S. Dept. Comm., Washington, D. C., 1964).

Peterson, R. H., and Anderson, J. M. (1969a). Influence of temperature change on spontaneous locomotor activity and oxygen consumption of Atlantic salmon, *Salmo salar,* acclimated to two temperatures. *J. Fisheries Res. Board Can.* **26**, 93–109.

Peterson, R. H., and Anderson, J. M. (1969b). Effects of temperature on brain tissue oxygen consumption in salmonid fishes. *Can. J. Zool.* **47**, 1345–1353.

Pitkow, R. B. (1960). Cold death in the guppy. *Biol. Bull.* **119**, 231–245.

Pitt, T. K., Garside, E. T., and Hepburn, R. L. (1956). Temperature selection of the carp (*Cyprinus carpio* Linn.) *Can. J. Zool.* **34**, 555–557.

Potts, D. C., and Morris, R. W. (1968). Some body fluid characteristics of the Antarctic fish, *Trematomus bernacchii. Marine Biol.* **1**, 269–276.

Precht, H. (1968). Der Einfluß "normaler" Temperaturen auf Lebensprozesse

bei wechselwarmen Tieren unter Ausschluß der Wachstums- und Entwick-
lungsprozesse. *Helgolaender Wiss. Meeresuntersuch.* **18**, 487–548.

Precht, H., Christophersen, J., and Hensel, H. (1955). "Temperatur und Leben."
Springer, Berlin.

Prosser, C. L., and Farhi, E. (1965). Effects of temperature on conditioned reflexes and on nerve conduction in fish. *Z. Vergleich. Physiol.* **50**, 91–101.

Prosser, C. L., Barr, L. M., Pinc, R. A., and Lauer, C. Y. (1957). Acclimation
of goldfish to low concentrations of oxygen. *Physiol. Zool.* **30**, 137–141.

Rao, G. M. M. (1967). Oxygen consumption of rainbow trout (*Salmo gairdneri*)
in relation to activity, salinity and temperature. Ph.D. Thesis, University of
Toronto (National Library of Canada, Canadian theses on microfilm, No. 1987).

Rao, G. M. M. (1968). Oxygen consumption of rainbow trout (*Salmo gairdneri*)
in relation to activity and salinity. *Can. J. Zool.* **46**, 781–786.

Rao, G. M. M. (1969). Effect of activity, salinity, and temperature on plasma
concentrations of rainbow trout. *Can. J. Zool.* **47**, 131–134.

Roberts, J. L. (1966). Systemic versus cellular acclimation to temperature by
poikilotherms. *Helgolaender Wiss. Meeresuntersuch.* **14**, 451–465.

Roots, B. I., and Prosser, C. L. (1962). Temperature acclimation and the nervous
system in fish. *J. Exptl. Biol.* **39**, 617–628.

Rubin, M. A. (1935). Thermal reception in fishes. *J. Gen. Physiol.* **18**, 643–647.

Ruhland, M. L. (1965). Etude comparative de la consommation d'oxygène chez
différentes espèces de poissons teléostéens. *Bull. Soc. Zool. France* **90**, 347–353.

Ruhland, M. L. (1967). Controle des pressions partielles d'oxygène au cours des
mésures de la consommation d'oxygène chez les poissons dans un appareil
enregistreur continu. *Bull. Soc. Zool. France* **92**, 787–792.

Ruhland, M. L., and Heusner, A. (1959). Technique d'enregistrement de faibles
consommations d'oxygène: Application aux poissons de petite taille. *Compt.
Rend. Soc. Biol.* **153**, 161–164.

Saunders, R. L. (1962). The irrigation of the gills in fishes. II. Efficiency of oxygen
uptake in relation to respiratory flow, activity and concentrations of oxygen
and carbon dioxide. *Can. J. Zool.* **40**, 817–862.

Scheier, A., and Cairns, J., Jr. (1966). Persistence of gill damage in *Lepomis
gibbosus* following a brief exposure to alkyl benzene sulfonate. *Notulae Naturae
(Acad. Nat. Sci. Phila.)* **391**, 1–7.

Schmein-Engberding, F. (1953). Die Vorzugstemperaturen einiger Knochenfische
und ihre physiologische Bedeutung. *Z. Fischerei* **2**, 125–155.

Scholander, P. F., Haugaard, N., and Irving, L. (1943). A volumetric respirometer
for aquatic animals. *Rev. Sci. Instr.* **14**, 48–51.

Segrem, N. P., and Hart, J. S. (1967). Oxygen supply and performance in
*Peromyscus.* Metabolic and circulatory responses to exercise. *Can. J. Physiol.
Pharmacol.* **45**, 531–541.

Shelford, V. E. (1929). "Laboratory and Field Ecology." Williams & Wilkins,
Baltimore, Maryland.

Shepard, M. P. (1955). Resistance and tolerance of young speckled trout
(*Salvelinus fontinalis*) to oxygen lack, with special reference to low oxygen
acclimation. *J. Fisheries Res. Board Can.* **12**, 387–446.

Silver, S. J., Warren, C. E., and Doudoroff, P. (1963). Dissolved oxygen requirements of developing steelhead trout and Chinook salmon embryos at different
water velocities. *Trans. Am. Fisheries Soc.* **92**, 327–343.

Smit, H. (1965). Some experiments on the oxygen consumption of goldfish

(*Carassius auratus* L.) in relation to swimming speed. *Can. J. Zool.* **43**, 623–633.

Smit, H. (1967). Influence of temperature on the rate of gastric juice secretion in the brown bullhead, *Ictalurus nebulosus*. *Comp. Biochem. Physiol.* **21**, 125–132.

Smith, L. S., and Newcomb, T. W. (1970). A modified version of the Blažka respirometer and exercise chamber for large fish. *J. Fisheries Res. Board Can.* **27**, 1321–1324.

Snyder, C. D. (1905). On the influence of temperature upon cardiac contraction and its relation to influence of temperature upon chemical reaction velocity. *Univ. Calif. Publ. Physiol.* **2**, 125–146.

Spaas, J. T. (1959). Contribution to the biology of some cultivated cichlidae. Temperature, acclimation, lethal limits and resistance in three cichlidae. *Biol. Jaarboek Konink. Natuurw. Genoot. Dodonaea Gent* **27**, 21–38.

Spoor, W. A. (1946). A quantitative study of the relationship between the activity and oxygen consumption of the goldfish and its application to the measurement of respiratory metabolism in fishes. *Biol. Bull.* **91**, 312–325.

Sprague, J. B. (1968). Avoidance reactions of salmonid fish to representative pollutants. *Water Res.* **2**, 23–24.

Sprague, J. B., and Ramsay, B. A. (1965). Lethal levels of mixed copper-zinc solutions for juvenile salmon. *J. Fisheries Res. Board Can.* **22**, 425–432.

Steen, J. B. (1963). Oxygen secretion in the swimbladder. *Proc. 5th Intern. Congr. Biochem., Moscow, 1956* pp. 621–630. Pergamon Press, Oxford.

Stevens, E. D., and Fry, F. E. J. (1971). Brain and muscle temperatures in ocean caught and captive skipjack tuna. *Comp. Biochem. Physiol.* **38A**, 203–211.

Sullivan, C. M. (1954). Temperature reception and responses in fish. *J. Fisheries Res. Board Can.* **11**, 153–170.

Sullivan, C. M., and Fisher, K. C. (1953). Seasonal fluctuations in the selected temperature of speckled trout, *Salvelinus fontinalis* (Mitchill). *J. Fisheries Res. Board Can.* **10**, 187–195.

Swift, D. R. (1964). The effect of temperature and oxygen on the growth rate of the Windermere char (*Salvelinus alpinus willughbii*). *Comp. Biochem. Physiol.* **12**, 179–183.

Tang, P. (1933). On the rate of oxygen consumption by tissues and lower organisms as a function of oxygen tension. *Quart. Rev. Biol.* **8**, 260–274.

Tåning, Å. V. (1952). Experimental study of meristic characters in fishes. *Biol. Rev.* **27**, 169–193.

Taylor, G. I. (1923). Experiments on the motion of solid bodies in rotating fluids. *Proc. Roy. Soc.* **A104**, 213–218.

Tyler, A. V. (1966). Some lethal temperature relations of two minnows of the genus *Chrosomus*. *Can. J. Zool.* **44**, 349–364.

Tytler, P. (1969). Relationship between oxygen consumption and swimming speed in the haddock, *Melanogrammus aeglefinus*. *Nature* **221**, 274–275.

Ushakov, B. P. (1968). Cellular resistance, adaptation to temperature and thermostability of somatic cells with special reference to marine animals. *Marine Biol.* **1**, 153–160.

van Dam, L. (1938). "On the Utilisation of Oxygen and Regulation of Breathing in Some Aquatic Animals," pp. 1–143. Volharding, Gröningen.

van't Hoff, J. H. (1884). Études de dynamique chimique. Amsterdam.

Verheijen, F. J. (1958). The mechanism of the trapping effect of artificial light sources upon animals. *Arch. Neerl. Zool.* **13**, 1–107.

Veselov, E. A. (1949). Effect of salinity of the environment on the rate of respiration in fish. *Zool. Zh.* **28**, 85–98.

von Bertalanffy, L. (1950). The theory of open systems in physics and biology. *Science* **111**, 23–29.

von Ledeburg, J. F. (1939). Der Sauerstoff als ökologischer Faktor. *Ergeb. Biol.* **16**, 173–261.

Wells, N. A. (1935). Variations in the respiratory metabolism of the Pacific killifish *Fundulus parvipinnis,* due to size, season and continued constant temperature. *Physiol. Zool.* **8**, 318–336.

Whitmore, C. M., Warren, C. E., and Doudoroff, P. (1960). Avoidance reactions of salmonid and centrarchid fishes to low oxygen concentrations. *Trans. Am. Fisheries Soc.* **89**, 17–26.

Whitworth, W. R. (1968). Effects of diurnal fluctuations of dissolved oxygen on the growth of brook trout. *J. Fisheries Res. Board Can.* **25**, 579–584.

Wikgren, B. J. (1953). Osmotic regulation in some aquatic animals with particular respect to temperature. *Acta Zool. Fenn.* **71**, 1–93.

Winberg, G. G. (1956). Intensivnost obmena i pichchevye potrebnosti ryb. *Nauch. Tr. Belorussk. Gos. Univ.* [for translation, see *Fisheries Res. Board Can., Transl. Ser.* **194** (1960)].

Wohlschlag, D. E. (1957). Differences in metabolic rates of migratory and resident freshwater forms of an Arctic whitefish. *Ecology* **38**, 502–510.

Wohlschlag, D. E. (1964). Respiratory metabolism and ecological characteristics of some fishes in McMurdo Sound, Antarctica. *In* "Biology of the Antarctic Seas" (M. O. Lee, ed.), Antarctic Res. Ser. No. 1, pp. 33–62. Am. Geophys. Union.

Wohlschlag, D. E., Cameron, J. N., and Cech, J. J., Jr. (1968). Seasonal changes in the respiratory metabolism of the pinfish (*Lagodon rhomboides*). *Contrib. Marine Sci.* **13**, 89–104.

Woodhead, P. M. J. (1964). The death of North Sea fish during the winter of 1962-63, particularly with reference to the sole, *Solea vulgaris. Helgolaender Wiss. Meeresuntersuch.* **10**, 283–300.

Wuhrmann, K., and Woker, H. (1948). Beiträge zur Toxikologie der Fische. II. Experimentelle Untersuchungen über die Ammoniak- und Blausäurevergiftung. *Schweiz. Z. Hydrol.* **11**, 210–244.

Zahn, M. (1962). Die Vorzugstemperaturen zweier Cypriniden und eines Cyprinodonten und die Adaptationstypen der Vorzugstemperatur bei Fischen. *Zool. Beitr.* [N.S.] **7**, 15–25.

Zahn, M. (1963). Jahreszeitliche Veränderungen der Vorzugstemperaturen von Scholle (*Pleuronectes platessa* Linne) und Bitterling (*Rhodeus sericeus* Pallas). *Verhandl. Deut. Zool. Ges. Muenchen* pp. 562–580.

# 2

# BIOCHEMICAL ADAPTATION TO THE ENVIRONMENT

*P. W. HOCHACHKA and G. N. SOMERO*

## I. INTRODUCTION

### Basic Problems of Metabolic Control in Poikilotherms

During a good portion of this century biochemistry went through a period of descriptive and largely empirical research. A major reversal of this approach began when the basic fabric of metabolic organization had become clear, and biochemists were faced with the problem of interpreting from a functional standpoint a vast body of empirical knowledge. To this end biochemical thinking came to draw heavily on the theories of evolution and physiological adaptation; consequently, issues of selective advantage and physiological significance, which previously had been ignored to a large extent, have become central to many areas of biochemical research.

The functional approach to biochemistry is perhaps best exemplified by the current theories of enzymic regulation; these concepts are based on ideas of physiological adaptation, and the contributions made by different enzymic reactions are considered in the context of cellular "needs" and "demands" (see Atkinson, 1965, 1966, 1968; Stadtman, 1968). When metabolism is viewed from a functional perspective, the importance of an individual enzymic reaction is not seen as catalysis per se but in terms of the role which the reaction plays as part of a tightly regulated network of metabolic transformations.

Current theories of metabolic (enzymic) regulation may be summarized as follows: The contribution which a given metabolic pathway makes to overall metabolism is largely dependent on the cell's needs for the products of that pathway. The activity of a pathway is tightly controlled by means of "on–off switches" on one or more of the enzymes functioning early in the pathway. In many instances the "switch" enzyme is the first enzyme which is unique to the pathway; this site of control obviously permits a highly efficient regulation of the pathway as a whole. The parameter which has generally been selected for the role of switch is enzyme–substrate (E–S) affinity, which is normally inversely proportional to the reciprocal of the apparent Michaelis constant ($K_m$) of substrate. Most known positive modulators (activators) increase E–S affinity; negative modulators have the opposite effect. These relationships are illustrated in Fig. 1. An important implication of E–S affinity modulation as a mechanism for controlling metabolic activity is that substrate levels in the cell must normally be well below saturating ($V_{max}$) concentrations, for $K_m$ changes can affect reaction rates only if the substrate

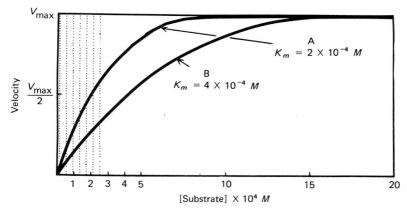

**Fig. 1.** The effect of varying enzyme–substrate affinity, defined as the reciprocal of the Michaelis constant ($K_m$), on enzyme activity. Enzymes A and B catalyze the same reaction and exhibit the same activity at saturating ($V_{max}$) concentration of substrate. The enzymes have a twofold difference in $K_m$. At physiological substrate concentrations (indicated by stippling) enzyme A is much more active than B. In terms of enzyme regulation theory, B may represent a deactivated state of the enzyme and A may represent an activated state. B may be converted to A by the binding of a positive modulator; conversely, A may be converted to B under the influence of a negative modulator.

concentration is low (Fig. 1). When substrate levels are low, it is apparent that small, e.g., twofold, changes in $K_m$ can lead to large changes in reaction velocity. Thus, $K_m$ modulation is seen as a highly efficient mechanism for governing rates of metabolic activity.

The basic problems of metabolic control for poikilothermic organisms arise from the fact that most of the regulatory functions described above are directly affected by changes in environmental parameters such as temperature; for example, E–S affinity may vary with temperature in complex manners (Hochachka and Somero, 1968; Somero and Hochachka, 1968, 1969; Somero, 1969a,b; Baldwin and Hochachka, 1970; Baldwin, 1971). For some enzymes, enzyme–modulator interactions are highly temperature sensitive; changes of this sort pose major problems for the control of metabolism (Behrisch and Hochachka, 1969a,b). In other cases, regulatory functions appear temperature independent (Somero and Hochachka, 1968; Somero, 1969a). The flow of carbon through metabolic branch points is particularly sensitive to thermal changes and can display apparently anomalous temperature characteristics (Hochachka, 1968a; Dean, 1969). Ion concentrations and, presumably, ion compartmentalizations may change in response to temperature variation (Hickman *et al.*, 1964; Heinicke and Houston, 1965). These findings indicate that the complex mechanisms which permit an organism to regulate closely its enzymic activity also render the organism highly vulnerable to the deleterious effects of sudden changes in environmental conditions. However, the fact that these enzymic functions are directly affected by temperature raises an important possibility: If the environmentally (thermally) induced changes in enzymic properties occur in an adaptive direction and at an adaptive rate, then it is possible that the organism may be able to make use, in a positive manner, of environmental changes which might appear deleterious on *a priori* grounds. The elaborate metabolic control mechanisms we have discussed therefore may serve as "raw material" for selection to "design" homeostatic mechanisms enabling the organism to adapt to environmental changes. We shall spend much of this essay discussing the manner in which poikilotherms can accomplish this process.

In this discussion it is important to bear in mind that the adaptive "strategies" employed by poikilotherms will vary considerably among different organisms; for example, if avenues of behavioral escape from harmful environmental circumstances are available, then the need for extensive biochemical adaptation is lessened. In addition, the range of adaptive capacities shown by an organism may be dependent on the complexity of its environment. Fishes living at near-constant temperatures in the Antarctic seas show less ability to acclimate their metabolism than do more eurythermal fishes such as trout or goldfish (Somero *et al.*, 1968). This complexity of the organism's adaptive responses may thus depend on the rate at which, and the extent to which, environmental parameters such as temperature, dissolved gases, and salinity fluctuate.

Considerations of this type have led physiologists to consider a number of different time-courses of environmental adaptation; we feel it may be useful to organize our discussion within a time-course framework of this type. At one extreme is evolutionary adaptation, a process likely requiring many generations for completion. An example of this process is the latitudinal adjustment of metabolic rates of fishes (Wohlschlag, 1964; Somero *et al.*, 1968; Hemmingsen *et al.*, 1969).

Similar physiological responses to temperature are commonly observed on a seasonal basis (Roberts, 1964, 1967). This adaptation process, which occurs over a time-course of days, weeks, or months, is termed "acclimation" or "acclimitization."

Finally, biologists have recently recognized that for at least some organisms and physiological processes, adaptation to the environment is immediate; for example, metabolic $Q_{10}$'s approximating unity have been reported for several intertidal organisms (Newell, 1966; Newell and Northcroft, 1967; S. Baldwin, 1968).

In the past few years, comparative biochemists have initiated an intensive investigation of the biochemical changes which are involved in each of these adaptation processes. This chapter will consider what progress has been made along these lines to date, and what avenues of future research seem most promising and relevant.

## II. IMMEDIATE EFFECTS OF TEMPERATURE ON ENZYMES

### A. Enzyme–Substrate Interactions

Our approach to the problem of enzymic mechanisms of temperature adaptation has been guided to a great extent by the findings of workers in the field of enzyme regulation. Although these workers have normally considered biochemical systems which function at essentially constant temperature, the elucidation of the most important parameters involved in regulation enzymic activity *in vivo* has suggested numerous questions for physiologists and biochemists concerned with organisms facing large variations in body temperatures.

The most important contribution of biochemists studying the regulatory functions of enzymes has been the elaboration of control mechanisms involving modulator induced changes in E–S affinity. As stated above, this parameter seems of crucial importance in the vast majority of enzyme regulatory processes. Thus it was of interest to determine how E–S affinity is influenced by temperature in the case of enzymes from poikilothermic organisms.

Our initial investigations of this question suggested the following important relationship: Decreases in temperature over most or all of a poikilotherm's range of physiological temperatures are accompanied by increases in E–S affinity. Phrased in the terminology of enzyme regulation theory, this statement says that, over most of a poikilotherm's range of habitat (body) temperature, decreases in temperature affect its enzymes in a manner analogous to positive modulators.

This "positive thermal modulation" of poikilothermic enzymes is well illustrated in the case of Alaskan king crab, *Paralithodes camtschatica*, phosphofructokinase (PFK); addition of the positive modulatory 5′-AMP and a 10°C decrease in temperature have analogous affects on the enzyme (Fig. 2). Similar temperature-dependent changes in E–S affinity have been found for a large number of enzymes from phylogenetically diverse poikilotherms. This list includes lungfish liver fructose-diphosphatase (FDPase) (Fig. 3); rainbow trout acetylcholinesterase (Baldwin and Hochachka, 1970) (Fig. 4); pyruvate kinases (PyKs)

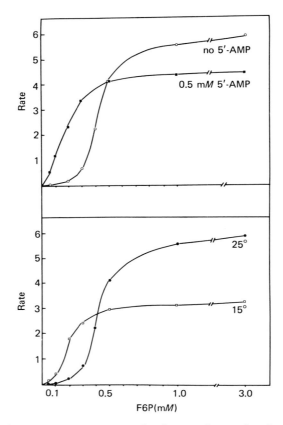

**Fig. 2.** Substrate saturation curves for king crab muscle phosphofructokinase at 15° and 25°C (lower panel) and at 25°C in the presence and absence of the positive modulator, 5′-AMP (upper panel). Note how addition of 5′-AMP and the decrease in assay temperature have analogous effects on PFK. Phosphofructokinase activity was assayed in 0.1 $M$ tris-HCl buffer, pH 8.0, 2 mM ATP, 5 mM MgSO₄, 0.3 mM NADH, 0.05 $M$ KCl, 1 mM dithiothreitol, excess coupling enzymes (aldolase, triosephosphate isomerase, and glycerolphosphate dehydrogenase), and varying amounts of F6P. Data from Freed (1971).

from different poikilotherms (Somero and Hochachka, 1968; Somero, 1969a,b) (Fig. 5); lactate dehydrogenase (LDHs) (Hochachka and Somero, 1968; Somero and Hochachka, 1969); goldfish choline acetyltransferases (Hebb *et al.*, 1969); and rainbow trout phosphofructokinases (PFKs) and fructosedephosphate aldolases (Alds) (Somero, 1970).

The relationship between temperature and E–S affinity may display two important characteristics. First, over most of the physiological temperature range, decreases in temperature promote increases in E–S

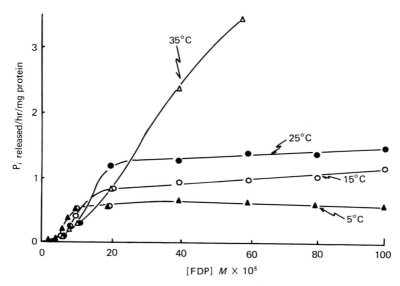

**Fig. 3.** Substrate saturation curves for lungfish FDPase at a series of assay temperatures. Note, in comparison with the theoretical saturation curves of Fig. 1, the similarity between the effects of positive modulators and decreases in temperatures. After Behrisch and Hochachka ( 1969b).

affinity. Second, for many enzymes, sharp increases in $K_m$ are observed at temperatures near the lower limits of the physiological range. Thus, at these lower temperatures, decreases in temperature function as negative modulators. The important biological consequences of this condition where temperature decreases slow the reactions by (a) reducing the kinetic energy of the reactants and (b) decreasing E–S affinity will be discussed in Section IV, B.

The extent to which E–S affinity increases can stabilize rates of enzymic activity as the temperature drops is documented in Tables I and II. If E–S affinity did not vary with temperature, then the $Q_{10}$'s of the reactions would be independent of substrate concentration. However, in all cases studied the $Q_{10}$ of the reaction is directly proportional to substrate concentration.

Another way of illustrating the significance of temperature-dependent changes in E–S affinity is shown in Fig. 6, which illustrates the effect of temperature on lungfish LDH activity. An important consequence of a differential change in $K_m$ and $V_{max}$ with temperature is that at low substrate concentrations, when $K_m$ is of primary importance in determining reaction velocity, catalytic rates are higher at lower temperatures than at higher temperatures (Fig. 6). When substrate concentrations are

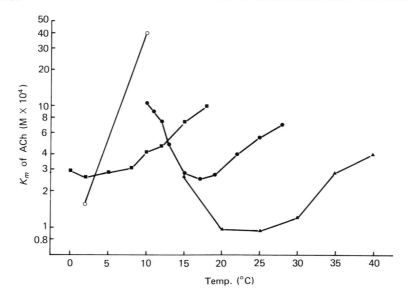

**Fig. 4.** The effect of temperature on the $K_m$ of acetylcholine for acetylcholin-esterase enzymes of the electric eel, *Electrophorus electricus* (▲); cold- (2°C) ac-climated trout (■); warm- (18°) acclimated trout (●); and *Trematomus borch-grevinki* (○). After Baldwin and Hochachka (1970) and Baldwin (1971).

raised above $K_m$ values, catalytic rates behave according to the Arrhenius relationship. The result of these $V_{max}$ and $K_m$ effects is a family of satura-tion curves for which the thermal optimum gradually shifts upward with increasing substrate concentration (Fig. 6).

It is evident from these findings that temperature dependence of an enzymic reaction will depend critically on the intracellular concentration of substrate. In most cases where intracellular substrate concentrations have been determined accurately, these levels have been less than or equal to the $K_m$ values of substrate for the different enzymes (Williamson *et al.*, 1967a,b; Hochachka *et al.*, 1971). Thus, temperature-dependent E–S affinity changes can be of major physiological significance to poikilo-thermic organisms.

The importance of substrate concentration in determining the thermal properties of enzymic systems has a number of further implications of biological importance. First, the temperature dependence of an enzymic reaction will change over time if the substrate level varies; we shall discuss this implication when we consider thermal effects on active and basal metabolism. Second, the temperature dependencies of different metabolic pathways may differ if certain pathways have relatively high

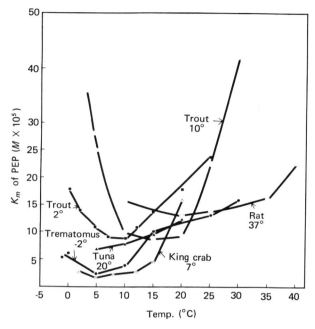

**Fig. 5.** The effect of temperature on the $K_m$ of phosphoenolpyruvate (PEP) of pyruvate kinase enzymes from differently adapted species. Modified after Somero (1969a).

levels of substrate available or if the temperature–$K_m$ responses of enzymes vary among pathways. These possibilities will also be considered in a later section.

## B. The Role of Enzyme Variants in Immediate Temperature Adaptation

For any given enzyme, rate stabilization resulting from increases in E–S affinity as the temperature drops is noted only over a certain range of temperatures. For an organism experiencing 10°–15°C changes in habitat temperature over daily or hourly time spans, it would seem advantageous to have two or more variants of a given enzyme in its tissues which, by acting together, could promote thermally independent enzymic function over a wider range of temperatures than would be possible if only a single form of the enzyme were present. We shall discuss several instances in which a number of enzyme variants would appear to broaden the range of temperatures over which enzymic activity

**Table I**

Temperature Coefficients ($Q_{10}$'s) of LDH Reactions as a Function
of Substrate (Pyruvate) Concentration[a]

| Pyruvate (mM) | King crab ($Q_{10}$) | Zooarcid ($Q_{10}$) | Shrimp ($Q_{10}$) |
|---|---|---|---|
| 2.0 | 1.8 | 1.7 | |
| 1.0 | 1.5 | 1.6 | |
| 0.5 | 1.5 | 1.4 | 2.3 |
| 0.2 | 1.2 | 1.4 | 1.9 |
| 0.1 | 1.2 | 1.2 | 1.8 |
| 0.05 | 0.8 | | |

[a] $Q_{10}$ values are for the temperature range 5°–15°C. From Somero (1969a).

can be held relatively independent of temperature. Biochemically, what is unique about these systems is that the enzyme variants detected in these kinetic studies very likely are not isoenzymes (isozymes) in the classic sense of being proteins with different primary structures.

One instance in which enzyme variants seem important in immediate compensation is in the case of Alaskan king crab pyruvate kinase (Somero, 1969b). Two kinetically distinct forms of the enzyme can be detected (Fig. 7). One PyK, termed "warm" PyK, exhibits maximal E–S affinity near 12°C; a second PyK ("cold" PyK) has highest E–S affinity near 5°C (Table III). The combined activities of these PyK variants yield highly stable rates of PyK activity over the king crab's entire range of habitat temperatures (approximately 4°–12°C) (Table II).

Our examination of the molecular basis of the two PyK activities

**Table II**

Temperature Coefficients ($Q_{10}$'s) of the King Crab Glucose-6-Phosphate (G6P)
Dehydrogenase, 6-Phosphogluconate (6PG) Dehydrogenase, and PyK
Reactions as Functions of Substrate Concentration[a]

| Substrate (mM) | G6P dehydrogenase ($Q_{10}$) | 6PG dehydrogenase ($Q_{10}$) | PyK ($Q_{10}$) |
|---|---|---|---|
| 0.5 | 3.6 | | 3.0 |
| 0.3 | | 3.8 | |
| 0.2 | 3.0 | 3.4 | 2.2 |
| 0.1 | 2.6 | 3.0 | |
| 0.05 | | 2.3 | 1.9 |
| 0.02 | | | 1.6 |
| 0.01 | | 1.9 | |

[a] All $Q_{10}$ values are for the temperature range 5°–15°C. From Somero (1969a).

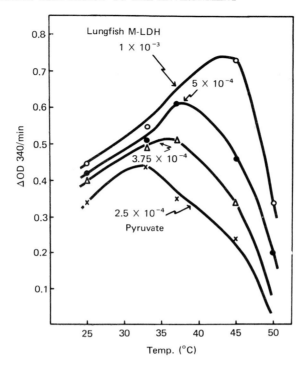

**Fig. 6.** Effect of temperature on lungfish muscle LDH activity at a series of pyruvate concentrations. From Hochachka and Somero ( 1968 ).

yielded a surprising result, for both activities seem a result of a single protein species. The two PyK variants are formed in a temperature-dependent interconversion reaction. As the temperature is lowered, warm PyK is formed at the expense of cold PyK; as temperature is raised, the opposite conversion occurs. This effect can be seen by comparing the relative contributions of the two PyK variants to $V_{max}$ activity at different temperature (Fig. 7). The biological significance of this interconversion is twofold. First, the interconversion forms increased quantities of the type of enzyme which functions well at a particular habitat temperature. Second, the warm variant of the enzyme exhibits sigmoidal saturation kinetics, whereas the cold PyK has hyperbolic kinetics. This fact allows us to extend our analogy between the effects of temperature decreases and the effects of positive enzyme modulators: Temperature decreases, like certain positive modulators, can activate enzymes by promoting allosteric interconversions between sigmoidal and hyperbolic stages of the same enzyme protein. Similarly, complex kinetics characterize the LDH

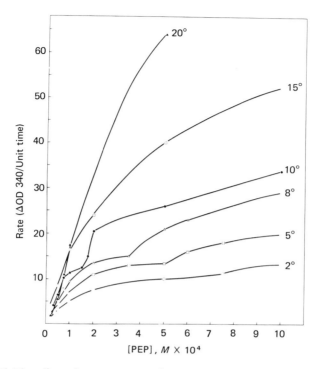

**Fig. 7.** The effect of temperature on king crab leg muscle pyruvate kinase activity at a series of substrate (PEP) concentrations. From Somero (1969b).

### Table III

The Effect of Temperature on the Apparent $K_m$ Value of Cold PyK and the $S_{0.5}$ Value of Warm PyK of the Alaskan King Crab[a]

| Temp. (°C) | Warm $S_{0.5}$ ($\times 10^4\,M$) | Cold $K_m$ ($\times 10^4\,M$) |
|:---:|:---:|:---:|
| 2 | 3.3 | 1.8 |
| 5 | 8.0 | 1.5 |
| 8 | 5.0 | 1.7 |
| 9 | 2.5 | 2.2 |
| 10 | 2.0 | 2.5 |
| 12 | 1.4 | 8.0 |
| 15 | 3.0 | [b] |
| 20 | 13.3 | [b] |

[a] Half-saturating substrate concentrations for sigmoidal enzymes are designated $S_{0.5}$. Note the similarity in the temperature–$K_m$ relationships of these two enzymes and the PyKs of warm- and cold- acclimated rainbow trout (Fig. 5). Data from Somero (1969b).

[b] Enzyme appears to be entirely converted to the warm conformation.

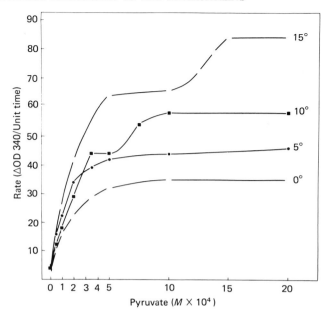

**Fig. 8.** The effect of temperature on the rate of king crab leg muscle LDH activity at a series of substrate (pyruvate) concentrations. Note that only one LDH variant is active at high temperatures under conditions of physiological pyruvate concentration. From Somero and Hochachka (1969).

systems of king crab (Fig. 8) and rainbow trout (Fig. 9). In both of these cases more than one type of LDH activity can be distinguished kinetically. For king crab LDH there appear to be two different LDH subunits operative in the system. Note that at higher temperatures only a single LDH is likely to contribute to pyruvate reduction at physiological substrate concentrations; the affinity of the second LDH is too low to enable it to function at physiological pyruvate levels, which are less than 0.1 mM (Somero and Hochachka, 1969). As the temperature is reduced, this low affinity LDH is activated by a sharp decrease in the apparent $K_m$ of pyruvate (Table IV). Consequently, at 5°C both LDH variants contribute to pyruvate metabolism, and the rate of LDH activity is higher at 5°C than at 10°C at physiological concentrations of pyruvate. In the rainbow trout system comparable changes occur. Again, at higher temperatures certain LDH variants would not function at physiological pyruvate levels. As the temperature is decreased, these "latently active" LDHs become functional through increased LDH–pyruvate affinity. At 10°C, the temperature to which the trout were acclimated, LDH activity is higher than at 15°C. These $Q_{10}$ relationships are further documented in Table V.

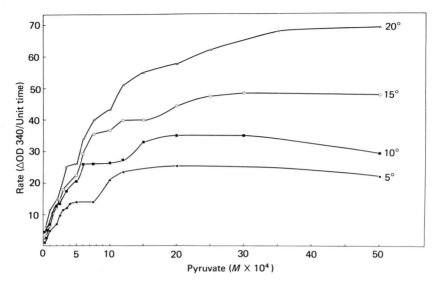

**Fig. 9.** The effect of temperature on rainbow trout epaxial muscle LDH activity at a series of substrate (pyruvate) concentrations. From Somero and Hochachka (1969).

The molecular basis for these LDH effects is unclear. Isozymes of LDH are present in both systems: Five isozyme bands were detected in the king crab (G. N. Somero and P. W. Hochachka, unpublished data), and ten were present in trout (Massaro and Markert, 1968), and these are assembled into tetramers in the usual manner (see Kaplan, 1964; Markert, 1968). However, complex saturation curves

**Table IV**

The Effect of Temperature on the $K_m$ of Pyruvate of
the Two King Crab LDH Variants[a]

| Temp. (°C) | $K_m$ of pyruvate | |
| | "Low-$K_m$" LDH (mM) | "High-$K_m$" LDH (mM) |
| --- | --- | --- |
| 0 | | 0.13[b] |
| 5 | | 0.08[b] |
| 10 | 0.30 | 0.70 |
| 15 | 0.33 | 1.20 |

[a] From Somero and Hochachka (1969).

[b] At these two lower temperature the $K_m$ values of the two LDHs are approximately equal (see Fig. 9).

**Table V**

$Q_{10}$ Values for the Rainbow Trout LDH Reaction at a Series
of Substrate Concentrations[a]

| Pyruvate (M) | ($Q_{10}$) | |
|---|---|---|
| | (5°–15°C) | (10°–20°C) |
| $5 \times 10^{-3}$ | 2.15 | 2.33 |
| $2 \times 10^{-3}$ | 1.74 | 1.66 |
| $1 \times 10^{-3}$ | 1.71 | 1.59 |
| $5 \times 10^{-4}$ | 1.61 | 1.27 |
| $2 \times 10^{-4}$ | 1.73 | 1.16 |
| $5 \times 10^{-5}$ | 1.36 | 1.00 |

[a] From Somero and Hochachka (1969).

of this type cannot be generated by adding two or more hyperbolic saturation curves together. Thus it seems unlikely that these kinetic effects result solely from the activities of several LDH isozymes displaying normal hyperbolic saturation curves. One explanation is that some LDHs exhibit sigmoidal saturation kinetics under certain conditions; this remains to be tested with purified isozymes. The formation of a sigmoidal type LDH might involve the formation of a "metastable" form of the molecule (see Nickerson and Day, 1969). Under certain conditions, e.g., high temperature, part of the population of LDH molecules may exist in a metastable state characterized by (1) sigmoidal kinetics and (2) low E–S affinity. Thus, this type of system may be comparable to the king crab PyK system discussed earlier.

## C. Temperature Effects on Active and Basal Metabolism

We have shown that the rate-stabilizing effects of temperature-dependent changes in E–S affinity are greatest when substrate concentrations are low (Tables I, II, and V). When substrate levels rise, as they might during strenuous activity, then the temperature dependence of enzymic reactions will also increase. This consideration suggests that the temperature dependence of whole organism metabolism may be a function of the metabolic rate per se. To the extent that the enzymic effects noted determine the thermal properties of whole organism metabolism, one would predict that metabolic $Q_{10}$'s will be lowest during basal metabolism and highest during active metabolism.

Because of the difficulties inherent in measuring the active and basal components of metabolism, there are few data available to test the above hypothesis. The strongest supporting evidence for a positive correlation between metabolic rate and metabolic $Q_{10}$ comes from studies of intertidal organisms, notably sedentary invertebrates and certain species of algae (Newell, 1966, 1967; Halcrow and Boyd, 1967; Newell and Northcroft, 1967; S. Baldwin, 1968). In all cases these organisms were found to exhibit metabolic $Q_{10}$'s approximating unity over the physiological temperature range. Since these organisms likely have only a limited active metabolism, the metabolic data obtained by these workers may be a fair approximation of true basal metabolic rates.

In the case of fishes, reliable separation of active and basal metabolism has been achieved in only a small number of instances (Fry, 1947, 1958; Brett, 1967). Where such data are available there does not appear to be a positive correlation between the metabolic rate and $Q_{10}$. Further investigation of the thermal properties of active and basal metabolism are thus needed to resolve this question.

## D. Temperature Effects on Regulatory Properties of Poikilothermic Enzymes

All of our discussion of the thermal behavior of enzyme systems has been based on a single kinetic parameter, enzyme–substrate affinity. Our predictions of relatively temperature-independent enzymic function are tenable only if we assume that the regulatory functions of poikilothermic enzymes are likewise relatively independent of temperature.

This latter condition does not appear to pertain in the case of homeothermic enzymes. For mammalian and bacterial enzymes, large and frequently differential effects of temperature on regulatory properties have been reported (Ingraham and Maaløe, 1967; Lowry et al., 1964; Helmreich and Cori, 1964; Taketa and Pogell, 1965; Iwatsuki and Okazaki, 1967). Intuitively, it was felt that the types of changes observed in these systems, as well as the magnitude of these changes, would be incompatible with survival in poikilothermic systems; for example, if the magnitude of the change in enzyme–modulator interaction observed for mammalian FDPase should occur in poikilothermic systems, then one would expect that gluconeogenesis would be completely "switched off" at low temperatures (Taketa and Pogell, 1965). A further problem for poikilotherms would appear to be the type of enzyme–modulator response which occurs at different temperatures. Thus, in the case of E. coli deoxythymidine kinase, deoxythymidine triphosphate inhibits the enzyme at temperatures above 30°C, while at lower temperatures it activates the enzyme. These findings plus similar data from other studies (see Lowry *et al.*, 1964) led us to expect that enzyme–modulator interactions in poikilothermic organisms would of necessity be largely unaffected by temperature. Our predictions have not been realized. It appears that just as enzyme–substrate interactions can be changed by temperature, so also specific enzyme–modulator and enzyme–cofactor interactions can be temperature sensitive (Behrisch and Hochachka, 1969a,b; Behrisch, 1969; Somero and Hochachka, 1968). However, even though certain regulatory functions of poikilothermic enzymes may be thermally sensitive, the overall regulation of any given enzyme may be highly independent of temperature (see Somero and Hochachka, 1968; Behrisch, 1969).

This conclusion is based on detailed studies of properties of enzymes at regulatory and branch points in metabolism. Fructosediphosphatase, for example, functions at a branch and regulatory site in the gluconeogenic and glycolytic pathways. The enzyme catalyzes the hydrolysis of FDP to F6P and $P_i$; in all other systems examined, negative modulation of FDPase by AMP appears to be the major means of controlling the activity of this reaction (Newsholme and Gevers, 1967). Thus when the energy charge of the cell is high (AMP levels are very low), FDPase activity is favored and gluconeogenesis is thereby promoted. When the energy charge of the cell is low, i.e., when AMP levels are high, the FDPase reaction is inhibited.

In our initial studies of trout liver FDPase we observed that the affinity of FDPase for AMP increases dramatically at lower temperatures. At 25°C the $K_i$ of AMP is approximately $80 \times 10^{-5} M$; at 15°C a value of $5 \times 10^{-5} M$ was found; and at 0°C the $K_i$ is $2.5 \times 10^{-5} M$. In other words, enzyme–AMP affinity is nearly 20–30 times greater at low temperatures than at higher temperatures (Behrisch and Hochachka, 1969a). Since the organism must be capable of maintaining gluconeogenic activity, and therefore FDPase activity, even at low temperatures (Hochachka, 1967), it seemed obvious that mechanisms must exist for FDPase activation—or at least reversal of AMP inhibition—at low temperatures.

From detailed consideration of the control of trout FDPase it has become evident that the increased efficiency of AMP inhibition at low temperatures can be counteracted by a number of mechanisms: (1) increasing cation concentrations, which decrease AMP site–site interactions; (2) decreasing hydrogen ion concentration, which lowers the $K_a$ values of the cations and increases the $V_{max}$ of the reaction; (3) decreasing values of the free $Mg^{2+}$/free $Mn^{2+}$ ratio, which effectively decreases the $K_a$ value of the cationic cofactor; (4) decreasing free $Ca^{2+}$, which lowers Ca inhibition; and (5) increasing FDP site–site interactions and decreasing AMP site–site interactions, which activates the enzyme at low temperatures (Behrisch and Hochachka, 1969a).

We will consider each of these mechanisms in turn.

(1) Previous studies indicated that $Mg^{2+}$ can reverse AMP inhibition of FDPase. In fish systems, however, the concentrations of $Mg^{2+}$ required to counteract AMP inhibition are quite high (Behrisch and Hochachka, 1969a) and therefore $Mg^{2+}$ activation would not seem to be a plausible control feature.

(2) A second means of regulating FDPase involves hydrogen ion control; increases in pH lead to marked changes in the $K_a$ values of $Mg^{2+}$ and $Mn^{2+}$ as well as to changes in $V_{max}$. The pH profile displays a steep section in the physiological range for both ions. Thus small changes in pH can lead to large changes in FDPase activity, a characteristic feature of many efficient regulatory metabolites. Recently, Trivedi and Danforth (1966) observed profound effects of pH on the PFK reaction; these pH effects could conceivably be coupled with pH effects on FDPase. A coupling of this sort would be analogous to the adenylate control of these two oppositely poised enzymes: Adenylate conditions which activate PFK inhibit FDPase and vice versa (Atkinson, 1966). Such a coupling of regulatory functions is obviously required in cells which possess both PFK and FDPase activities, for simultaneous activity of these two enzymes would lead to hydrolysis of ATP and, therefore, to a short circuit of energy metabolism.

(3) A third mechanism for activation of FDPase may also involve cation cofactors. In trout, lungfish, and salmon FDPases (Behrisch and Hochachka, 1969a,b; Behrisch, 1969), the saturation curves for $Mg^{2+}$, and usually $Mn^{2+}$ as well, are sigmoidal, indicating site–site interactions. The value of $n$, which is not an elementary kinetic parameter but rather a measure of the number of binding sites and their strength of interactions, varies between 1 and approximately 2.5 for $Mg^{2+}$ (Behrisch and Hochachka, 1969a) and is maximal at 15°C. If we assume that cellular concentrations of these cations are in the range of the apparent $K_a$ values, then it is evident that either cation could serve as a positive modulator, as well as a cofactor, of the enzyme.

(4) Another crucial aspect of cationic control of FDPase is the ratio of free $Mg^{2+}$/free $Mn^{2+}$ in the cell. For the three fish FDPases examined (rainbow trout, salmon, and lungfish), the $K_a$ of $Mn^{2+}$ is from 20- to 100-fold lower than the $K_a$ of $Mg^{2+}$. Thus, very small changes in the ratio of free $Mg^{2+}$/free $Mn^{2+}$ could lead to extremely large changes in the activity

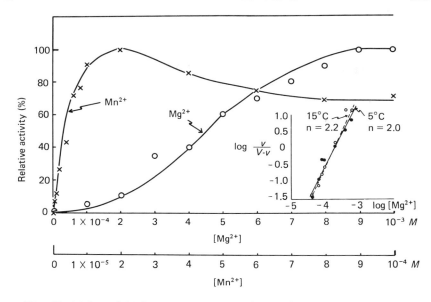

**Fig. 10.** $Mg^{2+}$ and $Mn^{2+}$ saturation curves for rainbow trout liver FDPase at pH 7.4 in 0.1 $M$ tris-HCl buffer, assayed at 15°C. The inset shows two Hill plots for the $Mg^{2+}$ saturation curve, at 5° and 15°C, with interaction coefficients of about 2.0. From unpublished data by Behrisch and Hochachka (1968).

of FDPase; for example, at $Mg^{2+}$ concentrations too low for detectable FDPase activity, $Mn^{2+}$ can fully activate the enzyme (Fig. 10).

(5) $Ca^{2+}$ functions as an inhibitor of fish FDPase reactions and is competitive with respect to both $Mg^{2+}$ and $Mn^{2+}$. Because the affinity of FDPase for $Mn^{2+}$ is higher than for $Mg^{2+}$ much higher concentrations of $Ca^{2+}$ are required to inhibit the enzyme in the presence of $Mn^{2+}$ than $Mg^{2+}$. Thus, $Ca^{2+}$ inhibition is also likely to be sensitive to the ratio of free $Mg^{2+}$/free $Mn^{2+}$.

(6) Perhaps the most significant aspect of cationic control of FDPase is related to the nature of AMP inhibition. As in the case of many regulatory enzymes, the AMP saturation curve for trout liver FDPase is sigmoidal, with $n$ values ranging from 1.0 to nearly 2.5. Significantly, the $n$ value is highest at about 15°C; by this criterion, AMP inhibition would therefore appear to be maximally efficient at this temperature. However, as noted above, cation activation is also most efficient at this temperature, and this effect would tend to reverse AMP inhibition. In addition, there is a definite drop in the value of $n$ for AMP as the cation concentration is increased; this effect would also reduce the in-

hibitory action of AMP. A similar reduction in AMP site–site interactions occurs at temperature below 15°C; this effect would likewise counteract increased FDPase-AMP affinity at lower temperatures.

(7) A final mechanism by which increased sensitivity to AMP inhibition at low temperatures can be reversed may involve the substrate, FDP. In previous studies, FDP has been implicated in the control of the interconversion of FDP to F6P by the complementary mechanisms of substrate inhibition of FDPase and product activation of PFK (Atkinson, 1966). Newsholme and Gevers (1967) argued, quite correctly, that substrate inhibition of FDPase is probably not of physiological significance because the concentrations of FDP required are high and the per cent inhibition is low. However, in fish liver FDPase, FDP itself can serve as a positive modulator as well as the substrate of the enzyme. In the case of trout liver FDPase, the FDP saturation curves are sigmoidal, with $n$ values of approximately 2 at elevated temperatures. At low temperatures there is a striking increase in the $n$ value of FDP ($n$ approximates 5), indicating an equally significant increase in the efficiency of FDP activation of the enzyme.

If any or all of these control mechanisms are, in fact, operative, one would predict that the FDPase reaction would be rather insensitive to temperature. This can indeed be observed. Within the usual biological temperature range the slopes of Arrhenius plots are quite low, particularly in the presence of manganese (Behrisch and Hochachka, 1969a; Behrisch, 1969). The optimal velocity for trout FDPase seems to be least dependent on temperature between about 10° and 20°C (Fig. 11); parallel studies of salmon liver FDPase indicated a $Q_{10}$ of 1.0 for the optimal velocity between 9° and 15°C (Behrisch, 1969). Since most of the above control mechanisms affect enzyme–ligand affinities or site–site interactions, the reaction rate is likely to be even more thermally insensitive at low substrate concentrations. Similar conclusions have been derived for lungfish liver FDPase (Behrisch and Hochachka, 1969b) and salmon liver FDPase (Behrisch, 1969).

In the case of PyK, which functions at another branch and control site in glycolysis and gluconeogenesis, the situation may be somewhat less complex. Pyruvate kinase catalyzes the conversion of phosphoenolpyruvate (PEP) to pyruvate according to the reaction: PEP + ADP = Pyr + ATP. The enzyme is activated in a "feed-forward" manner by FDP, inhibited by ATP, and is also subject to ionic modulation; $K^+$ and $Mg^{2+}$ activate the enzyme and $Ca^{2+}$ is considered to be an inhibitor.

The PyK isozyme found in rainbow trout muscle is highly activated

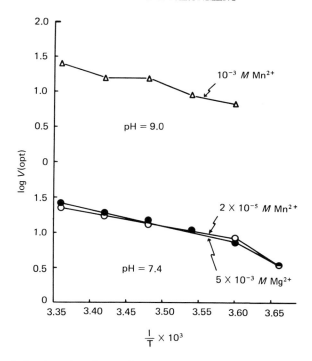

**Fig. 11.** Arrhenius plots for rainbow trout FDPase at pH 7.4 and 9.0. Values are calculated from reaction velocities under optimum conditions ($V_{opt}$). After Behrisch and Hochachka (1969a).

**Table VI**

FDP Activation of M-PyK from 10°–12°C Acclimated Rainbow
Trout at a Series of Temperatures[a]

| Temp. (°C) | Activation (%) |
|:---:|:---:|
| 7 | 190 |
| 10 | 200 |
| 13 | 230 |
| 15 | 240 |
| 16 | 240 |
| 17 | 240 |
| 20 | 240 |
| 25 | 210 |

[a] The FDP concentration was 0.1 m$M$. From Somero and Hochachka (1968).

by FDP, an activation which is effected primarily through a reduction in the $K_m$ of PEP (Somero and Hochachka, 1968). The PyK found in trout liver is distinguished from the muscle PyK by its insensitivity to FDP. For trout muscle PyK, the $K_a$ of FDP is approximately 2 $\mu M$ (at 15°C). We were unable to accurately estimate the temperature dependence of this $K_a$ value because it is so low. However, the activation observed by a given concentration of FDP was found to be remarkably independent of temperature through the trout's entire range of habitat temperatures (Table VI). This finding supported our initial hypothesis concerning temperature independence of regulatory functions. A similar temperature independence of FDP activation was noted for the muscle PyK of an Arctic Zooarcid fish (Somero, 1969a).

Support for our hypothesis was also obtained from studies of adenylate interactions with PyK. The $K_m$ of adenosine diphosphate (ADP) was independent of temperature, in sharp contrast to the $K_m$ of the other substrate, PEP (Fig. 12). In addition, ATP inhibition was found to be temperature independent (Somero and Hochachka, 1968). These findings suggested that PyK–adenylate interactions might exhibit a high degree of temperature insensitivity owing to the importance of adenylate control of enzymic functions (Atkinson and Walton, 1967; Atkinson and Fall, 1967; Atkinson, 1968).

Although the interactions of PyK with FDP, ADP, and ATP are all highly insensitive to temperature, the effects of different inorganic ions on PyK proved to be more complex. Several workers have demonstrated that $Ca^{2+}$ can be an important negative modulator of PyK under physiological conditions (Bygrave, 1966a,b, 1967; Nagata and Rasmussen, 1968). In the case of rainbow trout muscle PyK, $Ca^{2+}$ control of enzymic activity is highly complex. First, inhibition by high concentrations of $Ca^{2+}$ is relatively temperature independent. However, at low $Ca^{2+}$ concentrations activation of the enzyme is observed. This activation is greater at low temperatures and may be of significance in the regulation of PyK activity. Furthermore, it is possible that the activity of PyK may be critically dependent on the ratios of free $Ca^{2+}$/free $Mg^{2+}$ in the cytosol (Nagata and Rasmussen, 1968). In this connection, the $K_a$ of $Mg^{2+}$ varies with temperature in an irregular manner, with a minimal $K_a$ occurring near 15°C. The basis and the significance of these effects are not yet known.

Later studies of citrate synthase led to comparable conclusions. Citrate synthase in trout liver is regulated by ATP feedback inhibition. Adenosine triphosphate increases the apparent $K_m$ of acetyl-CoA with no major effect on the $V_{max}$ (Hochachka and Lewis, 1970). Over normal biological temperature ranges the $K_i$ values of ATP are thermally in-

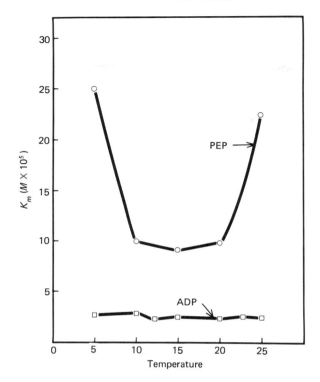

**Fig. 12.** The effect of temperature on the $K_m$ of substrate of PEP and ADP for muscle pyruvate kinase of 10°C-acclimated rainbow trout.

dependent, but in upper thermal extremes these values can change markedly.

Support for our hypothesis that regulatory properties are to some extent insensitive to temperature also comes from studies of control of glycolysis in king crab *Paralithodes* muscle preparations (Hochachka *et al.*, 1970). In these studies, it became apparent that the same control sites (hexokinase, phosphofructokinase, and pyruvate kinase) were operative in the regulation of glycolytic flux at all temperatures tested (1°, 8°, and 15°C) within the biological temperature range of the species. Whereas these studies support the proposition that overall regulatory integrity is maintained irrespective of temperature, they leave open the possibility that efficiency of control may vary with temperature. That efficiency of control at these and similar sites does change with temperature is evident in the kinetic studies and in studies of metabolic branch points.

## E. Temperature Effects on Carbon Flow through Branch Points in Metabolism

At the outset of this discussion it is important to stress that the action of temperature on multienzyme processes in poikilotherms has not been widely studied. The data which are available arise from studies of a small number of species, tissues, and metabolic processes. In a number of cases, however, the predictions stemming from the single enzyme studies discussed above appear to be realized. Thus, van Handel (1966) observed that glycogen synthesis can be entirely temperature insensitive over certain temperature ranges; at temperatures below this "plateau" range, the $Q_{10}$ is high (about 5.0). Similarly, Newell (1966) observed that the oxidation of pyruvate or succinate by mitochondrial preparations from several tissues and organisms can be independent of temperature over quite broad ranges. Gordon (1968) has reported that the temperature coefficients of oxygen consumption of red and white muscle from tuna are about 1.0 over the range of 5°–35°C. More recently, Dean (1969) has examined the action of temperature on acetate-1-$^{14}$C and palmitate-1-$^{14}$C oxidation in three tissues of the rainbow trout. In all cases, oxidation rates can show temperature coefficients as low as unity, at least over certain temperature ranges. Whatever the basic mechanisms underlying these effects, they seem to be unusually effective in the oxidation of palmitate by red and white muscle. Thus, between 5° and 18°C, palmitate oxidation rates exhibit a temperature coefficient less than one. Similar results are obtained in studies of acetate oxidation by white muscle of 18°-acclimated trout. In liver from these fish, acetate oxidation rates show almost complete thermal independence between 10° and 38°C. In liver, acetate incorporation into lipids also shows thermal independence over certain temperature ranges, albeit quite high $Q_{10}$ values are observed at other incubation temperatures.

In all of the above studies, the implicit assumption was made that the pathways concerned were functioning under conditions of saturating substrate concentrations.

Dean's studies are exceptional in this series in that exogenous substrate levels were low. In all the other studies cited, exogenous substrate concentrations were high, usually in the range 1–5 mM, and hence it was reasonable to assume that intracellular substrate concentrations were also in this high range. This need not be the case. All available evidence on probable intracellular concentrations of metabolites under similar experimental conditions suggest that concentrations are in the range of the $K_m$ values for the enzymes involved in their metabolism (Williamson *et al.*, 1967a,b; Hochachka *et al.*, 1970). Also, the distribution of metabolites within the different intracellular compartments is itself under regulation (see Chappell and Robinson, 1968, for example). Thus, the alternative assumption that substrate concentrations are low is more consistent with current information on *in vivo* substrate levels and with our studies on the thermal dependence of enzyme–substrate affinities of poikilothermic systems.

Interpreting the results of Newell, van Handel, Gordon, and Dean is difficult for these workers have not assessed the effect of temperature on branch pathways competing for common intermediates. It is clear that differential effects of temperature on branch point enzymes could lead to important changes in the relative activities of different metabolic pathways. We have examined this problem in an analysis of the action of temperature on the channeling of carbon through branch points in intermediary metabolism (Hochachka, 1968a).

Following glucose phosphorylation, three major pathways are present for the further metabolism of G6P: (1) classic Embden–Meyerhof glycolysis; (2) the hexose monophosphate pathway (pentose shunt), and (3) the biosynthetic pathway to glycogen. In liver slices of certain Amazon River fishes (*Symbranchus, Lepidosiren,* and *Electrophorus*), the pentose shunt and the pathway to glycogen become increasingly effective in competing for the common substrate, G6P, when temperature is raised from 22° to 38°C (Hochachka, 1968a). Glucose carbon flow to glycogen and through the pentose shunt is consequently strikingly raised; however, carbon flow through glycolysis is not increased at higher temperatures and may actually exhibit a decrease.

A number of enzymic processes probably account for these observations (Fig. 13):

(1) The thermal optimum for G6P dehydrogenase—the first enzyme of the pentose shunt—is high at all G6P levels; hence, this enzyme can compete for G6P more effectively at high temperatures.

(2) Enzyme–substrate affinities of LDH (Hochachka and Somero, 1968) and probably of other glycolytic enzymes increase as the temperature is decreased, thus thermally stabilizing glycolysis.

(3) The affinity of glycogen synthetase for the positive modulator G6P is apparently increased at high temperatures. Thus, even in the absence of large changes in G6P concentrations, high temperature will favor glycogen synthesis.

(4) Glycogen phosphorylase is held relatively inactive at high temperatures, possibly by increased affinity of this enzyme for its negative modulators G6P and ATP. This situation would also contribute to the thermal independence of glycolysis.

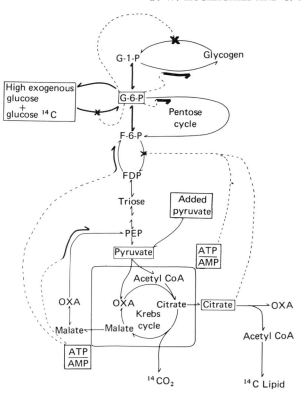

Fig. 13. Diagram of control of glucose metabolism in liver slices of *Symbranchus* and *Lepidosiren*. Broken lines connect effectors (boxed in metabolites) with the enzyme steps that they modulate. Effector activation is expressed with an arrow; effector inhibition is marked with a cross. Heavy arrows and crosses indicate that the effectiveness of modulation is dependent on temperature; presumably, appropriate enzyme–modulator affinities are increased at high temperatures. After Hochachka (1968a).

These four conditions are consistent with available information and supply minimal mechanisms for the observed action of temperature on the G6P branching point. In any event, it must be assumed that efficiency of control at any given site may vary with temperature, even though the same enzymic reactions serve as control sites (or valves) in any given pathway at all temperatures (Hochachka *et al.*, 1971).

Although further experiments are required for a more complete understanding of these events, the work nonetheless demonstrates that for relatively intact multienzyme systems the importance of temperature effects on enzyme–substrate and enzyme–modulator affinities is large. As a generalization, it seems reasonable to propose that carbon flow in any given pathway (glycolysis, the Krebs cycle, etc.) can be held independent of temperature by (1) changing sensitivity to modulation of key enzymes in that pathway, and (2) altering the capabilities of metabolic pathways to compete for common substrates by appropriate thermally dependent changes in enzyme–substrate affinities of key enzymes in the multienzyme sequence. Control mechanisms such as these would allow certain pathways to remain thermally independent, while other branching pathways, e.g., the pentose shunt, glycogen synthesis, and lipogenesis, could alter their activities greatly in response to temperature changes.

This type of differential thermal behavior of different metabolic pathways could be of major significance for poikilothermic organisms; for example, in tuna the highest and best-regulated intramuscular temperatures occur in the deep red muscle located midway along the bodies of these fishes (Carey and Teal, 1966). In white muscle and the more superficial red muscles, the temperature of the tissue is subject to rapid and large variation. Both ambient water temperature and activity would likely determine the precise temperature of these tissues. In a system such as this, in which large and nonpredictable temperature changes would be expected to occur, a good evolutionary "strategy" would appear to be the development of temperature-independent systems of energy-yielding reactions, which would enable the fish to have constant availability of energy for sudden work efforts such as those involved in prey capture and predator evasion. According to Gordon (1968) aerobic energy metabolism of tuna white muscle appears to conform to this pattern. Unfortunately, we do not know the thermal characteristics of other metabolic sequences in this tissue.

We conclude this discussion of immediate temperature effects with the following generalizations:

(i) Regulation of enzymic activity vis-à-vis changes in the metabolic state of the cell or changes in body temperature is accomplished by altering enzyme–substrate affinity. In the case of temperature adaptation, reductions in temperature over at least the upper range of physiological temperatures of the organism lead to increases in E–S affinity. Under these circumstances temperature decreases function analogously to positive modulators of enzymes, and rates of enzymic activity may be essentially independent of temperature at physiological substrate concentrations.

(ii) Although the efficiency of control at any given site may vary with temperature, the overall regulatory properties of poikilothermic enzymes seem to be highly insensitive to temperature changes.

## III. TEMPERATURE ACCLIMATION

### A. Enzymic Changes in the Process

Because of the influence of traditional biochemistry, which deals almost exclusively with mammalian and bacterial systems, the extent to which poikilothermic enzymes appear tailored for thermally independent function may seem somewhat surprising. However, as we have indicated, in only a few cases of those so far observed is the direct relationship between $K_m$ and temperature maintained over the entire biological temperature range of the organism. At temperatures below a critical minimum, the $K_m$ usually increases dramatically (Figs. 4 and 5). Thus, in the case of warm (18°) acclimated rainbow trout, the $K_m$ of brain acetylcholinesterase at 8°C is in the order of $10^{-2} M$ (Baldwin and Hochachka, 1970); the $K_m$ of muscle LDH at 5°C is about 0.1 $M$ (Hochachka and Somero, 1698); the $K_m$ of trout muscle PyK at 0°C extrapolates to PEP concentrations of over 0.4 m$M$ (Somero and Hochachka, 1968). These $K_m$ values may be as much as 100 times higher than probable physiological concentrations of substrate. This means that, over the lower range of physiological temperatures for the organism, these enzymes from warm-acclimated trout are highly inefficient and, indeed, except under unusually high substrate concentrations, are essentially inactive. Yet it is common knowledge that after a period of acclimation this species commonly thrives in waters at these low temperatures. How is this paradox resolved?

Many previous studies of cold acclimation imply, and sometimes explicitly suggest, that a basic mechanism of acclimation involves the production of higher quantities of enzymes in order to compensate for decreases in temperature. Thus, Ekberg (1962) noted that the activities of 6-phosphogluconate dehydrogenase (6PGDH) and FDP-aldolase increase during cold acclimation in carp. Jankowsky (1968) found that the activities of FDP-aldolase, malate dehydrogenase (MDH), and cytochrome oxidase increase in cold acclimation of the golden orfe. Baslow and Nigrelli (1964) reported complete or "perfect" compensation of brain acetylcholinesterase activity in *Fundulus*, and similar data are available for a variety of other enzymes (Freed, 1965; Caldwell, 1969). All these studies measured maximum catalytic activities; none determined the enzyme forms responsible for the activities in the different acclimation groups of test organisms. From a functional point of view, the selective advantage of producing increased amounts of inefficient or largely inactive enzymes is not evident. It appears, therefore, that increased enzyme synthesis as such is not always a sufficient mechanism for promoting the kinds of changes we observe in cold acclimation. When this does occur, the kinetic properties of the enzymes concerned probably are temperature insensitive over the entire biological range or perhaps exhibit the interconversion phenomena found in king crab pyruvate kinase (Fig. 7). The first evidence that qualitatively different enzymes might be synthesized during thermal acclimation came from studies on the role of LDH isozymes in thermal acclimation in goldfish (Hochachka, 1965) and later in trout (Hochachka, 1967; Hochachka and Somero, 1968). During cold acclimation in these organisms, new isozymes of LDH appear which differ kinetically from a noninducible set in having higher absolute affinities for substrate and having minimal $K_m$ values at lower temperatures (Hochachka and Somero, 1968). Essentially identical results are observed in the case of trout muscle PyK (Fig. 5). Muscle PyK from warm (10°–18°) acclimated trout shows a minimal $K_m$ at about 15°. With cold acclimation, a new muscle PyK is induced which shows minimal values at about 7°. In this case, the minimal value of the $K_m$ for both forms of the enzyme is about the same. J. Baldwin has identified two isozymes of acetylcholinesterase in trout brain (Fig. 4). Only one of these occurs in warm acclimated fish; the other occurs only during cold (2°) acclimation. Both isozymes are present when the trout are acclimated to an intermediate (12°) temperature. Again, the minimal $K_m$ values for the two isozymes are essentially identical, but the minimum occurs at about 18°C for the "warm" enzyme while the minimum occurs at about 2°C for the "cold" form of the enzyme (Baldwin and Hochachka, 1970). In the goldfish, choline acetyltransferase also appears to occur in two forms, the "warm" form of the enzyme having a higher absolute $K_m$ at low temperatures than the "cold" form (Hebb *et al.*, 1969). These data suggest that the crucial process in cold acclimation is not the biosynthesis of more of the same kinds of enzymes that are present in the warm-acclimated state, but rather the biosynthesis of new enzyme variants—perhaps in relatively larger quantities—which are better adapted for catalysis at low temperature.

This general conclusion has far-reaching implications.

(1) The time course of acclimation, which is classically accepted as ranging from about 1 week to several weeks (Brett, 1956; Prosser, 1958), may be synonymous with the time course of isozyme induction. In the case of acetylcholinesterase, Baldwin observed that following stepwise transfer from 18°C to 2°C water, the rainbow trout becomes highly immobile. Normal activity seems to return only after acclimation to the

low temperature and coincides with the appearance of a cold form of acetylcholinesterase. Clearly, there will be no sharp line in time dividing the cold- and warm-acclimated states, but in general the pattern of activity acclimation and isozyme induction should coincide rather closely.

Any factors or conditions which interfere with isozyme induction will interfere with thermal acclimation. This implication is difficult to test critically. Studies of fishes showing only minor acclimatory capacities might yield useful insight on this matter.

(2) The primary advantage of employing "better" isozymes in thermal acclimation—as opposed to producing altered quantities of a single enzyme species—is not $Q_{10}$ reduction (rate compensation) over the time-course of acclimation. The types of isozyme changes we have observed in the trout would not, in the absence of quantitative changes in enzyme levels, promote complete or "perfect" compensation. For PyK and acetylcholinesterase, the minimal $K_m$ values of the warm and cold isozymes are essentially equal (Figs. 4 and 5). If quantities of enzymes are not to be elevated in the cold, then complete rate compensation in activity of these enzymes would demand that the cold isozymes have drastically lower $K_m$ values than the warm enzymes. This situation does not seem to be general. The similar minimal $K_m$ values for warm and cold variants of a given enzyme suggest that a primary function of the isozyme changes is the production of enzymes with $K_m$ values in a range likely to be optimal for regulation of catalytic activity. Thus, at low temperatures, small changes in substrate concentration or small changes in $K_m$ can lead to large changes in the activities of cold forms of these enzymes ($A_4$ LDH in Fig. 14), a condition which would appear to be admirably suited to controlling the reaction.

In the case of warm variants of these enzymes at low temperatures, very large changes in substrate concentration are required to yield small changes in reaction rate ($C_4$ LDH in Fig. 14). This condition clearly is not one which allows efficient control of reaction rates. In evolutionary terms, it appears that there is a strong selection for enzymes having $K_m$ values allowing large changes in activity in response to physiological changes in substrate concentrations. This is reflected in the patterns of enzyme variants produced during acclimation and during evolutionary adaptation (see Hochachka and Somero, 1968; Somero, 1969a).

It is important to stress that the typical temperature–$K_m$ relationship is displayed by both cold and warm isozymes, but the minimal $K_m$ occurs at different temperatures. The usual thermal stability mechanism consequently is operative over different temperature ranges. Thus, temperature increases above about 5° can be compensated by concomitant

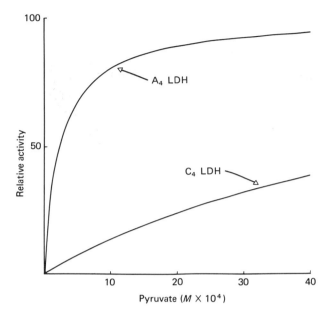

**Fig. 14.** The effect of varying levels of pyruvate on the activity of rainbow trout $A_4$ and $C_4$ LDHs at a temperature of 0°C, at which the $K_m$ values of the two enzymes differ approximately 20-fold. Probable physiological pyruvate concentrations are below 1 m$M$. Data from Hochachka and Somero (1968).

increases in $K_m$ of the cold form of AChE; temperatures above about 17° can be compensated in the case of the warm isozyme (Fig. 4).

One can visualize various kinds of isozyme systems induced during acclimation, which share in common the $K_m$ properties needed to retain control of catalytic activity. However, the effects that these various isozyme changes might have on acclimatory rate compensation might vary among isozymes, depending on how the absolute concentrations of the warm and cold isozymes vary among different enzymes. This finding parallels the conclusion made by Precht (1958) concerning physiological rate functions. Some 20 years earlier, he was led to categorize five kinds of acclimatory processes, ranging from overcompensation in the cold to a "paradoxical" situation in which the warm-acclimated function exhibited a higher rate. The above enzyme considerations would seem to provide a mechanistic basis for the types of changes discussed by Precht.

(3) Finally, for any given reaction, the induction of a new isozyme specifically adapted for function in the cold acclimation state may lead—in the absence of any changes in substrate or modulator concentrations—

to the establishment of new steady state conditions if the enzyme exhibits modified enzyme–substrate and/or enzyme–modulator interactions. This last implication is crucial for it is consistent with an important end result of the acclimation process—the appearance of new steady state conditions in various metabolic pathways and a general reorganization of cellular metabolism.

## B. Basic Metabolic Adjustments during Temperature Acclimation

Historically, the first indications of metabolic compensations for thermal changes in the environment were in studies of the activities and respiration of whole animals. Later experiments, utilizing somewhat more sophisticated methods, indicated that metabolic rate adjustments during acclimation are more complex; thus, some metabolic processes are activated to a large extent, others may remain unchanged, still others may be reduced in activity. That is, cold acclimation does not merely accelerate all rate processes and warm acclimation does not decelerate these processes. Rather, during acclimation the metabolism of the organism is fundamentally reorganized. The nature of metabolic reorganization during acclimation has been previously considered in some detail (Hochachka, 1967); hence, we shall at this point only summarize some of the salient features.

In general, the observations can be summarized as follows. In tissues of cold-acclimated fish, compared to tissues of warm-acclimated ones, (1) glycolysis rate is increased by up to 5-fold (Hochachka and Hayes, 1962); (2) the participation of the pentose shunt may be increased from negligible contributions to activities accounting for about 10% of glucose metabolism (Hochachka and Hayes, 1962); (3) the Krebs cycle may be decreased, unchanged, or possibly slightly increased (depending upon tissues and species) whereas fatty acid oxidation and electron transfer functions are characteristically increased (Freed, 1965; Hochachka and Hayes, 1962; Caldwell, 1969; Dean, 1969); (4) lipogenesis is activated, in some cases only by a small factor (Hochachka and Hayes, 1962; Dean, 1969), but in other cases the activation of synthesis of unique fatty acids may increase during cold acclimation by factors as high as 12-fold (Knipprath and Mead, 1968); (5) glycogen synthesis rate appears to be increased (Hochachka and Hayes, 1962); (6) protein synthesis rates appear to be generally higher during cold acclimation, at least in certain species and certain tissues (Das and Prosser, 1967; Haschemeyer, 1968, 1969a,b); (7) synthesis rates of nucleic acids (RNA in particular) and their turnover rates are higher in cold acclimated fishes (Das, 1967); and (8) the ionic microenvironment may alter during acclimation (Heinicke and Houston, 1965).

On examination of the evidence upon which the above summary rests, it is evident that in any given tissue not all of the above processes necessarily occur. Most of them probably occur in liver, in which metabolic organization is complex. In tissues such as gill, muscle, and brain, exergonic reactions are coupled to highly specialized work functions; hence, metabolic organization may be abbreviated. It is not surprising therefore that some of the above changes are not as evident in these tissues. The shunt, for example, is not important in the metabolism of brain or muscle; clearly, adjustments in its participation in glucose metabolism would not be relevant and indeed do not occur in these tissues during acclimation. The mechanisms by which participation of the above processes can change during acclimation have been considered in some detail previously (see Hochachka, 1967). However, the adaptive significance of metabolic reorganization has remained unexplained, and, indeed, is often avoided in discussions of thermal acclimation (see Hochachka, 1967; Precht, 1968).

## C. Biological Significance of Metabolic Reorganization

Empirically, we know that two kinds of processes often occur during acclimation: (a) compensatory adjustments in metabolic rate, which tend to free the organism from stringencies of the outer environment, and (b) biophysical and biochemical restructuring of many cellular and tissue components for operation under the new thermal regime imposed on the organism. Previous workers have, by and large, emphasized the first of these processes. We believe that it is the less fundamental, in the sense that it need not necessarily occur, and when it does, it is probably a consequence of the second process of rebuilding of the cell. Thus, the synthesis of new enzyme variants in relatively large amounts appears to promote (1) some stability in rate of enzyme-catalyzed reactions, and (2) maintenance of metabolic control. The synthesis of other kinds of proteins such as membrane proteins, ribosomal proteins, blood proteins, and specific transport proteins may also be an essential part of the restructuring process which occurs during acclimation.

Lipids, particularly membrane lipids, appear to play a particularly important role in acclimation. Although the literature here is vast (see Johnston and Roots, 1964; Knipprath and Mead, 1968; Roots, 1968), there is universal agreement that in responding to cold exposure, organisms tend to increase the degree of unsaturation of their fatty acids.

Earlier it was believed that by altering the degree of unsaturation of fatty acid chains, an organism is able to adjust lipid viscosity to a variety of thermal conditions. Recent data do not contradict this thesis but indicate that the adjustments in lipid composition are far more complex than would be required simply for viscosity regulation.

In addition to simple changes in saturation of fatty acids, very specific changes in lipid composition of different tissues occur during acclimation (Johnston and Roots, 1964; Roots, 1968; Knipprath and Mead, 1968). In the case of goldfish brain, total lipids tend to increase during cold acclimation, but the magnitude of this change is slight and may not be significant. Similarly, the total content of the major membrane-based phospholipids in goldfish CNS (choline glycerophosphatides and ethanolamine phosphatides) is not influenced by temperature acclimation, but the specific species of phospholipids that occur depend critically upon the acclimation state (Roots, 1968; Roots and Johnston, 1968). These complexities were recognized by Roots (1968); she suggested that the significance of these changes may relate to effects on membrane, particularly nerve membrane, functions (Roots, 1968; Johnston and Roots, 1964). That unsaturated fatty acids form expanded monomolecular films and cannot be as closely packed as saturated fatty acids has long been known; these conditions could lead to important adjustments in various membrane functions (transport, impulse transmission, electron transfer, etc.). Many enzymes, such as those of the electron transfer system, are critically dependent upon the membrane-lipid milieu in which they normally function; acclimation changes in membrane structure may be closely related to associated changes in the activities of these enzymes (Caldwell, 1969). It seems clear that just as unique isozyme systems appear to be requisite for survival at certain temperatures, unique membrane composition and membrane architecture are also basic to the acclimatory progress. Since many enzymes are membrane bound, it is possible that production of new membrane-based lipids and new isozymes during acclimation may be closely integrated aspects of a single rebuilding process.

Cellular restructuring during acclimation also appears to involve the ribosomes. We have found that ribosomes from cold and warm-acclimated rainbow trout have different melting temperatures ($T_m$) (Table VII; Somero and Gould-Somero, 1970). The molecular basis of this difference remains to be resolved. Interestingly, the cold trout ribosomes melt at a higher temperature than the warm ribosomes. Similarly, for the other eucaryotic ribosomes we have studied, there appears to be no correlation between the organism's adaptation temperature and the ribosomal melting temperature; these results contradict other reports

**Table VII**

Melting Temperatures for Liver Ribosomes from Organisms
Adapted to Different Temperatures[a]

| Species | Temp. (°C) | $T_m$ (°C) |
|---------|------------|------------|
| Rat | 37 | 53.4 |
| *Drosophila Melanogaster* | 27 | 51.5 |
| Salmon | 12 | 48.5 |
| Rainbow trout | 18 | 45.3[b] |
| Rainbow trout | 2 | 49.0[b] |
| *Trematomus bernacchii* | −1.9 | 50.4 |

[a] The melting buffer was 10 m$M$ tris/HCl, pH 7.8 (22°C), containing 5 m$M$ MgCl₂.
Temperatures given in parentheses are the approximate temperatures to which the organisms were adapted.

[b] The probability that the 18° and 2° trout melting temperatures are equal is less than 0.005%.

on bacterial (Pace and Campbell, 1967) and protozoan (Byfield *et al.*, 1969) ribosome melting where a positive correlation between ribosomal melting temperatures and lethal temperatures has been proposed.

Rebuilding processes undoubtedly require energy, reducing power, fundamental building blocks (amino acids and fatty acids), and the subcellular machinery necessary for biosynthetic processes. It therefore seems reasonable to propose that certain of the metabolic reorganizations found during thermal acclimation may be directed toward the effecting of cellular restructuring. In these terms, it seems necessary to distinguish between a set of changes which is at least semipermanent and a set of changes which is transitory and likely to disappear once cellular restructuring is completed.

Current data do not permit a clear distinction between these two sets of phenomena. Certainly, it seems most probable that the acclimatory isozyme changes which promote controlled catalysis will persist. Similarly, the restructuring of cellular membranes is likely to be part of the new steady state acclimation condition. However, other changes, notably certain of the metabolic reorganizations listed in Section III, C, may be somewhat transitory. Thus, metabolic activities associated largely with biosynthesis might be transitory events which are not characteristic of the final steady state. Increased pentose shunt activity and lipogenesis may fall into this category of transitory changes. Distinction between these two sets of phenomena could readily be made through study of the time-course of thermal acclimation. It should be added that this distinction between transitory and permanent acclimatory changes implies that both warm and cold acclimation will be characterized by certain common metabolic changes, and these might include increased rates of lipogenesis and protein biosynthesis, whereas other changes will be unique steady state characteristics of either the warm- or cold-acclimated condition.

Two final points should be considered in this discussion of temperature acclimation. First, it seems essential that attempts be made to define the metabolic needs or demands facing poikilotherms at different temperatures. Most studies of temperature acclimation appear to make the implicit assumption that adjusting to low temperature necessarily requires that the organism increase its metabolic rates to a large extent. References to "perfect" compensation indicate that the experimenter often assumes that the cold-acclimated organism ought to metabolize as rapidly as its warm-acclimated counterpart. However, there is no evidence to suggest that metabolic demands are the same at all temperatures; indeed, there are data suggesting that at low temperature maintenance metabolism is significantly reduced (Brett et al., 1969). At lower temperatures the likely reduction in rates of thermal inactivation of macromolecules, for example, might reduce the over-all metabolic demands. Second, it is important to determine whether or not all changes triggered by temperature change should be considered as temperature adaptations. Might certain changes which are triggered by changes in temperature really be related to seasonal processes, e.g., gametogenesis, and not temperature adaptation per se?

## IV. EVOLUTIONARY ADAPTATION

### A. Rate Compensation

As in the case of the temperature acclimation, two types of enzymic changes can be proposed to account for evolutionary adaptation to temperature in rate processes such as respiration (Peiss and Field, 1950; Scholander et al., 1953; Wohlschlag, 1964; Somero et al., 1968) and growth (Wohlschlag, 1961). First, cold-adapted species may possess higher concentrations of enzymes than warm-adapted species. Implicit in this "quantitative" hypothesis is the assumption that enzymes of warm- and cold-adapted organisms are qualitatively the same. We have indicated why this simple quantitative hypothesis is invalid in the case of thermal acclimation; identical arguments apply in the case of evolutionary adaptation. For example, the production of large quantities of LDH enzymes resembling warm trout LDH in an Antarctic fish like

*Trematomus bernacchii* would be an inefficient mechanism for promoting rate compensation. Further, the control of catalysis vis-à-vis changes in habitat temperature would be jeopardized by an adaptation of this sort. These shortcomings of the quantitative mechanism of enzymic adaptation are illustrated in Fig. 14, where the $A_4$ LDH can be analogized to *T. bernacchii* LDH and the $C_4$ LDH to warm trout LDH at a temperature of 0°C.

It is therefore apparent that evolutionary adaptation, like acclimation, depends upon "qualitative" changes in enzymes. In the case of evolutionary adaptation, the temperature at which the minimal $K_m$ occurs is shifted along the temperature axis (Figs. 4 and 5); thus, for most enzymes the temperatures of minimal $K_m$ are approximately the same as the species' minimal habitat temperatures (Hochachka and Somero, 1968; Somero, 1969a).

These $K_m$ changes have two important effects. First, the enzymic activities of differently adapted species have a built-in thermal stability over the range of temperatures the species-encounters in Nature. Second, examination of current available evidence (Somero, 1969a; Baldwin and Hochachka, 1970) will reveal that for all interspecific variants of a given enzyme the $K_m$ values at the adaptation temperatures of the different species are quite similar. This fact suggests that the absolute value of the $K_m$ of substrate is adjusted to enable optimal regulatory function by the enzyme; selection has not favored extremely low $K_m$ values in cold-adapted species. This situation is exactly the same as that observed for isozyme changes during thermal acclimation.

The fact that $K_m$ changes are not a good mechanism for effecting "perfect" or complete compensation in rates of enzyme function leads us to examine the role of activation energy ($E_a$) in this process. It has been proposed by several workers that $E_a$ may be an important parameter in evolutionary adaptation of enzymes to temperature (Vroman and Brown, 1963; Somero et al., 1968; Somero, 1969a). In systems where relatively limited amounts of thermal energy are present to drive metabolic reactions, it would seem advantageous to employ enzymes which have a relatively high efficiency in lowering the energy barriers to the reaction. In other words, if $E_a$ is an important factor in evolutionary adaptation to temperature, it should correlate positively with adaptation temperature.

Evaluating the significance of published values of activation energies is made difficult by the observation that Arrhenius plots are often nonlinear, frequently showing quite distinct breaks at critical temperature ranges (Somero and Hochachka, 1968; Massey et al., 1966). Hence, the temperature range over which activation energies are compared between homologous enzymes of different species can profoundly influence our conclusions concerning their significance. Also, the slopes of Arrhenius plots for a number of enzymes are known to depend critically upon the levels of substrates and modulators; indeed, such metabolites themselves appear to alter activation energies (Helmreich and Cori, 1964; Lowry et al., 1964). In the case of amino acid oxidases under certain assay conditions, Arrhenius plots can become Z-shaped (Koster and Veegers, 1968). Koster and Veegers' interpretation suggests a model which assumes a high and a low temperature form of the enzyme. Over certain temperature ranges, through which one form is converted to the other, the slopes of Arrhenius plots represent the sum of both activation energy and activation entropy (Koster and Veegers, 1968). Unfortunately, these kinds of effects have not been widely recognized by previous workers interested in comparative aspects of enzyme function; hence, one must approach with caution published $E_a$ values for homologous forms of the same enzyme from different species.

An additional uncertainty in interpreting activation energy effects involves the activation entropy of the reactions catalyzed by different variants of the same enzyme (see Wróblewski and Gregory, 1961). In order to predict accurately turnover numbers from activation energy data one must have knowledge of the activation entropy characteristic of the reaction. An example of the types of difficulties inherent in estimating turnover number from activation energy data without knowledge of entropy effects is found in studies of mammalian LDH isozymes. Plagemann et al. (1960) found that on the basis of activation energy differences rabbit $LDH_1$ should have a turnover number approximately 3300-fold higher than rabbit $LDH_5$. However, the experimentally determined turnover number for $LDH_1$ is only 3–4 times greater than that of $LDH_5$. Apparently the $LDH_5$ reaction is characterized by high activation energy and high activation entropy, and when both of these terms vary in the same direction their effects on reaction velocity tend to cancel each other out.

Activation energy measurements can, at best, provide circumstantial evidence for differences in turnover number. Unfortunately, the number of cases in which this latter parameter has been estimated for enzymes from differently adapted poikilotherms is small. However, when both activation energy and turnover number have been determined for an enzyme, a low activation energy value has been accompanied by a high turnover number (see, e.g., Cowey, 1967; Assaf and Graves, 1969).

Despite the uncertainty in estimation and interpretation of $E_a$ values, it does appear that $E_a$ correlates positively with adaptation temperature for certain enzymes. This relationship has been noted for succinic dehydrogenase (Vroman and Brown, 1963; Somero *et al.*, 1968), fructose-diphosphate aldolase (Kwon and Olcott, 1965), glyceraldehyde-3-phosphate dehydrogenase (Cowey, 1967), pyruvate kinase (Somero and Hochachka, 1968; Somero, 1969a), FDPase (Behrisch and Hochachka, 1969a,b), and muscle glycogen phosphorylase (Assaf and Graves, 1969). However, for other enzymes no relationship between these two parameters has been found [mollusk ribonucleases, Read (1964a,b); lactate dehydrogenase, Hochachka and Somero (1968) and Somero (1969a); acetylcholinesterase, Baldwin and Hochachka (1970)]. These data are summarized in Table VIII. While this variation among enzymes casts doubt as to the sensitivity of this parameter to selective pressure, there are several reasons why the importance, and the occurrence, of $E_a$ changes among enzymes might be expected to vary.

First, if an enzyme (PyK or FDPase) is rate limiting in a metabolic sequence, then selection for $E_a$ reduction during cold adaptation might be high relative to a case in which an enzyme is not likely to be rate limiting. (Both lactate dehydrogenase and ribonuclease would seem to fall into this latter category.) Second, the inherent activation energies of enzymic reactions differ. If the inherent $E_a$ of a reaction is high, then selection for a low $E_a$ during cold adaptation may be great.

Third, it is impossible for enzymes to be infinitely efficient, i.e., $E_a$ values for reactions can be lowered only to some finite value. In the case of enzymes such as AChE, in which cold and warm isozymes both exhibit very low $E_a$ characteristics (Baldwin and Hochachka, 1970), it seems probable that $E_a$ has been reduced about as far as possible.

Finally, the selective advantage of $E_a$ reduction may be related to parameters other than temperature; for example, in tuna, which have a high swimming velocity, selection for low $E_a$ values might be high relative to the case of a more sluggish animal.

The obvious selective advantage of low $E_a$ values for enzymes which function at low temperatures is documented in Table IX, by comparing the values of velocity constants for the PyK reactions ($K_x$) of different species relative to the *Trematomus* ($K_T$) PyK reaction. The rate constant for the *Trematomus* reaction (with $E_a = 10$ kcal/mole) may be approximately 10,000 times greater than that of the Zooarcid reaction, about $10^8$ times greater than that of the tuna reaction, and up to $10^{16}$ times greater than that of the rainbow trout reaction. It is evident that even small decreases in $E_a$ can lead to very significant increases in the velocity of the reaction; hence, the selective advantage of lowering $E_a$ during cold adaptation seems considerable. However, the differences in activation entropy among the homologous forms of the enzymes may cause major changes in the relative velocity constants (see above).

## Table VIII

Activation Energy Values for LDH, PyK, FDPase, GPDH,[a] FDP Ald, and Glycogen Phosphorylase Reactions Catalyzed by Enzymes from Differently Adapted Poikilotherms and Homeotherms[b]

| Enzyme | Organism | $E_a$ (kcal/ mole) | Temp. (°C) | Reference |
|---|---|---|---|---|
| LDH, $M_4$ | Lungfish | 13 | | Hochachka and Somero, 1968 |
| LDH, $M_4$ | Tuna | 11.2 | 15–30 | Hochachka and Somero, 1968 |
| LDH, $H_4$ | Tuna | 10.7 | 15–30 | Hochachka and Somero, 1968 |
| LDH, $A_4$ | Lake and brook | 10.8 | 15–30 | Hochachka and Somero, 1968 |
| LDH, $B_4$ | trout | 12.6 | 15–30 | Hochachka and Somero, 1968 |
| LDH, $C_4$ | | 22.1 | 15–30 | Hochachka and Somero, 1968 |
| LDH (mixed isozymes) | Rainbow trout | 12 | 15–30 | Hochachka and Somero, 1968 |
| LDH (mixed isozymes) | Zooarcid | 10 | 5–20 | Somero, 1969a |
| LDH (mixed isozymes) | King crab | 10 | 5–20 | Somero, 1969a |
| LDH (muscle) | *T. borchgrevinki* | 11 | 0–15 | Hochachka and Somero, 1968 |
| Glycogen Phosphorylase (muscle) | Rabbit | 21.2 | 0–30 | Assaf and Graves, 1969 |
| | Lobster | 15.9 | 0–30 | Assaf and Graves, 1969 |
| PyK (muscle) | Rat | 10 | 35–45 | Somero and Hochachka, 1968 |
| PyK (muscle) | Tuna | 20 | 5–20 | G. N. Somero, unpublished data |
| PyK (muscle) | Rainbow trout | 30 | 5–25 | Somero and Hochachka, 1968 |
| PyK (muscle) | Zooarcid | 15 | 5–20 | Somero, 1969a |
| PyK (muscle) | King crab | 12 | 5–20 | Somero, 1969a |
| PyK (muscle) | *T. bernacchii* | 10 | 1–10 | Somero, 1969a |
| FDPase (liver) | Rabbit | 16 | 2–46 | Behrisch and Hochachka, 1969b |
| FDPase (liver) | Lungfish | 17.6 | 25–35 | Behrisch and Hochachka, 1969b |
| | Lungfish | 10.5 | 15–25 | Behrisch and Hochackha, 1969b |
| | Lungfish | 11.0 | 5–15 | Behrisch and Hochachka, 1969b |
| FDPase (liver) | Rainbow trout | 6 | 5–25 | Behrisch and Hochachka, 1969a |
| | Rainbow trout | 9.5 | 15–25 | Behrisch and Hochachka, 1969a |
| GPDH (muscle) | Cod | 14.5 | 5–35 | Cowey, 1967 |
| GPDH (muscle) | Rabbit | 19.0 | 5–35 | Cowey, 1967 |
| GPDH (muscle) | Lobster | 14.5 | 5–35 | Cowey, 1967 |
| Ald (muscle) | Tuna | 4.2 | | Kwon and Olcott, 1965 |
| Ald (muscle) | *T. bernacchii* | 15.3 | 5–25 | G. N. Somero, unpublished data |
| Ald (muscle) | Rainbow trout | 15.7 | 5–25 | G. N. Somero, unpublished data |

[a] Here, GPDH indicates D-glyceraldehyde-3-phosphate dehydrogenase.

[b] Activation energy values were computed from the slopes of Arrhenius plots over the temperature ranges indicated.

**Table IX**

Comparisons of the Rate Constants of PyK Reactions Relative to the Rate
Constant of the *Trematomus* Reaction, Which Has the Lowest
$E_a$ Reported for the Reaction

The ratios of the rate constants were determined by the equation:

$$\log \frac{K_x}{K_T} = \frac{E_{a(Trematomus)} - E_{a_x}}{2.3 \times 1.98 \times \text{Temperature (°K)}}$$

$$\frac{\text{rate constant for } Trematomus \text{ PyK}}{\text{rate constant for PyKs of other species}}$$

| Ratio of rate constants | 0° | 10° |
|---|---|---|
| $k_{(T)}/k_{(crab)}$ | 40.6 | 35.6 |
| $k_{(T)}/k_{(Zooarcid)}$ | $1.5 \times 10^4$ | $7.5 \times 10^3$ |
| $k_{(T)}/k_{(tuna)}$ | $1.11 \times 10^8$ | $5.7 \times 10^7$ |
| $k_{(T)}/k_{(trout)}$ | $1.2 \times 10^{16}$ | $3.3 \times 10^{15}$ |

Finally, in those cases in which $E_a$ does not appear to be low in cold-adapted species, rate compensation may occur either by (1) increases in the quantities of enzymes and/or (2) increases in the steady state concentrations of appropriate substrates and modulators.

## B. Lethal Temperature Effects: A Possible Role of $K_m$ Changes in Establishing Thermal Tolerance Limits

The biochemical factors which establish thermal tolerance limits for poikilotherms are poorly understood. Changes in enzymes and lipids have been evoked as mechanisms of thermal death, but data supporting these hypotheses are sparse.

If the inactivation of enzymic activity is an important cause of thermal death, then the type of damage done to enzymes by temperature extremes is undoubtedly more suble than gross protein denaturation, the phenomenon which has received the most study (see Ushakov, 1967). For example, Antarctic fish of the genus *Trematomus*, adapted to −1.9°C, have an upper lethal temperature of 6°C (Somero and DeVries, 1967). Protein denaturation seems highly unlikely to account for thermal death in this case.

An alternate mechanism by which enzymic activity could be heat- or cold-inactivated is suggested by the data in Figs. 4, 5, and 6. Most enzymes exhibit sharp increases in $K_m$ at extremes of temperature. It seems likely that once the $K_m$ of an enzyme has increased beyond a certain value, the rate of the reaction may drop to a level which is lethal for the organism. Thus temperature extremes may inactivate an enzymic reaction even though no irreversible denaturation is done to the enzyme molecules per se. This mechanism of thermal death could be important at both extremes of temperature, although at lower temperatures the synergistic effects of reduced thermal energy *and* reduced E–S affinity would seem particularly important. One way of dramatizing the latter situation is to compute $Q_{10}$ values for warm trout isozymes at low temperatures. For warm pyruvate kinase of trout, $Q_{10}$ values at physiological PEP concentrations exceed 25 at temperatures below 7°C.

## V. TEMPERATURE ADAPTATION OF FISH HEMOGLOBINS

The temperature effects which have been observed for fish hemoglobins are strikingly analogous to the $K_m$ effects discussed in previous sections. The affinity of hemoglobin for oxygen, as measured by the half-saturating ($P_{50}$) concentration of oxygen, varies with temperature in a reciprocal manner, much like enzyme–substrate affinity. It is only fair to state that the discovery of the temperature–affinity relationship for hemoglobins by Krogh and Leitch (1919) predates the study of temperature-dependent changes in enzyme–substrate affinity by almost 50 years.

The biological effects of the temperature–$P_{50}$ relationship are very likely similar to those discussed for E–S affinity changes. On an evolutionary time scale, one finds a shifting of the $O_2$ saturation curve along the temperature axis (Fig. 15) in such a manner that each hemoglobin species is functional at the temperature to which the organism is evolutionally adapted. Thus, as in the case of E–S affinity, the protein (hemoglobin or enzyme) is capable of varying its function ($O_2$ transport or catalysis) in response to alterations in the concentrations of $O_2$ (or substrate) which are present.

Similar hemoglobin changes have been noted in differently acclimated fish (Grigg, 1969). However, Grigg reported that the oxygen-dissociation curve displacements which occur during acclimation do not result from the presence of different types of hemoglobin but rather from changes in the composition of the erythrocyte cytosol. It seems likely that changes in levels of organophosphate compounds (see Benesch and Benesch, 1969) might readily promote these saturation curve displacements; this hypothesis remains to be tested in fish. Recent data on seasonal changes in newt hemoglobins (Morpurgo *et al.*, 1970), which

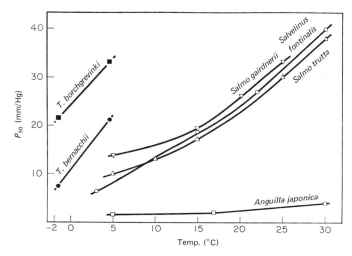

**Fig. 15.** The effect of temperature on the half-saturating ($P_{50}$) concentrations of oxygen for blood of differently adapted fishes. From Grigg (1967). Compare these data with Figs. 4 and 5.

show that the Bohr effect is strongly influenced by acclimation temperature, are also consistent with the hypothesis of small molecule modulation of hemoglobin properties.

## VI. GAS TENSIONS

### A. Anaerobiosis

It is not widely appreciated that some of the lower vertebrates, and even higher proportions of the invertebrates, are facultative anaerobes under certain circumstances; for example, during winter conditions the European carp often become "ice-locked" in small ponds which gradually become anaerobic and remain $O_2$ free for 2–3 months until the spring thaw. The carp show no apparent ill effects after this extreme exposure to anoxic conditions. Blažka (1958) was probably the first to recognize the fundamental consequences of this habit in carp. Unlike fishes such as the salmonids, which depend upon an aerobic metabolism, the carp do not accumulate an $O_2$ debt during anaerobiosis. Under similar conditions most vertebrates accumulate large amounts of lactic acid. In the carp, however, the usual end products of anaerobic breakdown of carbohydrates do not accumulate; rather, the organism accumulates large amounts of long-chain fatty acids (Blažka, 1958). At low temperatures, the amount of energy which can be obtained from the conversion of sugars to fatty acids (see chapter by Hochachka, Volume I, this treatise) apparently is adequate to meet both maintenance and active metabolic requirements.

When ambient $O_2$ tensions become low, goldfish similarly derive considerable energy for active and basal metabolism from anaerobic reactions. This partial anaerobiosis can be sustained for a long period. As oxygen concentrations near 15% of air saturation the goldfish sustains a respiratory quotient ($CO_2/O_2$) of about two for week-long periods (Kutty, 1968). Although pathways are unknown, it is clear that the goldfish has mechanisms for the production of metabolic $CO_2$ even in the complete absence of $O_2$ (Hochachka, 1961; Ekberg, 1962); at low temperatures these mechanisms apparently can adequately supply all the energy demands of the organism.

At this time it is difficult to ascertain the frequency of these anaerobic mechanisms among fishes. Coulter (1967) listed some 10 species of benthic fishes in Lake Tanganyika which appear to live in, or at least tolerate extended exposures to, deep, essentially anaerobic water. In the swamp waters of tropical regions, $O_2$ tensions often become critically reduced; one common adaptation to this condition has been the development of the air breathing habit in many of the fishes of the area (chapter by Johansen, Volume IV, this treatise). It would appear that other species in this area, which have not taken to air breathing, must rely heavily upon an anaerobic metabolism.

In studies of heat production by cichlid fishes, Morris (1968) has repeatedly observed that heat production rates exceed, by a factor of 1½–2-fold, rates which would be expected on the basis of $O_2$ consumption. In some cases, the rate of heat production is as much as 5-fold higher than theoretically expected values; this finding again indicates an unusually active anaerobic metabolism.

As far as we can ascertain, nothing is known of the mechanisms of anaerobic metabolism in fishes, and this is an area that is clearly in need of much further research. Reaction pathways of anaerobic metabolism are much better understood in various invertebrate organisms which are facultative anaerobes (Beuding and Saz, 1968; Ward and Schoefield, 1967) and in tissues such as the kidney in higher vertebrates which must supplant their aerobic metabolism with important anaerobic decarboxylations in order to support various maintenance and work functions (Cohen, 1968). Similar anaerobic decarboxylations may occur in fishes during exposure to anoxic waters.

## B. High $O_2$ Tensions

A primary function of the swim bladder in those fishes which use this organ in hydrostatic function appears to be the secretion of $O_2$ from the blood into the swim bladder, at times against exceedingly high concentration gradients. During gas deposition, lactic acid enters the blood circulating through the bladder epithelium. The pH of this blood drops to values approaching 1 pH unit lower than the pH of the blood entering the rete system. This pH change is brought about largely, if not solely, by lactic acid which is presumably produced as an end product of glycolysis in the swim bladder epithelium (Steen, 1963). These conditions raise two important problems: (1) high glycolytic rates are not normally expected in the presence of high concentrations of $O_2$ because of the Pasteur effect (inhibition of glycolysis by high $O_2$); and (2) the variability in intracellular pH may be expected to be high, being a function of the rate of $O_2$ secretion.

The Pasteur effect is brought about by the development of a high energy charge in the cell and the subsequent inhibition of the PFK reaction by high ATP concentrations. In swim bladder, the Pasteur effect is absent (Ball *et al.*, 1955), probably because mitochondrial metabolism is reduced (Steen, 1963). Also, it is possible that swim bladder epithelium possesses forms of PFK which are not sensitive to inhibition by high ATP.

We have little information on the pH responses of enzymes of the bladder epithelium. Swim bladder LDH, particularly at high substrate values, appears to be less sensitive to pH change than do other LDHs examined (Hochachka, 1968b); in this way, this particular enzyme appears to be well adapted for function in the microenvironment of the swim bladder epithelium. We do not know if the same is true for other enzymes.

## VII. ESTIVATION

The African lungfish *Protopterus* lives in quiet tropical swamp waters which are subject to seasonal drought. At the onset of the dry season, as the water level falls, the lungfish burrows into the semisolid muddy bottom and comes to lie some 1–1½ feet deep in the mud at the bottom of a burrow leading to the surface. When the surrounding mud dries, the mucous covering the fish hardens to form a thin, brown cocoon which is contiguous with the fish at essentially all points in contact with the mud. The tube of dried mucous is the only direct contact with the outer environment and allows the lungfish a channel for breathing. In estivation, the lungfish takes no food or water and excretes no waste nitrogen. Lungfish have been known to survive this condition for several years although usually the dry season lasts for only a few months. The South American lungfish *Lepidosiren*, faces a similar ecological situation. These conditions impose upon the lungfish a number of physiological problems which would appear to require specific and probably drastic biochemical adjustments. We can outline briefly at least four of these.

## A. Energy Sources

Since the estivating lungfish cannot take in food, all energy requirements must be fulfilled by the metabolism of endogenous reserves. On the basis of respiratory gas analyses, Homer Smith (1930) pointed out that the major endogenous fuel substances for metabolism during estivation were amino acids derived from body proteins. Major carbohydrate fuels were thought to be used up early in the estivation. It is clear from later studies that protein is the major endogenous fuel, but contrary to expectations, the reserves of carbohydrate and lipid are not used up early in the process. Instead, they are conserved and may actually accumulate somewhat during the estivation (Janssens, 1964). This situation may be general among fishes (Stimpson, 1965; see also Bellamy, 1968). However, it is not clear why fishes should differ in this way from mammals, nor are the mechanisms of protein mobilization understood.

## B. Maintenance of Carbohydrate and Lipid Reserves

We have no information whatever on the mechanisms by which the estivating lungfish is able to maintain its carbohydrate and lipid reserves during starvation. Presumably, turnover rates of lipid are reduced and fatty acid oxidation rates are balanced by fatty acid synthesis from carbohydrate precursors. The latter in turn are probably maintained by gluconeogenesis from amino acids (Janssens, 1964). However, this situation is complicated by the energy charge of the cell, which may be expected to be low under estivating conditions. If, as in other organisms (Newsholme and Gevers, 1967), the major control sites in gluconeogenesis are highly sensitive to ATP/AMP ratios in the liver, energy-depleted conditions would not favor the synthesis of carbohydrate. Since gluconeogenic rates appear to remain high during estivation, we initially postulated that regulatory enzymes in this pathway should be subject to control by mechanisms other than adenylate modulation (Behrisch and Hochachka, 1969b). Detailed examination of the enzyme, FDPase, which is an important control site in gluconeogenesis substantiates our prediction. In the case of the lungfish FDPase, AMP inhibition occurs, but the $K_i$ is some 3-fold higher than for trout liver FDPase; in addition, the flow of carbon through this bottleneck in the pathway can be effectively modulated by $Mg^{2+}$, $Mn^{2+}$, $H^+$, and FDP. These kinds of mechanisms can readily account for maintenance of gluconeogenesis during prolonged estivation (Behrisch and Hochachka, 1969b).

## C. The Problem of Urea Storage

In estivation, the nitrogenous end product of amino acid metabolism is urea. Because no excretion can take place, urea accumulates in the tissues. Similar increases in the concentration of urea in the tissues occur in response to dehydration in several amphibians (Balinsky et al., 1961; Scheer and Markel, 1962; Tercafs and Schoffeniels, 1962). Also, Rana cancrivora, a South Asian frog living in brackish water, concentrates urea within its tissues to remain in osmotic balance with the environment (Gordon et al., 1961; Schmidt-Nielsen and Lee, 1962). Estivation in the lungfish is rather comparable with dehydration in these amphibians in that these animals are unable to excrete their nitrogenous waste. However, the lungfish differs from the other species since normally its nitrogenous waste product is largely $NH_4^+$; the organism must therefore change the form of its nitrogenous end product as well as store it in the tissues. This changeover clearly involves changes in the relative activities of various enzymes of nitrogen metabolism (Janssens, 1964), but since the urea biosynthesis rate does not change noticeably during estivation, the changeover must be related to the control of $NH_4^+$ production. Because glutamine synthestase is not present in lungfish liver, $NH_4^+$ cannot be stored in the form of glutamine. Rather, Janssens and Cohen (1968) suggested that metabolite regulation of the activity of glutamate dehydrogenase to low levels can account for reduced $NH_4^+$ production during estivation. Since in vivo concentrations of the regulatory metabolites have not been estimated, it is clear that much further work on this aspect of metabolic control in the lungfish is required.

## D. Metabolic Depression during Estivation

From Homer Smith's initial studies of lungfish metabolism (1930) it was evident that the overall metabolic rate of the estivating lungfish is much reduced when compared with the nonestivating animal. Whereas this is a condition of estivation wherever it occurs in the animal kingdom, very little is known of the cellular mechanisms by which maintenance metabolic processes can be reduced to levels perhaps 1–2 orders of magnitude lower than normal basal rates. Since lipid and carbohydrate sources are clearly in abundance at this time, metabolic depression must lead to a great reduction in activities and/or amounts of enzymes involved in energy metabolism. Unfortunately, no information is available dealing with this aspect of lungfish metabolism. Solution of the problem should lead to much greater insight into the problems of estivation in general as well as to metabolic control in the lungfish specifically.

## VIII. PROSPECTS AND PROBLEMS FOR THE FUTURE

At the beginning of this essay the topic of "biochemical adaptation to the environment" was introduced by stressing the importance of the functional approach to biochemistry. We tried to illustrate the necessity of examining enzymic systems in an experimental context which (1) takes into account the physiological role of the reaction and (2) approximates as closely as possible the conditions experienced by the enzyme in the intracellular environment. In our own studies the most important difference in approach from previous studies has been the recognition that enzymic properties are critically dependent on the types of metabolites present and their concentrations. We have shown that the influence of temperature on metabolic reactions is grossly different at high (nonphysiological) and low (physiological) substrate concentrations. Thus, even such a simple change in experimental design as the use of non-saturating concentrations of substrate has led to important revisions in our understanding of temperature adaptation. To conclude, it might therefore be desirable to consider what other sorts of functional considerations might be fruitfully applied in our attempts to discover the biochemical mechanisms of environmental adaptation.

Perhaps the most logical extension of our discussion of the importance of using physiological substrate concentrations is to emphasize that enzyme concentrations may also have significant effects on the results of kinetic experiments. Vesell and co-workers (Wuntch *et al.*, 1970) have shown that the "classic" difference in substrate inhibition between muscle and heart LDHs does not occur at enzyme concentrations which approximate those in the cell. This observation suggests that whereas differences in substrate inhibition may be useful in characterizing isozymes of LDH, these differences may have no physiological significance. Although estimations of intracellular enzyme concentrations are difficult, it is nonetheless clear that *in vivo* enzyme concentrations are one to several orders of magnitude greater than the enzyme concentrations used in most assay systems (see, e.g., Srere, 1967, 1968, 1969).

Another uncertainty in enzyme studies arises from the difficulty in knowing whether the properties of an enzyme molecule free in solution differ from its properties when bound to other large molecules. In at least some cases, the bound and free forms of enzymes are known to have different kinetic properties. Hexokinase bound to mitochondria of the brain has a higher affinity for ATP than free hexokinase (Schwartz and Basford, 1967). In sea urchins, the pentose shunt may be controlled by releasing bound glucose-6-phosphate dehydrogenase, an effect which activates the enzyme (Isono and Yasumasu, 1968). The likelihood that effects of this sort are of general importance in metabolic regulation seems high. Recent observations that even such classically "soluble" enzymes as the glycolytic enzymes can bind reversibly with muscle protein (Arnold and Pette, 1968) suggest that enzyme binding phenomena must be included as a major consideration in the design and the interpretation of kinetic experiments.

Similar problems arise when we consider interactions among different enzyme molecules. Much as enzymes may interact with "structural" proteins, and thereby alter their kinetic properties, in some cases enzyme–enzyme interactions may greatly influence the kinetics of a system. One example of this effect is found in the case of enzymes involved in pyrimidine synthesis. The synthesis of carbamyl phosphate is an important branch point in metabolism, with one pathway leading to arginine and a second pathway to pyridine nucleotides (Stadtman, 1968). In the latter case, carbamyl phosphate synthetase is organized with the second enzyme in the pyrimidine pathway, aspartate transcarbamylase, into a single multifunctional enzyme complex which effectively channels carbon into this pathway (Lue and Kaplan, 1969). A second variant of carbamyl phosphate synthetase is found associated with the pathway to arginine.

An additional factor which may be important in determining the kinetic properties of enzymes is the possible existence of different "metastable" states of the same protein. The dogma of biochemistry states that the biologically active form of an enzyme is also the thermodynamically most stable state of the enzyme. Nickerson and Day (1969) questioned this assumption on kinetic grounds. They argued that, following the synthesis of a polypeptide chain, it is unreasonable to assume that each of approximately $10^{600}$ possible configurations of folding which are possible will be "tried" in order to find that one most stable conformation without first finding a conformation which is relatively stable at the particular temperature in question. Presumably there would be a continuous migration of initially synthesized protein over the energy barrier to a thermodynamically more stable form. The rate of migration may be expected to depend on the thermal energy available in the system. Thus, in poikilothermic systems, there is reason to believe that the conformation of a protein, or at least the distribution of different conformations, may be significantly affected by temperature. Our data from studies of king crab LDH and PyK and trout LDH (Figs. 7, 8, and 9) may represent instances where multiple "metastable" states of the enzymes are present. Gelb *et al.* (1970) have recently obtained similar data for glyceraldehyde-3-phosphate dehydrogenase from poikilotherms, and they concluded that metastable states of their enzymes may be involved in producing the complex kinetics they have observed.

In addition to the above factors which require careful study, there are several more effects which remain to be investigated vis-à-vis questions of environmental adaptation. Briefly, these may be listed as follows: (1) the influence of temperature on the aggregation–disaggregation of enzyme subunits—in at least some cases the aggregation of subunits is strongly temperature dependent (see Assaf and Graves, 1969); (2) the role of enzyme–phospholipid and enzyme–polysaccharide interactions in metabolic regulation; (3) the influence of temperature on cellular compartmentalization of metabolites and ions; and (4) the influence of the environment on intracellular pH.

The functional analysis of biochemical data has scarcely begun. Now that the first great functional achievement, namely, the elaboration of ideas of metabolic pathways which are tightly regulated, has been made, biochemists must determine how these different pathways are organized in time and space in the cellular environment. As Srere (1968) has stated:

> We shrink from considering the cell as it exists with redundant pathways and multiple interactions at each step. I am not suggesting an abandonment of studies on simplified systems, but I feel that we should also strive for some intermediate degree of complexity. What is really needed is another "Krebsian" step in biochemistry; an insight that enables us to advance conceptually to the next magnitude of complexity . . .

Progress in environmental biochemistry will be attendant on the pace with which biologists can rise to Srere's challenge.

## REFERENCES

Arnold, H., and Pette, D. (1968). Binding of glycolytic enzymes to structure proteins of the muscle. *European J. Biochem.* 6, 163–171.

Assaf, S. A., and Graves, D. J. (1969). Structural and catalytic properties of lobster muscle glycogen phosphorylase. *J. Biol. Chem.* **224**, 5544–5555.

Atkinson, D. E. (1965). Biological feedback control at the molecular level. *Science* **150**, 851–857.

Atkinson, D. E. (1966). Regulation of enzyme activity. *Ann. Rev. Biochem.* **35**, 85–124.

Atkinson, D. E. (1968). Citrate and the citrate cycle in the regulation of energy metabolism. *In* "Metabolic Roles of Citrate" (T. W. Goodwin, ed.), pp. 23–40. Academic Press, New York.

Atkinson, D. E., and Fall, L. (1967). Adenosine triphosphate conservation in biosynthetic regulation: *Escherichia coli* phosphoribosylpyruphosphate synthase. *J. Biol. Chem.* **242**, 3241–3242.

Atkinson, D. E., and Walton, G. M. (1967). Adenosine triphosphate conservation in metabolic regulation: rat liver citrate cleavage enzyme. *J. Biol. Chem.* **242**, 3239–3241.

Baldwin, J. (1971). Evolutionary adaptation of enzymes to temperature (in preparation).

Baldwin, J., and Hochachka, P. W. (1970). Functional significance of isoenzymes in thermal acclimation: Acetylcholinesterase from trout brain. *Biochem. J.* **116**, 883–887.

Baldwin, S. (1968). Manometric measurements of respiratory activity in *Acmaea digitalis* and *Acmaea scabra*. *Veliger* **11**, 79–82.

Balinsky, J. B., Cragg, M. M., and Baldwin, E. (1961). The adaptation of amphibian waste nitrogen excretion to dehydration. *Comp. Biochem. Physiol.* **3**, 236–244.

Ball, E. Q., Strittmatter, C. S., and Cooper, O. (1955). Metabolic studies on the gas gland of the swimbladder. *Biol. Bull.* **108**, 1–17.

Baslow, M. H., and Nigrelli, R. F. (1964). The effect of thermal acclimation on brain cholinesterase activity of the killifish, *Fundulus heteroclitus*. *Zoologica* **49**, 41–51.

Behrisch, H. W. (1969). Temperature and the regulation of enzyme activity in poikilotherms: Fructose diphosphatase from migrating salmon. *Biochem. J.* **115**, 687–696.

Behrisch, H. W., and Hochachka, P. W. (1969a). Temperature and the regulation of enzyme activity in poikilotherms: Properties of rainbow trout fructose diphosphatase. *Biochem. J.* **111**, 287–295.

Behrisch, H. W., and Hochachka, P. W. (1969b). Temperature and the regulation of enzyme activity in poikilotherms: Properties of lungfish fructose diphosphatase. *Biochem. J.* **112**, 601–607.

Bellamy, D. (1968). Metabolism of the red piranha (*Rooseveltiella nattereri*) in relation to feeding behaviour. *Comp. Biochem. Physiol.* **25**, 343–347.

Benesch, R., and Benesch, R. (1969). Intracellular organic phosphates as regulators of oxygen release by haemoglobin. *Nature* **221**, 618–622.

Beuding, E., and Saz, H. J. (1968). Pyruvate kinase and phosphoenolpyruvate carboxykinase activities of *Ascaris* muscle, *Hymenolepis diminuta*, and *Schistosoma mansoni*. *Comp. Biochem. Physiol.* **24**, 511–518.

Blažka, P. (1958). The anaerobic metabolism of fish. *Physiol. Zool.* **31**, 117–128.

Brett, J. R. (1956). Some principles in the thermal requirements of fishes. *Quart. Rev. Biol.* **31**, 75–87.

Brett, J. R. (1967). Swimming performance of sockeye salmon *Onchorynchus nerka* in relation to fatigue time and temperature. *J. Fisheries Res. Board Can.* **24**, 1731–1741.

Brett, J. R., Shelbourn, J. E., and Shoop, C. T. (1969). Growth rate and body composition of fingerling sockeye salmon, *Oncorhynchus nerka*, in relation to temperature and ration size. *J. Fisheries Res. Board Can.* **26**, 2363–2394.

Byfield, J. E., and Lee, Y. C., and Bennett, L. R. (1969). Thermal instability of *Tetrahymena* ribosomes: effects on protein synthesis. *Biochem. Biophys. Res. Comm.* **37**, 806–812.

Bygrave, F. L. (1966a). The effect of calcium ions on the glycolytic activity of Ehrlich ascites-tumour cells. *Biochem. J.* **101**, 480–487.

Bygrave, F. L. (1966b). Studies on the interaction of metal ions with pyruvate kinase from Ehrlich ascites-tumour cells and from rabbit muscle. *Biochem. J.* **101**, 488–494.

Bygrave, F. L. (1967). The ionic environment and metabolic control. *Nature* **214**, 667–671.

Caldwell, R. S. (1969). Thermal compensation of respiratory enzymes in tissues of the goldfish (*Carassius auratus* L.). *Comp. Biochem. Physiol.* **31**, 79–93.

Carey, F. G., and Teal, J. M. (1966). Heat conservation in tuna fish muscle. *Proc. Natl. Acad. Sci. U. S.* **56**, 1464–1469.

Chappell, J. B., and Robinson, B. H. (1968). Penetration of the mitochondrial membrane by tricarboxylic acid anions. *In* "Metabolic Roles of Citrate" (T. W. Goodwin, ed.), pp. 123–133. Academic Press, New York.

Cohen, J. J. (1968). Renal gaseous and substrate metabolism *in vivo:* Relationship to renal function. *Proc. Intern. Union Physiol. Sci.* **6**, 233–234.

Coulter, G. W. (1967). Low apparent oxygen requirements of deep water fishes in Lake Tanganyika. *Nature* **215**, 317–318.

Cowey, C. B. (1967). Comparative studies on the activity of D-glyceraldehyde-3-phosphate dehydrogenase from cold and warm-blooded animals with respect to temperature. *Comp. Biochem. Physiol.* **23**, 969–976.

Das, A. B. (1967). Biochemical changes in tissues of goldfish acclimated to high and low temperatures. II. Synthesis of protein and RNA of subcellular fractions and tissues composition. *Comp. Biochem. Physiol.* **21**, 469–485.

Das, A. B., and Prosser, C. L. (1967). Biochemical changes in tissues of goldfish acclimated to high and low temperatures. I. Protein synthesis. *Comp. Biochem. Physiol.* **21**, 449–467.

Dean, J. M. (1969). The metabolism of tissues of thermally acclimated trout (*Salmo gairdneri*). *Comp. Biochem. Physiol.* **29**, 185–196.

Ekberg, D. R. (1958). Respiration in tissues of goldfish adapted to high and low temperatures. *Biol. Bull.* **114**, 308–316.

Ekberg, D. R. (1962). Anaerobic and aerobic metabolism in gills of the crucian carp adapted to high and low temperatures. *Comp. Biochem. Physiol.* **5**, 123–128.

Evans, R. M., Purdie, F. C., and Hickman, C. P., Jr. (1962). The effect of temperature and photoperiod on the respiratory metabolism of rainbow trout (*Salmo gairdnerii*). *Can. J. Zoo.* **40**, 107–118.

Fluke, D. J., and Hochachka, P. W. (1965). Radiation indication of subunit activity of lactic dehydrogenase. *Radiation Res.* **26**, 395–402.

Freed, J. M. (1965). Changes in activity of cytochrome oxidase during adaptation of goldfish to different temperatures. *Comp. Biochem. Physiol.* **14**, 541–659.

Freed, J. M. (1971). Temperature effects on muscle phosphofructokinase of the Alaskan king crab *Paralithodes camtschatica*. *Comp. Biochim. Physiol.* (in press).

Fry, F. E. J. (1947). Effects of the environment on animal activity. *Publ. Ontario Fisheries Res. Lab.* **68**, 5–62.

Fry, F. E. J. (1958). Temperature compensation. *Ann. Rev. Physiol.* **20**, 207–224.

Gelb, W., Oliver, E., Brandts, J. F., and Nordin, J. H. (1970). Unusual kinetic transition in honeybee glyceraldehyde phosphate dehydrogenase. *Biochemistry* **9**, 3228–3235.

Gordon, M. S. (1968). Oxygen consumption of red and white muscles from tuna fishes. *Science* **159**, 87–90.

Gordon, M. S., Schmidt-Nielsen, K., and Kelly, H. M. (1961). Osmotic regulation in the crab-eating frog (*Rana cancrivora*). *J. Exptl. Biol.* **39**, 659–678.

Gordon, M. S., Amdur, B. H., and Scholander, P. F. (1962). Freezing resistance in some northern fishes. *Biol. Bull.* **122**, 52–62.

Grigg, G. C. (1967). Some respiratory properties of the blood of four species of Antarctic fishes. *Comp. Biochem. Physiol.* **23**, 139–148.

Grigg, G. C. (1969). Temperature-induced changes in the oxygen equilibrium curve of the blood of the brown bullhead *Ictalurus nebulosus*. *Comp. Biochem. Physiol.* **28**, 1203–1223.

Halcrow, K., and Boyd, C. M. (1967). The oxygen consumption and swimming activity of the amphipod *Gammarus oceanicus* at different temperatures. *Comp. Biochem. Physiol.* **23**, 233–242.

Haschemeyer, A. E. V. (1968). Compensation of liver protein synthesis in temperature acclimated toadfish, *Opsanus tau*. *Biol. Bull.* **134**, 130–140.

Haschemeyer, A. E. V. (1969a). Studies on the control of protein synthesis in low temperature acclimation. *Comp. Biochem. Physiol.* **28**, 535–552.

Haschemeyer, A. E. V. (1969b). Rates of polypeptide chain assembly in liver *in vivo*: Relation to the mechanism of temperature acclimation in *Opsanus tau*. *Proc. Natl. Acad. Sci. U. S.* **62**, 128–135.

Hebb, C., Morris, D., and Smith, M. W. (1969). Choline acetyltransferase activity in the brain of goldfish acclimated to different temperatures. *Comp. Biochem. Physiol.* **29**, 29–36.

Heninicke, E. A., and Houston, A. H. (1965). Effect of thermal acclimation and sublethal heat shock upon ionic goldfish, *Carassius auratus* L. *J. Fisheries Res. Board Can.* **22**, 1455–1476.

Helmreich, E., and Cori, C. R. (1964). The effects of pH and temperature on the kinetics of the phosphorylase reaction. *Proc. Natl. Acad. Sci. U. S.* **52**, 647–654.

Hemmingsen, E. A., Douglas, E. L., and Grigg, G. C. (1969). Oxygen consumption in an Antarctic hemoglobin-free fish, *Pagetopsis macropterus*, and in three species of *Notothenia*. *Comp. Biochem. Physiol.* **29**, 467–490.

Hickman, C. P., Jr., McNabb, R. A., Nelson, J. S., Van Breeman, E. D., and Comfort, D. (1964). Effect of cold acclimation on electrolyte distribution in rainbow trout (*Salmo gairdnerii*). *Can. J. Zool.* **42**, 577–597.

Hochachka, P. W. (1961). Glucose and acetate metabolism in fish. *Can. J. Biochem. Physiol.* **39**, 1937–1941.

Hochachka, P. W. (1965). Isoenzymes in metabolic adaptation of a poikilotherm: Subunit relationships in lactic dehydrogenases of goldfish. *Arch. Biochm. Biophys.* **111**, 96–103.

Hochachka, P. W. (1967). Organization of metabolism during temperature compensation. *In* "Molecular Mechanisms of Temperature Adaptation," Publ. No. 84, pp. 177–203. Am. Assoc. Advance. Sci., Washington, D. C.

Hochachka, P. W. (1968a). Action of temperature on branch points in glucose and acetate metabolism. *Comp. Biochem. Physiol.* **25**, 107–118.

Hochachka, P. W. (1968b). Lactate dehydrogenase function in *Electrophorus* swimbladder and in the lungfish lung. *Comp. Biochem. Physiol.* **27**, 613–615.

Hochachka, P. W., and Hayes, F. R. (1962). The effect of temperature acclimation on pathways of glucose metabolism in the trout. *Can. J. Zool.* **40**, 261–270.

Hochachka, P. W., and Lewis, J. K. (1970). The functional significance of enzyme variants in thermal acclimation: Trout liver citrate synthases. *J. Biol. Chem.* **245**, 6567–6573.

Hochachka, P. W., and Somero, G. N. (1968). The adaptation of enzymes to temperature. *Comp. Biochem. Physiol.* **27**, 659–668.

Hochachka, P. W., Freed, J. M., Somero, G. N., and Prosser, C. L. (1971). Control sites in glycolysis of crustacean muscle. *Intl. J. Biochem.* **2**, 125–130.

Ingraham, J. L., and Maaløe O. (1967). Cold-sensitive mutants and the minimum temperature of growth of bacteria. *In* "Molecular Mechanisms of Temperature Adaptation," Publ. No. 84, pp. 297–309. Am. Assoc. Advance. Sci., Washington, D. C.

Isono, N., and Yasumasu, I. (1968). Pathways of carbohydrate breakdown in sea urchin eggs. *Exptl. Cell Res.* **50**, 616–626.

Iwatsuki, N., and Okazaki, R. (1967). Mechanisms of regulation of deoxythymidine kinase of *Escherichia coli*. II. Effect of temperature on the enzyme activity and kinetics. *J. Mol. Biol.* **29**, 155–165.

Jankowsky, H. D. (1968). Versuche zur Adaptation der Fische in normalen Temperaturbereich. *Helgolaender Wiss. Meeresuntersuch.* **18**, 317–362.

Janssens, P. A. (1964). The metabolism of the aestivating African lungfish. *Comp. Biochem. Physiol.* **11**, 105–117.

Janssens, P. A., and Cohen, P. P. (1968). Nitrogen metabolism in the African lungfish. *Comp. Biochem. Physiol.* **24**, 879–886.

Johnston, P. V., and Roots, B. I. (1964). Brain lipid fatty acids and temperature acclimation. *Comp. Biochem. Physiol.* **11**, 303–310.

Kanungo, M. S., and Prosser, C. L. (1959). Physiological and biochemical adaptations of goldfish to cold and warm temperatures. II. Oxygen consumption of liver homogenate and oxidative phosphorylation of liver mitochondria. *J. Cellular Comp. Physiol.* **66**, Suppl. 1, 1–10.

Kaplan, N. O. (1964). Lactate dehydrogenase—structure and function. *Brookhaven Symp. Biol.* **17**, 131–153.

Kaplan, N. O. (1968). Nature of multiple molecular forms of enzymes. *Ann. N. Y. Acad. Sci.* **151**, 400–412.

Katzen, H. M., Soderman, D. D., and Cirillo, V. J. (1968). Tissue distribution and physiological significance of multiple forms of hexokinase. *Ann. N. Y. Acad. Sci.* **151**, 351–358.

Knipprath, W. G., and Mead, J. F. (1966). Influence of temperature on the fatty acid pattern of mosquitofish (*Gambusia affinis*) and guppies (*Lebistes reticulatus*). *Lipids* **1**, 113–117.

Knipprath, W. G., and Mead, J. F. (1968). The effect of environmental temperature on the fatty acid composition and on the *in vivo* incorporation of 1-$^{14}$C-acetate in goldfish (*Carassius auratus* L.). *Lipids* **3**, 121-128.

Koster, J. F., and Veeger, C. (1968). The relation between temperature inducible allosteric effects and the activation energies of amino-acid oxidases. *Biochim. Biophys. Acta* **167**, 48–63.

Krogh, A., and Leitch, I. (1919). The respiratory function of blood in fishes. *J. Physiol. (London)* **52**, 288–300.

Kutty, M. N. (1968). Respiratory quotients in goldfish and rainbow trout. *J. Fisheries Res. Board Can.* **25**, 1689–1728.

Kwon, T. W., and Olcott, H. S. (1965). Tuna muscle aldolase. I. Purification and properties. *Comp. Biochem. Physiol.* **15**, 7–16.

Licht, P. (1964). The temperature dependence of myosin-adenosinetriphosphatase and alkaline phosphatase in lizards. *Comp. Biochem. Physiol.* **12**, 331–340.

Licht, P. (1967). Thermal adaptation in the enzymes of lizards in relation to preferred body temperatures. *In* "Molecular Mechanisms of Temperature Adaptation," Publ. No. 84, pp. 131–145. Am. Assoc. Advance. Sci., Washington, D. C.

Lowry, O. H., Schulz, D. D., and Passonneau, J. V. (1964). Effects of adenylic acid on the kinetics of muscle phosphorylase-a. *J. Biol. Chem.* **239**, 1947–1953.

Lue, P. F., and Kaplan, J. G. (1969). The aspartate transcarbamylase and carbamylphosphate synthetase of yeast: A multi-functional enzyme complex. *Biochem. Biophys. Res. Commun.* **34**, 426–433.

Markert, C. L. (1968). The molecular basis for isozymes. *Ann. N. Y. Acad. Sci.* **151**, 14–40.

Massaro, E. J., and Markert, C. L. (1968). Isozyme patterns of Salmonid fishes: Evidence for multiple cistrons for lactate dehydrogenase polymers. *J. Exptl. Zool.* **168**, 223–238.

Massey, V., Curti, B., and Ganthers, H. (1966). A temperature-dependent conformational change in D-amino acid oxidase and its effect on catalysis. *J. Biol. Chem.* **241**, 2347–2357.

Morpurgo, C., Battaglia, P. A., and Leggio, T. (1970). Negative Bohr effect in newt haemolysates and its regulation. *Nature* **225**, 76–77.

Morris, R. (1968). Personal communication.

Nagata, N., and Rasmussen, H. (1968). Parathyroid hormone and renal cell metabolism. *Biochemistry* **7**, 3728–3733.

Newell, R. C. (1966). The effect of temperature on the metabolism of poikilotherms. *Nature* **212**, 427–428.

Newell, R. C. (1967). Oxidative activity of poikilotherm mitochondria as a function of temperature. *J. Zool. (London)* **151**, 299–311.

Newell, R. C., and Northcroft, H. R. (1967). A re-interpretation of the effect of temperature on the metabolism of certain marine invertebrates. *J. Zool. (London)* **151**, 277–298.

Newsholme, E. A., and Gevers, W. (1967). Control of glycolysis and gluconeogenesis in liver and kidney cortex. *Vitamins Hormones* **25**, 1–87.

Nickerson, K. W., and Day, R. A. (1969). Possible biological roles for metastable proteins. *Currents Mod. Biol.* **2**, 303–306.

Pace, B., and Campbell, L. L. (1967). Correlation of maximal growth temperature and ribosome heat stability. *Proc. Natl. Acad. Sci. U. S.* **57**, 1110–1116.

Peiss, C. N., and Field, J. (1950). The respiratory metabolism of excised tissues of warm-and cold-adapted fishes. *Biol. Bull.* **99**, 213–224.

Plagemann, P. G. W., Gregory, K. F., and Wróblewski, F. (1960). Die elektrophoretischtrennbaren Lactatdehydrogenasen des Saugetieres. III. Einfluss der

Temperatur auf die Lactatdehydrogenasen des Kaninchens. *Biochem. Z.* **334,** 37–48.

Precht, H. (1958). Concepts of the temperature adaptation of unchanging reaction systems of cold-blooded animals. *In* "Physiological Adaptation" (C. L. Prosser, ed.), pp. 50–78. Ronald Press, New York.

Precht, H. (1968). Der Einfluss "normaler" Temperaturen auf Lebensprozesse bei wechselwarmen Tieren unter Ausschluss der Wachstums- und Entwicklungs-prozesse. *Helgolaender Wiss. Meeresuntersuch.* **18,** 487–548.

Prosser, C. L. (1958). The nature of physiological adaptation. *In* "Physiological Adaptation" (C. L. Prosser, ed.), pp. 167–180. Ronald Press, New York.

Prosser, C. L. (1967). Molecular mechanisms of temperature adaptation in relation to speciation. *In* "Molecular Mechanisms of Temperature Adaptation," Publ. No. 84, pp. 351–376. Am. Assoc. Advance. Sci., Washington, D. C.

Read, K. R. (1964a). The temperature coefficients of ribonucleases from two species of gastropod molluscs from different thermal environments. *Biol. Bull.* **127,** 489–498.

Read, K. R. (1964b). Comparative biochemistry of adaptations of poikilotherms to the thermal environment. *Proc. Symp. Exptl. Marine Ecol., 1961.* Occasional Publ. No. 2, pp. 39–47. Graduate School of Oceanography, Univ. of Rhode Island.

Roberts, J. L. (1964). Metabolic responses of fresh-water sunfish to seasonal pho-toperiods and temperatures. *Helgolaender Wiss. Meeresuntersuch.* **14,** 451–465.

Roberts, J. L. (1967). Metabolic compensations for temperature in sunfish. *In* "Molecular Mechanisms of Temperature Adaptation," Publ. No. 84, pp. 245–262. Am. Association Advance. Sci., Washington, D. C.

Roots, B. I. (1968). Phospholipids of goldfish (*Carassius auratus* L.) brain: The influence of environmental temperature. *Comp. Biochem. Physiol.* **25,** 457–466.

Roots, B. I., and Johnston, P. V. (1968). Plasmalogens of the nervous system and environmental temperature. *Comp. Biochem. Physiol.* **26,** 553–560.

Scheer, B. T., and Markel, R. P. (1962). The effect of osmotic stress and hypophy-sectomy on blood and urine urea levels in frogs. *Comp. Biochem. Physiol.* **7,** 289–297.

Schmidt-Nielsen, K., and Lee, P. (1962). Kidney function in the crab-eating frog (*Rana cancrivora*). *J. Exptl. Biol.* **39,** 167–177.

Scholander, P. F., Flagg, W., Walters, V., and Irving, L. (1953). Climatic adapta-tion in Arctic and tropical poikilotherms. *Physiol. Zool.* **26,** 67–92.

Scholander, P. F., van Dam, L., Kanwisher, J. W., Hammel, H. T., and Gordon, M. S. (1957). Supercooling and osmoregulation in Arctic fish. *J. Cellular Comp. Physiol.* **49,** 5–24.

Schwartz, G. P., and Basford, R. E. (1967). The isolation and purification of solubilized hexokinase from bovine brain. *Biochemistry* **6,** 1070–1079.

Smith, M. W. (1930). Metabolism of the lungfish *Protopterus aethiopicus. J. Biol. Chem.* **88,** 97–130.

Somero, G. N. (1969a). Enzymic mechanisms of temperature compensation: Im-mediate and evolutionary effects of temperature on enzymes of aquatic poikilo-therms. *Am. Naturalist* **103,** 517–530.

Somero, G. N. (1969b). Pyruvate kinase variants of the Alaskan king crab: Evidence for a temperature-dependent interconversion between two forms have distinct- and adaptive-kinetic properties. *Biochem. J.* **114,** 237–241.

Somero, G. N. (1970). Unpublished observations.

Somero, G. N., and DeVries, A. L. (1967). Temperature tolerance of some Antarctic fishes. *Science* **156**, 257–258.

Somero, G. N., and Gould-Somero, M. (1970). Unpublished observations.

Somero, G. N., and Hochachka, P. W. (1968). The effect of temperature on catalytic and regulatory functions of pyruvate kinases of the rainbow trout and the Antarctic fish *Trematomus bernacchii*. *Biochem. J.* **110**, 395–400.

Somero, G. N., and Hochachka, P. W. (1969). The role of isoenzymes in immediate compensation to temperature. *Nature* **223**, 194–195.

Somero, G. N. Giese, A. C., and Wohlschlag, D. E. (1968). Cold adaptation of the Antarctic fish *Trematomus bernacchii*. *Comp. Biochem. Physiol.* **26**, 223–233.

Srere, P. A. (1967). Enzyme concentrations in tissues. *Science* **158**, 936–937.

Srere, P. A. (1968). Studies on purified citrate-enzymes: Metabolic interpretations. *In* "Metabolic Roles of Citrate" (T. W. Goodwin, ed.), pp. 11–21. Academic Press, New York.

Srere, P. A. (1969). Some complexities of metabolic regulation. *Biochem. Med.* **3**, 61–72.

Stadtman, E. R. (1968). The role of multiple enzymes in the regulation of branched metabolic pathways. *Ann. N. Y. Acad. Sci.* **151**, 516–530.

Steen, J. B. (1963). The physiology of the swimbladder of the eel, *Anguilla vulgaris*. III. The mechanism of gas secretion. *Acta Physiol. Scand.* **59**, 221–241.

Stimpson, J. H. (1965). Comparative aspects of the control of glycogen utilization in vertebrate liver. *Comp. Biochem. Physiol.* **15**, 187–197.

Taketa, K., and Pogell, B. M. (1965). Allosteric inhibition of rat liver fructose 1,6-diphosphatase by adenosine 5′-monophosphate, *J. Biol. Chem.* **240**, 651–662.

Tercafs, R. R., and Schoffeniels, E. (1962). Adaptations of amphibians to salt water. *Life Sci.* **1**, 19–24.

Trivedi, B., and Danforth, W. H. (1966). Effect on pH on the kinetics of frog muscle phosphofructokinase. *J. Biol. Chem.* **241**, 4110–4114.

Ushakov, B. P. (1967). Coupled evolutionary changes in protein thermostability. *In* "Molecular Mechanisms of Temperature Adaptation," Publ. No. 84, pp. 107–129. Am. Assoc. Advance. Sci., Washington, D. C.

van Handel, E. (1966). The thermal dependence of the rates of glycogen and triglyceride synthesis in the mosquito. *J. Exptl. Biol.* **44**, 523–528.

Vroman, H. E., and Brown, J. R. C. (1963). The effect of temperature on the activity of succinic dehydrogenase from livers of rats and frogs. *J. Cellular Comp. Physiol.* **61**, 129–131.

Ward, C. W. and Schoefield, P. J. (1967). Comparative activity and intracellular distribution of tricarboxylic acid cycle enzymes in *Haemonchus contortus* larvae and rat liver. *Comp. Biochem. Physiol.* **23**, 335–359.

Williamson, J. R., Cheung, W. Y., Coles, H. S., and Herczeg, B. E. (1967a). Glycolytic control mechanisms. IV. Kinetics of glycolytic intermediate changes during electrical discharge and recovery in the main organ of *Electrophorus electricus*. *J. Biol. Chem.* **242**, 5112–5118.

Williamson, J. R., Herczeg, B. E., Coles, H. S., and Cheung, W. Y. (1967b). Glycolytic control mechanisms. V. Kinetics of high energy phosphate intermediate changes during electrical discharge and recovery in the main organ of *Electrophorus electricus*. *J. Biol. Chem.* **242**, 5119–5124.

Wohlschlag, D. E. (1961). Growth of an Antarctic fish at freezing temperatures. *Copeia* pp. 11–18.

Wohlschlag, D. E. (1964). Respiratory metabolism and ecological characteristic of some fishes in McMurdo Sound, Antarctica. *In* "Biology of the Antarctic Seas" (M. O. Lee, ed.), Vol. I, pp. 33–62. Am. Geophys. Union, Washington, D. C.

Wróblewski, F., and Gregory, K. F. (1961). Lactic dehydrogenase isozymes and their distribution in normal tissues and plasma and in disease states. *Ann. N. Y. Acad. Sci.* **94**, 912–932.

Wuntch, T., Chen, R. F., and Vesell, E. S. (1970). Lactate dehydrogenase isozymes: Kinetic properties at high enzyme concentrations. *Science* **167**, 63–65.

# 3

## FREEZING RESISTANCE IN FISHES

*ARTHUR L. DeVRIES*

## I. INTRODUCTION

The occurrence of a large and varied fish fauna in the oceans of the polar regions illustrates how successfully fishes have been able to adapt to extremes of environmental stress. The mechanisms of cold adaptation which permit survival at low temperatures fall into two general categories: (1) mechanisms which permit survival per se at near-freezing temperatures and (2) mechanisms which lead to cold-adapted rates of activities in such physiological functions as respiration and growth. In this essay a particularly important example of the first class of cold adaptation mechanisms is considered, namely, the means by which fishes avoid freezing under environmental conditions where ice formation in the body fluids would most likely be favored. The analysis of this phenomenon will stress: (1) the nature of low temperature stresses in freshwater and marine habitats and (2) the varied adaptive responses—both behavioral and biochemical—which offer fishes avenues of escape from injury resulting from freezing.

## II. FREEZING AVOIDANCE IN FRESHWATER FISHES

### A. Freezing of Freshwater Streams and Lakes

With the onset of winter the temperatures of many temperate streams quickly drop to 0°C and freezing occurs at the surface. Once the water is frozen, heat exchange between the water and the colder air is greatly diminished and the temperature of the water below the ice remains near its freezing point. In cold weather it is only in the shallow parts of the streams that solid ice is likely to form all the way to the bottom. Fishes usually avoid these shallow habitats and spend their winters in the deep pools of shallow streams or in large rivers (Nikolsky, 1963).

The winter temperature regime of freshwater lakes is considerably different from that of streams. In a lake the entire water column cools to 4°C before freezing takes place. This phenomenon is explained by the fact that freshwater has a maximum density at 4°C, and in a lake which is being cooled from the surface the dense water sinks, thus setting up a convection system which mixes the entire lake. Only after the entire water column has reached 4°C does the surface water cool below freezing. Once the surface is frozen, heat loss from the water is greatly diminished because of the insulating properties of ice, and the temperature of the water column of most lakes remains near 4°C except immediately beneath the surface of the ice. It is only in the long severe winters of the mountainous temperate and polar regions that shallow lakes freeze to the bottom as a result of sustained heat loss from the surface.

### B. Freezing Avoidance through Habitat Selection

Most freshwater fishes are in little danger of freezing because the freezing point of freshwater is nearly 0.5°C above the freezing point of their body fluids. The freezing points of serum from freshwater fishes range from −0.50° to −0.65°C (Prosser and Brown, 1961; Black, 1957). Thus, even when the temperature of a stream or lake drops to its freezing point, there is still a comfortable margin between the freezing point of the fish and that of the water. Only in lakes which freeze to the bottom do fishes encounter conditions where they are likely to freeze. In these environments some fishes such as the Arctic black fish, *Dallia pectoralis*, and the crucian carp, *Carassius carassius*, overwinter in an inactive state and avoid freezing by burrowing into the warmer mud at the bottom of the frozen lake (Nikolsky, 1963). In some cases these fishes may be frozen into the mud or even into a lump of ice, yet they remain alive as long as their body fluids do not freeze (Kalabukhov, 1956) provided they are capable of anaerobic metabolism.

## III. FREEZING AVOIDANCE IN MARINE FISHES

### A. Freezing of Marine Environments

The range of temperatures encountered in the sea is notably small compared with the temperature extremes found in terresterial environments. The narrow temperature range of oceanic water results from continual mixing of the oceans and the high specific heat of water. The highest ocean temperatures, approximately 30°C, occur near the equator; the lowest temperatures, −1.7 to −1.9°C, which approximate the freezing point of seawater (which varies with salinity) occur in the high latitudes of the polar regions. In the coastal regions of the north Atlantic Ocean, freezing conditions also occur but only for short periods during the winter. In the Arctic, freezing conditions may last for most of the year and result in the formation of very thick ice. However, even in the high Arctic, summers are warm enough so that water temperatures rise, warming the surface waters, and some of the ice melts.

Only in parts of the Antarctic Ocean can freezing conditions be found throughout most of the year. Two such areas are the Ross Sea and the Weddell Sea which extend geographically into the high latitudes of the Antarctic Continent. The weather in these regions is cold throughout most of the year, and as a result the temperatures of these bodies of water are at their freezing points for long periods of time. These environmental conditions promote formation of a thick ice cover. Only for a brief period during the late austral summer, when solar radiation is at its greatest, does the seawater temperature in these regions rise slightly above the freezing point and a limited amount of melting occurs.

At these low water temperatures poikilothermic organisms are faced with the possibility of freezing. Invertebrates maintain body fluids which are slightly hyperosmotic to seawater. Consequently, these organisms are in little danger of freezing unless they become trapped in masses of ice crystals which eventually freeze into solid ice (Dayton *et al.*, 1969). In contrast to the invertebrates, most fishes have body fluids which are hypoosmotic to seawater and thus freeze at temperatures above the freezing point of seawater. The blood serum freezing points for a wide variety of marine fishes range from −0.5 to −0.8°C (Black, 1951) with the exceptions of *Latimeria chalumnae* and the elasmobranchs, whose sera are isomotic to seawater (Pickford and Grant, 1967; W. T. W. Potts and Parry, 1964).

## B. Survival by Means of Avoidance of Ice-Laden Seawater

### 1. ARCTIC FISHES

Scholander *et al.* (1957) described two groups of fishes inhabiting the fjords of northern Labrador which spend much of their lives at the freezing point of seawater. One group, consisting of *Boreogadus saida, Lycodes turneri, Liparis koefoedi, Gymnacanthus tricuspis,* and *Icelus spatula,* inhabits only the deep bottom water (200–300 meters) of the fjords where water temperatures are uniformly −1.7°C throughout the year (Fig. 1). The blood serum freezing points of members of this deep-water group range from −0.9°C in *L. koefoedi* to −1.0°C in *B. saida.* Since the water temperature remains a constant −1.7°C throughout the year, it is most likely that these fishes spend their entire lives with their body fluids supercooled by about 0.8°C. If members of this deep-water fauna are brought to the surface and put into ice-laden seawater at a temperature of −1.7°C, most of them immediately freeze. Interestingly, some tomcod, *B. saida,* can survive contact with ice at −1.7°C despite the fact that their blood freezes at −1.0°C. The basis of freezing resistance in these cases is not known.

Some shallow-water inhabitants of the Arctic Ocean avoid freezing by leaving the saltwater environment when water temperatures approach 0°C. The Arctic char, *Salvelinus alpinus,* migrates to freshwater streams and lakes where freshwater temperatures are always well above the freezing point of their body fluids (Andrews and Lear, 1956).

### 2. ANTARCTIC FISHES

The deep water of McMurdo Sound has a mean annual temperature of −1.86°C, and the temperature varies with season by only 0.2°C (Littlepage, 1965). DeVries (1970) described three Antarctic fishes which spend their entire lives in a supercooled state at depths of 500–600 meters in this Sound. Two of these fishes, a zoarcid, *Rhigophila dearborni,* and a liparid, *Liparis* sp., have blood serum freezing points of −1.54 and

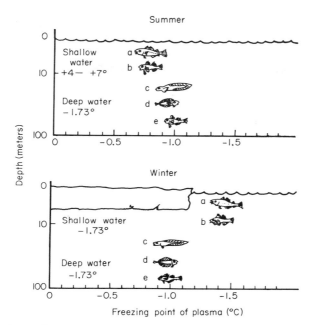

**Fig. 1.** Plasma freezing points of shallow-water fishes (a, *Gadus ogac;* b, *Myoxocephalus scorpius*) and benthic fishes (c, *Lycodes turneri;* d, *Liparis koefoedi;* e, *Gymnacanthus tricuspis*) in summer and winter. The position of the fishes on the abscissa indicates the freezing point of their plasma (modified from Scholander *et al.,* 1957).

−0.9°C (see Table I). Their body fluids are supercooled by 0.4° and 1.0°C, respectively. When caught in traps during the winter, these fishes cannot be raised through the ice-laden surface waters without freezing. However, in the summer when ice crystals are absent in the surface water because the water temperature is about 0.1°C warmer, these species can be raised to the surface without freezing. During the summer if these fishes are held at their environmental temperature of −1.86°C and a small quantity of fine ice crystals is added to their water, they freeze immediately. Another deep-water benthic fish, *Trematomus loennbergi*, whose serum freezes at −1.7°C, has also been observed to occasionally freeze when raised through the ice-laden surface water. In addition, when this species is maintained in an aquarium at its environmental temperature of −1.86°C it survives poorly. However, if this species is kept at −1.7°C, a temperature at which no ice formation occurs in refrigerated aquaria, no deaths from freezing are observed (Wohlschlag, 1964).

**Table I**

Freezing Points and Concentrations of Inorganic and Organic Solutes in the Sera of Several Fishes Inhabiting Cold Temperate, Arctic and Antarctic Waters

| Location / Species | Environmental parameters | Freezing point (°C) | Sodium (mmoles/liter) | Chloride (mmoles/liter) | Freezing point depression due to NaCl[a] (%) | Nonprotein nitrogen[b] (mg/100 ml of sera) | Urea nitrogen (mg/100 ml of sera) | α-Amino nitrogen (mg/100 ml of sera) | Carbohydrate (equivalents of glucose) (mg/100 ml of sera) | References |
|---|---|---|---|---|---|---|---|---|---|---|
| **Temperate** | | | | | | | | | | |
| *Fundulus heteroclitus* | 8–15 days acclimation at −1.5° | −0.83 | 208 | 171 | 78 | 33 | 57 | | 348[c] | Umminger, 1969a,b |
| **Arctic** | | | | | | | | | | |
| Summer | | | | | | | | | | |
| *Myoxocephalus scorpius* | 4–7° | −0.79 | | 200 | 86 | | | | | Scholander et al., 1957 |
| *Gadus ogac* | 4–7° | −0.79 | | 200 | 86 | | | | | Scholander et al., 1957 |
| Winter | | | | | | | | | | |
| *Myoxocephalus scorpius* | −1.7° (ice) | −1.25 | 216 | 234 | 61 | 130 | 40 | 15 | | Gordon et al., 1962 |
| *Gadus ogac* | −1.7° (ice) | −0.94 | 216 | 243 | 83 | 400 | 70 | 25 | | Gordon et al., 1962 |
| *Gadus ogac* | −1.7° (ice) | −1.47 | | 250 | 58 | | | | | Scholander et al., 1957 |
| **Antarctic** | | | | | | | | | | |
| **(Signy Island)** | | | | | | | | | | |
| Summer | | | | | | | | | | |
| *Notothenia neglecta* | | −0.92 | 248 | 212 | 85 | | 102 | | 470[d] | Smith, 1970 |
| Winter | | | | | | | | | | |
| *Notothenia neglecta* | −1.7° (ice) | −1.08 | 259 | 242 | 79 | | 221 | | 970[d] | Smith, 1970 |
| *Notothenia rossii* (Balleny Islands) | −1.7° (ice) | −1.06 | 237 | 238 | 77 | | 116 | | 600[d] | Smith, 1970 |
| Summer | | | | | | | | | | |
| *Notothenia kempi* | 1° | −0.84 | 191 | 191 | 77 | 120 | | | 121[e] | DeVries, 1970 |
| *Notothenia larseni* | 1° | −1.52 | 202 | 202 | 46 | 136 | | | 279[e] | DeVries, 1970 |
| **(Antarctic Peninsula)** | | | | | | | | | | |
| Summer | | | | | | | | | | |
| *Notothenia gibberifrons* | 0.0–1.0° | −1.89 | 219 | 219 | 39 | | | | | DeVries, 1972 |

| Species | Condition | Freezing point | | | | | | | Reference |
|---|---|---|---|---|---|---|---|---|---|
| Nothenia coriiceps | 0.0–1.0° | −1.61 | | 198 | 42 | | | | DeVries, 1972 |
| Chaenocephalus aceratus | 0.0–1.0° | −1.35 | | 209 | 53 | | | | DeVries, 1972 |
| Champsocephalus gunnari | 0.0–1.0° | −1.00 | | 168 | 56 | | | | DeVries, 1972 |
| Trematomus bernacchii | 0.0–1.0° | −1.81 | | 238 | 45 | | | | DeVries, 1972 |
| (McMurdo Sound) Summer and winter | | | | | | | | | |
| Liparis sp. | Deep water −1.86° (no ice) | −0.92 | | 237 | 93 | 54 | | | DeVries, 1970 |
| Rhigophila dearborni | Deep water −1.86° (no ice) | −1.52 | | 233 | 52 | 60 | | 88[e] | DeVries, 1970 |
| Trematous loennbergi | Deep water −1.86° (no ice) | −1.83 | | 233 | 43 | 274 | | | DeVries, 1970 |
| Trematomus bernacchii | Deep water −1.86° (no ice) | −1.87 | | 254 | 46 | 343 | | 587[e] | DeVries, 1970 |
| Trematomus hansoni | Deep water −1.86° (no ice) | −1.92 | | 258 | 46 | 375 | | 594[e] | DeVries, 1970 |
| Trematomus bernacchii | Shallow water −1.90° (ice) | −1.98 | | 254 | 44 | 481 | 52 | 880[e] | DeVries, 1970 |
| Trematomus hansoni | Shallow water −1.90° (ice) | −2.01 | | 259 | 44 | 480 | 12 | 838[e] | DeVries, 1970 |
| Trematomus borchgrevinki | Shallow water −1.90° (ice) | −2.07 | 274 | 235 | 42 | 504 | 61 | 831[e] | DeVries, 1970 |
| Trematomus hansoni | 30 days acclimation at +2° | −1.68 | | 206 | 42 | 345 | 20 | 625[e] | DeVries, 1968 |

[a] For calculations of the percentage of the freezing point depression resulting from sodium chloride, it was assumed that sodium was present in the sera at the same concentration as that of chloride in cases where sodium was not experimentally determined.

[b] Determinations were made on 10% trichloroacetic acid filtrates.

[c] Glucose was determined by glucose oxidase method.

[d] Reducing sugar by the method of Hagedorn and Jensen using glucose as a standard.

[e] Determinations for carbohydrate were made on 10% trichloroacetic acid filtrates using the phenol–sulfuric acid method in which polymers are hydrolyzed into their constituent residues. Therefore, these values represent both free hexoses as well as the hexoses present in the glycoproteins which are soluble in trichloroacetic acid. Glucose was used as a standard.

It is clear that as long as the deep-water Arctic and Antarctic fishes remain in their natural habitats where no ice crystals are present, the state of slight supercooling of their body fluids is stable enough to permit survival. In the past several years there have been several studies concerned with the deep-water fishes of McMurdo Sound (Wohlschlag, 1964; DeVries, 1968), and no one has ever captured these fishes in the shallow waters near shore. Thus it appears that these fishes are restricted to the ice-free deep water. Ice formation does not occur even at a relatively shallow depth of 200 meters because of the effect of hydrostatic pressure which at this depth is sufficient to lower the freezing point of water by approximately 0.15°C.

The fact that the deep-water fishes of the Arctic and Antarctic have not evolved a mechanism for prevention of freezing but rather survive only by avoiding ice in the deep water leads one to suggest that their adaptation to these cold environments is incomplete. Studies by Wohlschlag (1964) which have shown that the oxygen consumption of the Antarctic zoarcid fishes at environmental temperatures is much lower than that of the *Trematomus* fishes also indicate that the adaptation to the cold in the case of this deep-water fish is slight. In addition, it is worth noting that both the zoarcid and liparid fishes have affinities in the Arctic and temperate oceans (Wohlschlag, 1964) and that their appearance in the cold waters of the Antarctic may have taken place recently.

## C. Physiochemical Avoidance of Freezing in Marine Fishes

### 1. TEMPERATE FISHES

There are few studies concerned with the freezing of temperate marine fishes in their natural habitats. The reason for this is, in part, that water temperatures of the temperate oceans are usually well above freezing for most of the year and that well defined near-freezing marine habitats exist only in the shallow waters near the coastlines for short periods of time during the coldest part of the winter. Even then, thin ice and winter storms usually prevent access to likely areas where studies might be carried out.

Despite such short periods of low water temperatures in the temperate regions a few fishes have been studied such as the flounder, *Pseudopleuronectes americanus*, which spawns in the bottom waters of the Mystic River estuary in eastern Connecticut that reach −0.8°C during cold winters. Freezing experiments indicate that death occurs in this fish between −1.0° and −1.5°C, accompanied by ice formation in its tissues (Pearcy, 1961). The blood serum of such fishes freezes at −1.15°C which is in agreement with the temperature at which ice forms in the tissues. In the summer, however, the serum freezes at −0.63°C; this value is well within the range of serum freezing points of −0.5° to −0.8°C found in most teleosts (Black, 1957).

Both Scholander et al. (1957) and Umminger (1969a) have studied the killifish, *Fundulus heteroclitus*, at subzero temperatures. Although this fish can survive in a supercooled state at −1.5°C for long periods of time in the laboratory, it cannot do so if it comes into contact with ice for extended periods of time. Recent laboratory observations by Dr. B. L. Umminger have shown that this fish, despite the supercooled state of its body fluids, survives in aquaria containing large pieces of ice at −1.5°C because it avoids contact with the ice. This observation suggests that in their natural winter environment where ice is present on the surface of the water, the behavioral avoidance of ice may have survival value for *F. heteroclitus* (Umminger, 1970). Umminger (1969a,b) has made a thorough study of the serum of *F. heteroclitus* acclimated to temperatures between 20° and −1.5°C. The freezing point of the sera of killifish held at 20°C was −0.66°C. After acclimation for several days to −1.5°C, the freezing point dropped to −0.88°C. Although this slight increase in blood plasma osmolality in response to low temperatures does serve to lower the freezing point of the blood slightly, it is not clear whether the magnitude of change is great enough to have any significance in the prevention of freezing in nature. Studies of this fish in its natural winter habitat would most likely clarify this point.

## 2. ARCTIC FISHES

Of the two groups of Arctic fishes studied by Scholander et al. (1957), one group (composed of the fishes, *Gadus ogac* and *Myoxocephalus scorpius*) inhabits only the shallow waters of the Labrador fjords and experiences much warmer temperatures (4°–7°C) during the summer than the other group inhabiting the deep water (see Section III, B, 1). The blood plasma freezing points of −0.8°C for these shallow-water sculpins and fjord cod captured during the summer (Scholander et al., 1957; Gordon et al., 1962) are well within the range of those found in common temperate marine teleosts (Black, 1951). However, when winter water temperatures become low enough so that considerable ice formation occurs, these fishes increase the osmotic concentration of their blood until it freezes at slightly above the freezing point of the seawater in which they live. The summer and winter plasma freezing point data for these fishes are given in Table I. From these data it is clear that the response of the sculpins and fjord cod to low winter temperatures is one of lowering the freezing point of their body fluids as protection against freezing. It can be seen from Table I that the plasma freezing points of these fishes were lower during the winter of 1955 than during the winter of 1957. These differences between the freezing points are real according to Gordon et al. (1962) and probably reflect a response to lower temperatures of the first winter season. For this reason knowledge of the thermal history of the fishes utilized in both of these studies would have been of great interest. Raschack (1969) reported that the freezing points of the plasma of *M. scorpius* inhabiting the brackish water (salinity 12‰) of Kiel Bay in the Baltic Sea drops from —0.64° to —0.86°C in response to low winter temperature (—0.5°C) and that when this fish is acclimated to seawater of a salinity of 35‰ at —0.5°C the freezing point drops to —1.25°C.

### 3. Antarctic Fishes

As stated previously (Section III, A), the coldest waters of the world are found near the Antarctic Continent. Despite the low temperatures of the Antarctic Ocean, invertebrate organisms and fishes are relatively abundant (Dearborn, 1965; Norman, 1940; Andriashev, 1965, 1970) and the Antarctic fishes are known to have relatively high levels of metabolism as well as fast rates of growth (Wohlschlag, 1964; Hureau, 1963).

The waters of McMurdo Sound, the southernmost part of the Ross Sea, afford an excellent opportunity to carry out studies on cold-adapted fishes for several reasons: (1) the extensive ice cover forms a stable platform from which both physical and biological studies can be carried out during most of the year; (2) the waters of McMurdo Sound range from a few meters in depth near the shore to depths of a 1000 meters in the middle, forming a variety of habitats, some associated with the ice; and (3) the water temperatures of the sound are extremely cold and stable. The average water temperature is —1.87°C and temperature variation, either with season or depth, is of the order of only 0.2°C (Tressler and Ommundsen, 1962; Littlepage, 1965). The months inclusive of December through April are designated as the hydrographic summer season, and water temperatures during this period cluster around a mean —1.80°C. During the winter season, which includes the months of May through November, water temperatures cluster around a mean —1.90°C. Even though the temperature difference between the summer and winter hydrographic seasons is small, it is nevertheless important because McMurdo Sound water freezes at —1.90°C. During the winter months sea ice growth is rapid because of the cold air temperatures, and ice 1.5 meters in thickness is common in July. At certain times during the winter months, the presence of ice crystals is observed only in the 33 meters of surface water. Their absence in deeper water can be explained by the relationship which exists between the freezing point of water and pressure. For each 10-meter increase in depth, the freezing point of pure water decreases 0.0075°C (Montgomery, 1957).

In the surface water two types of ice crystals have been observed: (1) extremely small crystals which are invisible singly but, when present in large numbers, give the water the appearance of being filled with tiny reflective "needles"; and (2) large ice platelets up to 10 cm in diameter that form on lines suspended in the water (Littlepage, 1965). Aggregations of these same platelets are found as large irregular masses on the ocean bottom and have been quite appropriately termed "anchor ice" (Pearse, 1962). This anchor ice is found adhering loosely to the ocean floor to depths of 33 meters (Dayton *et al.*, 1969).

Over the deep water, ice platelets form a loosely aggregated layer on the underside of the solid sea ice. This layer begins forming in July, and by late November it is 3–4 meters thick. In December a small increase in seawater temperature causes the layer to melt, and by January it has disappeared.

From the above observations it is apparent that winter conditions in the surface waters of McMurdo Sound would be conducive to ice formation in most fishes. However, several fishes belonging to the family Nototheniidae are closely associated with this icy environment and, in fact, some even utilize it as part of their habitat. DeVries (1970) has often observed *T. bernacchii* and *T. hansoni* resting on masses of anchor ice, and *T. borchgrevinki*, a cryopelagic fish, swimming in the ice-laden surface waters beneath the annual ice. The latter species has also been observed to swim up into the holes and tunnels of the sub-ice platelet layer. Presumably the holes and tunnels within the matrix of this layer are a safe refuge, for only rarely has the Weddell seal, which preys on this fish, been observed to venture into the ice platelet layer and thrash about in search of fish. Frequently this fish rests on the platelets, most likely waiting for zooplankters to drift by.

*Trematomus bernacchii* and *T. hansoni* inhabit the deep as well as the shallow waters of the Sound; thus, these fishes can be studied in ice-laden and ice-free habitats at similar temperatures. The habitats of the fishes studied by DeVries (1968, 1970) are shown in Fig. 2, and the freezing points of their blood sera are given in Table I. As one would expect, *T. borchgrevinki*, living in waters where most ice formation occurs, has the lowest freezing point. The fact that populations of

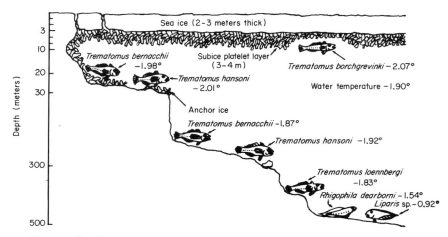

**Fig. 2.** Sketch of McMurdo Sound showing the habitats and blood serum freezing points of several of the *Trematomus* fishes and two deep-water fishes (a zoarcid and liparid) during the winter. The freezing points are in °C (redrawn from DeVries, 1970a, in "Antarctic Ecology").

*T. bernacchii* and *T. hansoni* living in the anchor ice zone have lower freezing points than individuals of these same two species living in the deep water of the Sound is of interest because the only readily apparent differences between the deep and shallow sites are ones of ice and pressure. The temperature differences between these sites appear to be insignificant (Littlepage, 1965); however, they should not be completely ignored. It should be noted that if the serum freezing points of the deep-water populations were measured under 30 atm of pressure, which is the hydrostatic pressure of their habitat at 300 meters, their freezing points would be identical to those of the shallow-water populations.

The waters of McMurdo Sound never rise above −1.5°C, even in the height of the austral summer. However, farther north the intrusion of sub-Antarctic waters into the northern part of the Ross Sea results in a slight warming. Near the Balleny Islands, which are located 1200 km north of McMurdo Sound near the Antarctic Circle, the temperature of the water column at a depth of 300 meters was +1.0°C during the summer of 1964 (DeVries, 1968). The serum freezing points of *Notothenia larseni* and *N. kempi* captured at this depth were −1.52° and −0.84°C, respectively (Table I). Because of the inaccessibility of this region to ships during the winter no seasonal studies have been attempted with these fishes.

The fishes inhabiting the waters adjacent to the Antarctic Peninsula, which is across the continent from the Ross Sea, experience water temperatures 2°C higher than those inhabiting McMurdo Sound. Serum freezing points have been determined for several of these summer fishes, including members of the Antarctic families, Nototheniidae and Chaenicthyidae (DeVries, 1969). In general, the freezing point data for these fishes (Table I) indicate that Antarctic fishes living in warmer water have higher serum freezing points than those of fishes living in the cold waters of McMurdo Sound. However, two exceptions are apparent, one being *Notothenia gibberifrons* and the other being *Trematomus bernacchii*. The latter fish also inhabits the cold water of McMurdo Sound, and it is evident that the warm temperatures (+1°C) of the waters of the Antarctic Peninsula have little effect on its freezing resistance since only a small rise in its serum freezing point is observed (Table I). Even in the case of *T. hansoni*, a closely related inhabitant of McMurdo Sound, warm acclimation for 30 days at +2°C causes only a 0.2°–0.3°C rise in its serum freezing point (Table I), which is in contrast to a 0.8°C rise observed with some of the Arctic fjord cod in response to warm summer temperatures (+4° to +7°C) (Scholander *et al.*, 1957; Gordon *et al.*, 1962). Smith (1970) has studied freezing resistance in Antarctic fishes inhabiting the waters near Signy Island, which is located in the extreme northern part of the Weddell Sea. In the summer, when water temperatures are warmer than those of McMurdo Sound, the blood serum freezing points of *Notothenia neglecta* and *N. rossii* were −1.1°C, almost 1.0°C higher than those of the *Trematomus* fishes of McMurdo Sound. During the winter, however, the serum freezing point of *N. neglecta* decreases only by 0.1°C despite the fact that the water temperatures are low enough so that considerable ice formation occurs.

## 4. SERUM FREEZING POINTS AS ESTIMATES OF RESISTANCE TO FREEZING

To determine whether serum freezing points are good estimates of the resistance to freezing for the Antarctic fishes, DeVries (1969, 1972) carried out experiments in which specimens of several of the species studied were subjected to progressively lower temperaturs in the presence of ice. The temperature at which each species froze was found to be the same as, or only slightly lower than, the serum freezing point (Table II), indicating that the serum freezing points are good estimates of freezing resistance for these fishes. These studies indicate that only a small degree of supercooling is possible in the presence of ice. Pearcy (1961) noted that the serum freezing points were in accord with the temperature at which *P. americanus* died from freezing in the presence of ice. Brett and Alderdice (1958) also reported that cultured chum salmon, *Oncorhyn-*

**Table II**

Comparison of Lower Incipient Lethal Temperatures in the Presence of Ice and
Freezing Point of Blood Sera for Several Antarctic Fishes

| Genus and species | Lower incipient lethal temperature (°C)[a] | Freezing point of blood serum (°C) |
|---|---|---|
| *Trematomus borchgrevinki* | −2.1 | −2.07 |
| *Trematomus bernacchii* | −2.0 | −1.98 |
| *Trematomus loennbergi* | −1.8 | −1.83 |
| *Rhigophila dearborni* | −1.6 | −1.52 |
| *Chaenocephalus aceratus* | −1.2 | −1.35 |
| *Notothenia coriiceps* | −1.6 to −1.7 | −1.61 |
| *Notothenia nudifrons* | −1.5 to −1.6 | −1.49 |

[a] Death was preceded by convulsions, and the temperature at which the fish lay
motionless on its back with operculars flared was taken as the temperature of freezing.

*chus keta* and sockeye salmon, *O. nerka,* are incapable of existing at temperatures below the freezing point of their blood and die from freezing at the same temperature at which their blood freezes.

On the other hand, the freezing points of plasma obtained from *M. scorpius* and *G. ogac* captured in the fjords of Labrador during the winter indicate that they are supercooled by 0.2°–0.4°C. Scholander *et al.* (1957) showed that some of these fishes could survive contact with ice for long periods of time at water temperatures of −1.73°C, the freezing point of seawater there. However, some of these same fishes, with identical plasma freezing points, were susceptible to freezing in the presence of ice. No one has offered a satisfactory explanation as to why some fish freeze and others do not in spite of their having identical plasma freezing points. Eliassen *et al.* (1960) found that the plasma freezing point of a boreal sculpin, *Cottus scorpius,* is −0.6°C when captured in 10°C water, but when acclimated at −1.5°C the plasma freezing point is reduced to −0.9°C. After acclimation at −1.5°C for 3 weeks *C. scorpius* is capable of surviving at water temperatures of −1.7°C in the presence of ice. However, if acclimation was limited to only a few days at −1.5°C, then extreme susceptibility to freezing was observed when these fish came into contact with ice at −1.85°C. Smith (1970) reported that *N. neglecta* spends the winter in ice-laden seawater at a temperature of −1.7°C, yet it does not freeze even though its blood serum freezing point is −1.1°C.

In the case of the Arctic fishes and two Antarctic fishes, *N. neglecta* and *N. rossii,* the freezing points which were reported by the investigators do not reflect the temperature at which the fishes freeze in the presence of ice. These are the only cases reported in the literature where fishes are able to survive in the presence of ice with their body fluids in a supercooled state.

In view of the fact that the integument of a fish does not form a barrier to the propagation of ice crystallization (Scholander et al., 1957), supercooling in the presence of ice is difficult to explain. Either the body fluids contain some solute which confers stability on the supercooled state or the method for determination of the freezing point does not accurately reflect the temperature at which ice formation can occur in the plasma leading to freezing of the fish. This latter point will be discussed in Section III, E, 6.

## D. Role of Small Solutes in Freezing Avoidance

### 1. INORGANIC IONS

In temperate teleosts sodium chloride is the principal electrolyte of the serum, and it is responsible for 80–90% of the blood osmolality. Potassium, calcium, urea, and the free amino acids account for much of the remainder. When temperate and boreal marine teleosts encounter low water temperatures, the concentration of sodium chloride in the blood serum increases (Eliassen et al., 1960; Pearcy, 1961; Woodhead and Woodhead, 1959; Gordon et al., 1962; Raschak, 1969; Umminger, 1969a). The extent of the increase in this electrolyte varies among species. For instance, when F. heteroclitus is transferred from 20° to −1.5°C water, the concentration of sodium chloride in the plasma increases by only 13% (Umminger, 1969a), whereas in P. americanus it increases by 18% (Pearcy, 1961). In M. scorpius taken from Kiel Bay in the Baltic Sea where the wintertime water temperatures are −0.5°C, the electrolyte content of the plasma is 20% over that found in the summer when water temperatures are around 10°C (Raschack, 1969). In the boreal cod, Gadus callarias, the chloride level in plasma taken from specimens captured at −1.5°C is 15% over that of those captured at +15°C (Eliassen et al., 1960). In the Antarctic fish, N. rossii, the serum concentration of sodium chloride in the winter is 15% over that of the summer (Smith, 1970).

With most temperate fishes the increases in plasma osmolality associated with low temperatures only partially result from increases in sodium chloride. Pearcy (1961) reported that in P. americanus sodium chloride accounts for 83% of the serum osmolality in the summer, whereas in the winter it accounts for only 57%. In other words, when the freezing point depression of the plasma is increased from 0.63° to 1.10°C, about

0.4°C of the increase results from solutes other than sodium chloride. In *M. scorpius* taken from the brackish water of Kiel Bay in the Baltic Sea the increase in plasma osmolality (−0.64° to −0.86°C) associated with low temperatures (−0.5°C) also only partially results from increases in electrolytes, and the remainder is attributed to nondissociated organic compounds. In *Taurulus bubalis*, a long-spined sea scorpion, the increase in plasma osmolality associated with low temperatures is not as great as that of *M. scorpius* and results exclusively from inorganic electrolytes (Raschack, 1969).

Increases in the serum levels of sodium chloride in response to cold acclimation in many temperate and boreal fishes have often been attributed to the breakdown of their osmoregulatory ability (Woodhead, 1964; Doudoroff, 1945). However, with many Arctic and Antarctic fishes living in permanently near-freezing habitats, the levels of sodium chloride are higher than those of temperate fishes and in fact show no variation (DeVries, 1968), or only a little variation with season (Smith, 1970). In addition, the levels of sodium chloride in the blood of the Arctic fishes *M. scorpius* and *G. ogac* show a natural seasonal variation with increases observed in the levels of sodium chloride as well as in nondissociated organic compounds during the winter when water temperatures are low (Scholander *et al.*, 1957; Raschack, 1969). With these Arctic fishes sodium chloride accounts for 87% of the blood osmolality in the summer, while in the winter it accounts for only 62% in the sculpin and 79% in the fjord cod (Gordon *et al.*, 1962). In addition, analyses for potassium ion during winter indicate that it is not present at concentrations much higher than those found in the plasma of temperate marine fishes which inhabit warmer waters. Thus, since the proportion of the total blood osmolality accounted for by these electrolytes is much less at low temperatures it seems unlikely that osmoregulatory failure is involved.

Studies of fishes in the Antarctic indicate that sodium chloride accounts for slightly less than half of the serum osmolality in those fishes showing the greatest resistance to freezing, while it accounts for slightly more than half in those showing only a moderate resistance to freezing (Table I). For instance, in *T. borchgrevinki*, which lives in the coldest water where ice is most abundant, sodium chloride accounts for only 42% of the serum osmolality. In the cases of *T. bernacchii* and *T. hansoni*, this electrolyte accounts for 44% when they inhabit the ice-laden shallow waters and 46% when inhabiting the deep waters where ice is absent. In *Chaenocephalus aceratus*, a hemoglobinless fish whose serum freezes at −1.3°C (Table I), sodium chloride accounts for 55% of the serum osmolality.

The concentrations of potassium, magnesium, and calcium ions have not been determined in the blood of the Antarctic fishes, except in the serum of *Notothenia neglecta* and *N. rossii* where potassium levels are not exceptionally high (Smith, 1970). In the Arctic sculpin and fjord cod the concentration of potassium ion is about the same as that for teleosts living in warmer temperate waters (Gordon *et al.*, 1962). One would not expect high levels of these electrolytes because of the importance of their ratios in intermediary metabolism and in the propagation of electrical impulses along neurons (Mahler and Cordes, 1966; Prosser and Brown, 1961). Thus it is clear that the inorganic electrolytes account for a much smaller fraction of the serum osmolality in many of the polar and boreal fishes than they do in temperate and tropical fishes.

## 2. ORGANIC SOLUTES

Since inorganic ions have been found to account for so little of the plasma osmolality in some of the cold-adapted fishes of the temperate and polar regions, the presence of high concentrations of osmotically active organic solutes has been investigated (DeVries, 1968; Umminger, 1969b; Raschack, 1969). Even with the sera from cold-acclimated *F. heteroclitus*, where most of the increase in plasma osmolality results from inorganic ions, part of the increase resulted from elevated levels of free glucose. Umminger (1969b) speculated that the primary role of the 430% increase in the level of glucose is the prevention of spontaneous nucleation in the absence of ice. However, the high levels of glucose provide no protection against nucleation if external ice is encountered (Umminger, 1970). No conclusive evidence supporting this thesis has been put forward at this time. It has been suggested that the increase in serum osmolality not resulting from sodium chloride in *P. americanus* during the winter results from some organic solute; however, no compound has been identified (Pearcy, 1961). In the plasma from Arctic sculpins and fjord cod the concentrations of organic solutes commonly found in the blood of teleosts are not extraordinarily high. However, the levels of nonprotein nitrogen are two and four times higher in the sculpin and cod, respectively, than in temperate marine teleosts. Since the nonprotein nitrogen in the body fluids of most organisms can be attributed to small nitrogen containing compounds, it was reasonable for Gordon *et al.* (1962) to postulate that there would be more than enough solute to account for the high serum osmotic concentration if all of the nonprotein nitrogen was present in molecules containing only one nitrogen atom per molecule. However, despite their systematic analyses of the serum, the high level of nonprotein nitrogen could not be correlated with high serum levels of small nitrogen-containing compounds such as urea, free amino acids, purines, pyrimidines, and amines (Table I). Concentrations of other small solutes such as reducing sugars, alcohols, and simple lipids were not significantly high either.

Examination of the osmotic role played by the salts and organic compounds identified in the serum revealed that they could supply enough solute to account for only 68% of the wintertime serum osmotic concentration in the sculpin. It was speculated that the remaining 32% of the wintertime osmotic concentration resulted from an "antifreeze" compound whose presence was associated with the high nonprotein nitrogen level. However, despite exhaustive analyses of the serum, no compound with antifreeze properties was identified. As with the populations of *M. scorpius* in the fjords of northern Labrador, the concentration of electrolytes in the wintertime populations inhabitating Kiel Bay in the Baltic Sea accounts for only 73% of the plasma osmolality as contrasted to 78% in the summer. This wintertime increase in osmolality was thus interpreted as partly resulting from an increase in nondissociated organic compounds which, however, were not identified (Raschack, 1969).

In the Antarctic fishes high levels of nonprotein nitrogen have also

### Table III

Freezing Points of Whole and Dialyzed Sera from Several Antarctic Fishes

| | Freezing point (°C) | |
|---|---|---|
| Genus and species | Whole sera | Dialyzed sera[a] |
| *Trematomus borchgrevinki* | −2.07 | −0.60 to −0.70 |
| *Trematomus hansoni* | −2.01 | −0.58 to −0.63 |
| *Trematomus bernacchii* | −1.90 | −0.50 to −0.58 |
| *Notothenia gibberifrons* | −1.89 | −0.53 to −0.62 |
| *Notothenia coriiceps* | −1.61 | −0.48 |
| *Notothenia larseni* | −1.67 | −0.37 |
| *Notothenia nudifrons* | −1.49 | −0.29 |
| *Chaenocephalus aceratus* | −1.35 | −0.22 |
| *Champsocephalus gunnari* | −1.00 | −0.10 |
| *Notothenia kempi* | −0.84 | −0.03 |
| *Raja* sp. | −2.00 | −0.00 |

[a] The samples of sera were dialyzed 48 hr against running distilled water and contained no inorganic ions or low molecular weight organic solutes. The majority of the freezing point depression of the dialyzed sera results from the nondialyzable glycoproteins with antifreeze properties.

been found. As in the case of the Arctic fishes, these high levels are not resulting from elevated concentrations of urea or free amino acids (Table I). In the Antarctic fishes the majority of the nonprotein nitrogen has been shown to be associated with macromolecular solutes (DeVries and Wohlschlag, 1969; DeVries, 1970). For several Antarctic fishes a positive correlation exists between the concentrations of nonprotein nitrogen in the sera (Table I) and the depressions of the freezing point of dialyzed sera (Table III). This correlation is explained by the fact that the macromolecular solutes which depress the freezing point of dialyzed sera are nitrogen-containing compounds which are soluble in trichloracetic acid. It is quite possible that the high levels of nonprotein nitrogen found in the plasma of the Arctic sculpins and fjord cod during the winter season (Scholander et al., 1957) likewise reflect the presence of glycoproteins with antifreeze properties. This is an obvious area for further research.

## E. Role of Macromolecular Solutes in the Avoidance of Freezing

A new approach to the problem of determining the amount of serum osmolality resulting from small solutes such as inorganic ions and low molecular weight organic solutes was employed by DeVries (1972). This approach involved removal of the small serum components by dialysis. As an example, when the serum of the temperate black perch, *Embiotoca jacksoni*, which has a freezing point of −0.7°C, is dialyzed against distilled water for 48 hr, the freezing point rises to 0.01°C, indicating almost all the depression of the freezing point of whole serum results from low molecular weight solutes. In contrast to the perch, the serum of the Antarctic fish, *T. borchgrevinki*, has a freezing point of −2.1°C and after dialysis still has a freezing point of −0.6°C, which is about 30% of the total serum freezing point depression. This experiment indicates that 70% of the serum freezing point depression results from low molecular weight solutes such as sodium chloride, potassium, urea, and glucose and that the remainder results from macromolecular solutes present in the protein fraction of the serum.

Removal of the serum proteins through heat precipitation has shown that the majority of the solutes responsible for the depression of the freezing point of dialyzed serum are large molecular weight substances which have been identified as a group of glycoproteins that contain only two types of amino acid and two sugar residues (DeVries, 1969; DeVries et al., 1970). These compounds with antifreeze properties have been termed "freezing-point-depressing" glycoproteins. They are composed of alanine, threonine, $N$-acetylgalactosamine, and galactose (see Section III, E, 4).

### 1. CONCENTRATIONS OF GLYCOPROTEINS IN DIALYZED SERA OF ANTARCTIC FISHES

The dialysis technique has been used to estimate the fraction of the serum-freezing-point depression which results from the glycoproteins in the Antarctic fishes. In Table III freezing points are given for whole and dialyzed sera for several Antarctic fishes. The data show that those fishes whose sera have the lowest freezing points before dialysis also have the lowest freezing points after dialysis, thus indicating the presence of high concentrations of the glycoproteins. It should be noted that the freezing point of the serum from *N. kempi* is similar to that of most temperate marine teleosts and the high freezing point of its dialyzed serum indicates that almost no freezing-point-depressing glycoproteins are present. The serum of *T. borchgrevinki* has the lowest serum freezing point of any Antarctic fish and also has the greatest amount of glycoproteins present in its serum, as evidenced by the low freezing point of its dialyzed serum.

### 2. ISOLATION OF THE FREEZING-POINT-DEPRESSING GLYCOPROTEINS FROM SERA

The freezing-point-depressing glycoproteins have been isolated from the blood sera of several Antarctic fishes (DeVries, 1969) but have been studied in detail only in the case of *T. borchgrevinki* (DeVries et al., 1970). These glycoproteins also occur in the pericardial fluid but have not yet been isolated from the tissues (DeVries, 1972). The fact that the glycoproteins have not been isolated from the tissues does not mean that they are not present, because it is extremely difficult to separate small amounts of glycoproteins from the large amounts of lipid and cellular debris which are present in tissue extracts. The glycoproteins present in the sera of *T. borchgrevinki* have been isolated and purified using DEAE-cellulose column chromatography. A total of eight glycoproteins have been isolated and have been identified on the basis of analytical acrylamide gel electrophoresis (Fig. 3). These glycoproteins have been assigned numbers beginning from the cathode and proceeding toward the anode. Most of the glycoproteins have been recovered after the purification procedures as mixtures of glycoproteins 3, 4, and 5, and mixtures of 7 and 8. Measurements of the freezing points of aqueous solutions of the individual glycoproteins indicate that only glycoproteins

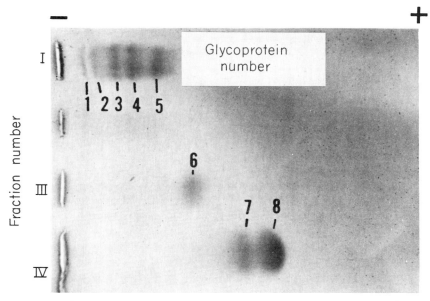

**Fig. 3.** Electrophoretogram of glycoproteins isolated from the sera of *T. borch-grevinki* using DEAE-cellulose chromatography. Electrophoresis was carried out at pH 8.6 using a sodium borate buffer and the acrylamide gel was stained for carbohydrate with concentrated sulfuric acid and α-naphthol (taken from DeVries *et al.*, 1970, *J. Biol. Chem.* **245**, 2901, with permission of the copyright owner).

1–5 depress the freezing point of water more than expected on the basis of the number of particles in solution and therefore form nonideal solutions. These glycoproteins are referred to as active glycoproteins. Solutions of glycoproteins 6, 7, and 8 form ideal solutions as in the cases of solutions of galactose and sodium chloride (Fig. 4).

## 3. FREEZING-POINT-DEPRESSING PROPERTIES OF THE GLYCOPROTEINS

The properties of a solution that depend upon the number of particles in solution and not on the kinds of particles are called its "colligative properties." These properties are a manifestation of a common phenomenon; i.e., solute molecules decrease the tendency of water molecules to escape from one phase to another or from one solution to another. The colligative properties of solutions are (1) the osmotic pressure, (2) the vapor pressure lowering, (3) the boiling point elevation, and (4) the freezing point depression. The relationship that exists between the freezing point and molal concentration of solutions of galactose and sodium chloride are shown in Fig. 4. The fact that the curve shown for

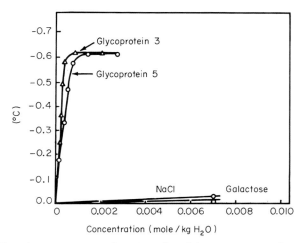

**Fig. 4.** Freezing points as a function of molal concentration for aqueous solutions of galactose, sodium chloride, and glycoproteins 3 and 5. Freezing points were determined using a Fiske osmometer.

sodium chloride has a slope almost twice that of galactose clearly illustrates that the colligative properties are dependent upon the number and not upon the kind of particles in solution. The fact that sodium chloride ionizes into two particles at the same molal concentration as galactose explains why the former has twice the effect on the freezing point of water as the latter, which is a much larger molecule.

The freezing points of aqueous solutions of glycoproteins 3, 4, and 5 have been determined at several concentrations and are shown in Fig. 5 (DeVries *et al.*, 1970). It should be noted that on a weight basis each glycoprotein has the same freezing point lowering capacity. On a weight basis, at low concentrations these three glycoproteins are slightly more effective than sodium chloride with regard to their capacity for lowering the freezing point of water. However, at high concentrations they are less effective.

Since the lowering of the freezing point of a solution is dependent upon the number of particles present in solution, a much more meaningful illustration of the freezing-point-depressing properties of the relatively large molecular weight glycoproteins can be made by plotting freezing points as function of molal concentrations. Such a plot shows that the glycoproteins are about 500 times more effective than galactose in depressing the freezing point of water (Fig. 4).

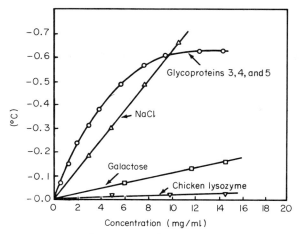

**Fig. 5.** Freezing points of aqueous solutions of chicken lysozyme, galactose, sodium chloride, and the glycoproteins which were isolated from the sera of *T. borchgrevinki*. One curve is shown for glycoproteins 3, 4, and 5 because on a weight basis they have the same freezing point depressing capacities. The freezing points were determined with a Fiske osmometer.

## 4. PHYSICAL AND CHEMICAL PROPERTIES OF THE FREEZING-POINT-DEPRESSING GLYCOPROTEINS

The molecular weight of a homogeneous unknown compound can theoretically be calculated from data from any one of the colligative properties of its solution provided no association occurs within the range of concentrations used for the determinations. If one calculates the molecular weight of the glycoproteins from the freezing point data in Fig. 5, a very low value of 15 g/mole is obtained. This value is not in accord with those shown in Table IV, which were obtained from sedimentation equilibrium determinations on the analytical ultracentrifuge and osmotic pressure measurements. Data from these methods indicate that glycoprotein 3 has a molecular weight of 21,500 g; glycoprotein 4, 17,000 g; and glycoprotein 5, 10,500 g. From the sedimentation equilibrium studies, osmotic pressure determinations and the fact that the glycoproteins will not pass through dialysis tubing, it is clear that they are relatively large molecules. Such data, however, give no information about the molecular size or shape of molecules. A molecular weight of 78,000 g has been determined for a mixture of the active glycoproteins, using Sephadex gel filtration (DeVries, 1968). Since the behavior of a molecule on Sephadex depends on its Stokes radius rather than on its

**Table IV**

Physical Properties of Glycoproteins Isolated from the Serum of *T. borchgrevinki*

| Sample | Molecular weight (g/mole) | | Shape and structure | |
| | Sedimentation equilibrium | Membrane osmometry | Viscosity[a] [n] (cm³/g) | Circular dichroism (24°) |
| --- | --- | --- | --- | --- |
| Glycoprotein 3[b] | 21,500 | | 22[c] | Random coil |
| Glycoprotein 4[b] | 17,000 | | | without |
| Glycoprotein 5[b] | 10,500 | 11,000 | | α-helical or |
| Glycoprotein 8 | 2,600 | | 5 | β structure |

[a] The intrinsic viscosity of β-lactoglobulin, a spherical molecule is 4 cm³/g.

[b] Glycoproteins 3, 4, and 5 will not pass through 3/32 in. Visking dialysis tubing.

[c] The viscosity was determined on a mixture of glycoproteins 3 and 4 at 0.5°C.

true molecular weight, the high value of 78,000 g indicates that these molecules have expanded structures. Studies of the viscosity of the glycoproteins yielded an intrinsic viscosity of 20 cm³/g for a mixture of glycoproteins 3 and 4. The intrinsic viscosity of a spherical molecule such as β-lactoglobulin is 4.0 cm³/g (Table IV). Thus the study on viscosity also indicates that the glycoproteins have expanded structures. Circular dichroism studies show that the glycoproteins are random coils lacking α-helical or β structure.

The chemical composition has been determined for each of the active glycoproteins and they contain only alanine, threonine, N-acetylgalactosamine, and galactose (Table V). All of the active glycoproteins have identical compositions, and these four residues account for about 96% of the total weight of the molecule. The structure is the same for all

**Table V**

Chemical Composition of Glycoproteins Isolated from the Sera of
*T. borchgrevinki* and *D. mawsoni*

| Constituents (residues/10,000 g glycoprotein) | *T. borchgrevinki* | | | | *D. mawsoni* | |
| | 3[a] | 4 | 5 | 8 | 3–5 | 7, 8 |
| --- | --- | --- | --- | --- | --- | --- |
| Threonine | 14.6 | 14.7 | 14.5 | 11.5 | 14.1 | 11.2 |
| Alanine | 32.5 | 32.2 | 32.6 | 25.2 | 31.4 | 25.0 |
| Proline | 0 | 0 | 0 | 6.55 | 0 | 5.5 |
| N-Acetyl-galactosamine | 14.7 | 14.4 | 14.2 | 14.0 | 14.0 | 13.6 |
| Galactose | 17.5 | 17 | 17 | 16.5 | 17.5 | 17 |

[a] Glycoproteins are designated by their electrophoretic band numbers.

the active glycoproteins and is one of a polypeptide of alanine and threonine to which disaccharides are attached. The disaccharides are composed of galactose and N-acetylgalactosamine and are linked to every threonine through a glycosidic linkage involving carbon-1 of N-acetyl-galactosamine. Studies of the primary structure of the active glycoproteins indicate that they are composed of a basic repeating unit which is shown in Fig. 6 (DeVries et al., 1970; DeVries, 1972). The inactive gly-coproteins, which account for 80% of the glycoprotein fraction of the serum, have roughly the same composition as the active glycoprotein with the exception that a small amount of proline is present. They are smaller in size and do not have antifreeze activity. The function of these inactive glycoproteins is not clear. The possibility exists that they might serve as intermediates in the biosynthesis of the active glycoproteins.

Analyses of serum glycoproteins isolated from *Dissostichus mawsoni*, *T. bernacchii*, *N. coriiceps*, *N. gibberifrons*, and *C. aceratus* indicate that their glycoproteins have compositions identical to those isolated from *T. borchgrevinki*. Preliminary evidence based on electrophoretic data indicates that the active glycoproteins isolated from the sera of *C. aceratus* and the *Notothenia* fishes have molecular weights which differ from those of the glycoproteins isolated from the serum of *T. borch-grevinki* (DeVries, 1972). However, conclusive evidence must come from determinations of the molecular weight using sedimentation equilibrium ultracentrifugation.

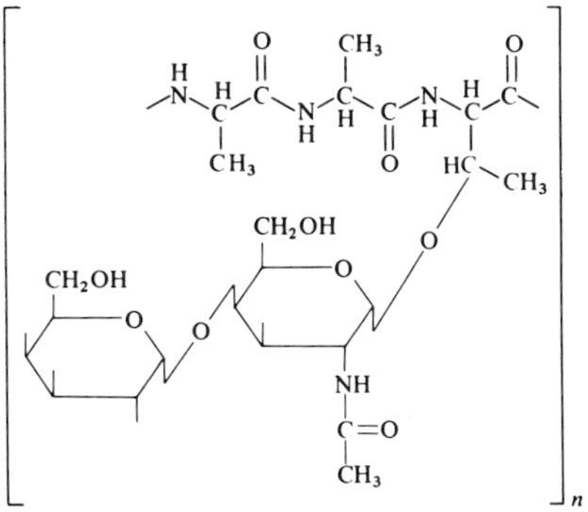

Fig. 6. Repeating structural unit of the glycoproteins with antifreeze properties. The structure is that of a polypeptide chain composed of repeating units of alanyl-alanyl-threonine to which disaccharides composed of galactosyl-(1,4)-N-acetyl galactosamine are attached glycosidically through the hydroxyl group of every threonine. The glycoproteins which differ in size are thought to be composed of different numbers of this repeating unit. It has not been determined whether the glycosidic linkages of the carbohydrate moiety are $\alpha$ or $\beta$.

## 5. Mechanism of Action of Freezing-Point-Depressing Glycoproteins

*a. Effect of Chemical Modification and Enzymic Degradation on the Activity of the Glycoproteins.* To determine how the structure of a protein is related to its function, protein chemists often alter the structure of a protein by chemical modifications and enzymic degradation and study the effects on its function. This approach has been used with some success in attempts to elucidate the mechanism by which the glycoproteins so effectively lower the freezing point of water. For instance, if only 30% of the hydroxyl groups of the sugar residues are acetylated, the freezing-point-depressing properties of the glycoproteins are destroyed (Komatsu *et al.*, 1970). However, if the acetyl groups are removed then most of the activity is restored. Periodate oxidation results in the elimination of carbon-3 of the galactose ring as formic acid, and as a result freezing-point-depressing activity is also lost.

It is known that borate can form complexes with the hydroxyl groups of many sugar residues and does so much more readily with those containing *cis*-hydroxyls than those containing *trans*-hydroxyls (Zittle, 1952). In the presence of 0.15 M sodium borate the freezing-point-depressing properties of the active glycoproteins are completely lost (DeVries, 1972); however, the freezing-point-depressing activity of the glycoproteins can be completely restored if the borate is removed by dialysis against distilled water. Presumably the borate forms a reversible complex with the *cis*-hydroxyls of carbons-3 and -4 of the galactose residue. The relationship which exists between the amount of borate and the freezing-point-depressing activity of the glycoproteins is shown in Fig. 7. This experiment clearly implicates the hydroxyl groups of the sugar residues, most likely those of the galactose residues, as being involved in the freezing-point-lowering activity. Such experiments employing chemical modifications clearly indicate that the carbohydrate moieties of the glycoproteins are necessary for their function.

Studies by DeVries (1968) and Komatsu *et al.* (1970) indicate that the intact polypeptide backbone of the glycoproteins is also essential for function. In fact, splitting of only two or three peptide bonds is all that is necessary to destroy all activity. Thus, it is clear that the integrity of the molecule is highly important.

*b. Effect of Salt on the Activity of the Glycoproteins.* The relationship between freezing points and the molality of the solutions of the glycoproteins indicates that they do not depress the freezing point of water on the basis of the number of particles in solution. When the freezing-point-depressing activities of the glycoproteins were determined in the

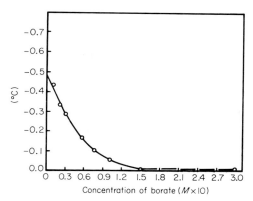

**Fig. 7.** Freezing points of aqueous solutions of glycoprotein 5 (4.5 mg/ml) in the presence of sodium borate showing the amount of borate which reversibly inactivates the glycoproteins with antifreeze activity. The contribution of the sodium borate to the freezing point depression has been subtracted for each of the freezing point determinations which were made with Fiske osmometer.

presence of 0.05 $M$ sodium chloride, which is one-fifth the concentration of this electrolyte in the serum of *T. borchgrevinki*, it was found that the sodium chloride had no effect on the ability of the glycoproteins to lower the freezing point of water. In fact the freezing point depressions resulting from the sodium chloride and the glycoproteins are additive (Fig. 8).

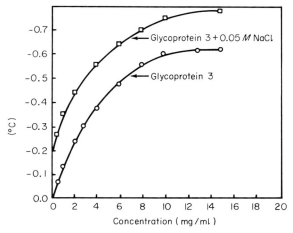

**Fig. 8.** Freezing points of aqueous solutions of glycoprotein 3 in the presence of 0.05 $M$ sodium chloride as a function of glycoprotein concentration. The freezing points were determined using a Fiske osmometer.

Any explanation of the mechanism of action of the freezing-point-depressing property of the glycoproteins must take into account the fact that large volumes of water are affected. The attenuation of the freezing-point-lowering capacity at relatively low concentrations (Fig. 5) would lead one to suggest that the volume of water influenced by each glycoprotein molecule is large. In order to influence large volumes of water, the architecture of the glycoproteins must be such that a maximum amount of interaction can occur between the glycoprotein and water. Maximal interaction would require that the glycoproteins have expanded structures, as the experimental data indicate, rather than compact ones (Table IV).

In an attempt to gain some insight into how expanded molecules can interact with water, it is helpful to consider the nonideal behavior of solutions of certain extended long chain, flexible polymers composed of repeating residues. The osmotic effects observed with these polymers can be explained only on the basis that each repeating residue acts as if it were almost entirely independent of its neighbor (Alexander and Johnson, 1953; Brey, 1958). If the glycoproteins described in this paper are polymers in which the component residues act independently of their neighboring residues, one finds there are enough residues to account for only 30% of the freezing-point-depressing activity associated with the glycoproteins on a weight basis. That such a mechanism is not involved is clearly indicated by experiments in which enzymic hydrolysis by proteolytic enzymes destroys all activity with cleavage of only a few peptide bonds (Komatsu et al., 1970). If the above mechanism were involved, no change in freezing point would be observed upon limited hydrolysis.

Many flexible polymers contain considerable solvent within the domain of their coils. Much of this solvent is trapped and moves wherever the molecule moves (Tanford, 1963). Assuming the glycoproteins are expanded molecules, it is possible that water is likewise immobilized within their structure. There are a few instances in which the freezing points of such immobilized water is lower than expected. The solvent trapped within synthetic gels consisting of polyvinyl alcohol and polyacrylic acid has a freezing point 1–2 degrees lower than that of the swelling buffer (Bloch et al., 1963). A similar, structurally based mechanism may explain why the glycoproteins lower the freezing point of water. Although a hypothesis of structural ordering of water making it less available for the type of molecular ordering essential to ice formation seems attractive, the possibility still exists that the glycoproteins prevent macroscopic ice crystal formation by interacting with microscopic ice crystal nuclei in such a way so as to prevent their growth.

The hydroxyl groups of the galactose could well be involved in binding the glycoproteins in a monomolecular layer to ice crystal nuclei thus preventing them from growing into crystals. Elucidation of the mechanism by which these glycoproteins prevent freezing presents a challenging problem which awaits further research.

## 6. THE RELEVANCE OF FREEZING POINTS OF BODY FLUIDS IN RELATION TO FREEZING RESISTANCE

In the studies made by DeVries (1968, 1972) the freezing points were determined using a Fiske osmometer. In this device a sample of serum is supercooled, then frozen by means of a mechanical shock, and the temperature of the ice–water mixture is measured with an accurate thermister. The degree of supercooling is limited so that only about 4% of the solution is in the form of ice after freezing. In the case of distilled water the temperature as measured by the osmometer after freezing is $0°C$; however, the temperature of a solution of sodium chloride frozen in the osmometer will be lower than the true freezing point of the solution because pure solvent freezes out leaving behind a more concentrated solution which will have a lower freezing point. To circumvent this difficulty the osmometer is calibrated with standard solutions of sodium chloride for which freezing points are already known. With this method the assumption is made that the unknowns behave in the same way as the sodium chloride standards. This appears to be the case with biological fluids such as urine and human serum but not with the serum of some of the Antarctic fishes.

When the freezing points of solutions of the glycoproteins were determined by measuring the temperature at which one small ice crystal in a small capillary tube filled with a solution of the glycoproteins begins to increase in size, it was found that they are approximately the same as those obtained using the Fiske osmometer (see Fig. 9, DeVries, 1972). However, the melting point of this small ice crystal is not the same as the temperature of incipient crystal growth, nor is it even close to it. In fact, the melting point is very close to the freezing point of water. If one measures the temperatures at which a small crystal of ice grows and melts in a dilute solution of sodium chloride, one finds that the temperatures for these two processes are the same, which is what is expected for solvent–solute system in thermodynamic equilibrium. The method of Ramsay and Brown (1955) for determining freezing points of solutions is based on this principle. However, in the case of solutions of the glycoproteins it appears that the ice–glycoprotein solution forms a system which is not in thermodynamic equilibrium.

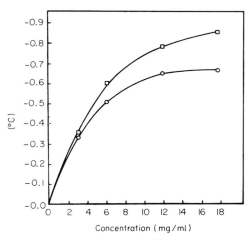

**Fig. 9.** Comparison of freezing points of solutions of a mixture of glycoproteins 3, 4, and 5 determined with the Fiske osmometer ( ○ ) and the temperatures of incipient ice crystal growth of the same solutions ( □ ).

The possibility exists that the effect of the glycoproteins on ice formation may be a kinetic phenomenon, i.e., that its effect is one of slowing down the formation of ice crystals in the blood. It has been suggested by several workers that the mechanism of action of some cryoprotective agents is one of "poisoning" ice crystal nuclei thus preventing their growth (Karow, 1969; Lusena, 1955). If the glycoproteins were enhancing survival in the Antarctic fishes by slowing the rate of ice crystal formation in their blood, then one would expect that the ice which did eventually form would have to be melted at some "thermal site" within the body of the fish or else by elevation of the body temperature of the whole fish. D. C. Potts and Morris (1968) and Morris (1970) have proposed that thermogenesis plays a role in the survival of the *T. bernacchii* by keeping their hypotonic urine, which freezes at —1.0°C, from turning to ice in the bladder. However, other studies (DeVries, 1968, 1972) indicate that the deep body temperatures of these fishes are only 0.02°–0.05°C above the temperature of the water which surrounds the fish even when the fish is actively swimming. Smith (1970) also reported that the deep body temperatures of *N. neglecta* and *N. rossi* are within 0.1°C of the temperature of water.

Therefore in view of the high melting points of solutions of the glycoproteins, which exist in the serum, it is unlikely that the fishes of Antarctica can tolerate any ice formation in their blood, no matter how slowly it occurs, because if ice did form it would never melt, especially in McMurdo Sound where the water temperature is never higher than the melting point of the blood.

As pointed out previously, the freezing points of the body fluids of the Antarctic fishes studied by DeVries (1968, 1972) were in agreement with the temperature at which the live fish froze (Section III, C, 4). However, this was not the case in those studied by Smith (1970) who showed that *N. neglecta* was supercooled by approximately 0.8°C in the presence of ice. The conflicting aspects of these two studies can be partially resolved if one considers the techniques by which the freezing points were determined. In the case of the Antarctic fishes from Signy Island (Smith, 1970) the freezing points were determined by the method of Ramsay and Brown (1955) which employs the temperature at which the last ice crystal melts as the freezing point. Scholander *et al.* (1957) noted that the freezing points and melting points of the serum of the Arctic winter fishes were 0.1°C apart and therefore used the difference between the two as the freezing point. In the case of the Antarctic fishes from McMurdo Sound and the Antarctic Peninsula, the freezing points of the serum were determined using a Fiske osmometer (DeVries, 1968, 1972) and agree with the temperature at which the live fish freezes. However, these freezing points probably do not represent the freezing point of serum in thermodynamic equilibrium and therefore cannot truly be called freezing points. In fact the freezing points determined by the Fiske osmometer may be more accurately a measure of the degree to which the Antarctic blood serum can be consistently supercooled. A solution to these apparently contradictory findings will most likely become apparent when the mechanism of action of the glycoproteins with antifreeze properties is elucidated.

## REFERENCES

Alexander, A. E., and Johnson, P. (1953). "Principles of Polymer Chemistry." Cornell Univ. Press, Ithaca, New York.

Andrews, C. W., and Lear, E. (1956). The biology of the Arctic char (*Salvelinus alpinus* L.) in northern Labrador. *J. Fisheries Res. Board Can.* 13, 843–860.

Andriashev, A. P. (1965). A general review of the Antarctic fish fauna. *In* "Bigeography and Ecology in Antarctica" (J. van Mieghem, ed.), pp. 491–550. Junk Publ., The Hague.

Andriashev, A. P. (1970). Cryopelagic fishes of the Arctic and Antarctic and their significance in polar ecosystems. *In* "Antarctic Ecology" (M. W. Holdgate, ed.), Vol. 1, pp. 297–304. Academic Press, New York.

Black, V. S. (1951). Some aspects of the physiology of fish. II. Osmotic regulation in teleost fishes. *Univ. Toronto Biol. Ser.* 59, No. 71, 53–89.

Black, V. S. (1957). Excretion and osmoregulation. *In* "The Physiology of Fishes" (M. E. Brown, ed.), Vol. 1, pp. 163–205. Academic Press, New York.

Bloch, R., Walters, D. H., and Kuhn, W. (1963). Structurally caused freezing point depression of biological tissues. *J. Gen. Physiol.* **46**, 605–615.

Brett, J. R., and Alderdice, D. F. (1958). The resistance of cultured young chum and sockeye salmon to temperatures below 0°C. *J. Fisheries Res. Board Can.* **15**, 805–813.

Brey, W. S., Jr. (1958). "Principles of Physical Chemistry," pp. 306–308. Appleton, New York.

Dayton, P. K., Robilliard, G. A., and DeVries, A. L. (1969). Anchor ice formation in McMurdo Sound, Antarctica, and its biological effects. *Science* **163**, 273–274.

Dearborn, J. H. (1965). Ecological and faunistic investigations of the marine benthos at McMurdo Sound, Antarctica. Doctoral dissertation, Stanford University, Stanford, California.

DeVries, A. L. (1968). Freezing resistance in some antarctic fishes. Doctoral dissertation, Stanford University, Stanford, California.

DeVries, A. L. (1969). Freezing resistance in fishes of the Antarctic peninsula. *Antarctic J. U. S.* **4**, 104–105.

DeVries, A. L. (1970). Freezing resistance in Antarctic fishes. *In* "Antarctic Ecology" (M. W. Holdgate, ed.), Vol. 1, pp. 320–328. Academic Press, New York.

DeVries, A. L. (1972). Glycoproteins in thermophobic organisms. *In* "Biochemical Adaptation," (F. P. Conte, ed.) (in press).

DeVries, A. L., and Wohlschlag, D. E. (1969). Freezing resistance in some Antarctic fishes. *Science* **163**, 1073–1075.

DeVries, A. L., Komatsu, S. K., and Feeney, R. E. (1970). Chemical and physical properties of freezing-point-depressing glycoproteins from Antarctic fishes. *J. Biol. Chem.* **245**, 2901–2908.

Doudoroff, P. (1945). The resistance and acclimatization of marine fishes to temperature changes. II. Experiments with *Fundulus* and *Antherinops*. *Biol. Bull.* **88**, 194–206.

Eliassen, E., Leivestad, H., and Moller, D. (1960). Effect of low temperatures on the freezing point of plasma and on the potassium/sodium ratio in muscles of some boreal and subarctic fishes. *Arbok Univ. Bergen, Mat.-Nat., Ser:* No. 14, 1–24.

Gordon, M. S., Amdur, B. H., and Scholander, P. F. (1962). Freezing resistance in some northern fishes. *Biol. Bull.* **122**, 52–62.

Hureau, J. C. (1964). Contribution à la connaissance de *Trematomus bernacchii Antarctic Biol., 1st Symp.*, Bouleneger. *Paris, 1962* pp. 481–487.

Kalabukhov, N. I. (1956). "The Hibernation of Animals." Gorki State Univ. Press, Charkov, U.S.S.R. (in Russian).

Karow, A. M. (1969). Cryoprotectants—a new class of drugs. *J. Pharm. Pharmacol.* **21**, 209–223.

Komatsu, S. K., DeVries, A. L., and Feeney, R. E. (1970). Studies of the structure of freezing-point depressing glycoproteins from an Antarctic fish. *J. Biol. Chem.* **245**, 2909–2913.

Littlepage, J. L. (1965). Oceanographic investigations in McMurdo Sound, Antarctica. *In* "Biology of the Antarctic Seas" (M. O. Lee, ed.), Vol. II, pp. 1–37. Am. Geophys. Union, Washington, D. C.

Lusena, C. V. (1955). Ice propagation in systems of biological interest. III. Effect of solutes on nucleation and growth of ice crystals. *Arch. Biochem. Biophys.* **57**, 277–284.

Mahler, H. R., and Cordes, E. H. (1966). "Biological Chemistry." Harper, New York.

Montgomery, R. B. (1957). Oceanographic data. In "American Institute of Physics Handbook" (D. E. Gray, ed.), Part 2, p. 117. McGraw-Hill, New York.

Morris, R. W. (1970). Thermogenesis and its possible survival value in fishes. In "Antarctic Ecology" (M. W. Holdgate, ed.), Vol. 1, pp. 337–343. Academic Press, New York.

Nikolsky, G. V. (1963). "The Ecology of Fishes." Academic Press, New York.

Norman, J. R. (1940). Coast fishes, 3, the Antarctic zone. 'Discovery' Rept. 18, 3–104.

Pearcy, W. G. (1961). Seasonal changes in osmotic pressure of flounder sera. Science 139, 193–194.

Pearse, J. S. (1962). Letter to editor. Sci. Am. 207, 12.

Pickford, G. E., and Grant, F. B. (1967). Serum osmolality in the coelacanth, Latimeria chalumnae: Urea retention and ion regulation. Science 155, 568–570.

Potts, D. C., and Morris, R. W. (1968). Some body fluid characteristics of an Antarctic fish, Trematomus bernacchii. Marine Biol. 1, 269–276.

Potts, W. T. W., and Parry, G. (1964). "Osmotic and Ionic Regulation in Animals," Vol. 19, p. 171. Pergamon Press, Oxford.

Prosser, C. L., and Brown, F. A., Jr. (1961). "Comparative Animal Physiology," 2nd ed., pp. 57–80. Saunders, Philadelphia, Pennsylvania.

Ramsay, J. A., and Brown, R. H. J. (1955). Simplified apparatus and procedure for freezing-point determinations upon small volumes of fluid. J. Sci. Instr. 32, 372–375.

Raschack, M. (1969). Untersuchungen uber osmo- und elektrolytregulation bei knochenfischen aus der ostsee. Intern. Rev. Ges. Hydrobiol. Hydrog. 54, 423–462.

Scholander, P. F., van Dam, L., Kanwisher, J. W., Hammel, H. T., and Gordon, M. S. (1957). Supercooling and osmoregulation in Arctic fish. J. Cellular Comp. Physiol. 49, 5–24.

Smith, R. N. (1970). The biochemistry of freezing resistance of some Antarctic fish. In "Antarctic Ecology" (M. W. Holdgate, ed.), Vol. 1, pp. 329–336. Academic Press, New York.

Tanford, C. (1963). "Physical Chemistry of Macromolecules," pp. 344–346. Wiley, New York.

Tressler, W. L., and Ommundsen, A. M. (1962). "Seasonal Oceanographic Studies in McMurdo Sound, Antarctica," Tech. Rept. TR–125, p. 141, U. S. Navy Hydrog. Office.

Umminger, B. L. (1969a). Physiological studies on supercooled killifish (Fundulus heteroclitus). I. Serum inorganic constituents in relation to osmotic and ionic regulation at subzero temperatures. J. Exptl. Zool. 172, 283–302.

Umminger, B. L. (1969b). Physiological studies on supercooled killifish (Fundulus heteroclitus). II. Serum organic constituents and the problem of supercooling. J. Exptl. Zool. 172, 409–424.

Umminger, B. L. (1970). Personal communication.

Wohlschlag, D. E. (1964). Respiratory metabolism and ecological characteristics of some fishes in McMurdo Sound, Antarctica. In "Biology of the Antarctic Seas" (M. O. Lee, ed.), Vol. I, pp. 33–62. Am. Geophys. Union, Washington, D. C.

Woodhead, P. M. J. (1964). The death of North Sea fish during the winter of 1962–63, particularly with reference to the sole, Solea vulgaris. Helgolaender Wiss. Meeresuntersuch. 10, 283–300.

Woodhead, P. M. J., and Woodhead, A. D. (1959). The effects of low temperature on the physiology and distribution of the cod, *Gadus morhua* L., in the Barents Sea. *Proc. Zool. Soc. London* **133**, 181–199.

Zittle, C. A. (1952). Reaction of borate with substances of biological interest. *Advan. Enzymol.* **12**, 439–527.

4

# LEARNING AND MEMORY*

*HENRY GLEITMAN and PAUL ROZIN*

## I. INTRODUCTION

Fish as a group are by far the most active of the poikilothermic verte-brates, and, if nothing else, this high output of behavior makes them suitable candidates for the study of learning. Many investigators have been impressed by the rapidity of learning in fish and the general similar-

* This work supported in part by U. S. Public Health Service Grant MH 10629-05 and National Science Foundation Grant GB 8013.

ity of this learning to that seen in the presumably more sophisticated mammals. In addition, fish are capable of some of the most impressive adaptive specialized learning feats in the animal kingdom; witness, for example, migration in salmon.

Possibly the first formal experiment on learning in fish was performed by Mobius on pike in 1873 and repeated by Triplett on perch in 1901. In Triplett's classic work, perch were placed in a tank with their minnow prey but were separated from the minnows by a glass partition. After repeatedly crashing into the glass in vain attempts to reach the minnows, the perch learned not to chase the minnows. When the glass partition was removed, the minnows were able to swim near the perch unmolested. The inhibition of attack in this case is apparently specific to the prey used in training: Perch first trained to avoid the minnows banged en-thusiastically into the glass partition when angleworms were now dropped behind it. This simple but highly illustrative experiment has been followed by an enormous number of reports of learning in fish. We can only hope to describe representative samples of this work and to highlight major trends in research of the past and present.

Research on learning and memory in fish has followed two separate traditions. One, the biological or naturalistic, has addressed itself pri-marily to the role of learning in the natural life of fish and has thus organized itself around such functional problems as feeding, migration, and the like. The other tradition concerns the learning and memory capacities of fish determined under experimental laboratory conditions, viewed against the backdrop of traditional learning theory, and related to a comparative psychology of learning. (An offshoot of this second approach has been concerned with the role of brain structures in learning and memory.) Unfortunately, these two basic lines of research have interacted very little. This has had a serious effect on the development of a comparative psychology of learning. The experimental psychologists have often forced the fish into standardized forms of apparatus developed for mammals and may thus have prevented some of the most salient plastic features of fish behavior from expressing themselves. On the other hand, the biologically oriented investigators have all too rarely employed the sophistication in experimental design and techniques developed by experimental psychologists; thus, their conclusions sometimes rest on shaky empirical foundations. These two different traditions in research are reflected in the organization of the important review of fish learning by Thorpe (1963). Research in fish has been motivated either by a desire to understand fish and their achievements (typically in the naturalistic tradition) or to use the fish as a convenient preparation to study some basic issue or phenomenon in the vertebrates or indeed the entire animal kingdom (typical of the physiological or behavioral laboratory approach).

The first two sections of this chapter deal with learning in fish. In these, we first consider the largely descriptive naturalistic evidence, by functional area (e.g., feeding and migration), and then the more programmatic and analytic "learning–theoretical" approach. The second section concerns the physiological or neural aspects of memory and learning and deals with memory and consolidation, the neural substrate of learning, interocular transfer, and cold block of learning. (For convenience, we have considered both physiological and psychological aspects of memory in the same section.) The final section of the chapter reviews representative examples of the use of learning as a tool to investigate other aspects of fish behavior.[*]

## II. LEARNING—THE NATURALISTIC TRADITION

Careful observation of almost any species of fish is likely to suggest some cases of learning in the natural environment. Indeed, many of the examples of fish learning in nature come from incidental observations reported in studies not explicitly devoted to learning. As a result, we may have overlooked many unique and unusual examples and only hope that we can provide a representative sample. In one article, "The social behavior of the jewel fish, *Hemichromis bimaculatus*, Gill," G. K. Noble and Curtis (1939) give some indication of the many roles which learning may play and the importance of learning in the normal behavior of at least this one species. Considering only one broad category of behavior, the authors demonstrated that learning played an important role in recognition of the male by the female, of home territory, of eggs, of young, and of parents by the young. Of course, in this highly evolved cichlid fish, learning may be more predominant than in other groups of fish; still, it is very instructive that these authors found suggestions of an involvement of learning in virtually every major function they studied.

### A. Migration and Orientation

The migration of salmon may well be the single most impressive behavioral accomplishment of any fish. Learning appears to be involved

---

[*] Alternative reviews of fish learning and memory can be found in the chapter on fish in Thorpe's "Learning and Instinct in Animals" (1963) and many of the chapters in Ingle's "The Central Nervous System and Fish Behavior" (1968b). The Thorpe review in particular has been a valuable source for the naturalistic section of the chapter.

as a part of the phenomenon, and is also useful as a tool in the study of sensory factors in migration.

There is no doubt about the migration phenomenon. Several species of salmon migrate from their home stream to the open ocean when fairly young (one to a few years of age) and then return to their home stream after a period of years (for reviews, see Harden-Jones, 1968; Hasler, 1966). In addition, there are many well-documented examples of spawning migrations of land-locked salmon from lakes into streams (Harden-Jones, 1968). Marking experiments have demonstrated unequivocally that salmon migrate from their home streams into parts of the ocean up to over a thousand miles from their river's mouth and that salmon identified in the open ocean return to streams to spawn (Harden-Jones, 1968). Of course, the demonstration that these returning fish are actually returning to their own home stream requires marking the fish before downstream migration, catching and marking the same fish again in the open ocean, and then finally identifying the fish once more in its home stream. Considering the low percentage of returns from the open ocean to freshwater, it is not surprising that there are virtually no direct data of this sort. Three such cases have been reported (Harden-Jones, 1968). In any case, this point bears only upon the migration from the ocean to the river mouth: That salmon can find their way from the river mouth to their home stream is almost indisputable. From our point of view, this is the critical phenomenon, since it is here that learning mechanisms seem to be involved. The mechanisms of open sea migration are poorly understood at this time; possible roles for learning in this behavior will be discussed in Section II, A, 2.

## 1. Return of Salmon from River Mouth to Home Stream—The Odor Hypothesis

Since our goal here is to discuss learning and memory in fish (and not migration per se), we will first consider the evidence that learning is involved in the phenomenon before asking what it is that is learned. We can ask then whether adult fish recognize their home stream by virtue of their genetic make-up or because of their early experience in it, without knowing what it is that they recognize. In fact, historically, there was much evidence which pointed to learning of home-stream characteristics before anyone had a good idea of what those characteristics might be.

The critical experiment involves transplanting salmon eggs or fry from their home stream to some other stream. Will the fish return to their true home (genetic determination) or to their adopted home (learning)? In such transplantation experiments traps are set at both the "adopted" stream and at the "ancestral" home stream (or streams near the adopted stream) for a period of a few years following seaward migration and catches of the distinctively marked transplants are recorded. A series of such studies is reviewed by Harden-Jones (1968). Most of the studies gave results that support the learning hypothesis: Most of the animals return to their adopted home. While some of the studies gave inconclusive results (in two such experiments there were essentially no returns to either stream), no experiment ever gave results to support the genetic determination hypothesis.

Two particularly significant studies are by White and Huntsman (1938) and Donaldson and Allen (1957). White and Huntsman transplanted fry of *Salmo salar* (originally from the Chaleur Bay area) to the east branch of the Apple River (Nova Scotia). Since this branch had been previously dammed it contained virtually no native salmon; thus, observation of the transplanted fry and their migration was particularly easy. The south branch of the river supported a run of native salmon. The transplanted smolts were trapped and marked during their descent to the sea. Barriers and traps were placed across both branches of the Apple River during the return migration, and more than 90% of the marked fish that were trapped were found in the east branch, their adopted home. The salmon apparently learned something about the distinctive features of their new home, although this study does not demonstrate that they preferred their adopted stream to their ancestral stream. The Donaldson and Allen (1957) study offers extremely impressive data. Salmon, (*O. kisutch*) fingerlings (over 1 year old), were transferred from the Soos Creek Hatchery to either the University of Washington School of Fisheries Hatchery or Issaquah Hatchery, in two groups, each containing 36,000 fish. The groups at each new "home" were differentially marked by fin clipping. They migrated to the sea a few months later. Upon return, 2 or 3 years later, all but one of the 195 marked fish identified returned to its adopted home: of the 71 Issaquah fish caught, 70 were caught at Issaquah, one at the university fishery, and none at the ancestral home; all 124 university hatchery fish were caught at the university hatchery. The Issaquah fish passed the university hatchery site both on the downstream and upstream migration.

This study and a report by Carlin (1963, cited by Hasler, 1966) indicate that the critical learning of the identity of the home stream may take place over a relatively short period of time rather late in development at the smolt stage. Carlin found that fish transplanted as smolts returned to the site of transplantation and not to the stream where they were hatched and raised. [On the other hand, Harden-Jones (1968,

p. 267) cites a study in which some transplanted sockeye fingerlings returned to their site of early rearing, bypassing the release area, where
they had spent a few months prior to migration. Some of these dis-
crepancies may represent species differences.] The learning process here
involved seems to occur at a critical time period and produces long-lasting
effects; for these reasons it has sometimes been described as *imprinting*
(e.g., Hasler, 1966).

What has the salmon learned that permits it to identify its home
stream and return to it? Largely as a result of the work of Hasler and
his colleagues (see Hasler, 1966, or Hasler's chapter in this treatise),
it is possible to answer this question with some degree of assurance.

Hasler and his colleagues have formulated the "odor hypothesis" to
explain the freshwater phase of migration. They suggest that "the salmon
identifies the stream of its birth by a characteristic odor, imprinted in
the salmon as a fry or fingerling"; the fish presumably follows some sort
of odor gradient while swimming upstream. This hypothesis assumes
that (a) each stream has a characteristic and stable odor; (b) the fish
can detect this odor and discriminate it from odors in related streams;
(c) the fish, in swimming upstream, is guided by this odor; (d) the
home-stream odor "memory" is acquired by an imprinting process and
retained over a period of years; and (e) there is some kind of odor
gradient that can guide the salmon from river mouth to home stream.

Walker and Hasler (1949) provided strong evidence on stream odors
and their recognition [steps (a) and (b)] by training minnows (which
are more easily dealt with in the laboratory than salmon) to discriminate
between water specimens washed in one or another plant species. Their
technique involved discriminative instrumental learning with food as
the positive and electric shock as the negative reinforcer. A later study
(Hasler and Wisby, 1951) used the same technique to show that blunt-
nosed minnows can discriminate between waters from different streams,
a discrimination abolished by destruction of the olfactory capsule. Are
characteristic stream odors stable enough so that the salmon can utilize
them in their travels? They evidently are. Hasler and Wisby (1951)
showed that fish trained to respond to a given stream odor would still
respond when presented with water taken from the same stream but
during a different season. Another indication that home-stream recognition is chemically mediated is the fact that migrating salmon show increased excitability when exposed to water from their home lake rather
than from a neighboring lake (Fagerlund *et al.*, 1963). Similarly, there
is evidence of markedly enhanced electrical activity in the olfactory bulbs
of adult spawning salmon when their olfactory sacs are infused with

home waters (Hara *et al.*, 1965). This response is specific to water from their particular spawning site (Ueda *et al.*, 1967); salmon from different spawning sites each show a specific response to water from their site. Add to this the fact that salmon migrating upstream do not behave consistently at forks when the olfactory pits are plugged (Hasler, 1966) and we have strong evidence that olfactory cues serve as guiding influences in upstream migration [step (c) above].

Whether other sensory factors may also be involved in homing and migration in the salmon studied or whether the olfactory imprinting mechanism proposed accounts for all freshwater homing in salmon is yet to be determined. There is evidence that temperature gradients guide lakeward migration of young rainbow trout (Northcote, 1969), and that genetic differences in stocks of sockeye salmon influence differences in lakeward migrations (Raleigh, 1967). Brannon (1967) has provided impressive evidence for genetic control of lakeward migration in sockeye fry. Eggs from a stock that normally migrates upstream into a lake nursery area and eggs from a downstream migrating stock were raised under identical conditions. The resulting fry showed strong tendencies to swim in the same direction as their parents. About 80% of the experimentally reared fish, faced with an upstream–downstream choice in the laboratory, chose the direction characteristic of their stock. Hybrids showed intermediate behavior. In short, migration mechanisms are likely to differ in different situations or in different species. However, the olfactory imprinting hypothesis is the only precise formulation that can account for much of the data on freshwater homing of adult salmon. Most of the alternative stimulus dimensions suggested ($CO_2$ level and temperature gradients) do not seem to provide enough inherent variations to allow specific identification of a large number of distinct home sites and routes. The odor hypothesis could be established conclusively if salmon were imprinted to an arbitrary "decoy" odor in their home stream or hatchery and returned to a different stream labeled with this odor a few years later. Hasler (1966) is attempting this experiment now, but of course it could easily fail even if the odor hypothesis is true (see discussion of gradients below).

Implicit in the odor hypothesis is the assumption that salmon can remember the home-stream odor over a period of years [step (d) above]. Despite the absence of any direct laboratory evidence for such long-lasting memory in fish (Hasler, 1966) the assumption of such memorial capacities in salmon is not unreasonable considering the convincing arguments for the odor hypothesis which implies it. We know fairly little of how this memory is laid down or stored, but there is some evidence for peripheral storage in the olfactory bulbs (Hara *et al.*, 1965). Further, some recent studies indicate that puromycin and other metabolic inhibitors, introduced intracranially, can temporarily suppress the electrophysiological response to home waters (Oshima *et al.*, 1969; see section on memory).

But there are still some further problems. At the present time, the odor hypothesis cannot satisfactorily explain the full migration from river mouth to home stream. After all, the home-stream waters must be enormously diluted by the time they reach the river mouth; under the circumstances, detecting the home-stream odor would be an exceedingly difficult task. Estimates of water flow in the Fraser River system suggest that home water concentration could be below one part in one thousand at the river mouth (Harden-Jones, 1968), considerably lower than any demonstrated thresholds for detection of home water by salmon (Idler *et al.*, 1961; Ueda *et al.*, 1967). We must therefore assume either that the salmon has remarkable olfactory acuity in the face of masking odors or that other mechanisms are at work. One possible mechanism, suggested by Harden-Jones (1968), is multiple imprinting: The salmon remembers ("imprints" upon) the distinctive odors at a number of points on his downstream migration. The return voyage is then seen as a series of subvoyages with the salmon reading out the series of imprinted odors, in reverse order, as "subgoals." Such a system will obviously magnify the odor concentrations and gradients. There is very little evidence for or against this hypothesis. Harden-Jones (1968, p. 268) cites a study by Carlin showing that a smaller percentage of Atlantic salmon smolts released into the sea near their river mouth returned to their home stream than fish released from points further upstream. This, of course, suggests that experiences gained on the downstream journey are utilized during the return journey.

There is one study on record (Harden-Jones, 1968, p. 267) in which some transplanted sockeye fingerlings returned to the site of their early rearing, swimming past the site of release. In this case the returning salmon negotiated a stretch of at least 10 miles of presumably unfamiliar waters. This distance may have been too small to permit a distinction between the various imprinting hypotheses. Still, this study raises some interesting questions about the actual identification of the spawning area. Is it possible that the odor tracking brings the fish into the appropriate watershed and that visual as well as olfactory cues (perhaps both imprinted) then contribute to guide the fish to the specific spawning area?

We can think of one further explanation, within the odor hypothesis, for the identification of the home-stream odor at the river mouth. If we assume that the odor of neighboring streams in the same river system is more similar than distant streams in different river systems, it would be possible to argue that at every choice point, the fish simply chooses that stream which smells most like its home stream. Presumably, its home river outlet would smell more like its home stream than other river outlets; stimulus generalization will then take care of the rest. If this were the case, no further demands have to be made on either the odor hypothesis or the salmon olfactory system. In a sense, this hypothesis makes greater demands on the environment (the similarity in odor of neighboring streams) and less on the fish. In contrast to dilution in laboratory experiments, dilution as it occurs in the actual river system may be with water of an odor that is similar to home-stream water. It is interesting in this regard that Ueda *et al.* (1967) recorded weak responses from the olfactory bulbs of salmon when exposed to water from a bypassed tributary near the spawning site. Although this result is certainly consistent with the sequential imprinting hypothesis, it is absolutely essential for the generalization hypothesis. On the other hand, Idler *et al.* (1961) performed some laboratory studies which showed that sockeye salmon migrating into Great Central or Cultus lakes gave a positive response (school disruption and increased swimming speed) to water from some but not all streams flowing into their respective lakes. Water that produced no such response came from streams that were not inhabited by sockeye salmon. In this case at least, neighboring streams seem to have quite different odors. The critical experiment here would be to demonstrate that salmon, before migrating downstream (or salmon beginning their return voyage after having been flown to the river mouth), will show varying levels of positive response to water taken from various tributaries along their projected route while showing no such positive response to other waters. This might be done electrophysiologically or in a laboratory training experiment of the sort done by Hasler and his colleagues. Alternatively, a generalization experiment of this sort might be performed on purely freshwater species to demonstrate an environmental basis for the generalization hypothesis.

In summary, we feel that there is very strong evidence that at least some species of salmon "imprint" on their home-stream odor and that this odor serves as a principal source of guidance during the freshwater phase of their return migration. The use of additional (perhaps visual) sensory cues is quite possible, particularly for the identification of the home breeding grounds themselves, but this possibility has not as yet been investigated. At this point, we still do not know how the salmon tracks his home-stream odor from the river mouth and whether visual or other cues aid him in this task.

## 2. SUN-COMPASS ORIENTATION

Though much of the research on sun-compass orientation in fish has its impetus from attempts to explain open sea migration in salmon, almost all work has been done on freshwater species, since their movements are easier to track and they are more adaptable to the laboratory environment (for a general review, see Hasler, 1966).

Sun-compass reactions, probably at varying levels of sophistication, have been demonstrated in a number of species (Hasler, 1966; Winn et al., 1964). According to Gerking (1959), homing has been reported in 21 fish species, and many of these cases probably involve some sort of sun orientation. The phenomenon in fish is in many respects quite similar to that in bees and birds and will be discussed here only as it relates to learning.

The standard technique for the study of sun-compass reactions in fish is a circular chamber, with 16 channels extending radially outward from a central core (Hasler, 1966). Figure 1 presents a diagram of this apparatus. Fish (centrarchids, such as the pumpkinseed sunfish) can be trained to swim in a particular direction (e.g., north) at a particular time of day. They then maintain this direction when tested at different times of the day by altering the angle they take from the sun. Careful precautions are taken to prevent "landmark" orientation by rotating the experimental chamber between trials (Hasler, 1956).

It is apparent that fish are capable of sophisticated sun-compass orientation including precise compensation for changes in the sun azimuth position, time of sunrise, and sun altitude; they clearly possess an impressive, highly structured navigational system. But even if we assume that this system is primarily predetermined genetically [since fish raised in artificial indoor environments show prompt compensation for sun movements, etc., when placed outdoors (Schwassman and Hasler, 1964)] it is obvious that the system requires environmental information to work properly. Environmental cues must play an important role in calibrating the system. This is a kind of "learning" in which experience produces long-term changes in behavior while the usual sort of contingencies and associated trappings of learning are absent. The organism evidently experiences certain environmental changes and "remembers" them (that is to say, takes account of them in its actions for long periods measured in days or weeks following these single events).

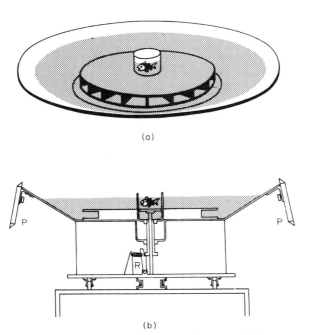

(a)

(b)

**Fig. 1.** Tank for training fish to a compass direction: (a) as seen from above showing the hiding boxes; (b) side view showing periscopes (P) for indirect observation and the release lever (R) to permit the cage to be recessed by remote control when fish is released. From Hasler (1968). Copyright (1968) by the University of Chicago Press.

For example, changing the onset of illumination (sunrise) produces gradual (over days) changes in the fish's chronometric navigation resulting in an orientation adapted to the newly phased day cycle. Furthermore, although inexperienced fish (sunfish, *Lepomis cyanellus*) raised in artificial light cycles show prompt accurate compensation when placed at a given latitude, experienced fish that have become accustomed to the true sun cycle at a given latitude, tend to retain their original, inappropriate compensation pattern when switched to a new latitude. They gradually readjust or "decondition" this original orientation and adopt an appropriate new one (Schwassman and Hasler, 1964). Experience of one particular pattern of sun movement evidently produces some lasting effects.

The direction of the sun's movement may be another feature of the sun's arc that has to be environmentally calibrated in some fish. In some tropical cichlids (Braemer, 1960; Hasler and Schwassman, 1960), there is evidence for compensation in both clockwise and counterclockwise directions (the sun in the northern hemisphere runs clockwise and in the southern hemisphere counterclockwise). It appears that appropriate direction of compensation may also be learned.

What emerges is a picture of an elaborate sun-arc projection mechanism, genetically determined, but requiring some definite information about the sun—its time of rise, its altitude, and, in some cases, its direction. This environmental information not only calibrates the arc but also produces rather long-lasting effects which lead to somewhat sluggish changes in the navigational mechanisms when different sets of environmental cues are presented.

Hoar (1958) performed some experiments on a few species of Pacific salmon that may well relate to migration, but their exact relevance to the actual achievements of the fish in their natural habitat is difficult to assess. He found that, if a group of downstream migrating juvenile salmon were placed in a circular trough, then all fish would swim in the same direction after a few minutes, that is, clockwise or counterclockwise. Further, the fish would usually maintain this direction for periods up to 24 hr. Chum salmon showed this effect most strikingly, maintaining their original direction after a period of 20–40 hr out of the testing environment. If it is assumed that the selection of the original direction was arbitrary and accidental, then we are dealing with a phenomenon in memory.

## B. Homing and Territorial Recognition

Homing and territorial activities in fish are well known. Such behavior patterns (Gerking, 1959) may be based upon the kinds of mechanisms already discussed. In addition, specific learning of visual or other landmarks may also be involved. For example, it is obvious that fish that establish territories learn certain identifying features of these territories. Thus jewel fish return to their nest site even if the nest has been removed, and they show disruption in behavior if the color of a flowerpot in their territory is altered (G. K. Noble and Curtis, 1939). Numerous other studies yield similar testimony (see Thorpe, 1963, for a general review). Williams (1957) reported that if he could locate and identify the same individual fish (wooly sculpin, *Clinocottus analis*, or opaleye, *Girella nigricans*) in a tide pool in the area he studied on two consecutive days, the likelihood was 0.80 that it would be in the same tide pool on these two days. Considering that the fish wander away from these pools during high tide, there is likely to be some sort of memory for the terrain or, possibly, a learning of specific routes. Hasler (1956) directly demonstrated that *Phoxinus* used multiple small incidental markings for orientation in a circular metal tank in sun-compass type of experiments. Only consistent rotation of the tank during training induced the fish to use sun-compass rather than local landmarks to maintain orientation.

The most suggestive example of landmark learning comes from work by Aronson (1951) on the goby, *Bathygobius soporator*. These fish tend, as did those of Williams (1957), to be found in the same tide pool from day to day and can be observed to jump from one pool to another during low tide. The remarkable feature of this performance is that in almost all of the 51 jumps observed (by 18 fish) the movement was accurately aimed to take the fish into an adjacent tide pool (see Fig. 2). Random jumping would result in the fish hitting rocks most of the time. The nature of the tide pools made it impossible for the fish to see directly into the next tide pool, and Aronson presents data that seem to defy explanation in terms of open sea or sun orientation. The fish could hardly be sensing water nearby since many jumps were into tide pools that happened to be dry at the time. Significantly few errors were observed, and it was not possible to get fish transplanted into unfamiliar tide pools to jump out, even when they were severely prodded. Fish in familiar tide pools would jump spontaneously or when prodded. These observations weaken a trial-and-error interpretation. Aronson also found that removing fish from a pool for 2 weeks and then replacing them did not diminish in any way their ability to jump accurately. These data can be interpreted as indicating a remarkable instance of latent landmark learning: The jumpers may have learned the terrain and arrangement of tide pools in their home area, presumably by swimming over the area during high tide. However, further replication and investigation of this phenomenon may reveal a simpler mechanism.

## C. Timing Mechanisms

The operation of an accurate, entrainable time sense or clock can be inferred from data on sun-compass orientation. Davis (1963) and Davis and Bardach (1965) have explicitly conditioned swimming activity to a time cue. In the first experiments (Davis, 1963) bass, *Micropterus salmoides*, and bluegills, *Lepomis m. microhirus*, were fed at light onset in a 12-hr-light, 12-hr-dark cycle. Within 10–20 days, clear hyperactivity appeared in the fish during the last 1–3 hr of darkness. This activity peak could be phase-shifted by simultaneous switching of light cycle and feeding with characteristic readjustment times of a few days. Since both species involved are more active in the light, it is possible to interpret

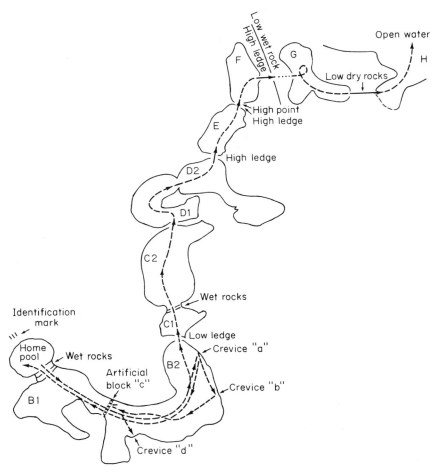

**Fig. 2.** Schematic outline of the tide pools, paths, and jumps of one goby. From Aronson (1951).

this result as a direct effect of light: The hyperactivity is an anticipation of light onset rather than of feeding. This seems particularly likely, since placing the feeding in the middle of the light cycle did not eliminate the predawn peak.

Davis and Bardach (1965) recognized this qualification and were able to demonstrate circadian anticipations of a daily feeding time under constant illumination in the killifish, *Fundulus hetericlitus*, which seem more likely to be entrained or affected by food than by light stimuli. Again, the development of the circadian activity response was rapid.

Formally, this demonstration has the same paradigm as temporal conditioned reflexes of the Pavlovian type (e.g., food presented every 30 min resulting in anticipatory salivation). It differs from this classical learning phenomenon in that the interval selected (24 hr) may be the only one that could be effective. However, it is possible, with difficulty and time, to train goldfish to discriminate time intervals of the order of minutes (Rozin, 1964).

## D. Feeding and Mimicry

Included among the fishes are species with the widest possible variety of feeding habits from prototypical predators such as pike, some sharks, and barracuda, through omnivores such as goldfish, to herbivores. On *a priori* grounds it would be indeed surprising if learning and memory did not play a role in the identification of suitable food, particularly among those species that seem to feed on a wide variety of items. Furthermore, a significant number of examples of apparent mimicry or warning coloration have been reported in fish or the prey of fish (Cott, 1940; Randal and Randal, 1960; Wickler, 1968). The issue from the point of view of learning is whether the conspicuous or mimicked colors or forms are effective because experience within the individual lifetime of a predator leads him to avoid a particular species after one or a few unpleasant experiences with it. This has been demonstrated in birds (e.g., see Wickler, 1968, for a review). For example, it is possible that fish learn to avoid attacking or ingesting one of the varieties of poisonous fish (Cott, 1940) and subsequently generalize this avoidance to the mimics of poisonous creatures (Randal and Randal, 1960).

The cleaner wrasse, *Labroides dimidiatus*, which eats small crustacea and other organisms from the surface of cooperating "customer" fish, is mimicked by the blenny, *Aspidontus tueniatus* (Wickler, 1968). The blenny, in fact, extracts pieces of flesh from the customer and takes advantage of the customer's quiet and placid stance in the presence of an apparent cleaner. Wickler suggests that the customer, or "host," learns to submit to the cleaner and very likely learns to discriminate it from the harmful mimic. Older fish tend to avoid the mimic more, and this apparently results from experience since adults deprived of experience with mimics confuse the cleaners and the mimics. Fish customers kept with mimics will subsequently avoid cleaners. It is possible to observe a fish that accepts mimics, is maimed on a few occasions, and then becomes wary of them.

More direct experimental evidence for a role for learning in food preferences is provided in an ancient study by Reighard (1908). Reighard hypothesized that the conspicuous coloration of coral reef fish might be warning coloration and set out to demonstrate that the gray snapper, *Lutianus griseus*, which preys on coral reef fish, could learn to form an association between colors and disagreeable qualities and could retain these associations. Although Reighard could not demonstrate that any of the bright-colored fish actually found in the reefs were avoided by the gray snapper, he was remarkably successful in establishing the avoidance in an experimental situation. He employed dead, usually dyed, sardines (*Atherina laticeps*) as an acceptable prey, and worked primarily with a tank of 150 gray snappers in the laboratory but also with some natural populations. He made the sardines aversive by the exotic expedient of sewing the tentacles of a medusa into the mouth of some sardines, and found that such adulterated sardines would be initially accepted by snappers but subsequently were rapidly rejected. He was able to teach a group of 150 snappers to completely avoid red-dyed sardines with tentacles. After a modest number of trials with these, the snappers were tested with untentacled sardines: They readily accepted those that were undyed and completely avoided those that were red. Twenty days later, the colony still avoided red sardines (and to a lesser extent blue and yellow ones) but readily accepted sardines that were not dyed. Although this experiment lacks some of the controls now accepted as standard (a series of experiments and controls was run on the same laboratory colony), it nonetheless appears valid and furthermore suggests the existence of the rapid kind of learning that would be the only useful kind in situations involving potentially damaging prey.

Aquarists know that fish, like rats, respond reluctantly to new foods and must be adapted to them over a period of time measured in minutes to days. Miller (1963) found that a few species of sunfish will not accept new foods for a few days and gradually familiarize themselves with new foods by sampling them, tasting, and spitting out, until eventually the food is accepted. Beukema (1968) has studied the feeding of sticklebacks, *Gasterosteus aculeatus*, and has confirmed the phenomenon of adaptation to new prey and greatly extended our knowledge of the mechanism of this adaptation and its manifestations. He finds that sticklebacks cruising around in a multiple, hexagonal-unit maze initially fail to attack a new prey (e.g., *Drosophila* larvae or worms) when they enter a compartment containing it. After about 10 such encounters with the prey, the probability that they will "recognize" and ingest it increases, leveling off after about 50 encounters. As acceptance of the new prey increases, the stickleback seems to recognize the prey at greater distances (development of a search image) and to adopt more systematic and efficient search patterns in swimming through the maze. When a highly palatable prey is introduced in one compartment of the maze, the stickleback is less likely to attack less palatable prey in other compartments. This effect remains for a period even after the highly palatable prey is removed from the situation. In the Beukema experiments, which seem to combine much of the best of both traditions in the study of learning, prey recognition, search pattern, and acceptance of prey are all adaptively influenced by experience.

Tugendhat (1960) administered electric shock to sticklebacks either for entering a feeding area in a tank or for actually grasping prey. She found that despite the great differences in contingencies produced by the two shocks both had very similar effects, notably, much less time spent feeding. Although less time was spent feeding by fish shocked for grasping both types of fish showed an increased efficiency of feeding (percent of general feeding time actually spent eating, rather than fixating and grasping prey, etc.). In a related experiment, Myer and Ricci (1968) have recently demonstrated a food avoidance response in goldfish in a typical laboratory learning situation.

Finally, it is likely that many if not most fish show moderately sophisticated forms of food intake regulation. One feature of such regulations is caloric compensation: When the caloric density of food is diluted, the organism gradually increases bulk intake to hold calories constant. Such behavior has been demonstrated in the goldfish (Rozin and Mayer, 1961a). Insofar as this modification in intake involves change in rate of ingestion or size of meals, it must involve some sort of learning. In many ways, it seems analogous to the sun-compass situation, since it involves a highly structured built-in system which receives calibration signals from the environment.

## E. Social Behavior

### 1. INDIVIDUAL AND SPECIES RECOGNITION

Evidence for a significant role for learning in social behavior appears incidentally in many studies and is well reviewed by Thorpe (1963). Establishment of relatively stable dominance hierarchies has been reported for a number of species (e.g., Braddock, 1945, in platies; Newman, 1956, in trout). For example, if four *Platypoceilus maculatus* (Braddock, 1945) are placed together in the same tank, a stable hierarchy is established in from 30 min to 12 days. This means that fish respond consistently to particular conspecifics. It is hard to imagine the development of such behavior in the absence of individual recognition, which obviously depends upon the establishment of a memory system. Male jewel fish are able to recognize their mates (G. K. Noble and Curtis, 1939). The male does not attack his mate even while attacking an approaching group of fish of which she is one; he will reject other female conspecifics when these are presented (in the absence of the mate); and, when faced with a choice of his mate and another similar looking female (placed in neighboring tanks), will approach his mate. The authors found that visual stimuli from the head were most critical in this identification. By lacquering one side of the head, they arranged a situation in which the male would recognize his mate from her normal side and attack her from the lacquered side.

There is clear evidence for significant learning factors in species recognition in cichlids (G. K. Noble and Curtis, 1939; Baerends and Baerends-van Roon, 1950). For example, experienced pairs of adult jewel fish, *Hemichromis bimaculatus*, recognize their young: If their eggs are replaced by eggs of a different species, the young will be eaten when they hatch. However, if the same switch is performed on a pair breeding for the first time, they will accept the young of different species. Such a pair, if subsequently allowed to breed normally, may eat their normal young. Apparently, the fish learn to recognize their young on the first breeding, and this learning remains stable. This interesting phenomenon has sometimes been described as imprinting of the parents upon the young.

Again, both of the major studies on cichlid behavior (G. K. Noble and Curtis, 1939; Baerends and Baerends-van Roon, 1950) offer evidence that learning is involved in following by the young. Tendencies to follow objects of particular colors (which differ in different species) can be modified by cross-rearing or raising in isolation. For example, normal young jewel fish tend to follow moving red discs (the brooding adult is reddish) but may be shifted toward a black disc preference by being raised with the darker colored *Cichlasoma*.

## 2. Schooling

Relatively little can be said about the role of learning in schooling, although many authors have assumed that it plays a substantial role both in explaining the adaptive value of schooling and understanding its development (e.g., Thorpe, 1963). (Also, see Shaw, 1970, for a full discussion and for an excellent review of schooling.) It has been suggested that one of the possible explanations of the adaptive advantages of schooling is that it facilitates learning. O'Connell (1960) has demonstrated conditioning of a school of sardines to a light-food contingency, with the conditioned response (CR) an increased speed and a tightening up of the school. Other results also suggest better learning performance in grouped than isolated fish (see Shaw, 1970). Welty (1934) performed a number of laboratory experiments on learning in grouped vs. isolated goldfish, all suggesting facilitation of learning in groups. He showed that placing a naive fish in a maze with a trained fish will facilitate acquisition of the maze by the naive fish and that watching (through a glass partition) a fish acquire the response of swimming through a hole to food facilitates acquisition of this response by the observing fish. This result could be accounted for, as Welty recognizes, in terms of familiarization with the situation, resulting in inhibition of fear responses, although it may also result from direct acquisition of information relevant to the task. Appropriate analysis on this point has not been performed. In general, experiments showing faster acquisition in groups must be treated cautiously: One general account of this is simply that in a group the fastest learning fish lifts the whole group's rate, since the slower learners may tend to follow him. Also, in Welty's experiments, a food reward was given when the first fish entered the reward compartment, and olfactory or visual cues relating to food could then have lured the other fish. Such reservations do not reflect on adaptive explanations of schooling as facilitating learning. The general principle that "many eyes are better than one," offered to account for groupings in animals, may hold here for learning: cohesion allows the best learner, as it were, to raise the performance of the group.

The role of learning in the development and mechanism of schooling itself has been studied by Shaw (1970). Since the behavioral mechanisms involved in schooling are not well understood, it is not easy to delineate a role for learning. Schooling usually develops gradually in nature, but this may be a purely maturational phenomenon; for example, completion of innervation of the lateral line organs correlates with the development of parallel orientation in young schooling fish (Shaw, 1970). Furthermore, a number of studies have reported the rapid appearance of normal schooling when fish raised in isolation are placed with conspecifics. Recent research by Shaw indicates that isolates placed together after 20 days of age school promptly, whereas isolates placed together before this age do not. However, the schooling of isolates was slightly abnormal: The fish kept readjusting position and reorienting. Shaw suggests that this irregularity is a result of poor integration of approach and withdrawal tendencies (which figure prominently in her analysis of the dynamics and development of schooling) and that the social environment may facilitate this proper integration. This facilitation could be of a general permissive sort, or specific types of learning might occur.

## III. LEARNING—THE TRADITION OF "LEARNING THEORY"

There is by now a very considerable literature on learning in fish conducted within the context of more or less classical learning theories and built upon evidence generated by experimental procedures developed in the learning theorist's laboratory. To students working within this tradition the fish as such is of only secondary interest; the fish's achievements and capabilities are of concern only to the extent that they illuminate some general propositions about the learning process itself. To them the fish is a tool, a kind of behavioral "preparation" provided by nature, whose behavior can sharpen our understanding about behavior in general much in the way in which a decorticated cat yields information about the functioning of the intact nervous system.

Typically such investigators are interested in differences between fish and other (vertebrate) classes: "What is it that the rat and the bird can learn that the fish cannot," is the kind of question often asked here. These investigators will quickly grant that "the fish" is a somewhat overambitious category, considering that this group comprises some 18,000 species, for their studies sample at most a few dozen of these (e.g., Herter, 1953) and more typically are based upon the behavior of but two or three. [For example, the work of Bitterman, Gonzalez, and their associates, and that of Mackintosh and Sutherland. For respresentative reviews, see Bitterman (1968) and Mackintosh (1969a).] Even so they point with some justification to the rather remarkable similarities in the behavior of the few species within the group of teleost fish (most often goldfish and the African mouthbreeder) and to the difference between either of these and the learning achievements found in several birds (primarily pigeons, also chickens) and mammals (usually the rat).

Setting aside the possibility that species within the class might differ markedly, in what sense can the fish be considered a preparation? The decorticated cat lacks a cortex; what does the fish lack that can give significance to whatever it is that it cannot learn? Some American investigators occasionally speak as if they wish to study the evolution of learning (e.g., Bitterman, 1964a). The fish is considered as "lower" on the phylogenetic ladder than the bird which is somehow "below" the mammals; if we find some orderly progression in learning capacities as we ascend the ladder we may gain some understanding of what the "higher" learning processes are all about. Such a formulation is of doubtful value and is severely limited since we are obviously not studying the rat's forebears when we study birds or the highly specialized present-day teleost fish. Given the organisms studied, there is little basis for constructing an evolution of learning (see Hodos and Campbell, 1969). Finally, there is some evidence which suggests that the octopus performs midway between the fish and the rat on the very tasks that had been used to differentiate these two (Mackintosh and Mackintosh, 1964); obviously the attempts to link behavioral differences obtained thus far to phylogenetic position are on rather shaky ground.

But phylogenetic comparisons of learning ability may be of considerable interest even apart from evolutionary considerations (and without explicit reference to known differences in neurological structure) for the program of comparative studies of learning can be stated more modestly. Suppose there are several phenomena in the learning laboratory that, at least on the face of it, are not obviously related to each other: If all of them occur in one group of animals and none of them in another, we have some good reason to suppose that these various phenomena are somehow linked to each other (e.g., share a common mechanism). Such correlations between diverse laboratory phenomena may then suggest a theoretical approach that can profitably connect them (e.g., Mackintosh, 1969a). Whether this strategy is fruitful will simply depend upon (a) the discovery of correlations between learning phenomena and animal groups, and (b) the development of theoretical models that can account for these correlations. In terms of such models, these correlations might be interpreted as pointing to a common mechanism; on the other hand, they might be interpreted as the result of common ecological problems shared by the various species.

We will now turn to a survey of the evidence as it bears on these matters. First of all, we shall consider the empirical questions. In what ways does the fish learn like other vertebrates, in what ways does it not? Since the main body of learning theory has been largely built upon the laboratory accomplishments of rats and pigeons, the comparisons will center primarily upon similarities and differences between the fish and these. For convenience, we will follow Bitterman's practice of using the term "the fish" to describe characteristics of two (admittedly quite different) fish species, but we have serious reservations about the use of this general term, as stated above and in Section III, C.

## A. Similarities

There is no doubt that the fish performs admirably in many learning tasks traditionally presented to birds and lower mammals. Typically these are tasks that are sometimes thought to involve "simple" learning processes—habituation, classical conditioning, instrumental conditioning, and the like. Unquestionably fish are capable of all of these. Nor is there any evidence to suggest that—allowing for sensory and motor differences and variations in motivational conditions—the basic phenomena which usually characterize these learning categories in rats or birds (e.g., rate of learning, extinction, generalization, and the like) are noticeably different when we turn to fish. Consider the evidence in the following section.

## 1. CLASSICAL CONDITIONING

There is no doubt that fish can be classically conditioned, that is, conditioned to respond to a conditioned stimulus (CS) when the presentation of the unconditioned stimulus (UCS) is not contingent upon the performance of the response. Among the first studies in this area were those of McDonald (1922), Froloff (1925), and Bull (1928). In these studies the unconditioned stimulus was shock or food, the unconditioned responses were flight or food-ingesting reactions. A wide variety of stimuli was successfully used as CS: color, sound, temperature changes, variations in the salinity of the water, touch, smell, taste, etc. (Bull, 1957). Reviews of the earlier work can be found in Bull (1957) and Herter (1953). More recent investigators have extended the classical conditioning paradigm to a broader range of unconditioned responses, among them general activity (e.g., Horner et al., 1960; see Fig. 3), electric organ discharge in mormyrids (Mandriota et al., 1965), aggressive display in Siamese fighting fish (Adler and Hogan, 1963; Thompson and Sturm, 1965a), changes in respiration (Kellogg and Spanovick, 1953), and heart rate (McCleary and Bernstein, 1959; McCleary, 1960). While some of the earlier studies are open to the charge that their results might be owing to instrumental rather than (or perhaps, in addition to) classical conditioning, this hypothesis is hard to maintain when the unconditioned responses are such things as heart rate or respiration changes.

For at least the grosser phenomena of classical conditioning, there is no evidence that points to major differences between conditioning as it occurs in fish and as it appears in other animals. Thus we find rates of conditioning and extinction that are of roughly the same order (e.g., Voronin, 1962); we find generalization and discrimination effects similar to those found elsewhere (e.g., Yarczower and Bitterman, 1965); we find evidence for sensitization (Harlow, 1939) and even some suggestion of higher order conditioning (Sanders, 1940).

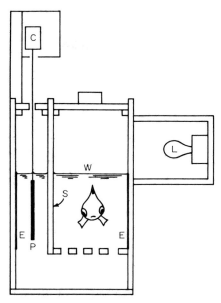

Fig. 3. A classical conditioning situation for the fish. L, lamp, the onset of which serves as the CS; E, electrode; P, paddle; S, slatted wall; W, water level; and C, phonograph cartridge. From Bitterman (1966).

One parametric difference has been debated in the literature. A study by M. Noble *et al.* (1959) using mollies had raised the possibility that the "optimum" CS–UCS interval is of the order of 2 sec for fish, compared to the optimum of about ½ sec usually claimed for mammals (Kimble, 1961). A further study by M. Noble and Adams (1963) ruled out the possibility that this effect resulted from differences in body temperature: the effect of the CS–UCS interval was the same for mollies at a temperature of 75°F as for those at 90°F. However, later work by Bitterman and his associates (e.g., Klinman and Bitterman, 1963; Behrend and Bitterman, 1964; Bitterman, 1964b; Bitterman, 1965) makes a strong case for supposing that the original difference was based on an artifact by pointing to a serious methodological problem in the study by M. Noble *et al.:* The intervals used during test trials were identical to those used during CS–UCS pairings, thus confounding the effect of training interval with opportunity to respond. After all, an animal with a CS–UCS interval of 10 sec has 10 times more opportunity to respond than another animal whose CS–UCS interval is 1 sec. If test trials (when CS is presented alone) are identical to those used during training (when CS and UCS are paired), and if the measure is probability of response, then we can hardly obtain a pure measure of the effect of this interval during training. Bitterman and his associates have shown that when opportunity to respond on test trials is equalized, the effect described by M. Noble *et al.* disappears; the latency of the response is shortest with the 0.5-sec training interval and rises as the interval increases. Considering Bitterman's critique, it is not too clear what might be meant by an optimum CS–UCS interval in any case; to the extent, however, that the term is meaningful there is little reason to suppose that fish differ from other animals in this regard.*

## 2. INSTRUMENTAL CONDITIONING

It is equally clear that fish can be instrumentally conditioned, that is, trained to perform a response upon which some reinforcing event is contingent. One of the earliest experiments reported is that of Triplett (1901) who kept a perch and some minnows in the same aquarium separated by a glass partition; after numerous collisions with the glass, the perch refrained from attacking the smaller fish even after the partition was removed. Many other studies testify to the same capacity for instrumental learning in other species of fish, both in aversive and appetitive situations. Much of the earlier work is summarized by Bull (1957) and by Herter (1953). Some Russian studies are described in Voronin (1962).

Most of the early experiments concentrated on orientation and locomotor responses; more recently, the area of study has been enlarged and manipulative responses such as lever pressing have commanded more attention. The pattern of recent investigation has been seriously affected by the general trend toward instrumentation in animal learning studies. Two devices to study instrumental learning in fish have become increasingly popular. One is an aquatic analog of the "Skinner box" so widely used in the study of rats and pigeons: The fish (usually a goldfish, or an African mouthbreeder) lunges at a visual target attached to a lever and thus gains a food reward (Haralson and Bitterman, 1950; Longo and Bitterman, 1959; Hogan and Rozin, 1962; see Fig. 4). The other is a shuttle box or, rather, a shuttle tank for fish, patterned after that used in the study of escape and avoidance learning in rats and dogs (Horner *et al.*, 1961; see Fig. 5).

Both appetitive and aversive reinforcers have proved effective in in-

* A further argument against the view that the "optimum interval" differs in different vertebrate classes comes from the demonstration that this optimum interval within one and the same subject depends upon the response system being studied (see Vandercar and Schneiderman, 1967, comparing nictitating membrane and heart-rate conditioning in rabbits).

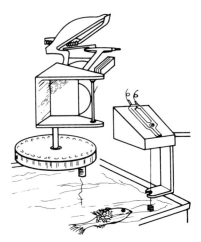

**Fig. 4.** Automatic device for the study of operant behavior in fish. From Longo and Bitterman (1959).

strumental conditioning of various fish species. On the appetitive side, food has been the reinforcer of convenience, but others have been found quite successful: thus Van Sommers showed that a brief exposure to aerated water reinforced a locomotor response in oxygen-deprived gold-fish (Van Sommers, 1962), and Rozin and Mayer (1961b) found that a squirt of cold water could reinforce lever pressing in a high tem-

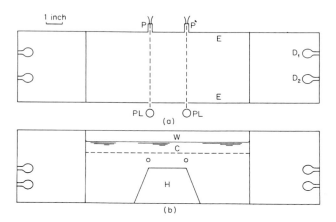

**Fig. 5.** Two views of a shuttle box for the fish: (a) plan and (b) side view: P, photocell; PL, photocell lamp; E, electrode; $D_1$, $D_2$, colored lamps, the onset of which serves as the CS; W, water level; C, ceiling of the animal's chamber; H, hurdle. From Bitterman (1966).

perature stressed goldfish. Of considerable interest is the finding that male Siamese fighting fish will perform an instrumental response that is reinforced by a releasing stimulus, the sight of another male (Thompson, 1963; Thompson and Sturm, 1965b).

On the aversive side, all of the major experimental paradigms have been utilized and with success, almost invariably with electric shock as the aversive stimulus. Both escape and avoidance learning have been demonstrated with several species of fish (Wodinsky et al., 1962), with primary attention paid to avoidance learning. Active avoidance is readily obtained, especially with locomotor responses; the fish learn, whether the fear-producing stimulus is exteroceptive (as in the usual shuttle tank procedure, e.g., Wodinsky et al., 1962) or whether it is response-produced (as in Sidman avoidance, e.g., Behrend and Bitterman, 1963). Passive avoidance (that is, suppression by punishment) can also be obtained in goldfish (Geller, 1963); not surprisingly, some related phenomena of classical conditioning such as "conditioned suppression" have been found as well (Geller, 1963, 1964).

Many of the functional relationships that have been discovered in the instrumental conditioning of homotherms have also been found in the fish. This holds true whether the process is studied using discrete trials or whether it is considered in the context of the free operant. Considering the inevitable differences in a host of experimental variables (e.g., shock intensity, CS–UCS interval, intertrial interval, effortfulness of the response) the results do not seem too dissimilar. Various other phenomena, well established in the rat or the pigeon, have also been shown to hold for fish. On the appetitive side we might list among others secondary reinforcement (Salzinger et al., 1968), stimulus generalization, contrast, and peak shift (Ames and Yarczower, 1965). On the aversive side we might mention the well-known Kamin effect, a U-shaped function relating the retention of avoidance to retention interval: In a recent experiment Pinckney (1966) has shown this effect in goldfish with the trough of the retention curve falling at 24 hr.

## B. Differences

The preceding discussion has shown that, for tasks that involve "simple" learning processes, few systematic differences show up between fish and those most commonly studied representatives of homotherms, rats, and pigeons. The situation changes when we turn to more complex problems. Generally speaking we find that systematic differences are obtained if the specific task requirements of the problem vary from moment to moment and if the "solution" of the problem requires either an integration over the various individual experiences (for example, in probability learning) or a differentiation between them (for example, in habit reversal). Most of the work in this rubric has been contributed by two groups of investigators: Bitterman, Gonzalez, and their co-workers on the one hand and Mackintosh and Sutherland on the other. For general reviews of results and overall orientation see Bitterman (1968) and Mackintosh (1969a). Both groups of investigators have concentrated upon a few critical problem areas in which they both have shown systematic differences between fish and other animals: habit reversal, the effect of changes in reinforcement pattern (including such matters as the partial reinforcement effect, the effect of amount of reward upon resistance to extinction, and the depression effect), and probability learning. We now turn to these studies in some detail.

## 1. HABIT REVERSAL

In a serial habit-reversal experiment, an animal is trained on a given discrimination (say, black plus vs. white minus), then given the reverse of this (now, white plus vs. black minus), and then the reverse of that, and so on. Mammals that have been given such sets of problems (mostly monkeys, cats, and rats), as well as pigeons, reliably improve as reversals proceed; in fact, after many such reversals some animals may approach a one-trial solution. [Some representative studies using rats and pigeons are North (1950), Gatling (1952), and Gonzalez et al. (1966).] This pattern of results contrasts sharply with that obtained with the fish. Several studies using goldfish or African mouthbreeders showed no evidence of progressive improvement over a series of successive reversals—sometimes as many as 168—in either visual or spatial discrimination (Bitterman et al., 1958; Behrend et al., 1965); Warren (1960) obtained similar results on paradise fish. Figure 6 presents the results of two habit-reversal studies: one, on rats, using a horizontal–vertical discrimination, the other, on African mouthbreeders, using a red–green discrimination (Bitterman, 1968). As the figure shows, the error curves start to deviate after a few reversals: Those for the rats decline, those for the fish stay the same.

The difference in habit-reversal performance in fish on the one hand and rats or pigeons on the other is indeed dramatic. This sharp discrepancy is not lessened by the fact that two recent studies have shown that given very special conditions fish can improve somewhat in progressive reversal problems. But this improvement was hardly of the order usually shown by rats and pigeons, and to obtain even this limited effect the fish's task had to be greatly simplified.

Setterington and Bishop (1967) trained African mouthbreeders on

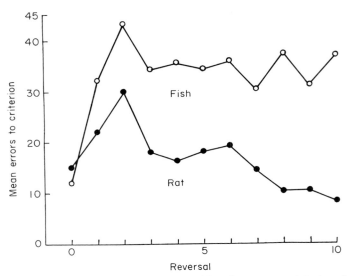

**Fig. 6.** The performance of rats and mouthbreeders in an original problem and 10 subsequent reversals. The rats were trained in a horizontal–vertical discrimination and the fish in a red–green discrimination. From Bitterman (1968). Copyright (1968) by the University of Chicago Press.

a simple spatial discrimination with an unlimited correction procedure: After each incorrect response the targets were withdrawn from the tank for 2 sec and then presented again, a procedure that was repeated until the correct response was finally made. A 2-sec intertrial interval followed the correct response. Under these circumstances, progressive improvement was found both for initial errors (made on the very first choice on a given trial) and repetitive errors (errors persisted in on any given trial after an initial error was made). Behrend and Bitterman (1967) repeated some features of Setterington and Bishop's study; they failed to find the decline in initial errors but did find significant improvement in repetitive errors over reversals. It appears that given optimum conditions some fish can show some minimal improvement over reversals; the critical feature is evidently a short time interval between trials.

A rather similar pattern of results appears in experiments in which an instrumental response is repeatedly conditioned, extinguished, and reconditioned. It is well known that in this situation the rat's resistance to extinction declines quite sharply (Perkins and Cacioppo, 1950). The same is true of the goldfish and the African mouthbreeder but with some important differences. To begin with, the fish's asymptotic resistance to extinction is much higher than that of the rat. There is another difference that may be more important. Consider a rat given successive sessions of extinction and reconditioning: Its resistance to extinction is essentially unaffected by the number of trials given on each reconditioning session. Not so for the fish. The more trials (and thus, of course, reinforcements) it receives during a reconditioning session, the longer it takes to extinguish during the extinction session thereafter [Gonzalez et al., 1961, 1962a, 1967a (Fig. 7 presents a graphic comparison of rats and fish in this situation)]. Similar results are described by Voronin (1962), whose account of a number of comparative studies of conditioning conducted in Soviet laboratories indicates a fairly orderly progression in what he calls the "lability of the nervous system"; baboons and dogs reached a criterion of one-trial extinction in relatively few experimental sessions, tortoises and fish reached it in very many sessions, and rabbits and birds fell in between.

What accounts for the relatively greater rigidity of the fish in experimental situations that require repeated reversals of prior reaction tendencies? Put another way, what is it that the fish lacks (or possesses to a lesser degree) which is present in the rat or the pigeon? It appears that to understand the fish, we must understand the rat and the pigeon and must now ask what it is that accounts for progressive improve-

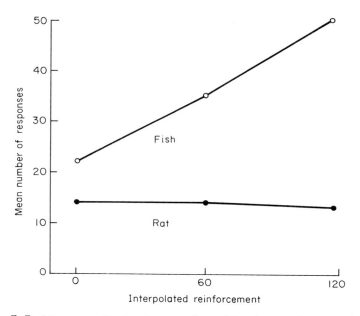

Fig. 7. Resistance to extinction in rat and mouthbreeder as a function of amount of interpolated reinforcement (Gonzalez et al., 1961). From Bitterman (1968). Copyright (1968) by the University of Chicago Press.

ment in habit reversal in these animals. Several interpretations of these and related effects have been offered.

*a. Response Strategies.* One possibility is the discovery of a general response rule and its adoption as a strategy: For example, the subject might learn to shift the response as soon as the reward conditions change ("win-stay, lose-shift"). Warren (1965) points out that while rhesus monkeys do adopt such a strategy, cats do not. Warren's monkeys showed impressive transfer from spatial training to a set of object discrimination problems; his cats, on the other hand, showed no such transfer even though they manifested the usual improvement during reversal training. A study by Mackintosh *et al.* (1968) makes a similar point for the rat. These authors showed that training on several position reversals did not improve the animals' performance on a subsequent series of brightness reversals: This renders such notions as a general "set to reverse" rather inappropriate. In another experiment, the same authors showed that experience on a serial reversal task did not impair performance in a subsequent probability learning situation: a two-choice discrimination in which one stimulus was reinforced 75% of the time, the other 25% of the time. If anything, the subjects "maximized" even more readily than without such training (that is, they chose the "majority stimulus" 100% of the time, thus maximizing their overall gain). A win-stay, lose-shift strategy in a probability learning task would necessarily lead to matching (that is, choosing the majority stimulus 75% of the time, thus matching the subject's choices to the reinforcement probabilities). This being so, such a strategy is probably not the factor that accounts for progressive improvement in habit reversal in rats and pigeons.

*b. Attentional Models.* A very different interpretation derives from a two-stage model of discrimination learning developed by Sutherland (1964), extended by Mackintosh (1969a,b), and put into formal terms by Lovejoy (1968). These authors argue that in learning a discrimination an animal learns not only which of the several possible stimulus dimensions (position, brightness, form, or whatever) is relevant to the problem at hand but also which value of the given dimension (e.g., black vs. white) is appropriate. According to the model, both dimensions ("analyzers") and responses to particular values ("response attachments") are acquired in an incremental fashion; what may vary from species to species is the rate at which analyzers and response attachments are formed, reach asymptote, and are extinguished. In principle, the model could account for improvement in habit reversal given that the relevant dimension stays the same and only the response attachment need be altered (Mackintosh and Mackintosh, 1964). Of course, it is true that the analyzer may extinguish when the reinforcement conditions are altered. Mackintosh (1969a) suggests that this is precisely what happens to rats in the early stages of reversal learning thus accounting for initial negative transfer; given that the analyzer has extinguished while the previous response attachments have not, what is carried over from before cannot help but must hinder. However, after a few reversals the analyzer resists extinction and is in fact strengthened, for the same analyzer is reinforced on both components of the reversal. The result is that the rate of response attachment (and thus improvement in reversal learning) increases. Mackintosh argues further that in fish the analyzers are not as stable as they are in rats, thus accounting for the fact that fish are very much inferior to rats in serial reversal learning.

Evidence relevant to this interpretation of habit reversal in fish and other animals comes from two areas that have occupied the attention of two-stage theorists: the effect of overtraining and a comparison of reversal and nonreversal shifts.

Many studies have found an overlearning-reversal effect in rats: Rats trained in a discrimination for many trials beyond criterion perform better on the reversal of this discrimination than do rats trained to criterion only (for general reviews, see Mackintosh, 1965b; Sutherland and Mackintosh, 1970; Sperling, 1965). Such an effect does not occur in fish. In fact, both Warren (1960), using paradise fish, and Mackintosh *et al.* (1966), using goldfish, found that in these animals overtraining *retards* reversal, although the deleterious effect of overtraining was not statistically significant in the latter study. The two-stage model can account for the beneficial effect of overtraining in rats on the assumption that under ordinary circumstances response attachment proceeds more quickly than acquisition of the relevant analyzer during the original learning: Being more weakly established the analyzer will then extinguish more rapidly in the course of reversal. If the animal is overtrained, the analyzer is strengthened and its subsequent extinction is retarded; under these conditions the old response attachments can be broken more readily and new ones can be substituted for them (Lovejoy, 1968). To account for the failure of overtraining to benefit reversal in fish, Mackintosh *et al.* postulate that in fish analyzers and response attachments grow at the same rate, thus further trials strengthen both equally with little or no effect upon reversal.

Granted that in fish overtraining does not benefit reversal learning, we nevertheless feel that two-stage theory is on rather shaky ground in drawing upon this fact as evidence for its general position. There is something rather contrived in the assumptions that must be made about the critical parameters: To account for the fish's poor performance on habit reversal we must assume that the analyzer is unstable (while it is more stable in the rat), while to account for the absence of an overtraining effect we must assume that the analyzer and the response attachments grow at the same rate (while in rats the analyzer grows more slowly). More important is the fact that the conditions under which the overtraining-reversal effect appears are still quite unclear, even in the rat (Warren, 1965). It seems that the effect is determined by a multitude of factors other than species differences (for example, reward magnitude); and it is still an open question whether two-stage theory can account for such factors by appropriate manipulations of its parameters.

A more direct test of two-stage analysis of habit reversal comes from a study by Schade and Bitterman (1966), who trained pigeons, African mouthbreeders, and goldfish to discriminate between stimuli that varied along two dimensions (e.g., position and color). On any one problem, one dimension was relevant, the other irrelevant. The subjects were trained on a long series of problems which required them sometimes to shift dimensions (nonreversal shift) and sometimes to reverse within a dimension (reversal). The results showed progressive improvement for the pigeons, both for reversal and for nonreversal shifts. The fish showed no improvement of any kind, either for reversal or nonreversal shifts. The attentional model asserts that an increase in the strength of one analyzer implies a decrease in the strength of all others. The results obtained by Schade and Bitterman do not bear out the model, which should predict a decrease in dimensional flexibility when there is progressive improvement in habit reversal given that the model tries to explain this improvement by the ever-increasing strength of the analyzer.

This is not to argue against the importance of dimensional set in discrimination learning. For rats and pigeons there is little doubt that cues are selectively attended and selectively ignored (Mackintosh, 1965b). Similar evidence exists for the fish. Mackintosh et al. (1966) trained goldfish on two successive discrimination problems with stimuli that varied along two dimensions (orientation and brightness). One dimension was relevant during the first discrimination; the other was relevant during the second. They found that overtraining on the first problem hindered acquisition on the second, quite predictable if we but assume that strengthening one analyzer will retard the acquisition of the other. Thus selective attention is not at issue, whether in rats, in birds, or in fish. What is at issue is whether the two-stage model can account for the progressive improvements in habit reversal (and the related overtraining-reversal effect) found in rats and in pigeons. What is further at issue is whether this model can also account for the fact that the fish does very poorly at habit reversal and shows no trace of an over-training-reversal effect.

 *c. Progressive Improvement as Forgetting.* A recent paper by Gonzalez *et al.* (1967c) proposes that the explanation of progressive improvement in habit reversal lies in the development of proactive inhibition. Proactive inhibition is typically studied using the following experimental paradigm:

|  | Short retention interval | Long retention interval |
|---|---|---|
| Control groups | $R_2 - R_2$ | $R_2 ------ R_2$ |
| Experimental groups | $R_1 R_2 - R_2$ | $R_1 R_2 ------ R_2$ |

The control groups merely learn $R_2$ and are tested for the retention of this response some time thereafter (the length of this "retention interval" is here indicated by the dashes). The experimental groups first learn $R_1$ (a response incompatible with $R_2$), then learn $R_2$ and are tested for its retention over intervals equivalent to those used for the controls. Proactive inhibition (PI) is indicated by inferior retention of $R_2$ in an experimental group as compared to the retention of $R_2$ in a control group with equal retention interval. Proactive inhibition (that is, the difference between experimental and control performance) is known to increase with the retention interval and with the strength of the previously learned competing response; it has been demonstrated repeatedly for both men (Underwood, 1948) and rats (S. Maier and Gleitman, 1967).

Gonzalez *et al.* (1967c) argue that with increasing series of reversals the animal develops ever-increasing proactive interference for whatever it has learned. Each reversal session acts as a proactive inhibitor on the sessions that follow. Normally, poor retention of the last training session would hinder performance on the subsequent sessions, but it does not here where we are dealing with a series of reversals. Remembering the last problem would impede the acquisition of its reversal; forgetting should be of help. Thus, increasing PI should generate progressive improvement.

In support of their view, Gonzalez *et al.* (1967c) present data of a habit-reversal experiment using pigeons and goldfish as subjects, red and green targets as stimuli, with daily sessions and reversals occurring on every second day. The usual analysis (by overall error scores) produced no surprise: marked improvement for the pigeons, none for the goldfish. More important for the authors' position is their analysis of the day-to-day retention scores. They compared the response pattern of the last five trials of each day with the first five trials of the day following. Consider three experimental sessions, S-1, S-2, and S-3, such that S-1 and S-2 represent two successive sessions on the same problem while S-3 represents the first day of its reversal. Compare performance on S-2 with that on S-3. Here, the subjects were reinforced for changing their response pattern, and thus a difference here could well be ascribed to habit-reversal improvement rather than to a retention loss. But similar comparisons were also made between the performance found on S-1 and S-2, two days within each discrimination; here a difference would be a pure measure of retention loss. Pigeons showed retention loss as measured by both indices, while goldfish showed none in either. Further, the retention loss in pigeons, while very variable, increased with progressive reversals. Gonzalez *et al.* (1967c) ascribe their results to PI increasing with increasing degrees of prior interference. When present (as in pigeons) this causes improvement in habit reversal; when absent (as in fish) there will be no such effect. In short, the fish fails to improve over a series of reversal problems because it cannot forget the last discrimination it has learned.

Several arguments can be leveled against this interpretation:

(1) As Sutherland and Mackintosh point out (1970) there is a problem with the method used to score retention loss. Gonzalez *et al.* (1967c) employ difference scores which necessarily depend upon the baselines achieved during the last five trials of any given day. Forgetting scores could only reach a maximum if base line performance was high. For pigeons these scores did increase as reversal training progressed. Was this owing to increased forgetting or to increases in base line performance as the animals mastered the reversals?

(2) Gonzalez *et al.* (1967a) assert that PI (and by implication, forgetting) is more pronounced in pigeons than in fish. As it happens, it is rather difficult to demonstrate PI in the pigeon: Both Kehoe (1963) and Gleitman and Kosiba (1967) failed to do so, the latter using an experimental procedure that had always produced PI in the rat. As for fish, Gleitman *et al.* (1970) have shown considerable retention losses of an avoidance response over a period of 4 weeks, which appear of the same order as forgetting rates found in rats.

(3) Most important, the PI interpretation has difficulty in explaining the essence of the habit-reversal effect: There is improvement in the error score from the first problem to the last. The PI theory can perhaps explain why the animal no longer suffers from the deleterious effect of the previous reversals. Can it explain why the subject eventually does better than he did on the very first problem, before he ever had learned any competing responses? Gonzalez *et al.* (1967a) might answer that the subject brings prior competing responses into the situation before he ever starts on the experiment; whether this is a plausible argument is certainly debatable.

*d. Habit Reversal and Changes in Brain Structure.* Two intriguing studies have attempted to relate habit-reversal improvements to brain structure. The first used the ablation method and showed that adult rats subjected to extreme decortication as infants showed no improvement in habit reversal: They became, in effect, like fish in this respect (Gonzalez *et al.*, 1964). A more recent study by Bresler and Bitterman (1968) reversed the logic of the ablation studies. If rats with diminished brain tissue act like fish in a habit-reversal study, will fish with "augmented" brain tissues act like rats? Bresler and Bitterman tried to supplement the brain tissue of *Tilapia macrocephala* by grafting donor embryonic tissue in the prospective tectal tissue of the embryo host. Upon reaching adulthood, six survivors were trained in a habit-reversal experiment. Of these six, two showed no brain abnormalities upon later examination; the other four showed various degrees of tectal thickening. The two normal animals showed no improvement in habit reversal; of the other four, two showed evidence of improvement in habit reversal, and two showed unusually rapid learning scores.

There may be a tendency to write off the result of the Bresler and Bitterman study because it appears in principle unlikely that a crude manipulation such as implantation of tissue at prospective tectal sites would produce an improvement in the operation of the highly organized and finely tuned nervous system. It is somewhat equivalent to expecting that deletion of a gene locus would improve an organism's survival, or that throwing a wrench at a computer would improve its operation. (Of course, once in a while it probably does.) However, it is just possible that the unlikely has happened; certainly the result suggests it. Under the circumstances, serious and careful examination of this paper is mandatory.

Quite apart from its results, the Bresler-Bitterman paper is based on the following reasoning:

(1) The authors assume that meaningful supplementation (that is, in relation to nervous organization and behavior) of portions of the brain is possible by means of grafting techniques. This assumption is certainly debatable, if only on the grounds that tampering with any highly organized system is unlikely to improve its operation, even if new tissue is in some sense incorporated. It is true enough that the evidence from gross anatomy presented in the paper does suggest an orderly integration of the grafted tissue in the four fish studied, but, while the gross sections shown are very impressive, no information is provided about the microscopic structure of the "new" tecta. Of course, the data on neural development and reorganization in the tectum (Sperry, 1965) suggest a remarkable amount of plasticity and organ-

ization in the tectum, so that it is not inconceivable that the "new" tectum could be an organized whole.

(2) The authors assume that supplementation at tectal sites would produce an improvement in performance and increased adaptibility. Offhand, there is little evidence for or against the idea that, granted that supplementation might produce improvement, this should be manifested at tectal sites. There is little evidence for the role of the tectum in learning and memory. Contrary to the statements of Bresler and Bitterman, the tectum is not homologous with the mammalian cortex. In fact, it might be disputed whether it is analogous to the cortex, so that this heuristic plausibility argument offered by the authors can hardly be maintained.

The issue is undoubtedly important. One would wish that future studies might provide more detailed information about the structure of such aberrant brains. One would also hope that the behavior of such supplemented fish could be examined in a wide variety of learning situations, especially including other tasks in which Bitterman, Gonzalez, Mackintosh, and others have found significant differences between fish and homotherms. If the results hold up upon replication and extension, they obviously buttress the argument that habit reversal taps a learning process that is somehow "higher" and may well have revolutionary implications for our understanding of the role of brain structures in the determination of such "higher" functions.

## 2. The Effect of Inconsistent Reinforcement

Few phenomena of instrumental conditioning are as widely known as the partial reinforcement effect (PRE). Animals that are reinforced on every trial during acquisition training will extinguish more rapidly than those reinforced only intermittently. This effect occurs in humans, monkeys, pigs, dogs, rats, and pigeons. Does it also occur in fish?

To answer this question we must first distinguish between two kinds of partial reinforcement effects: those that occur with massed trials, and those that occur when trials are substantially spaced. The importance of this distinction was first brought into prominence by Sheffield (1949) who tried to explain the PRE as due to sensory carryover. She believed that sensory remnants (traces) of the previous trial are carried into the trial (or perhaps, trials) following and then become part of the total stimulus operative during that trial. If the response is reinforced on a given trial, it will become conditioned to all stimulus components operative on that trial, including the stimulus traces of previous trials.

If all trials are reinforced, then the sensory carryover to which the response becomes conditioned is that characteristic of a positive trial (e.g., particles of food still in the mouth). If some of the trials are not reinforced, then the opposite condition holds: Some of the stimulus traces that become conditioned to the response will be those characteristic of a negative trial (e.g., a persistent frustrative reaction). During extinction, all trials are negative. Given inconsistent reinforcement during training, the aftereffects of these trials (that is, of course, all extinction trials) will have been conditioned to the response. On the other hand, nothing of the sort occurred during continuous reinforcement for there the animal never encountered a negative trial. The partial reinforcement effect follows.

Almost by definition, sensory carryover (as opposed to memorial representations) can last only for a rather short time interval. If this is so, the PRE should occur only if trials are massed. Sheffield performed a test of this hypothesis on rats and found supporting results (1949). These results were subsequently challenged by various investigators, several of whom found clearcut evidence that rats yield a solid PRE even when the intertrial interval is a full day (e.g., Weinstock, 1954). The demonstration of a spaced-trial PRE was generally taken as conclusive evidence against a theory of the effect based upon sensory carryover. In retrospect, however, this verdict may have been premature, for a recent analysis by Gonzalez and Bitterman (1969) makes a good case for supposing that the processes which underlie the massed-trial and the spaced-trial PRE are quite different.

Gonzalez and Bitterman (1969) studied the spaced-trials PRE in rats, varying reward magnitude. Whether trials were massed or spaced, the PRE was greater with large reward. But this effect came about in different ways. When trials were massed, increasing reward magnitude led to an increase in the resistance to extinction of the partially reinforced group (with little effect on the control)—hence, greater PRE. When trials were spaced, increasing reward magnitude affected the consistently rewarded group; its resistance to extinction dropped markedly while that of the partially reinforced group showed little change—again an increase of PRE, but, at least on the face of it, for rather different reasons. Gonzalez and Bitterman (1969) suggest that the massed-trial PRE results from a carryover process quite similar to the one proposed by Sheffield. The spaced-trial PRE according to them is essentially a negative contrast effect, akin to the "depression effect" studied by Crespi (1942): the animal's reward magnitude drops from some acquisition level to zero during extinction, and zero "seems like even less" in contrast to what was obtained before, so the animal overreacts. The greater the reward value during acquisition the greater the drop in extinction, and thus the greater the contrast. This effect occurs only for the consistently reinforced groups. The partially reinforced groups have already experienced zero magnitudes and thus suffer no contrast. In a sense then there is no "genuine" PRE when trials are spaced: It is not that inconsistent reward increases resistance to extinction but rather that consistent reward decreases it because of the contrast phenomenon.

Let us turn to the data as they pertain to fish:

*a. PRE When Trials Are Massed.* Many of the earlier studies questioned the reality of the PRE (whether spaced or massed) in fish, or at least hedged its appearance by a multitude of qualifications and restrictions. Thus, for example, it was initially asserted that the massed-trial PRE in fish would occur only if the consistent and inconsistent groups were given the same number of reinforcements rather than being given (as is usually the case in experiments with rats) the same number of trials (e.g., Gonzalez *et al.*, 1962a, 1967b). More recent studies indicate that the situation is simpler than this. Fish will show a massed-trial PRE (with trials equated) if the inconsistent group is trained with long sequences of unreinforced trials and/or if large reward magnitudes are used (Gonzalez *et al.*, 1965; Gonzalez and Bitterman, 1967). An example of such an effect is shown in Fig. 8. The effects of these variables upon PRE is of course well documented in the rat literature: increasing

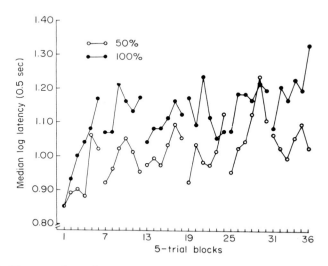

**Fig. 8.** The partial reinforcement effect in a high-reward equated-trials experiment with the goldfish (Gonzalez and Bitterman, 1967). Extinction is plotted in terms of the latency of response on successive five-trial blocks. From Bitterman (1968). Copyright (1968) by the University of Chicago Press.

either run length of unrewarded trials (Capaldi, 1967) or reward magnitude (Hulse, 1958) increases PRE. It appears then that earlier failures to obtain unqualified PRE effects with massed trials in the fish were probably caused by a faulty choice of the specific experimental conditions. As things stand now, there is no reason to believe that (minor parametric matters aside) the essential phenomena of the massed-trial PRE are substantially different in fish and homotherms. Assuming a Sheffield-type theory of the massed-trial PRE, we can then conclude that fish respond to sensory carryover from previous trials and that these aftereffects can become part of the stimulus complex to which the animals' responses are conditioned.

 *b. PRE When Trials Are Spaced.* To date, there have been rather few studies that bear on this issue. Still, the available evidence suggests that when intertrial intervals are long the PRE and related effects may be absent for the fish. Thus, Gonzalez *et al.* (1965) trained African mouthbreeders to strike a target for food, giving them one trial per day. Trials were equaled, and there were two inconsistent groups, one with a Gellerman order of positive and negative trials (in which the maximum run of unreinforced trials is three) and one in which the runs of unreinforced trials were more extended. Figure 9 shows mean number of trials to successively more severe criteria of extinction. As the figure indicates, resistance is if anything greatest for the consistent group and clearly greater for the group that had short rather than long runs of unreinforced trials during training.

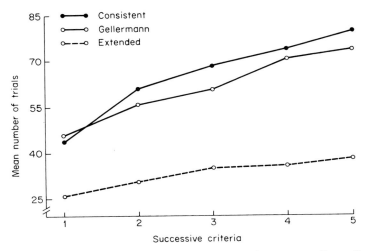

**Fig. 9.** Trials to successively more severe criteria of extinction. From Gonzalez *et al.* (1965).

As we have seen previously, there is some evidence which suggests that the spaced-trial PRE in rats is ultimately a result of a negative contrast effect suffered by the consistent group during extinction. Assuming this interpretation and given that African mouthbreeders (Gonzalez *et al.*, 1965) and goldfish (Schutz and Bitterman, 1969) show no spaced-trial PRE, one would similarly expect them not to show negative contrast. Several studies have come up with precisely this result. Lowes and Bitterman (1967) performed an experiment patterned after the classic study of Crespi on rats (1942). Goldfish were trained to strike a target for food and rewarded with either few or many worms. As in rats, performance was more effective the larger the reward magnitude. After 22 trials, half of the animals in each of the groups were continued on the original regimen, the other half were switched to the other reward magnitude. Figure 10 shows the results. There was no trace of a depression effect. Quite the contrary: Fish that were switched downward continued to respond at about the level of their prior performance, while those switched upward gradually approached their new asymptote.

A related phenomenon concerns resistance to extinction as a function of reward magnitude during training (here we are dealing with consis-

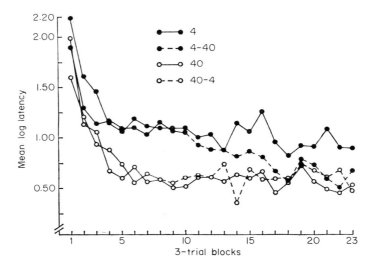

**Fig. 10.** Mean log latency of a simple instrumental response in the goldfish as a function of amount of reward. One group was rewarded throughout with 4 worms; a second, with 40 worms; a third was shifted from 4 to 40 worms; and a fourth, from 40 to 4 worms. Postshift performance is shown by broken lines. From Lowes and Bitterman (1967). Copyright (1967) by the American Association for the Advancement of Science.

tently reinforced animals only). With rats, there is a contrast effect: After a few minutes, the extinction curve of the high-reward group dips below that of the low-reward group. With fish, there is no trace of contrast: As Fig. 11 shows, the high-reward group outperforms the low-reward group throughout the entire extinction period (Gonzalez *et al.*, 1967c).

Assuming the interpretation offered by Gonzalez and Bitterman, the critical finding centers upon the fish's response to shifts in reward magnitude. The rat quickly adjusts its performance to the reward level it obtained on the last trial (or perhaps on the last few trials), often overshooting and undershooting the appropriate performance asymptote as it "contrasts" the reward it receives now with that to which it has previously become accustomed. Such effects in the rat suggest that the animal has acquired some "knowledge about" the reward and that it can refer to this representation and can differentiate it from other representations (that is, can distinguish the recent reward level from that obtained much before). Not so the fish. As Bitterman points out (1968), the fish's reaction is precisely that which would follow from the simplest version of stimulus–response (S–R) reinforcement theory (Hull, 1943). Does this mean that the fish cannot form "representations" in the sense in which the rat surely can? Or, perhaps, that fish cannot distinguish between the more recent and the more remote representations? Bitterman suggests that the fish is an "S–R reinforcement animal," a suggestion

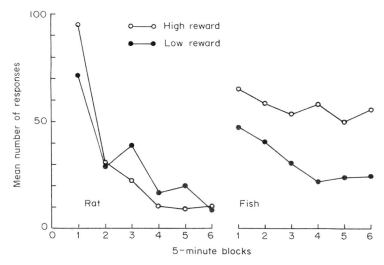

**Fig. 11.** Resistance to extinction in rat and goldfish as a function of amount of reward (Gonzalez *et al.*, 1967b). From Bitterman (1968). Copyright (1968) by the University of Chicago Press.

that would be more helpful if we knew what we meant by a "non-S–R reinforcement animal."

## 3. PROBABILITY LEARNING

In the customary two-choice discrimination experiment, one cue is always, and the other cue is never, associated with reinforcement. In a probability learning problem, the association is not so clear-cut: one cue will be "correct" for some proportion of the trials, while the other will be correct for the remaining trials. The subject may respond by *maximizing:* He may always choose the "majority stimulus" (that is, the most probable stimulus), a strategy that assures him of the greatest number of reinforcements. On the other hand, he may respond by *matching:* He may distribute his choices between the two stimuli according to the ratio of reinforcements accorded to each of them. For our purpose, the critical studies are those which employed a procedure that guarantees that the subject learns which cue is correct on every trial whether that cue is chosen or not; after an error the subject may correct himself (correction method) or in effect may be led to the correct cue by the experimenter (guidance method).*

The literature indicates that in this experimental situation fish perform quite differently from rats and from pigeons. The asymptotic performance of rats tends toward maximizing (e.g., Bitterman *et al.,* 1958; Uhl, 1963). This does not hold for the fish. Both goldfish and African mouthbreeders match the reinforcement ratios even after many

---

* Several further points should be mentioned briefly. (1) Conclusions about asymptotic performance can only be made if the subject has been given a long series of trials. A subject may be matching; on the other hand, he may only be passing through the matching level on his way towards maximizing. (2) The empirical generalizations here described hold for correction and guidance methods only. The results are different when the noncorrection method is employed in which the trial is over after the subject's choice, whatever its outcome. In the noncorrection method the subject never learns what "might have happened" had he chosen the other alternative. With this method the performance of most subjects quickly rises to maximizing asymptotes; this holds for rats and also for fish (e.g., Bitterman *et al.,* 1958). (3) On the face of it maximizing would seem to be the more intelligent approach to a probability learning situation, and, indeed, in a rough sort of way the data fit a simplistic notion of a phylogenic intellectual hierarchy: Monkeys and rats maximize, fish and cockroaches match (Warren, 1965). There is at least one problem with this simple and straightforward ordering: Humans also match. It is evident that matching or maximizing may be caused by very different factors depending upon the subject and upon the experimental situation. In humans, a critical factor is the subject's interpretation of the task: Does he believe it is truly random, does he feel that there is a way in which he can outguess the situation? [For a theoretical account of probability learning in humans, see Estes (1964).]

trials on the discrimination. They choose the majority stimulus by roughly the proportion that this is rewarded, but their choices show no sequential dependency (e.g., staying with the winner). "The fish chooses on each trial as though it were consulting a table of random numbers" (Bitterman, 1968). Results of this kind are reported by Bitterman *et al.* (1958) and Behrend and Bitterman (1961, 1966); a thorough discussion of this literature may be found in Mackintosh (1969a,b). The results of a representative study are presented in Fig. 12.

What accounts for the difference in the performance of these animals? Two-stage theorists have tried to provide an answer. According to them, maximizing depends upon an animal's maintaining its attention to the relevant cue dimension. They suggest that when an animal does not choose the majority stimulus this is caused not by a preference for the minority stimulus but rather by failure to attend to the relevant cue dimension (Mackintosh, 1969b).

Consider a black–white discrimination with black positive on 75% of the trials and white positive on the remaining 25%. An animal that maintains its brightness analyzer will continue to approach black on most of the trials. After all, the response attachment to black far outweighs that to white. But if the brightness analyzer weakens, other analyzers can take over. Suppose an animal has just experienced a trial on which white (which happened to be presented on the left side) was reinforced. The animal may now switch to a right–left analyzer with subsequent impairment of its performance.

According to two-stage theory, "failures of attention" account for the fact that even the rat does not really achieve true maximization—the usual asymptotes are of the order of 90%, rarely much above this. But in

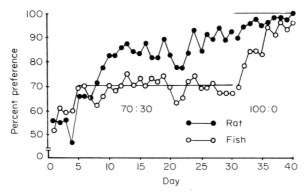

**Fig. 12.** Preference of fish and rat for the more frequently reinforced stimulus in 70:30 and 100:0 visual problems. From Bitterman *et al.* (1958).

the rat the analyzer is still maintained to a considerable degree: The animal can learn to attend to a particular cue dimension even if reward is inconsistent. The fish is not so proficient in this regard. It abandons the relevant analyzer as reward becomes inconsistent, and ever more so the higher the proportion of minority stimulus reinforcements. Trends toward maximizing and matching are the behavioral results.

Mackintosh discusses various lines of evidence which support this interpretation. For example, he points out that in rats pretraining which strengthens or weakens a particular analyzer will facilitate or hinder (that is, lead to maximizing or matching, respectively) subsequent probability learning in which the relevant cues involve the same analyzer (Mackintosh, 1969b). It is probably too early to evaluate this analysis. Thus far, only a very few of its empirical implications have been tested (Mackintosh and Holgate, 1968); in fact, since the theory has been stated rather informally, at least to date, its precise empirical implications are not always apparent. Mackintosh believes that this position can account for differences both in serial habit reversal and in probability learning in roughly similar terms: Competent performance in both of these (that is, improvement in the one and maximizing in the other) depends upon the strengthening and maintenance of attention to the relevant cue dimension in the face of changing reinforcement conditions. It is still debatable whether the two-stage model can account for results in both of these areas while using the same parametric assumptions throughout.

4. COMPLEX PROCESSES—DELAYED REACTION, DETOUR BEHAVIOR
   AND "INSIGHT"

There is a final group of phenomena generally classified together under the vague heading of "complex processes." These refer to effects which—at least to some earlier students of learning—suggested the operation of "higher processes." They include the delayed response, detour, and short-cut behavior, "insight" as manifested by unusually broad and appropriate transfer, and perhaps some species of latent learning. There has been rather little in the way of systematic work on any of these phenomena in the fish, although some attention has been paid to detour behavior and to delayed response.

Thorpe (1963) reviews several studies which establish the ability of various species of fish to detour around obstacles and through holes. There is evidently no doubt that fish can acquire more efficient spatial patterns of navigating through various environments. What is at issue is whether all of these accomplishments can be explained as instances of trial-and-error learning. Thorpe seems to feel that at least some successful solutions reported in the literature are "insightful." ". . . the law of effect, at least in the form then usually accepted, was inadequate to account for these results: for movements which had led to success in earlier trials were by no means necessarily 'stamped in' . . ." (Thorpe, 1963, p. 310). In this conclusion he leans heavily upon several studies performed by von Schiller (1949), who trained minnows in various detours around glass obstacles. These animals acquired a generalized movement pattern that was established and maintained despite considerable alterations of the cues. The solution was often very rapid and the final response smooth and continuous; under the circumstances this form of problem solving seemed to merit the term *insightful*. This conclusion was probably premature. A later study by Munn (1958) has thrown serious doubt upon this interpretation. Munn repeated Schiller's experiment in its essentials and showed that what the animals had learned primarily was to abandon a direct approach to the lure.

Experimental attention has also been paid to another phenomenon occasionally subsumed under "complex processes"—the delayed response. Several decades ago, the delayed response was generally taken as an index of behavioral complexity, and it was widely believed that "higher" animals could delay longer (and with fewer motor mediators) than "lower" ones. By now, this view is out of favor. Given the proper conditions rats can delay for hours (N. R. F. Maier, 1929), and the delayed response of monkeys may break down after a few minutes (Gleitman *et al.*, 1963). Even so, it would be interesting to see what delays can be found for the fish. As it happens, there is no conclusive evidence. Von Schiller (1948) claimed that minnows could delay for several seconds; however, an attempted replication by Munn (1958) was unsuccessful.

## C. Some Matters of Interpretation

### 1. ARE ALL THE EFFECTS THE RESULT OF A COMMON FACTOR?

Let us summarize the respects in which the fish's behavior in the learning laboratory differs from that found in rats and pigeons:

(1) Unlike rats and pigeons, the fish shows little or no improvement in serial habit reversal. A related effect concerns the fish's comparative rigidity in repeated conditioning-extinction sessions.

(2) While rats (and, to a lesser extent, pigeons) tend to maximize in probability learning situations, fish tend to match.

(3) Rats show a partial reinforcement effect whether trials are massed or spaced; fish show the effect only with massed trials. It appears

that the critical feature is a negative contrast effect in response to change in reward magnitude—rats show it; fish do not.

We have considered various interpretations of these separate effects. We must now ask: Can we find one central theme that links these effects together?

First, consider the attentional approach (Mackintosh, 1969a) which claims to provide a qualitative account of two of the phenomena (habit reversal and probability learning). As we have seen, this approach is not without its difficulties; for example, its analysis of habit reversal is seriously questioned by the results of Schade and Bitterman (1966). But quite apart from this, there seems no plausible way in which the attention mechanism can be extended to account for the fish's different reaction to shifts in reward magnitude. From this point of view, one factor could obviously not account for all the differences.

Another possibility is suggested ·by Bitterman (1968), who notes that the fish acts in neat accordance with a simple S–R reinforcement theory, of the kind proposed by Hull (1943). After all, some of the most telling evidence against Hullian theory comes precisely from those phenomena not found in the fish—habit reversal, spaced trials PRE, rapid readjustment, and contrast effects when confronted by changes in reward magnitude. (It is tempting to speculate about what might have happened had the fish been the primary subject studied in learning laboratories instead of the rat; for all we know, simple S–R reinforcement theory might still be the dominant theoretical position.) But does this mean that the mechanisms which underlie the fish's behavior are well described by a Hullian scheme based on simple S–R connections forged by reinforcement? Even granting Bitterman's view that the fish is an S–R animal, it is still difficult to identify the difference(s) between fish and homotherms: We cannot really tell what it is that the fish lacks until we know what it is the rat (and the pigeon) possesses.

One might suggest that the rat is able to order its experiences over time in a far more complex way than is the fish. It may be better able to differentiate them from each other and to act on the basis of the more recent event (e.g., improvement in habit reversal and response to shifts in reward magnitude). It may also be more capable of integrating a set of events over time, extracting an overall pattern (e.g., maximizing in a probability learning task). Unfortunately, these comments are at best descriptive and at worst too vague to provide much of a hint telling us how to proceed.

Under the circumstances, we must conclude that the common factor that underlies the various effects is still undiscovered; for all we know, no such common factor exists. Of course, the lure of such a common factor is still very powerful and one cannot quite give up the hope that once discovered it will give us a clue to what "complex learning" (perhaps intelligence is a better term) is all about. It is unlikely that this hope will become reality but, even if it should, we ought not to forget that, given the tremendous specialization of the teleosts and given further their enormous distance from any ancestor common both to them and to present mammals, any serious discussion of the evolution of intelligence is surely premature.

## 2. NATURALISTIC AND LEARNING THEORETIC APPROACHES COMPARED

It is quite clear that the naturalistic approach and the learning theory approaches have developed in mutual isolation. This is unfortunate because the strengths and weaknesses of these two orientations are in many ways complementary. For example, the naturalistic approach is primarily concerned with the animal's achievements in its normal habitat, while the learning theorist's approach is directed toward the discovery of basic processes or mechanisms (of learning or motivation) that he believes will reveal themselves in some idealized behavioral situation.

There is a further difference in the style of the actual investigations. Typically, the experiments performed in the learning laboratory are more carefully controlled with regard to variables involved in learning. The biologically oriented investigator controls sensory factors and a whole host of physiological variables with the utmost care, but he is typically less careful in the control of past experience and of cues that might influence his animals as he studies their "learning ability." With regard to learning, the learning theorist is far more of a skeptic: He needs more proof before he concludes that learning has actually occurred, and very much more proof yet before he will entertain the hypothesis that this learning process is of a "higher" sort (e.g., reasoning and insight). Lloyd Morgan's famous canon is but one early instance of this skepticism: "In no case may we interpret an action as the outcome of the exercise of a higher psychical faculty, if it can be interpreted as the outcome of the exercise of one which stands lower in the psychological scale" (quoted in Boring, 1950).

*a. Problems or Limitations of the Naturalistic Approach.* It is important to realize that few of the zoologically oriented investigators are primarily interested in learning as such; instead they focus upon particular functions that are of special importance to their subjects, e.g., migration or species recognition. The failure of much of this work to be more than descriptive or suggestive with respect to learning is partly a consequence of this orientation. However, some of the most interesting learning phenomena have been uncovered in the naturalistic framework and have remained as yet unanalyzed. For example, the work of Aronson (1951) suggesting an unusual ability for latent learning in the goby has not been followed up, although published almost 20 years ago. This phenomenon, if replicated and carefully analyzed, could contribute significantly to our understanding of memory. Similarly, the work on migration poses fascinating problems in learning and memory. The orientation of this work has been by and large toward the analysis of sensory factors (and for good reason). Hasler has made some promising beginnings in the study of the role of memory in migration, but it would indeed be useful if some investigators whose primary interest focuses upon learning and memory were to bring this interest to bear on such problems as the time course of "imprinting" of the home-stream odor, the conditions which interfere with the memory for this odor, and the role of reinforcement (perhaps the increasing strength of home-stream odor) in guiding the salmon back to his home stream.

Unfortunately, many of the phenomena described by naturalistic studies are often disregarded by psychologists working within the traditions of the learning laboratory. To the extent that this attitude is based on the lack of some important controls in certain of these studies, it is quite understandable. But the psychologist of learning has reservations that go beyond this. Extending Lloyd Morgan's canon, he tends to think of learning as an explanation of the last resort, to be accepted after all the "simpler" mechanisms (e.g., sensory effects) have been ruled out. In our view, this bias is unjustified. Biologically speaking, why should we prefer the hypothesis that a particular behavior pattern is genetically determined to the hypothesis that it is acquired through learning? Until the actual evidence comes in, both views are equally plausible. Since learning abilities of some sort are probably present in almost all multicellular animals, it would seem that the relative merits of an instinctive or a learned solution to a particular problem would depend primarily on the specific adaptive requirements of the situation and the developmental history of the species in question. Boring makes the point succinctly: ". . . nature is notoriously prodigal; why should we interpret only parsimoniously?" (Boring, 1950, p. 474).

*b. Problems or Limitations of the Learning Theory Approach.* As we have noted, the recent decade has seen an enormous increase in the comparative study of learning undertaken within the traditions of the learning laboratory. This increase has been highlighted by considerable ingenuity in the design and instrumentation of experiments coupled with imaginative theoretical formulations. But the ultimate success of this enterprise rests on the adequacy of the theoretical framework within which it is embedded. Naturalistic studies have a certain face validity: Migration, species recognition, and schooling play an obvious role in the actual life of various fish and have a clear adaptive significance to the species. No comparable face validity can be claimed for the studies coming out of the learning laboratory. Whether goldfish are capable of habit reversal or of a spaced-trials PRE is an issue the significance of which must be assessed by considerations that go beyond the particular phenomena in question.

Consider the goals of the enterprise. Some (by no means all) comparative psychologists of learning feel that the comparative psychology of learning should be able to provide, first, an accurate description of the differences in learning and related functions that hold across different animal groups, and, second, the use of this description to provide an account of the evolution of learning.

Have these two aims been met? We think not. A serious discussion of the evolution of intelligence seems clearly premature given the material at hand: Virtually all our data come from two teleosts, goldfish and mouthbreeders, hardly a fair sample of such an enormous group of animals comprising more than 18,000 species. Granted that these teleost fish are very different; still, they can hardly be classified as primitive, and they are surely far off the main line of evolution to mammals. Equally important perhaps, these animals have been studied in a context and in experimental situations very much different from those in which they evolved. This is not to say that such studies will not help us learn something important about learning but rather that they will not help us learn much about evolution. [For a further discussion of the shortcomings of present-day comparative psychology in its attempts to deal with the evolution of behavior, see Hodos and Campbell (1969).] Similar arguments make it clear that we do not as yet have the data to characterize the differences between fish and mammals, each taken as a group. We may know something about the goldfish and the mouthbreeder, but it seems rather premature to talk about *the fish*. At the same time, we must certainly note the remarkable similarities of these two species when studied in the laboratory situation.

As Mackintosh (1969a) has pointed out, however, the comparative study of learning has another purpose, which is generally considered to be the primary one and is not subject to the criticisms we have just raised. This is the use of different species as convenient preparations to study basic processes of learning and memory. Granted that the goldfish is not an appropriate representative of fish as a group, granted also that it (no more than any other modern teleost) is hardly in the mammalian ancestral line, the differences between it and another organism may still help us to understand something about the learning process as such. If two phenomena (say, contrast effects and spaced-trials PRE) both occur in one organism and both fail to occur in another, does this not tell us something about the mechanisms that might underlie these two? This argument is very persuasive, but even so we have some serious misgivings. For there is a further problem with the learning-theoretic approach as applied comparatively, a problem that lies in the very nature of the comparative question. What does it mean to say that some species has a capacity (e.g., habit reversal) which another species does not have? Let us consider the learning-theoretic approach with this question in mind.

Recent students of learning (particularly Bitterman, Gonzalez, and their colleagues) have made a strong case against the rat-centered outlook which had dominated learning theory during previous decades and are embarked on a vigorous program of investigation embracing many other species. But the fact is that as they turn from the study of one species to the next, their techniques remain essentially unchanged. Of course the apparatus in which the fish is studied is aquatic, is modified to suit the sensory and motor capacities of the animal, and delivers rewards appropriate to the fish. But, in its essentials, the boxes for fish and rats and pigeons are all built according to the same theoretical plan. This procedure does indeed allow for easier interspecies comparison. But one might well argue that we distort our notions of animal behavior by forcing these various species into the same experimental mold.

We must certainly grant that Bitterman, Gonzalez, and their colleagues have been especially careful, in some respects, about concluding that there are species differences. They are keenly aware of sensory and motivational differences, and by various parametric variations they attempt to circumvent such limitations. Nor do they concern themselves with simple quantitative features such as rate of learning but instead look for qualitative differences (e.g., habit reversal). But they do assume that given "appropriate" stimuli, responses and reinforcers, given also some invariance across different levels of motivation, it is then possible to make meaningful comparisons across species. This conclusion is at least debatable.

Consider the "response" the animal is rewarded for. Students of learning sometimes talk as if virtually all responses were interchangeable, equally joined to any stimulus, equally strengthened by any reinforcer. This is probably not the case. Various lines of evidence are accumulating which suggest that the relation between response and reinforcer is not altogether arbitrary (e.g., Breland and Breland, 1966). An interesting example is provided by Sevenster (1968) who trained sticklebacks to bite a rod with presentation of a female as reward. Acquisition rate was slow because the reward released incompatible behavior sequences: In sticklebacks, the sight of a female inhibits biting. Hogan and Rozin (1958) ran into similar difficulties when they tried to train male Siamese fighting fish, *Betta splendens*, to press a lever with the sight of its mirror image as reward; after more than a year of effort, they finally gave up. It is not the case that the sight of the mirror image is an inadequate re-inforcer: When a locomotor response is used, whether it is breaking a light beam (Thompson, 1963; Hogan, 1967) or swimming down a "runway" (Hogan, 1967), instrumental conditioning proceeds very well. What is evidently critical is the relationship between the particular response and the particular reinforcer.

[Work by Hogan and his colleagues suggests the possibility that mirror image and food rewards may be functionally different in other respects as well. Mirror-image rewards led to faster extinction rates than did food (Hogan, 1967); furthermore, no PRE was obtained when the reinforcement was the mirror image (Hogan et al., 1969). It is very possible that these differences will ultimately be shown to be a function of effective reward magnitude. On the other hand, the very fact that food and mirror image differ substantially in their functional relations to various response classes should make us more wary about dismissing further differences between these two rewards by prematurely subsuming them under the old, familiar categories.]

A similar point holds for the relation between conditioned and un-conditioned stimuli. It appears that some associations are formed more readily than others. For example, Garcia et al. (1968) showed that in rats visual or auditory stimuli are more easily conditioned to peripheral pain, while gustatory stimuli are more easily conditioned to internal malaise produced by X-rays.

Perhaps one may argue that all such problems of situational speci-ficity apply only to a "lower" level: the determination of an adequate stimulus, an appropriate response, an effective reinforcer. For these one might perhaps grant some situational specificity. But the student of learning would propose that once it is established that the species is motivated, that the cues are clearly perceived, that the manipulandum is appropriate, and that different levels of drive do not alter the essential shape of the behavioral curves, then the pattern of results can be compared across situations and across species. Put another way, he argues that once a fish has been trained to strike a lever for food reward in the presence of a light signal one can then ask meaningfully whether it is capable of habit reversal. But we believe that it is at least an open question whether habit reversal (or the Crespi contrast effect, or probability matching) is a phenomenon that is any less situation specific than the conditionability of a particular response.

Comparative psychologists of learning tend to regard such phenomena as essentially all-or-none: Some species show them and others do not, and this quite independent of the situational conditions.* But we should not forget that as far as fish are concerned such interpretations are necessarily based upon null effects—failure to find improvement in habit reversal, failure to find a contrast effect, and so on. One may want to conclude that the failure to find such effects in several species of fish (given the presence of this effect in rats and pigeons) indicates a difference in underlying mechanisms. Alternatively, one might conclude that the failure to find an effect simply resulted from the particular conditions under which it was sought.

In a sense, the burden of proof would seem to fall upon those who favor the second alternative. What we want to do here is to argue that this alternative is still open and is in fact quite plausible. We believe that an animal's capabilities are best determined if that animal is studied in situations where such capabilities would have adaptive value for its species. It is not clear that the various devices in which fish have been studied in the learning laboratory fit this condition. After all the salmon exhibits remarkable memorial feats with regard to stream odors; would we have discovered comparable capacities in its retention of a visual discrimination in a standard fish box? Improvement at habit reversal might well occur in fish given a situation in which such improvement is critical to the survival of the species, perhaps, in learning the location of food sources.

Whether these comments constitute genuine limitations of the learning-theoretic approach is as yet an open question. But at least on one point we are quite certain: The limited interchange between the learning-theoretic and the naturalistic approach has been of detriment to both.

## IV. MEMORY

Recent discussions of memory generally start with a distinction between short- and long-term memory since enough evidence has accumulated to suggest that we may be dealing with two essentially different

---

* Not all investigators in this area subscribe to an all-or-none view of habit reversal or probability learning performance. While Bitterman and his colleagues tend towards this view, Mackintosh explicitly rejects it (Mackintosh, 1969a).

processes. We will follow this distinction as we consider studies of memory in fish.

## A. Short-Term Memory in Fish

A critical line of evidence for the existence of a separate short-term memory system comes from studies on consolidation in which the subject suffers some severe trauma which leads to memory deficits (sometimes impermanent) for experiences just previous to the trauma. The trauma (e.g., electroconvulsive shock) presumably interferes with a consolidation process whereby the trace "establishes itself," that is, is transferred to the much more stable long-term store. Some of the most active research along these theoretical lines has been performed on goldfish in the laboratories of Agranoff, Davis, and their colleagues (see Agranoff and Davis, 1968).

### 1. MAJOR METHODS

Agranoff and Davis employ the same basic technique throughout. Goldfish were trained to cross back and forth in a standard fish shuttle tank to avoid an electric shock signalized by a light. The CS–UCS interval was usually 20 sec, the number of trials per day 20–30. Training was for one day only; retention was tested over 10 trials 4 days later by a relearning method. The effect of experimental procedures (e.g., electroconvulsive shock and injection of puromycin) was assessed by first "predicting" the retention score the animal would have achieved had there been no experimental manipulation and then comparing this theoretical score with the retention score actually obtained. Reference to a large control group allows calculation of the predicted score (the score the subject "should have" obtained had there been no further experimental manipulations) given the subject's performance on Day 1. Retention is expressed as the difference between predicted and obtained scores; thus, a score of —2.00 would mean that the number of avoidances on the relearning day was two less than it would have been without experimental manipulation.

The main manipulation employed by Agranoff and Davis was intracranial injection of puromycin. However, other amnesic agents such as electroconvulsive shock and KCl were also used. Previous investigators (e.g., Flexner et al., 1963) have shown amnesic effects following puromycin treatment in mammals, effects which are greater the shorter the interval between training and puromycin injection. These and other investigators were particularly interested in the fact that puromycin is a substance which inhibits protein synthesis and related its disruption of consolidation to various biochemical theories of memory. The search for the biochemical or molecular correlates of memory has indeed been the major impetus behind the Agranoff and Davis program but, quite apart from this, their work has important implications for the possible relationship between short-term and long-term storage systems, whatever their biological basis may ultimately turn out to be.

## 2. MAJOR FINDINGS

*a. Retention Decrements.* Puromycin injected 1 min after acquisition produces serious decrements in retention (Agranoff *et al.*, 1965). Other amnesic agents employed with similar effects include electroconvulsive shock (Davis *et al.*, 1965a), acetoxycycloheximide (Agranoff *et al.*, 1966), actinomycin (Agranoff *et al.*, 1967), KCl (Davis and Klinger, 1969), and possibly heat narcosis (Cerf and Otis, 1958). Not surprisingly, the effect of puromycin injection is greater, the higher the dose (Agranoff *et al.*, 1965), complete amnesia resulting from 170–210 $\mu$g of puromycin administered immediately after training.

*b. Consolidation Intervals.* The amnesic effect is greater the smaller the interval between training and puromycin injection. This of course is the critical finding, suggesting a disruption of trace consolidation. Agranoff *et al.* (1965) found that in their situation the "consolidation interval" was about 30 min: Amnesic effects were obtained if the interval between training and injection was 30 min but not much longer. Somewhat different consolidation intervals were obtained when other training procedures were employed (Davis, 1968; see discussion below) or when different amnesic agents were utilized (Davis *et al.*, 1965a). The longest consolidation interval was obtained with KCl: Intracranial injection of KCl has amnesic effects even when injected 18 hr after training (Davis and Klinger, 1969). Finally, there is some evidence to indicate that the consolidation interval is longer at lower temperatures (Davis *et al.*, 1965a).

*c. Effects on Acquisition.* Puromycin does not seem to interfere markedly with acquisition, as opposed to retention. Injection of puromycin 1 or 20 min before the onset of training is followed by normal acquisition and a significant decrement in retention (Agranoff *et al.*, 1965).

*d. A "Pure" Curve of Short-Term Memory.* Davis and Agranoff (1966) argue that since immediate post-training injection of high doses of puromycin results in complete amnesia 3 days later, all that is left of the fish's learning experience is in the short-term store. Retention tests after various intervals following training *and* immediate puromycin injection thus provide a forgetting curve of short-term memory. The authors report a gradual decay starting some time after 6 hr and extending past 48 hr. A somewhat steeper forgetting curve was obtained using KCl rather than puromycin (Davis and Klinger, 1969).

*e. Trigger Effect.* The consolidation interval can be significantly extended by leaving the fish in the shuttle box during the interval between completion of training and administration of puromycin. This phenomenon, called the *trigger effect* (Davis and Agranoff, 1966; Davis, 1968), suggests that consolidation is inhibited by stimuli associated with the learning situation—in effect, the memory trace stays on in the short-term store while still "aroused" by the perceptual context in which it was first acquired.

*f. Extensions of Consolidation Interval (Intertrial Environment Effect).* Davis and Klinger (1969) report the important finding that amnesia can be produced by puromycin, KCl, or acetoxycycloheximide injected as long as 24 hr after training (that is, vastly beyond the usual consolidation interval), if, just prior to injection, the fish are placed in the intertrial environment (ITE) (that is, the shuttle tank without CS and UCS; see Table I). Appropriate controls indicate that this phenomenon cannot be discounted as a simple effect of the drug upon test performance. The ITE effect appears to be sensitive to manipulations of the training-test intervals. For example, using puromycin, no ITE effect was apparent at the usual 4-day interval, but an effect appeared with an 8-day interval (Table I).

*g. Biochemical Correlates.* Agranoff and Davis have searched for

### Table I
#### Procedures and Results for ITE Studies[a]

| Time since training | | | | Performance (difference score) |
|---|---|---|---|---|
| 0 hr | 1 Day | 4 Days | 8 Days | |
| KCl | | Test | | −3.22 |
| | KCl | Test | | −0.30 |
| | ITE | Test | | +0.02 |
| | ITE then KCl | Test | | −3.76 |
| | ITE then puromycin | Test | | +0.83 |
| | ITE then puromycin | | Test | −1.66 |
| | Puromycin | | Test | −0.04 |

[a] ITE refers to intertrial environment (placing fish in deactivated experimental box). Difference scores refer to differences between test performance and predicted control result (from Davis and Klinger, 1969).

correlations between amnesic disruption and biochemical effects on brain metabolism produced by the amnesic agent. They have discovered that substances which block protein synthesis (puromycin and acetoxy-cycloheximide) or RNA synthesis (actinomycin) act as amnesic agents. In some cases, they have been able to show a correlation between inhibition of protein synthesis and amnesic disruption (Agranoff *et al.*, 1966); in other instances a meaningful correlation did not emerge (Agranoff *et al.*, 1965).

It is certainly possible that these various substances are acting purely as traumatic agents and that their specific biochemical effects have no particular relevance to the understanding of memory. There is abundant evidence to show that traumatic treatment of almost any sort can seriously interfere with ongoing, labile processes such as memory formation. After all, the specific action of KCl and electroconvulsive shock is probably not identical to that of the protein or RNA inhibitors; still, as Agranoff and Davis themselves have shown, they produce quite similar results. So far then, the work is only a beginning, no matter how exciting. One might eventually hope to look for specific alterations in the tissues participating in memory formation, and indeed a start has been made along these lines (Shashoua, 1968).

## 3. Implications of the Results

The results presented here have far reaching implications for our conception of the memory system. This makes it all the more important to ask some questions about procedural details, generality of results, and interpretation.

*a. Some Questions of Procedure.* It should be pointed out that most of the effects reported by Agranoff and Davis are quite robust despite considerable variability in some aspects of the data [e.g., large variations in acquisition scores attributable in part to seasonal effects; Agranoff and Davis (1968)]. However, we do not feel as sanguine about the trigger effect. The experiment here involves two procedures: maintaining the fish in the (deactivated) shuttle tank for an interval after training and injecting puromycin immediately after that interval. For the trigger effect to be real it must be larger than the summed effects of these two component procedures considered separately. Davis (1968) did run the two appropriate controls in which the two component treatments were administered separately. Both control groups did show some effect. The sum of the two effects is almost as large as that obtained for the trigger group (the respective difference scores with a trigger interval of 3½ hr are —2.2 and —2.7), a difference that is almost surely within range of sampling error. Davis based his claim for the phenomenon upon a comparison between the retention scores of the trigger group and those of the usual large, standard control group; this comparison yielded a statistically significant result. We have some reservations about Agranoff and Davis' practice of testing most of their experimental effects against a standard control rather than against one trained as part of the same experiment; since most of their phenomena are quite sizable this point is in general not relevant. In the case of the trigger effect, however, this objection seems serious enough.

b. *Generality of the Findings.* There is an obvious advantage in using the essentially identical experimental procedure throughout an entire research program, but such an approach is not without some disadvantages. Consider the specific test intervals used by Davis and Agranoff. Four days is the standard interval almost always used in their laboratory; a test at 8 days was conducted in but one recent study (Davis and Klinger, 1969) and then only if a 4-day test failed to yield an effect. Since we know that avoidance learning in particular is very sensitive to manipulations of intersession intervals (e.g., Kamin, 1957) even some modest parametric variations would be very helpful.

Perhaps the most serious limitation of the generality of the whole set of phenomena here described concerns their application to forms of learning other than avoidance. Agranoff and Davis imply that their work bears on the nature of the *memory trace;* presumably they expect that their effect would hold regardless of what the trace is "about" (aversive or appetitive learning, classical or instrumental conditioning, etc.). There is some evidence that not all forms of learning are equally disrupted by puromycin injection.

Potts and Bitterman (1967), using a discriminated avoidance in goldfish, demonstrated that immediate post-training injection of puromycin led to a decrement in the *overall* level of performance during the retention test but with little loss of the discrimination: While the absolute number of responses declined considerably, the ratio of responses during CS+ to that during CS— remained unchanged. Avoidance conditioning is usually considered as being composed of two components—classically conditioned fear and instrumentally conditioned escape (Rescorla and Solomon, 1967). The findings of Potts and Bitterman suggest that puromycin acts primarily to reduce the level of motivation based upon the classically conditioned fear component.

Considering the outcome of the Potts and Bitterman study (1967) further studies of the puromycin effect in other learning situations (e.g., appetitive) are clearly necessary. No systematic work has been performed in this area. However, two recent studies represent beginnings along these lines. Oshima *et al.* (1969) found that the characteristic electrophysiological response to home waters in salmon disappeared 4–7 hr after intracranial administration of puromycin or other metabolic inhibitors. A partial restoration of the response was apparent at 9–28 hr postinjection. At this point, it is not clear to what extent this interference can be described as acting upon memory rather than some form of sensory integration. The partial recovery by 9–28 hr suggests that the effect may be temporary. Therefore, if we are dealing with a memory-blocking effect, the agents seem to be acting on the retrieval process. It would certainly be surprising if the storage (as opposed to the retrieval) of a memory some few years old were easily disturbed by metabolic inhibitors.

In a rather unorthodox study, Shashoua (1968) found a disruptive effect of puromycin upon an adaptive motor pattern in goldfish: The fish had to compensate for small floats attached to their undersides which first caused them to swim upside down, a compensation that became much more rapid and efficient after one session. The interpretation of the puromycin effect here obtained is not too clear, since we cannot be sure that the fish's motor compensation represents an example of true (or at least, conventionally considered) learning.

*c. Some Problems of Interpretation.* Throughout this discussion we have used the terms *amnesia* or *memorial disruption* as though it were clearly established that the basic effects were indeed upon memory. The actual story may be more complicated than this. There is a sizable literature on memory consolidation in mammals (mostly on rats or mice subjected to electroconvulsive shock) with experimental paradigms similar to those used by Agranoff and Davis. This literature reveals considerable disagreement about the mechanisms responsible for the deficits on the retention test. Various alternatives have been suggested.

First, the effects may indeed reflect a disruption of the memory consolidation process. This still leaves many further questions quite open. For example, we may still ask whether they are limited to short-term memory and whether the various traumatic agents (e.g., KCl and puromycin) wreak their havoc by similar mechanisms.

A second interpretation holds that the effects are the result of some additional learning with the so-called amnesic agent serving as a UCS for some competing response which interferes with the original response pattern (e.g., Lewis and Maher, 1965). A related interpretation suggests that some "amnesic agents" act as punishers (Coons and Miller, 1960). There is certainly evidence that shows that electroconvulsive shock (ECS) does act as an aversive stimulus in rats (McGaugh and

4. LEARNING AND MEMORY

Madsen, 1964; McGaugh, 1965). On the other hand, it does appear that some part of the disruptive effect is genuinely amnesic and not easily attributable to the role of ECS as a UCS for defensive or other incompatible responses (e.g., McGaugh, 1965; Quartermain *et al.*, 1965).

In the case of the various traumatic agents used on fish, we are probably fairly safe in ruling out most versions of a competing response theory. Since the traumatic agent is administered only once and often at least 30 min after training (an administration that takes place outside of the training situation), it is hard to imagine how an association could be formed between the training situation and the traumatic agent.

On the other hand, interference might contribute to some of the effects. A case in point is the ITE. According to Davis and Klinger (1969), a few traumatic agents will have an effect as long as 24 hr after training, if just prior to injection the fish are placed in the intertrial environment. This finding is not easy to interpret in the context of a memorial theory. One might suggest that the intertrial experience rearouses memories of the training situation and that the amnesic agent somehow blocks later retrieval, but this line of argument obviously raises far more questions than it solves. As Davis and Klinger themselves suggest, such findings may be encompassed by the competing response hypothesis.

A third interpretation of the phenomena is that they represent a (presumably transient) performance effect, analogous to that created by a change in drive level or a dose of a tranquillizing drug administered just before a test session. This interpretation is all the more plausible when we are dealing with escape or avoidance learning, for here much hinges on the animal's level of fear. It is well known, for example, that the retention of avoidance learning in rats is a U-shaped function of interval since training: Performance worsens immediately after training, reaching its nadir at 1–4 hr. Thereafter, it improves again, continuing to rise until somewhere between 24 hr and 19 days (Kamin, 1957). Kamin attributes this effect to an increase in fear which "incubates" over time. A similar effect has recently been demonstrated with goldfish who avoid more competently if tested 168 hr as compared to 24 hr after training (Pinckney, 1966; see Fig. 13). Considerations of this sort are all the more relevant to the studies of Agranoff, Davis, and their associates since almost all of these studies use the same interval between training and test. Thus, at least some of the retention decrements they report might be performance effects caused by changes in fear incubation, shifting the curve of the Kamin effect along the time axis. The ITE effect could be a performance decrement of this sort: Returning the animal to the shuttle box surely arouses fear and could interact with the Kamin effect.

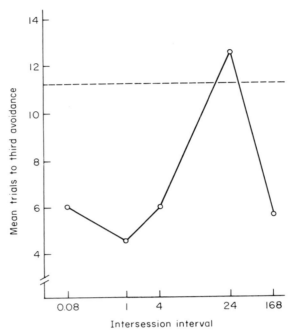

**Fig. 13.** Relearning of an avoidance as a function of interval since training. From Pinckney (1966).

It should be noted that the ITE effect does depend on the training-test interval (see Table I).

To summarize, the work of Agranoff, Davis, and their associates is presently at the forefront in the psychophysiology of memory. Their accomplishments are very impressive, and they were achieved in part by the strategy of concentrating upon a narrowly defined experimental situation. But this very concentration has the necessary drawback of limiting the generality of their findings. One of their empirical conclusions (the trigger effect) is not firmly established as yet, and in some cases alternative (that is, nonmemorial) interpretations have not yet been ruled out. Most of these criticisms would be answered by variations from a fairly standardized experimental procedure; for example, using more than one retention interval, employing learning situations with reinforcement contingencies other than avoidance (especially those based upon appetitive rather than aversive reinforcement) and with different response requirements (in which some animals are trained to respond while others are trained not to respond). We would guess [and Davis and Klinger (1969) have implied] that such variations would show that some of the phenomena under discussion are in fact caused by more than one factor. We think that the primary factor will indeed prove to be memorial, but its exact relevance to the understanding of storage and retrieval processes will not be known until it is isolated from those factors that are not.

## B. Long-Term Memory in Fish

What accounts for forgetting when the interval between training and test is long enough so that we are sure that we are beyond the range of short-term memory? There are rather few studies on long-term forgetting in fish or for that matter in other animals. In part this comparative neglect has resulted from a widespread conviction that habits are essentially permanent (assuming only that they are well protected from interference by further learning): Most workers in the area have been so impressed by the fact that learned patterns persist that few have asked whether they persist in full. Some recent studies have shown that this view is false. There is little doubt that rats forget various instrumental responses over a 1- or 2-month interval after training (e.g., Gleitman and Steinman, 1963); similar intervals produce serious declines in the avoidance performance of goldfish (Gleitman et al., 1970).

What produces forgetting of long-term memories? Two major theoretical approaches have been proposed, one based on the concept of interference, the other on that of decay. The overwhelming preponderance of studies has concentrated upon interference, mostly in the context of human rote learning (e.g., Underwood, 1957). This position asserts that forgetting is essentially a species of negative transfer: Access to one set of memories is interfered with by another set learned either before (proactive inhibition) or after (retroactive inhibition). The alternative position, which argues that memories decay over time, has been largely neglected. Interestingly enough, the few studies which seriously consider this second alternative have been conducted upon fish, capitalizing upon their poikilothermy and considerable learning capacity.

The decay theory of forgetting asserts that the memory trace (or perhaps, the access to this trace) becomes degraded by an unknown process that operates over time. Whatever the process, it is thought to be part and parcel of the normal biological life functioning of the organism. If so, then whatever speeds up the overall life processes (of which the hypothesized decay process is somehow a part) must necessarily increase forgetting; whatever slows these down will decrease it.

Several studies have tried to provide an empirical base for decay theory by varying the temperature (and thus the metabolic rate) of poikilotherms during retention intervals after learning. Not surprisingly, the fish has been the organism of choice in such experiments.

The classic study in this area is by French (1942). Goldfish were taught a four-unit maze, "rewarded" by escape from bright light and the company of another fish in the goal box. Following acquisition the animals were placed in small containers and gradually brought to one of three temperatures: 28°, 16° (the training temperature), or 4°C. They were left at these temperatures for 20 hr, then adjusted back to 16°C, and finally retrained to criterion. Relearning was more rapid the lower the temperature during the retention interval (see Fig. 14). French argues that this cannot be because lower temperatures facilitate learning. Quite the contrary: naive fish, after 20 hr at 4°C, learn the maze more slowly. This control also rules out the possibility that the retention effects might be artifacts of temperature acclimation, a point raised by Jones (1945). Jones points out that fish held at a lower temperature during the

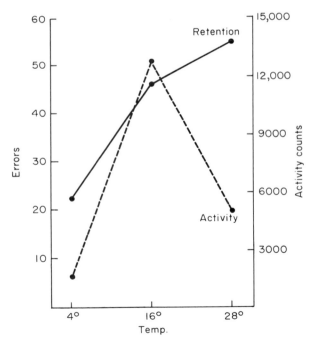

**Fig. 14.** Comparison of retention with amount of gross activity during the period of forgetting. From J. W. French (1942). The effect of temperature on the retention of a maze habit in fish. *J. Exptl. Psychol.* **31,** 85. Copyright (1942) by the American Psychological Association, and reproduced by permission.

retention interval would be expected to show a higher metabolic rate when returned to the standard temperature than would fish held at a higher temperature. Given that the naive animals in French's study performed most poorly after an interval at 4°C, Jones' criticism does not seem to apply.

Can this effect be attributed to increased activity level during retention intervals spent at higher temperatures? If so, it might be a function of interference after all: More active animals perform more responses and perhaps learn more incompatible reactions. French monitored the activity level of his animals and his results suggest that activity level is not the critical factor. His fish were more active at 16° than at 28°C; forgetting, however, was greater after they had been stored at 28° than after 16°C. Since temperature accelerates metabolic rate, French concludes that decay of memory traces accounts for the effect: Decay is accelerated in the fish with higher metabolic rate.

Later investigations have obtained results that are rather more equivocal. Erickson (1956) performed a modified version of French's experiment, using the paradise fish, *Macropodus opercularis* (as a surface breather, this fish would not be subject to anoxia caused by small volumes of overheated water). He found that a sojourn at 10°C retarded forgetting, but he could not show any effect for any of the other four temperatures (in five steps from 15° to 30°C) used during the interval. Given his findings on activity level at different temperatures, his retention effect could well be a side effect of activity.

Some recent experiments by the authors (Gleitman *et al.*, 1970) were designed to provide a more systematic test of a temperature retention effect. After all, French's learning situation was a bit unorthodox: The reinforcement condition was somewhat obscure (escape from confinement and the company of another goldfish in the goal chamber) and the retention interval was quite small (22 hr). There was also the possibility that long-term retention effects were confounded by consolidation phenomena: French's experimental animals were brought to the high temperature within 30 min after learning, a period that may be within the range of the consolidation process in fish. For these reasons, Gleitman *et al.* used goldfish in an avoidance shuttle tank in an experimental paradigm patterned after French's. The fish were trained to a moderate criterion in a shuttle tank and then extinguished 1 day, 4 weeks, or 8 weeks after acquisition. The 4- and 8-week retention intervals were spent in a tank at either 25°–26°C (the training and test temperature) or 33°C. To avoid any possible disruption of consolidation processes the "hot groups" remained at the lower temperatures for 2 days before transfer to the warmer temperature; to avoid complications with re-adaptation they were brought back to the training temperature 2 days before test. To control for the effects of prior temperature on learning and performance, two groups of naive fish were maintained for 8 weeks at 25°C to 26°C and 33°C, respectively, and then trained on the avoidance task.

The results of a first study were in line with French's findings: Retention deteriorated with increasing interval and with increasing temperature during that interval. The results of a second study which used several temperature levels during the interval were not in agreement: The temperature effect on retention was virtually absent (limited to the first test trial only and quickly declining as trials progressed). To date, we have not been able to resolve the inconsistency in the results of these two experiments.

Some related support for the decay theory comes from two studies by Rensch and Dücker. These authors considered the effect of chlorpromazine during the retention interval (Rensch and Dücker, 1966). Two groups of fish (*Carassius auratus*) were trained to criterion on a visual form discrimination. Following acquisition, retention tests were administered every 12 days over a 108-day interval. The experimental fish spent the retention interval in water treated with chlorpromazine. To avoid drug effects upon performance, the experimental animals were placed in drug-free water for 3 days prior to each retention test. The results show better retention for the animals treated with chlorpromazine (see Fig. 15).

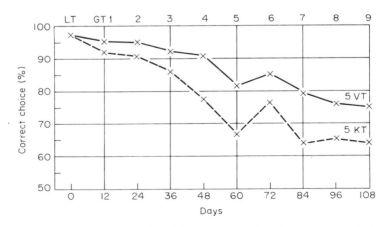

Fig. 15. Forgetting curves for fish treated with chlorpromazine during the retention interval (VT) and for control fish (KT). The ordinance indicates percent correct choice; the abscissa indicates days. From B. Rensch and G. Dücker (1966). Verzögerung des Vergessens erlernter visueller Aufgaben bei Tieren durch Chlorpromazin. *Arch. Ges. Physiol.* **289**, 200–214. Springer-Verlag, Berlin-Heidelberg-New York.

A related finding concerns the effect of darkness during the retention interval upon the retention of a visual habit. Dücker and Rensch (1968) trained two groups of goldfish to criterion on a visual form discrimination. Following acquisition, retention tests were administered every 12 days over an interval of 84 days. The experimental animals were kept in total darkness, the controls in constant illumination. The dark-kept fish were exposed to the controls' illumination level for 3 hr prior to each retention test. The results show better retention for the dark-kept animals.

Rensch and Dücker interpret both the chlorpromazine and the darkness effect as support for the decay position. They argue that the critical factor in trace breakdown is overall neural activity; whatever lowers the general metabolic level or perhaps the overall excitatory state of the relevant system (e.g., lowered visual input during darkness given that the relevant trace is visual) should slow up the forgetting process.

It is worth noting that interference theory can also account for the slowing down of the forgetting rate by chlorpromazine, darkness, or for that matter cold temperatures, on the assumption that all of these minimize learning (and thus later interference) during the retention interval. This hypothesis is difficult to establish and equally difficult to disconfirm. To establish it one must show not only that the factor which accelerates forgetting (e.g., heat or light) speeds up learning during the retention interval but also that the learned patterns which are acquired during this interval are such as to interfere with the habit the retention of which we are testing. To disconfirm the interference interpretation one must show that heat, light, and so on have no effect upon rate of learning during the retention interval. One approach might be to make learning during the retention interval virtually impossible (e.g., by anesthesia); should the temperature effect, for example, still hold under these conditions, an interference hypothesis becomes rather implausible. Such a study is far from easy to execute. (The present authors have tried unsuccessfully to develop an appropriate technique for over a year.) However, until an experiment of this kind is actually performed, one would guess that both decay and interference theories will continue to coexist for the foreseeable future.

## V. PHYSIOLOGICAL MECHANISMS

This section will present highlights of some recent approaches to the physiology of learning in fish: (a) localization of function, (b) inter-ocular transfer, and (c) cold block and temperature acclimation.

## A. Localization of Function

Fish have not infrequently been the subjects of experiments which attempt to relate specific brain areas to learning or memory capacity. [General reviews may be found in Aronson (1963) and in Healey (1957).] In spite of the fact that at least until recently, the connection between forebrain function and memory has been at best debatable (for reviews, see Janzen, 1933; Aronson, 1963; Aronson and Kaplan, 1968), most investigators have continued to study the effects of forebrain ablation on learning while neglecting to consider more caudal areas which are almost surely more critical to learning and memory. At the same time, the interest in the fish forebrain is perfectly understandable. To begin with, it is a structure that is the homolog of some areas (e.g., limbic system) which are of great importance for learning and motivational functions in mammals. But furthermore the fish forebrain is also homologous with the tissue that gives rise to the cerebral cortex in mammals.

What is the effect of forebrain removal in fish? The earlier literature suggested little or no disturbance of learning and memory (see Aronson, 1963). More recent studies, however, show a different and rather consistent pattern. Studying goldfish or *Tilapia*, Hainsworth *et al.* (1967), Savage (1968), and Aronson and Kaplan (1968) have found massive defects in the acquisition and retention of avoidance performance after forebrain removal. Considering the widely accepted two-factor approach to avoidance learning (e.g., Rescorla and Solomon, 1967), it seems natural to ask whether the ablation interferes with the classically or the instrumentally conditioned responses or with both. The answer is probably not simple. Classical conditioning as such is unimpaired in the forebrainless goldfish (Aronson and Kaplan, 1968). Nor is there any evidence that the instrumental component is lacking: Hainsworth *et al.* (1967) report that forebrainless fish, showing severe avoidance deficits, do still escape with normal latencies once the shock begins. The deficit in avoidance learning may not be caused by the absence of either of its two components but rather by some inability to link these up. A recent experiment by Overmier and Curnow (1969) demonstrates perfectly normal classical conditioning with an electric shock UCS in forebrainless goldfish and thus supports the notion of a defect in the interaction of classical and instrumental stages.

On the other hand, the deficit may be broad and very generalized. According to Aronson and Kaplan (1968) forebrain removal seems to depress many functions but to eliminate few. They suggest that the fish forebrain serves primarily as a modulator of other (lower) brain centers and that it functions in regulating arousal or awareness. They marshal several lines of evidence to support this contention. Thus, learning deficits are not usually found in classical conditioning or in simple instrumental tasks where performance would be facilitated by behavioral consistency. On some tasks which actually put a premium on rigidity, forebrainless fish do better than normal (e.g., learning a Y-maze with position cues; Ingle, 1965a). Perhaps the matter was put most succinctly by Janzen (1933) who described the forebrainless fish as "lacking in initiative."

As we have mentioned already, there has been but little work on the roles of lower centers in learning in the fish. Some promising beginnings have been made by Regestein (1968), who showed that unilateral hypothalamic lesions affect avoidance responses when the CS is presented to the contralateral eye.

## B. Interocular Transfer

Fish have been useful subjects for the study of interocular transfer since they offer the advantages of considerable learning ability coupled with complete contralateral projection in the visual systems. The neat interocular transfer paradigm has been applied to fish, first, to elucidate the role of the tectum and the neural cross-connections in the establishment of learning and memory and, second, as a tool to study coding in the visual system on the assumption that intertectal and other cross-connections may act as filters which selectively pass or attenuate certain components of the visual input.

Following an initial demonstration of interocular transfer in fish by Sperry and Clark (1949), McCleary was able to obtain clear evidence for interocular transfer of a cardiac deceleration CR to light paired with shock (McCleary, 1960) using a technique developed by McCleary and Bernstein (1959). On the other hand, interocular transfer was quite weak when an avoidance situation was employed. McCleary noted that fish that failed to avoid the CS on transfer trials nonetheless showed an agitated response to that stimulus (thrashing around and changes in respiratory rate); this suggests that the classical "fear" components of the response did transfer. Further studies showed that interocular transfer of avoidance conditioning ultimately depended upon one factor—whether the untrained eye was open or occluded during the initial training. While classical (cardiac) interocular transfer could be easily obtained in either case, avoidance only transferred (and not

completely even then) when the untrained eye was open during training. The critical dimensions are certainly not clear. Apparently the
visuomotor coordination requirements of the task somehow add further
restrictions on the possibility of interocular transfer.

Utilizing a "go, no-go" avoidance situation with goldfish (Fig. 16),
in which the animals must swim forward when given one stimulus but
not when given the other, Ingle has further elaborated these intriguing
findings. [For a more detailed review of this work see Ingle (1968a).]

Under appropriate conditions, the left and right visual systems can
operate independently since opposite discriminations (either simultaneous or successive) can be easily established in the two eyes. But this
does not mean that what enters the "untrained" eye while the other is
"trained" is of no consequence. Quite the contrary. In essential accord
with McCleary, Ingle found interocular transfer when the untrained
eye was unoccluded but not stimulated in any systematic way. On the
other hand, he did not find such transfer when the untrained eye was
stimulated by a five-dot pattern while the trained eye was exposed to
the critical stimuli, horizontal vs. vertical stripes. This suggests that
input into the contralateral hemisphere somehow masks input arriving
via the commissure. That the commissural input is relatively impoverished
in formation is suggested by other work of Ingle (1965a) discussed
below. But some intermediate state of integration is normally operative:
that such integration can and does occur is suggested by experiments in
which Ingle successfully conditioned goldfish in a situation where comparison of input from the two eyes was necessary for a correct response
(e.g., positive stimulus: the same stimulus in both eyes; negative stimulus:
a different stimulus in each eye).

The interocular transfer approach has provided information that
bears on the comparison of fish and mammalian learning much in the

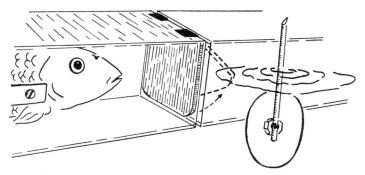

Fig. 16. Active avoidance apparatus for the study of interocular transfer. Fish
are trained to swim through swinging door when the striped disc is lowered into
the water. From Ingle (1968b). Copyright (1968) by the University of Chicago Press.

manner of Bitterman and Mackintosh. Ingle has performed two experiments which suggest that attentional processes are not critical in determining learning in fish, a view quite consistent with that taken by Mackintosh. One eye was exposed to the critical discriminative stimuli (horizontal vs. vertical), while the other eye was presented with irrelevant horizontal stripes on every trial. During training, all stimuli were black. After criterion was reached, Ingle introduced red horizontal stripes on the untrained side. When introduced on either positive or negative trials, these novel stimuli had a disinhibitory effect (Ingle, 1968a). This suggests that although the irrelevant eye was "tuned out" during training it could be "tuned in" again by a novel stimulus.

A more direct test of attentional factors was performed in a further study (Ingle, 1969). Goldfish were first trained on a horizontal-vertical discrimination. After criterion was reached, redundant color cues were added to the horizontal and vertical stripes on five or six unreinforced trials (e.g., red horizontal vs. green vertical). Tested on color alone, 10 fish showed 58/60 "correct" responses. Considering this enormous effectiveness of color cues, the orientation analyzer was obviously not very dominant. It appears that attentional processes in fish are weak and labile.

Much of Ingle's work has been directed at the properties of the visual system as such, quite apart from issues that bear upon the mechanisms of learning and memory. He has demonstrated loss of information in cross-tectal transfer by showing that easy discriminations transfer nicely while more difficult ones do not (Ingle, 1965b).

Interocular generalization tests have also revealed the "mirror-image phenomenon." Stimuli polarized in the anterior–posterior plane are generalized to their mirror images: thus an E with its vertical at the anterior side generalizes to an E whose vertical is on the posterior side if tested on the other eye. Ingle has developed this point and discussed the limitations and significance of mirror-image transfer (Ingle, 1968b). McCleary and Longfellow (1961), along somewhat similar lines, have asked whether interocular equivalence was learned or innately given. They showed clear interocular transfer (in the go, no-go avoidance task) with patterns presented to the noncorresponding parts of the two retinae.

## C. Cold Block of Learning and Temperature Acclimation

Over the last 6 years, Prosser and his colleagues (reviewed in Prosser and Nagai, 1968) have examined conditioning in fish as a possible index of temperature acclimation in the nervous system. They have indeed found a sensitive measure and, at the same time, have made promising beginnings in the study of many physiological aspects of conditioning.

Using the technique of cold block, Roots and Prosser (1962) and Prosser and Farhi (1965) demonstrated a neat hierarchy of sensitivities: Conditioned decreases of respiratory rate (see Fig. 17) or shuttle box avoidance conditioning were both blocked at a higher temperature than were reflexes, which in turn were blocked at a higher temperature than was peripheral nerve conduction. Conditioning showed typical temperature acclimation effects with blocking temperature rising about 5°C for each 10°C increase in the adaptation and conditioning temperature. Similarly, the minimum temperature at which conditioning could be accomplished rises with adaptation temperature (Prosser and Farhi, 1965). Clearly temperature adaptation includes alterations in the capacity for central nervous system integration.

The conditioning procedure employed here deserves attention because it is extremely effective. Goldfish are presented with a light, followed by a single electric shock. Within 20 trials a clear suppression of respiration is observed when the light is presented alone (see Fig. 17). When the UCS is omitted, the CR lasts about the length of the CS–UCS interval— an interesting temporal conditioning effect. The usual control procedures for pseudoconditioning (shock alone, etc.) show that true conditioning is involved.

By using this conditioning procedure, it was possible to show that adaptation occurs more quickly in moving from low to high temperatures

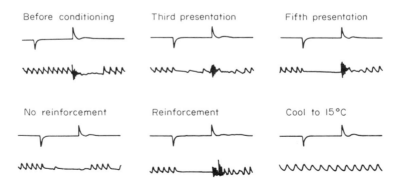

Fig. 17. Selected records from one experiment during conditioning of inhibition of respiration of a goldfish acclimated and conditioned at 25°C and blocked on cooling to 15°C. Upper records signal onset of light and cessation of light with simultaneous electric shock. Lower records indicate respiratory movements of operculum. From C. L. Prosser and E. Farhi (1965). Effects of temperature on conditioned reflexes and on nerve conduction in fish. Z. Vergleich. Physiol. 50, 91–101. Springer-Verlag, Berlin-Heidelberg-New York.

than in moving in the opposite direction: the development of the ability to condition at low temperatures took days.

More recently, Prosser (1965) developed a technique for recording conditioned electrical changes in the tectum. Changes can be recorded, during light-shock conditioning, in the slow potentials in the deep layers of the tectum (Prosser, 1965), and the appearance of a second response to light onset was observed after many light-shock pairings (Prosser and Nagai, 1968). These effects do not appear in the electroretinogram. Again, it was shown that cooling selectively blocks the conditioned response, demonstrating once more the increased temperature sensitivity or lability of conditioning effects. Interestingly, at temperatures where conditioning was cold-blocked (e.g., 5°C), extinction was ineffective; presentation of light in the absence of shock at this temperature did not diminish response to light when tested at the training temperature.

The detailed analysis of tectal-evoked potentials (Prosser, 1965; Prosser and Nagai, 1968) and their changes during conditioning could shed light on exactly what is being conditioned in the nervous system. Such an analysis is dependent, of course, on a complete understanding of the genesis of the various components of the tectal response.

## VI. LEARNING IN FISH AS A TOOL TO STUDY OTHER ASPECTS OF BEHAVIOR

Learning has often been used as simply a tool, not in an effort to understand the learning process itself but rather as a means to investigate other aspects of the behavior of physiology of fish. We have seen many examples of this already; e.g., Hasler's studies of the role of stream odors in salmon migration often employ the techniques of discriminative learning. In this case, the focus is on the fish and its behavior; in other cases both the technique (learning) and the subject (fish) are used as tools with the focus upon some general aspect of behavior (e.g., color vision) that is most conveniently studied in this manner. One reason for choosing the fish as subject in such studies is the animal's greater "simplicity" as compared to mammals. The absence of a highly developed telencephalon and the presence of a well-developed hypothalamus make the fish a particularly attractive subject for certain problems. Since both positive and negative reinforcement effects can be produced by intracranial stimulation in the goldfish (Boyd and Gardner, 1962), the study of drive-reward systems in the simpler fish brain is very appealing. In addition to these and other neurological advantages (see Ingle, 1965a), fish present the additional attraction of being the only group of verte-brate poikilotherms that are generally active (Rozin, 1968). Thus, when poikilothermy allows a greater range in relevant experimental manipula-tions, fish tend to be the organisms of choice. This feature has been especially utilized in the study of memory and learning as already described above (Prosser and Nagai, 1968; French, 1942; Gleitman *et al.*, 1970; Davis *et al.*, 1965a). The general use of poikilothermy in studying behavior has been discussed by Rozin (1968).

We now present a brief review of some recent applications of learning in fish to other problems in behavior. This review will be highly selective and is designed to indicate the range of phenomena which have been studied and the variety of learning techniques that have been applied.

## A. Sensory Discrimination and Capacity

Many of the experiments using fish and learning techniques are concerned with the measurement of various sensory capacities of fish, utilizing discrimination procedures in either classical or operant condi-tioning. Both Bull (1957) and Herter (1953) have performed extensive series of experiments in which they employed conditioning techniques to determine absolute and differential thresholds, and Tavolga and his colleagues (e.g., Tavolga and Wodinsky, 1963) have been studying auditory capacities in fish for many years using similar techniques, particularly avoidance training. We have already described the use of such methods in the investigation of sun-compass reactions and olfactory discrimination of stream odors (see Hasler, 1966).

Yaeger (1967) has used sophisticated learning techniques to obtain spectral sensitivity and saturation functions from goldfish as part of a systematic study of color vision in these animals. In the spectral satura-tion study, the fish was presented with two levers: one was trans-illuminated by white light, the other by white light to which was added a small amount of monochromatic light (the two were balanced for brightness). Yaeger obtained thresholds for discrimination by reward-ing the fish with food for pressing the lever containing the light mix-ture and systematically varying the wavelength of the added mono-chromatic light. When a trial was over the fish had to press a third lever located at the opposite end of the tank; only thus would the stimuli light up again to start a new trial. This additional refinement probably increased accuracy, for the fish was now more likely to "attend" to the stimuli.

Ingle has performed several sophisticated studies on movement per-ception and interocular equivalence in goldfish by using a simple discriminated avoidance task (Ingle, 1968b; see also Ingle's article in this volume). He has also employed the interocular transfer problem to study the relevant dimensions of visual coding in fish; by determining what kind of information can be transferred from one tectum to the other, and under what conditions, he has gained insight into both the nature of visual coding and, more directly, into the organization of binocular vision in fish (Ingle, 1968a). Similarly, Sutherland (1968), in part of a wide-ranging research program on form perception, has employed a simple two-choice discrimination apparatus to study shape discrimination in goldfish, discovering interesting similarities across species and phyla, and contributing considerably to our understanding of the relevant dimensions along which shape is coded by fish and other organisms. His technique is simple: An organism (e.g., fish, rat, or octopus) is trained on a shape discrimination (e.g., line 1 of Fig. 18) and is then tested for transfer to other shapes (lines 2–5, Fig. 18). His results give clues to the nature of stimulus equivalence in his subjects. For example, his work to date suggests that goldfish rely more heavily on points and

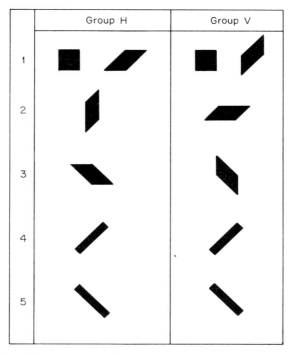

**Fig. 18.** Animals are taught to discriminate between the square and quadrilateral in the first line. The stimuli in lines 2–5 are used in generalization tests. From Sutherland (1968). Copyright (1968) by the University of Chicago Press.

less heavily on orientation contours than do rats, and that goldfish tend to recognize figures on the basis of their upper halves.

## B. Motivational Processes

"Motivation" in fish has not been studied as intensively as one might expect considering the broad spectrum of motivated behavior shown by these animals (see Baerends, 1957, or chapter by Baerends in this treatise). A few investigators have taken advantage of the fact that fish show many highly stereotyped instinctive behavior sequences and also learn quite easily. Adler and Hogan (1963) demonstrated that a more or less "typical" instinctive response display in the male Siamese fighting fish to the sight of another male (as measured by gill cover extension) could be classically conditioned to a neutral stimulus and could also be eliminated from the fish's repertoire if this response was punished by electric shock whenever it was elicited by the normal releasing stimulus. This study emphasized the lability of instinctive responses. Thompson and Sturm (1965a) confirmed the classical conditioning effect and extended it to other components of display, noting that conditioning occurred at different rates for different aspects of the display response. Thompson (1963) demonstrated that sight of a male Siamese fighting fish could serve as a reinforcer for an instrumental response in another male *Betta* (the response was breaking a light beam). He measured the rate of the response which produced the releasing stimulus and used this response rate to assess the effectiveness of a variety of stimuli, colors, movement, etc. (Thompson, 1963; Thompson and Sturm, 1965b). This work suggests an important linkage between ethological theory and learning theory by integrating the notion of reinforcement with that of consummatory behavior: It confirms the idea that appetitive behavior may be shaped and rewarded by appropriate reinforcements.

Only a few studies have used learning to investigate the "homeostatic" drives (e.g., thirst and hunger). This is rather surprising, considering that in mammals learning techniques are used so very extensively to study these problems. One might have expected that since water intake is not ordinarily a behavioral problem for fish, correspondingly more attention might have been paid to hunger mechanisms with the troubling hunger–thirst interactions no longer present. Using operant techniques by which fish were trained to press a lever for food pellets (see Fig. 19), the daily food intake patterns (e.g., lever-pressing patterns) of goldfish have been determined, as well as their food intake,

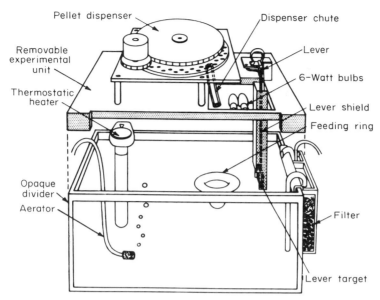

Fig. 19. Diagram of the apparatus used to measure food intake in the goldfish. The fish remained in his home tank, and the apparatus was moved from tank to tank. From Rozin and Mayer (1964).

as a function of temperature change, caloric dilution of their food, and increased work requirements (Rozin and Mayer, 1961a, 1964). These experiments show fish to possess rather sophisticated caloric regulation. Coupled with the fact that fish show the self-stimulation phenomenon (Boyd and Gardner, 1962), and that they have a well-developed hypothalamus, the opportunity for further investigation is manifest.

Fish can be trained to lever-press for a thermal reward. When placed in a hot tank, goldfish will respond for squirts of cold water which transiently lower the temperature in their tank by about 0.5°C (Rozin and Mayer, 1961b; Rozin, 1968). It is evidently possible to turn the fish into a behavioral homotherm: When allowed to regulate their water temperature through operant responding, fish maintained the temperature within a narrow range (see Fig. 20). Van Sommers (1962), using an ingeniously designed piece of apparatus (see Fig. 21), similarly demonstrated that oxygenated water could serve as a reinforcement for a fish held in deoxygenated water. Goldfish learned to break a light beam in the presence of a red light in order to get oxygenated water and to avoid breaking the same beam in the presence of a green

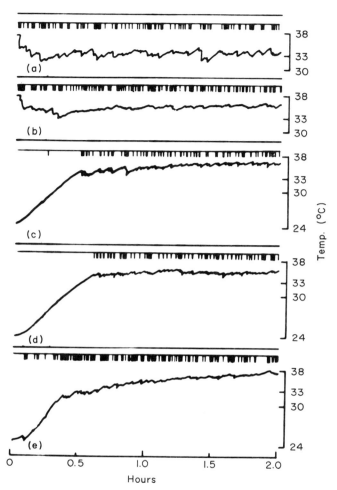

**Fig. 20.** Some typical records of individual goldfish thermoregulation sessions. In each panel, the top line indicates lever presses, as downward deflections, and the bottom line graphs temperature in the small container in °C. (a) Fish SG 106, session begins at 38°C, double reinforcement (2 sec). (b) Fish SG 106, session begins at 38°C, standard reinforcement (1 sec). (c) Fish SG 106, session begins at 25°C, standard reinforcement (1 sec). (d) Fish SG 111, session begins at 25°C, double reinforcement (2 sec). (e) Fish SG 110, session begins at 25°C, standard reinforcement (1 sec). From Rozin (1968). Copyright (1968) by the University of Chicago Press.

light to get this same reward. This procedure neatly rules out possible artifacts produced by effects of oxygenation upon activity level. Since they are potentially very sensitive and easily automated, both thermal and respiratory reinforcement effects could well be used for long-term

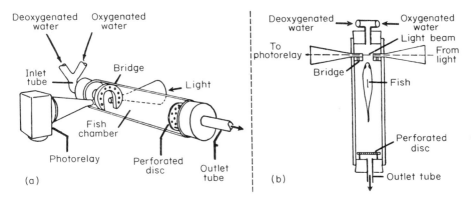

Fig. 21. Experimental chamber used to study respiratory reinforcement in goldfish: (a) general view and (b) plan (Van Sommers, 1962). Copyright (1962) by the American Association for the Advancement of Science.

studies of adaptation and of the operation of thermoregulatory and respiratory systems.

Fish have been shown to move to preferred positions in salinity gradients and to indicate salinity preferences in choice situations (see Baggerman, 1959; McInerney, 1964). The problem of establishing unequivocally that fish will regulate salinity by instrumentally learned responses is aggravated by the changes that occur in general activity levels with changing salinity. Such variations in activity may be a function not only of absolute salinity levels but also of rates of change of salinity. One strategy for controlling for this activity, adopted with the goldfish by Van Sommers (1969), is to record on punched tape all responses and salinity changes in the learning situation. These data are then used to reproduce exactly the same conditions of salinity in control periods. With this procedure it is possible to demonstrate that the behavior of the fish is not merely a result of changing salinity but depends to some degree upon the effectiveness of the behavior in changing the salinity, that is, that the behavior is instrumentally learned.

## C. Biological Rhythms

Learning techniques have proved useful in this area also. For example, feeding rhythms can be easily measured through the mediation of an operant (lever pressing) response for food reinforcement (Rozin and Mayer, 1961a). Learning techniques have also been used here to help disentangle certain theoretical issues. Rozin showed that the performance of goldfish on a fixed-interval schedule (FI-1) is in its essentials independent of temperature (Rozin, 1964). The animals were trained to press a lever for food, but the lever-press would be followed by delivery of food only once every minute. Under these conditions rats learn not to respond immediately after reinforcement but to respond more readily as the interval comes to its close (Skinner, 1938), suggesting that something like a temporal discrimination has been acquired. Rozin's study showed that this effect ("FI-scallop") can be found in goldfish, although typically only after many training sessions. The important point is that this discrimination is temperature-independent: goldfish trained at one temperature show the same effect when the temperature is lowered substantially. This suggests that some endogenous timer is at work even over these short intervals, as in longer (e.g., circadian) periods. It is sometimes proposed that circadian intervals are controlled by subtle cyclical environmental cues (e.g., Brown, 1962), but this argument can hardly hold for intervals of only 1 min. Rozin found that the *absolute* response rate dropped at lower temperatures but the FI-scallop (that is, the temporal discrimination) was unchanged. This suggests that the fish did not time himself by reference to some form of behavioral pacing.

## D. Regeneration Processes

Learning techniques have been applied elegantly to problems of neural integration and regeneration (Arora and Sperry, 1963). Here the regenerative capabilities of the fish (shared with the amphibia) and its learning ability (shared in many respects with the mammals) make it the ideal subject. Arora and Sperry asked the question: Will the coding of color and the connections that must mediate this be restored if the optic nerve is first cut and then regenerates? They trained *Astronotus ocellatus* to jump out of the water to obtain food from a wire suspended over the water. Two "targets" were available on each trial, a different color presented on each. Color was the cue indicating which wire was baited with real food and which with imitation (sponge). Once criterion was achieved, the optic nerves were severed. Following regeneration, the fish clearly demonstrated a preference for the color reinforced prior to surgery. Similar results have been obtained with retention of an olfactory discrimination following section and regeneration of the olfactory tracts (von Baumgarten and Miessner, 1968).

ACKNOWLEDGMENTS

The authors are indebted to Dr. Nicholas J. Mackintosh, Dr. Peter Van Sommers, and Dr. Larry Stein for generous advice and criticism. They also want to thank Alcine Potts and Elisabeth Rozin.

REFERENCES

Adler, N., and Hogan, J. A. (1963). Classical conditioning and punishment of an instinctive response in *Betta splendens*. *Animal Behaviour* **11**, 351–354.

Agranoff, B. W., and Davis, R. E. (1968). The use of fishes in studies on memory formation. *In* "The Central Nervous System and Fish Behavior" (D. J. Ingle, ed.), pp. 193–201. Univ. of Chicago Press, Chicago, Illinois.

Agranoff, B. W., Davis, R. E., and Brink, J. J. (1965). Memory fixation in the goldfish. *Proc. Natl. Acad. Sci. U. S.* **54**, 788–793.

Agranoff, B. W., Davis, R. E., and Brink, J. J. (1966). Chemical studies on memory fixation in goldfish. *Brain Res.* **1**, 303–309.

Agranoff, B. W., Davis, R. E., Casola, L., and Lim, R. (1967). Actinomycin D blocks formation of memory of shock-avoidance in goldfish. *Science* **159**, 1600–1601.

Ames, L., and Yarczower, M. (1965). Some effects of wavelength discrimination on stimulus generalization in the goldfish. *Psychonomic Sci.* 3, 311–312.

Aronson, L. R. (1951). Orientation and jumping behavior in the gobiid fish *Bathygobius soporator*. *Am. Museum Novitates* **1486**, 1–22.

Aronson, L. R. (1963). The central nervous system of sharks and bony fishes with special reference to sensory and integrative mechanisms. *In* "Sharks and Survival" (P. W. Gilbert, ed.), pp. 165–241. Heath, Boston, Massachusetts.

Aronson, L. R., and Kaplan, H. (1968). Function of the teleostean forebrain. *In* "The Central Nervous System and Fish Behavior" (D. J. Ingle, ed.), pp. 107–125. Univ. of Chicago Press, Chicago, Illinois.

Arora, H. L., and Sperry, R. W. (1963). Color discrimination after optic nerve regeneration in the fish, *Astronotus ocellatus*. *Develop. Biol.* **7**, 234–243.

Baerends, G. P. (1957). The ethological analysis of fish behavior. *In* "The Physiology of Fishes" (M. E. Brown, ed.), Vol. 2, pp. 229–269. Academic Press, New York.

Baerends, G. P., and Baerends-van Roon, J. M. (1950). An introduction to the study of the ethology of Cichlid fishes. *Behaviour Suppl.* **1**, 1–243.

Baggerman, B. (1959). The role of external factors and hormones in migration of sticklebacks and juvenile salmon. *In* "Comparative Endocrinology" (A. Gorbman, ed.), pp. 24–37. Wiley, New York.

Behrend, E. R., and Bitterman, M. E. (1961). Probability-matching in the fish. *Am. J. Psychol.* **74**, 542–551.

Behrend, E. R., and Bitterman, M. E. (1963). Sidman avoidance in the fish. *J. Exptl. Anal. Behav.* **6**, 47–52.

Behrend, E. R., and Bitterman, M. E. (1964). Avoidance-conditioning in the fish: Further studies of the CS-US interval. *Am. J. Psychol.* **77**, 15–28.

Behrend, E. R., and Bitterman, M. E. (1966). Probability-matching in the goldfish. *Psychonomic Sci.* **6**, 327–328.

Behrend, E. R., and Bitterman, M. E. (1967). Further experiments on habit reversal in the fish. *Psychonomic Sci.* **8**, 363–364.

Behrend, E. R., Domesick, V. B., and Bitterman, M. E. (1965). Habit reversal in the fish. *J. Comp. Physiol. Psychol.* **60**, 407–411.

Beukema, J. J. (1968). Predation by the three-spined stickleback (*Gasterosteus aculeatus* L.): The influence of hunger and experience. *Behaviour* **31**, 1–126.

Bitterman, M. E. (1964a). The evolution of intelligence. *Sci. Am.* **212**, 92–100.

Bitterman, M. E. (1964b). Classical conditioning in the goldfish as a function of the CS-UCS interval. *J. Comp. Physiol. Psychol.* **58**, 359–366.

Bitterman, M. E. (1965) The CS-US interval in classical and avoidance conditioning. In "Classical Conditioning" (W. F. Prokasy, ed.), pp. 1–19. Appleton, New York.

Bitterman, M. E. (1966). Animal learning. In "Experimental Methods and Instrumentation in Psychology" (J. B. Sidowsky, ed.), pp. 451–484. McGraw-Hill, New York.

Bitterman, M. E. (1968). Comparative studies of learning in the fish. In "The Central Nervous System and Fish Behavior" (D. J. Ingle, ed.), pp. 257–270. Univ. of Chicago Press, Illinois.

Bitterman, M. E., Wodinsky, J., and Candland, D. K. (1958). Some comparative psychology. Am. J. Psychol. 71, 94–110.

Boring, E. G. (1950). "A History of Experimental Psychology." Appleton, New York.

Boyd, E. S., and Gardner, L. C. (1962). Positive and negative reinforcement from intracranial stimulation of a teleost. Science 136, 648–649.

Braddock, J. C. (1945). Some aspects of the dominance-subordination relationship in the fish Platypoecilus maculatus. Physiol. Zool. 18, 176–195.

Braemer, W. (1960). A critical review of the sun-azimuth hypothesis. Cold Spring Harbor Symp. Quant. Biol. 25, 413–427.

Brannon, E. L. (1967). Genetic control of migrating behavior of newly emerged sockeye salmon fry. Intern. Pacific Salmon Fisheries Comm. Progr. Rept. 16.

Breland, K., and Breland M. (1966). "Animal Behaviour." Macmillan, New York.

Bresler, D. E., and Bitterman, M. E. (1968). Learning in fish with transplanted brain tissue. Science 163, 590–592.

Brown, F. A. (1962). Extrinsic rhythmicality: A reference frame for biological rhythms under so-called constant conditions. Ann. N. Y. Acad. Sci. 98, 775–787.

Bull, H. O. (1928). Studies on conditioned responses in fishes. I. J. Marine Biol. Assoc. 15, 485–533.

Bull, H. O. (1957). Conditioned responses. In "The Physiology of Fishes" (M. E. Brown, ed.), Vol. 2, pp. 211–228. Academic Press. New York.

Capaldi, E. J. (1967). A sequential hypothesis of instrumental learning. In "Advances in the Psychology of Learning and Motivation Research and Theory" (K. W. Spence and J. T. Spence, eds.), Vol. 1, pp. 67–156. Academic Press, New York.

Cerf, J. A., and Otis, L. S. (1958). Heat narcosis and its effects on retention of a learned behavior in the goldfish. Am. Psychologist 13, 419.

Coons, E. E., and Miller, N. E. (1960). Conflict versus consolidation of memory traces to explain "retrograde amnesia" produced by ECS. J. Comp. Physiol. Psychol. 53, 524–531.

Cott, H. B. (1940). "Adaptive Coloration in Animals." Methuen, London.

Crespi, L. P. (1942). Quantitative variation of incentive and performance in the white rat. Am. J. Psychol. 55, 467–517.

Davis, R. E. (1963). Daily "predawn" peak of locomotion in bluegill and largemouth bass. Animal Behaviour 12, 272–283.

Davis, R. E. (1968). Environment control of memory fixation in goldfish. J. Comp. Physiol. Psychol. 65, 72–78.

Davis, R. E., and Agranoff, B. W. (1966). Stages of memory formation in goldfish: Evidence for an environmental trigger. Proc. Natl. Acad. Sci. U. S. 55, 555–559.

Davis, R. E., and Bardach, J. E. (1965). Time co-ordinated prefeeding activity in fish. Animal Behaviour 13, 154–162.

Davis, R. E., and Klinger, P. D. (1969). Environmental control of amnesic effects of various agents in goldfish. *Physiol. Behav.* **4**, 269–271.

Davis, R. E., Bright, P. J., and Agranoff, B. W. (1965a). Effect of ECS and puromycin on memory in fish. *J. Comp. Physiol. Psychol.* **60**, 162–166.

Davis, R. E., Klinger, P. D., and Agranoff, B. W. (1965b). Automated training and recording of a light-tracking response in fish. *J. Exptl. Anal. Behav.* **8**, 353–355.

Donaldson, R., and Allen, G. H. (1957). Return of silver salmon, *Oncorhynchus kisutch* (Walbaum) to point of release. *Trans. Am. Fisheries Soc.* **87**, 13–22.

Dücker, G., and Rensch, B. (1968). Verzögerung des Vergessens erlernter visueller Aufgaben bei Fischen durch Dunkel haltung. *Arch. Ges. Physiol.* **301**, 1–6.

Erickson, R. P. (1956). The effect of temperature-induced activity upon retention in the paradise fish. M. A. thesis Brown University (unpublished).

Estes, W. K. (1964). Probability learning. *In* "Categories of Human Learning" (A. W. Melton, ed.), pp. 90–128. Academic Press, New York.

Fagerlund, U. H. M., McBride, J. R., Smith, M., and Tomlinson, N. (1963). Olfactory perception in migrating salmon. III. Stimulants for adult sockeye salmon (*Oncorhynchus nerka*) in home stream waters. *J. Fisheries Res. Board Can.* **20**, 1457–1463.

Flexner, J. B., Flexner, L. B., and Stellar, E. (1963). Memory in mice as affected by intracerebal puromycin. *Science* **141**, 57–59.

French, J. W. (1942). The effect of temperature on the retention of a maze habit in fish. *J. Exptl. Psychol.* **31**, 85.

Froloff, J. P. (1925). Bedingte Reflexe bei Fischen. I. *Pflug. Arch. Ges. Physiol.* **209**, 261–271.

Garcia, J., and Ervin, F. R. (1968). Gustatory-visceral and telereceptor-cutaneous conditioning—adaptation in internal and external milieus. *Commun. Behav. Biol.* **1**, Part A, 389–415.

Gatling, F. (1952). The effect of repeated stimulus reversals on learning in the rat. *J. Comp. Physiol. Psychol.* **45**, 347–351.

Geller, I. (1963). Conditioned "anxiety" and effects of punishment on operant behavior of goldfish. *Science* **141**, 351–353.

Geller, I. (1964). Conditioned suppression in goldfish as a function of shock-reinforcement schedule. *J. Exptl. Anal. Behav.* **7**, 345–359.

Gerking, S. D. (1959). The restricted movement of fish populations. *Biol. Rev.* **34**, 221–242.

Gleitman, H., and Kosiba, R. (1967). Failure to find proactive inhibition in the pigeon. Undergraduate research paper, University of Pennsylvania (unpublished).

Gleitman, H., and Steinman, F. (1963). The retention of runway performance as a function of proactive interference. *J. Comp. Physiol. Psychol.* **56**, 834–838.

Gleitman, H., Wilson, W. A., Jr., Herman, M. M., and Rescorla, R. A. (1963). Massing and within-delay position as factors in delayed-response performance. *J. Comp. Physiol. Psychol.* **56**, 445–451.

Gleitman, H., Rozin, P., Potts, A., and Holmes, P. (1970). Forgetting in goldfish as a function of temperature during the retention interval. In preparation.

Gonzalez, R. C., and Bitterman, M. E. (1967). Partial reinforcement effect in the goldfish as a function of amount of reward. *J. Comp. Physiol. Psychol.* **64**, 163–167.

Gonzalez, R. C., and Bitterman, M. E. (1969). The spaced-trials PRE as a function of contrast. *J. Comp. Physiol. Psychol.* **67**, 94–103.

Gonzalez, R. C., Eskin, R. M., and Bitterman, M. E. (1961). Alternating and random partial reinforcement in the fish with some observations on asymptotic resistance to extinction. *Am. J. Psychol.* **74**, 561–568.

Gonzalez, R. C., Eskin, R. M., and Bitterman, M. E. (1962a). Extinction in the fish after partial and consistent reinforcement with number of reinforcements equated. *J. Comp. Physiol. Psychol.* **55**, 381–386.

Gonzalez, R. C., Milstein, S., and Bitterman, M. E. (1962b). Classical conditioning in the fish: Further studies of partial reinforcement. *Am. J. Psychol.* **75**, 421–428.

Gonzalez, R. C., Eskin, R. M., and Bitterman, M. E. (1963). Further experiments on partial reinforcement in the fish. *Am. J. Psychol.* **76**, 366–375.

Gonzalez, R. C., Roberts, W. A., and Bitterman, M. E. (1964). Learning in adult rats extensively decorticated in infancy. *Am. J. Psychol.* **77**, 547–562.

Gonzalez, R. C., Behrend, E. R., and Bitterman, M. E. (1965). Partial reinforcement in the fish: Experiments with spaced trials and partial delay. *Am. J. Psychol.* **78**, 198–207.

Gonzalez, R. C., Berger, B. D., and Bitterman, M. E. (1966). Improvement in habit-reversal as a function of amount of training per reversal and other variables. *Am. J. Psychol.* **79**, 517–530.

Gonzalez, R. C., Holmes, N. K., and Bitterman, M. E. (1967a). Asymptotic resistance to extinction in fish and rat as a function of interpolated retraining. *J. Comp. Physiol. Psychol.* **63**, 342–344.

Gonzalez, R. C., Holmes, N. K., and Bitterman, M. E. (1967b). Resistance to extinction in the goldfish as a function of frequency and amount of reward. *Am. J. Psychol.* **80**, 269–275.

Gonzalez, R. C., Behrend, E. R., and Bitterman, M. E. (1967c). Reversal learning and forgetting in bird and fish. *Science* **158**, 519–521.

Hainsworth, F. R., Overmier, J. B., and Snowdon, C. T. (1967). Specific and permanent deficits in instrumental avoidance responding following forebrain ablation in the goldfish. *J. Comp. Physiol. Psychol.* **63**, 111–116.

Hara, T. J., Ueda, K., and Gorbman, A. (1965). Electroencephalographic studies of homing salmon. *Science* **149**, 884–885.

Haralson, J. V., and Bitterman, M. E. (1950). A lever-depression apparatus for the study of learning in fish. *Am. J. Psychol.* **63**, 250–256.

Harden-Jones, F. R. (1968). "Fish Migration." St. Martin's Press, New York.

Harlow, H. F. (1939). Forward conditioning, backward conditioning and pseudo-conditioning in the goldfish. *J. Genet. Psychol.* **55**, 49–58.

Hasler, A. D. (1956). Influence of environmental reference points on learned orientation in fish (*Phoxinus*). *Z. Vergleich. Physiol.* **38**, 303–310.

Hasler, A. D. (1966). "Under Water Guideposts. Homing of Salmon." Univ. of Wisconsin Press, Madison, Wisconsin.

Hasler, A. D. (1968). Memory in homing of migratory fishes. *In* "The Central Nervous System and Fish Behavior" (D. J. Ingle, ed.), pp. 247–255. Univ. of Chicago Press, Chicago, Illinois.

Hasler, A. D., and Schwassmann, H. O. (1960). Sun orientation in fish at different latitudes. *Cold Spring Harbor Symp. Quant. Biol.* **25**, 429–441.

Hasler, A. D., and Wisby, W. J. (1951). Discrimination of stream odors by fishes and relation to parent stream behavior. *Am. Naturalist* **85**, 223–238.

Healey, E. G. (1957). The nervous system. *In* "The Physiology of Fishes" (M. E. Brown, ed.), Vol. 2, pp. 1–119. Academic Press, New York.

Herter, K. (1953). "Die Fisch Dressuren and ihre Sinnesphysiologische Grundlagen." Akademie Verlag, Berlin.

Hoar, W. S. (1958). Rapid learning of a constant course by travelling schools of juvenile Pacific salmon. *J. Fisheries Res. Board Can.* **15**, 251–274.

Hodos, W., and Campbell, C. G. B. (1969). Scala Naturae: Why there is no theory in comparative psychology. *Psychol. Rev.* **76**, 337–350.

Hogan, J. A. (1967). Fighting and reinforcement in the Siamese fighting fish. *J. Comp. Physiol. Psychol.* **64**, 356–359.

Hogan, J. A., and Rozin, P. (1958). Unpublished observations.

Hogan, J. A., and Rozin, P. (1962). An improved mechanical fish-lever. *Am. J. Psychol.* **75**, 307–308.

Hogan, J. A., Kleist, S., and Hutchings, S. L. (1969). Display and food as reinforcers in the Siamese fighting fish (*Betta splendens*). Unpublished manuscript.

Horner, J. L., Longo, N., and Bitterman, M. E. (1960). A classical conditioning technique for small aquatic animals. *Am. J. Psychol.* **73**, 623–626.

Horner, J. L., Longo, N., and Bitterman, M. E. (1961). A shuttle box for fish and a control circuit of general applicability. *Am. J. Psychol.* **74**, 114–120.

Hull, C. L. (1943). "Principles of Behavior." Appleton, New York.

Hulse, S. H., Jr. (1958). Amount and percentage of reinforcement and duration of goal confinement in conditioning and extinction. *J. Exptl. Psychol.* **56**, 48–57.

Idler, D. R., McBride, J. R., Jonas, R. E. E., and Tomlinson, N. (1961) Olfactory perception in migrating salmon. II. Studies on a laboratory bio-assay for home-stream water and mammalian repellent. *Can. J. Biochem. Physiol.* **39**, 1575–1584.

Ingle, D. J. (1965a). The use of the fish in neuropsychology. *Perspectives Biol. Med.* **8**, 241–260.

Ingle, D. J. (1965b). Interocular transfer in goldfish: Color easier than pattern. *Science* **149**, 1000–1002.

Ingle, D. J. (1968a). Interocular integration of visual learning by goldfish. *Brain, Behav. Evolution* **1**, 58–85.

Ingle, D. J. (1968b). Spatial dimensions of vision in fish. *In* "The Central Nervous System and Fish Behavior" (D. J. Ingle, ed.), pp. 51–59. The University of Chicago, Chicago, Illinois.

Ingle, D. J. (1969). Errorless transfer between color and pattern discriminations in goldfish. Unpublished manuscript.

Janzen, W. (1933). Untersuchungen über Grosshirnfunktionen der Goldfisches (*Carassius auratus*). *Zool. Jahrb. Abt. Allg. Zool. Physiol. Tiere* **52**, 591–628.

Jones, F. N. (1945). An alternative explanation of the effect of temperature upon retention in the goldfish. *J. Exptl. Psychol.* **35**, 76–79.

Kamin, L. J. (1957). The retention of an incompletely learned avoidance response. *J. Comp. Physiol. Psychol.* **50**, 457–460.

Kehoe, J. (1963). Effects of prior and interpolated learning on retention in pigeons. *J. Exptl. Psychol.* **65**, 537–545.

Kellogg, W. N., and Spanovick, P. (1953). Respiratory changes during the conditioning of fish. *J. Comp. Physiol. Psychol.* **46**, 124–128.

Kimble, G. A. (1961). "Hilgard and Marquis' Conditioning and Learning." Appleton, New York.

Klinman, C. S., and Bitterman, M. E. (1963). Classical conditioning in the fish: The CS-US interval. *J. Comp. Physiol. Psychol.* **56**, 578–583.

Lewis, D. J., and Maher, B. A. (1965). Neural consolidation and electroconvulsive shock. *Science* **144**, 182–183.

Longo, N., and Bitterman, M. E. (1959). Improved apparatus for the study of learning in fish. *Am. J. Psychol.* **72**, 616–620.

Lorz, H. W., and Northcote, T. G. (1965). Factors affecting stream location, and timing and intensity of entry by spawning Kokanee (oncorhynchus nerka) into an inlet of Nicola Lake, British Columbia. *J. Fisheries Res. Board Can.* **22**, 665–687.

Lovejoy, E. (1968). "Attention in Discrimination." Holden-Day, San Francisco, California.

Lowes, G., and Bitterman, M. E. (1967). Reward and learning in the goldfish. *Science* **157**, 455–457.

McCleary, R. A. (1960). Type of response as a factor in interocular transfer in the fish. *J. Comp. Physiol. Psychol.* **53**, 311–321.

McCleary, R. A., Bernstein, J. J. (1959). A unique method for control of brightness cues in study of color vision in fish. *Physiol. Zool.* **32**, 284–292.

McCleary, R. A., and Longfellow, L. A. (1961). Interocular transfer of pattern discrimination without prior binocular experience. *Science* **134**, 1418–1419.

McDonald, H. E. (1922). Ability of *Pimephales notatus* to form associations with sound vibrations. *J. Comp. Physiol. Psychol.* **2**, 191–193.

McGaugh, J. L. (1965). Facilitation and impairment of memory storage processes. *In* "The Anatomy of Memory" (D. P. Kimble, ed.), Vol. 1, pp. 240–291. Science and Behavior Books, Inc., Palo Alto, California.

McGaugh, J. L., and Madsen, M. C. (1964). Amnesic and punishing effects of electroconvulsive shock. *Science* **144**, 182–183.

McInerney, J. E. (1964). Salinity preference: An orientation mechanism in salmon migration. *J. Fisheries Res. Board Can.* **21**, 995–1018.

Mackintosh, N. J. (1965a). Overtraining, extinction, and reversal in rats and chicks. *J. Comp. Physiol. Psychol.* **59**, 31–36.

Mackintosh, N. J. (1965b). Selective attention in animal discrimination learning. *Psychol. Bull.* **64**, 124–150.

Mackintosh, N. J. (1969a). Comparative studies of reversal and probability learning: Rats, birds, and fish. *In* "Animal Discrimination Learning" (R. Gilbert and N. S. Sutherland, eds.), pp. 175–185. Academic Press, New York.

Mackintosh, N. J. (1969b). Attention and probability learning. *In* "Attention: Contemporary Studies and Analyses" (D. Mostofsky, ed.). Appleton, New York.

Mackintosh, N. J., and Holgate, V. (1968). Effects of inconsistent reinforcement on reversal and nonreversal shifts. *J. Exptl. Psychol.* **76**, 154–159.

Mackintosh, N. J., and Mackintosh, J. (1964). Performance of *Octopus* over a series of reversals of a simultaneous discrimination. *Animal Behaviour* **12**, 321–324.

Mackintosh, N. J., Mackintosh, J., Salfriel-Jorne, O., and Sutherland N. S. (1966). Overtraining reversal, and extinction in the goldfish. *Animal Behaviour* **14**, 314–318.

Mackintosh, N. J., McGonigle, B., Holgate, V., and Vanderver, V. (1968). Factors underlying improvement in serial reversal learning. *Can. J. Psychol.* **22**, 85–95,

Maier, N. R. F., (1929). Delayed reaction and memory in rats. *J. Genet. Psychol.* **36**, 538–550.

Maier, S., and Gleitman, H. (1967). Proactive interference in rats. *Psychonomic Sci.* **7**, 25–26.

Mandriota, F. J., Thompson, R. L., and Bennett, M.V.L.(1965). Classical conditioning of electric organ discharge rate in mormyrids. *Science* **150**, 1740–1742.

Miller, H. C. (1963). The behavior of the pumpkinseed sunfish, *Lepomis gibbosus*

(Linneaus), with notes on the behavior of other species of *Lepomis* and the pigmy sunfish, *Elassoma everglade*. *Behaviour* **12**, 88–151.

Munn, N. L. (1958). The question of insight and delayed reaction in fish. *J. Comp. Physiol. Psychol.* **51**, 92–97.

Myer, J. S., and Ricci, D. (1968). Delay of punishment gradients in the goldfish. *J. Comp. Physiol. Psychol.* **66**, 417–421.

Neisser, U. (1966). "Cognitive Psychology." Appleton, New York.

Newman, M. A. (1956). Social behavior and interspecific competition in two trout species. *Physiol. Zool.* **29**, 64–81.

Noble, G. K., and Curtis, B. (1939). The social behavior of the jewel fish, *Hemichromis bimaculatus*, Gill. *Bull. Am. Museum Nat. Hist.* **76**, 1–46.

Noble, M., and Adams, C. K. (1963). The effect of length of CS-US interval as a function of body temperature in a cold-blooded animal. *J. Gen. Psychol.* **69**, 197–201.

Noble, M., Gruender, A., and Meyer, D. R. (1959). Conditioning in fish (*Mollienisia* sp.) as a function of the interval between CS and US. *J. Comp. Physiol. Psychol.* **52**, 236–239.

North, A. J. (1950). Improvement in successive discrimination reversals. *J. Comp. Physiol. Psychol.* **43**, 442–460.

Northcote, T. G. (1969). Lakeward migration of young rainbow trout (*Salmo gairdneri*) in the Upper Lardeau River, British Columbia. *J. Fisheries Res. Board Can.* **26**, 33–45.

O'Connell, C. P. (1960). Use of fish school for conditioned response experiments. *Animal Behaviour* **8**, 225–227.

Oshima, K., Gorbman, A., and Shimada, H. (1969). Memory-blocking agents: Effects on olfactory discrimination in homing salmon. *Science* **165**, 86–88.

Overmier, J. B., and Curnow, P. F. (1969). Classical conditioning, pseudoconditioning and sensitization in "normal" and forebrainless goldfish. *J. Comp. Physiol. Psychol.* **68**, 193–198.

Perkins, C. C., Jr. and Cacioppo, A. J. (1950). The effect of intermittent reinforcement on the change in extinction rate following successive reconditionings. *J. Exptl. Psychol.* **40**, 794–801.

Pinckney, G. A. (1966). The Kamin effect in fish. *Psychonomic Sci.* **4**, 387–388.

Potts, A., and Bitterman, M. E. (1967). Puromycin and retention in the goldfish. *Science* **158**, 1594–1596.

Prosser, C. L. (1965). Electrical responses of fish optic tectum to visual stimulation: Modification by cooling and conditioning. *Z. Vergleich. Physiol.* **50**, 102–118.

Prosser, C. L., and Farhi, E. (1965). Effects of temperature on conditioned reflexes and on nerve conduction in fish. *Z. Vergleich. Physiol.* **50**, 91–101.

Prosser, C. L., and Nagai, T. (1968). Effects of temperature on conditioning in goldfish. *In* "The Central Nervous System and Fish Behavior" (D. J. Ingle, ed.), pp. 171–180. Univ. of Chicago Press, Chicago, Illinois.

Quartermain, D., Paolino, R. M., and Miller, N. E. (1965). A brief temporal gradient of retrograde amnesia independent of situational change. *Science* **149**, 1116–1118.

Raleigh, R. F. (1967). Genetic control in the lakeward migrations of sockeye salmon (oncorhynchus nerka) fry. *J. Fisheries Res. Board Can.* **24**, 2613–2622.

Randal, J. E., and Randal, H. A. (1960). Examples of mimicry and protective resemblance in tropical marine fishes. *Bull. Marine Sci. Gulf Carribean* **10**, 444–480.

Regestein, Q. (1968). Some monocular emotional effects of unilateral hypothalamic lesions in goldfish. *In* "The Central Nervous System and Fish Behavior" (D. J. Ingle, ed.), pp. 139–144. Univ. of Chicago Press, Chicago, Illinois.

Reighard, J. (1908). An experimental field-study of warning coloration in coral-reef fishes. *Carnegie Inst. Wash. Papers Tortugas Lab, Dep. Marine Biol.* No. 9.

Rensch, B., and Dücker, G. (1966). Verzögerung des Vergessens erlernter visueller Aufgaben bei Tieren durch Chlorpromazin. *Arch. Ges. Physiol.* **289**, 200–214.

Rescorla, R. A., and Solomon, R. L. (1967). Two-process learning theory: Relationships between Pavlovian conditioning and instrumental learning. *Psychol. Rev.* **74**, 151–182.

Roots, B. I., and Prosser, C. L. (1962). Temperature acclimation and the nervous system in fish. *J. Exptl. Biol.* **39**, 617–629.

Rozin, P. (1964). Temperature independence of an arbitrary temporal discrimination in the goldfish. *Science* **149**, 561–563.

Rozin, P. (1968). The use of poikilothermy in the analysis of behavior. *In* "The Central Nervous System and Fish Behavior" (D. J. Ingle, ed.), pp. 181–192. Univ. of Chicago Press, Chicago, Illinois.

Rozin, P., and Mayer, J. (1961a). Regulation of food intake in the goldfish. *Am. J. Physiol.* **201**, 968–974.

Rozin, P., and Mayer, J. (1961b). Thermal reinforcement and thermo-regulatory behavior in the goldfish, *Carassius auratus. Science* **134**, 942–943.

Rozin, P., and Mayer, J. (1964). Some factors influencing short-term food intake of the goldfish. *Am. J. Physiol.* **206**, 1430–1436.

Salzinger, K., Freimark, S. J., Fairhurst, S. P., and Wolkoff, F. D. (1968). Conditioned reinforcement in the goldfish. *Science* **160**, 1471–1472.

Sanders, F. K. (1940). Second-order olfactory and visual learning in the optic tectum of the goldfish. *J. Exptl. Biol.* **17**, 416–433.

Savage, G. E. (1968). Function of the forebrain in the memory system of the fish. *In* "The Central Nervous System and Fish Behavior" (D. J. Ingle, ed.), pp. 127–138. Univ. of Chicago Press, Chicago, Illinois.

Schade, A. F., and Bitterman, M. E. (1966). Improvement in habit reversal as related to dimensional set. *J. Comp. Physiol. Psychol.* **62**, 43–48.

Schutz, S. L., and Bitterman, M. E. (1969). Spaced-trials partial reinforcement and resistance to extinction in the goldfish. *J. Comp. Physiol. Psychol.* **68**, 126–128.

Schwassmann, H. O., and Hasler, A. D. (1964). The role of the sun's altitude in sun orientation of fish. *Physiol. Zool.* **37**, 163–178.

Setterington, R. G., and Bishop, H. E. (1967). Habit reversal improvement in the fish. *Psychonomic Sci.* **7**, 41–42.

Sevenster, P. (1968). Motivation and learning in sticklebacks. *In* "The Central Nervous System and Fish Behavior" (D. J. Ingle, ed.), pp. 233–245, Univ. of Chicago Press, Chicago, Illinois.

Shashoua, V. E. (1968). The relation of RNA metabolism in the brain to learning in the goldfish. *In* "The Central Nervous System and Fish Behavior" (D. J. Ingle, ed.), pp. 203–213. Univ. of Chicago Press, Chicago, Illinois.

Shaw, E. (1970). Schooling in fishes: Critique and review. *In* "The Development and Evolution of Behavior" (L. Aronson *et al.*, eds.), pp. 452–480. Freeman, San Francisco, California.

Sheffield, V. F. (1949). Extinction as a function of partial reinforcement and distribution of practice. *J. Exptl. Psychol.* **39**, 511–526.

Skinner, B. F. (1938). "The Behavior of Organisms." Appleton, New York.

Sperling, S. E. (1965). Reversal learning and resistance to extinction: A review of the rat literature. *Psychol. Bull.* **63**, 281–297.

Sperry, R. W. (1965). Embryogenesis of behavioral nerve nets. *In* "Organogenesis" (R. L. DeHaan and H. Ursprung, eds.), pp. 161–186. Holt, New York.

Sperry, R. W., and Clark, E. (1949). Interocular transfer of visual discrimination habits in a teleost fish. *Physiol. Zool.* **22**, 372–378.

Sutherland, N. S. (1964). The learning of discrimination by animals. *Endeavour* **23**, 148–152.

Sutherland, N. S. (1968). Shape discrimination in the goldfish. *In* "The Central Nervous System and Fish Behavior" (D. J. Ingle, ed.), pp. 35–50. Univ. of Chicago Press, Chicago, Illinois.

Sutherland, N. S., and Mackintosh, N. J. (1970). "Stimulus Analyzing Mechanisms." Academic Press, New York (in preparation).

Tavolga, W. N., and Wodinsky, J. (1963). Auditory capacities in fishes. Pure tone thresholds in nine species of marine teleosts. *Bull. Am. Museum Nat. Hist.* **126**, 177–240.

Thompson, T. I. (1963). Visual reinforcement in Siamese fighting fish. *Science* **141**, 55–57.

Thompson, T. I., and Sturm, T. (1965a). Classical conditioning of aggressive display in Siamese fighting fish. *J. Exptl. Anal. Behav.* **8**, 397–404.

Thompson, T. I., and Sturm, T. (1965b). Visual-reinforcer color and operant behavior in Siamese fighting fish. *J. Exptl. Anal. Behav.* **8**, 341–346.

Thorpe, W. H. (1963). "Learning and Instinct in Animals." Methuen, London.

Triplett, N. (1901). The educability of the Perch. *Am. J. Psychol.* **12**, 354–360.

Tugendhat, B. (1960). The normal feeding behavior of the three-spined stickleback (*Gasterosteus aculeatus* L.). *Behaviour* **15**, 284–318.

Ueda, K., Hara, T. J., and Gorbman, A. (1967). Electroencephalographic studies on olfactory discrimination in adult spawning salmon. *Comp. Biochem. Physiol.* **21**, 133–143.

Uhl, C. N. (1963). Two-choice probability learning in the rat as a function of incentive, probability of reinforcement, and training procedure. *J. Exptl. Psychol.* **66**, 443–449.

Underwood, B. J. (1948). Retroactive and proactive inhibition after five and forty-eight hours. *J. Exptl. Psychol.* **38**, 29–38.

Underwood, B. J. (1957). Interference and forgetting. *Psychol. Rev.* **67**, 73–95.

Vandercar, D. H., and Schneiderman, N. (1967). Interstimulus interval functions in different response systems during classical discrimination conditioning of rabbits. *Psychonomic Sci.* **9**, 9–10.

Van Sommers, P. (1962). Oxygen-motivated behavior in the goldfish, *Carassius auratus*. *Science* **137**, 678–679.

Van Sommers, P. (1969). Personal communication.

von Baumgarten, R. J., and Miessner, H. J. (1968). Regeneration in teleost olfactory system. *In* "The Central Nervous System and Fish Behavior" (D. J. Ingle, ed.), pp. 101–105. Univ. of Chicago Press, Chicago, Illinois.

von Schiller, P. (1948). Analysis of detour behavior, learning of roundabout pathways in fish. *J. Comp. Physiol. Psychol.* **41**, 233–238.

von Schiller, P. (1949). Delayed response in the minnow (*Phoxinus laevis*). *J. Comp. Physiol. Psychol.* **42**, 463–475.

Voronin, L. G. (1962). Some results of comparative-physiological investigations of higher nervous activity. *Psychol. Bull.* **59**, 161–195.

Walker, T. J., and Hasler, A. D. (1949). Detection and discrimination of odors of aquatic plants by the bluntnose minnow (*Hyborhynchus natatus Raf.*). *Physiol. Zool.* **22**, 45–63.

Warren, J. M. (1960). Reversal learning by paradise fish (*Macropodus opercularis*). *J. Comp. Physiol. Psychol.* **53**, 376–378.

Warren, J. M. (1965). The comparative psychology of learning. *Ann. Rev. Psychol.* **16**, 95–118.

Weinstock, S. (1954). Resistance to extinction of a running response following partial reinforcement under widely spaced trials. *J. Comp. Physiol. Psychol.* **47**, 318–322.

Welty, J. C. (1934). Experiments in group behavior of fishes. *Physiol. Zool.* **7**, 85–128.

White, H. C., and Huntsman, A. G. (1938). Is local behaviour in Salmon heritable? *J. Fisheries Res. Board Can.* **4**, 1–18.

Wickler, W. (1968). "Mimicry in plants and animals." McGraw-Hill, New York.

Williams, G. C. (1957). Homing behavior of California rocky shore fishes. *Univ. Calif.* (*Berkeley*) *Publ. Zool.* **59**, 249–284.

Winn, H. E., Salmon, M., and Roberts, N. (1964). Sun-compass orientation by parrot fishes. *Z. Tierpsychol.* **21**, 798–812.

Wodinsky, J., Behrend, E. R., and Bitterman, M. E. (1962). Avoidance-conditioning in two species of fish. *Animal Behaviour* **10**, 76–78.

Yaeger, D. (1967). Behavioral measures and theoretical analysis of spectral sensitivity and spectral saturation in the goldfish, *Carassius auratus*. *Vision Res.* **7**, 707–727.

Yarczower, M., and Bitterman, M. E. (1965). Stimulus generalization in the goldfish. *In* "Stimulus Generalization" (D. Mostofsky, ed.), pp. 179–192. Stanford Univ. Press, Stanford, California.

# 5

# THE ETHOLOGICAL ANALYSIS OF FISH BEHAVIOR

*GERARD P. BAERENDS*

## I. INTRODUCTION

Ethology is the study of all aspects of behavior using biological methods. These aspects can be considered under five categories: the description, the causation, the ontogeny, the function or survival value, and the evolution. Since this is a book on physiology the emphasis shall here be laid on the causation of behavior. However, independently of the aspect we want to emphasize, we must always start with a description of the behavior concerned. Moreover, problems of causation can often be better approached against a background of knowledge about the function of the behavioral patterns and can sometimes benefit from ideas on their evolution. When concentrating on one aspect it is of advantage to remain open minded to the others, provided one always carefully avoids mixing up problems and answers of different categories.

What is the typical ethological contribution to the study of the machinery underlying behavior? Behavior is usually a very complicated phenomenon through which the animal is capable of adjusting its various functions to a constant or changing environment. Physiologists have mainly approached this phenomenon by starting at the bottom with detailed studies of relatively simple behavioral components, e.g., reflexes, locomotory or vegetative automatisms, and influences of definite parts of the nervous system or of the endocrine system. By isolating smaller components, the physiologist can reduce the number of intervening variables, and thus increase the exactness of his work. However, one cannot understand the functioning of an entire machinery from a detailed study of only the parts. To obtain an insight in how these parts work together, one must also start from the entity and analyze it downward to the level of the simple parts. This is what causal ethological analysis is trying to do: to split up a behavioral complex, as it appears to the observer, into units—often of different levels of integration—with circumscribed tasks and sufficiently restricted to be an acceptable challenge for attack by the physiologist with his methods. As a matter of fact, there is still a considerable gap between the level of integration of behavioral mechanisms reached by the physiologists with his studies of component parts and that in which the ethologist has gained insight by systematically breaking up complex behavior sequences. Nevertheless, progress is being made from both sides and there is an increasing number of studies in which ethologists, physiologists, and anatomists have met each other and now work in close cooperation.

Causal ethological analysis uses the "black box approach." The black box is the machinery inside the animal through which behavior, that is muscular and glandular activity (the output) is produced, under the influence of sensory stimuli (the input). By observing the output in great detail under different, often experimentally manipulated, input conditions the ethologist can predict the presence of mechanisms in the black box and define them on the basis of the functions he found them to fulfill. In doing this the ethologist may use the common knowledge that the black box contains receptors and effectors, but for the rest he should try to refrain from all temptations to be guided in his analysis by his knowledge of possible physiological mechanisms. His concern is not where a mechanism is situated, of which anatomical and physiological components it consists, or how it functions, but rather his findings should be formulated in such a way to appeal to the physiologist and warrant an unbiased approach by the latter. The ethologist should be aware of the danger of rigid theories or concepts and continuously adapt his hypotheses to newly obtained evidence.

The causal ethological analysis can only be carried sufficiently far when the input and output of the black box is measured quantitatively so that the relationship between input and output can, for different qualities and quantities, be subjected to quantitative analytical methods. The interesting effects seldom have an all-or-none character, but are usually only released in quantitative investigations. Quantitative work is also necessary to cancel the influence of intervening variables that are always active in complex behavioral phenomena and that may mask definite input–output relations if not systematically and quantitatively studied. It is therefore of great importance to make the behavioral phenomena measurable.

As a part of biology, ethology uses scientific methods. This implies that only those behavioral phenomena that are observable and measurable can be used in this analysis. Since there is no way of obtaining qualitative or quantitative knowledge about subjective feelings in animals —independent of whether such feelings exist or not—the subjective aspects have to be left out of consideration. It is because of this that the term "ethology" was adopted for "objective" behavioral research to differentiate it from animal psychology where, particularly in Europe until a few decades ago, this viewpoint was generally rejected.

Ethologists differ from the American comparative psychologists in that the former are primarily zoologists interested in a great many aspects of the animal, whereas the latter have largely concentrated their research on a study of the mechanisms of learning, irrespective of the species used as a subject.

In this chapter the picture of ethology will be developed insofar as possible with examples from studies of fish. Most of these results are supported by studies on other animals, particularly birds, but often mammals also. For readers interested in a more general survey of the present state of behavioral research along ethological lines the books of Marler and Hamilton (1966), Eibl-Eibesfeldt (1967), Manning (1967), and Hinde (1970) are recommended.

## II. FUNCTIONS OF FISH BEHAVIOR

As a basis for the causal considerations in the following sections we shall give in this section a short survey of the different functions fish behavior can serve and of some principles governing these functions.

In the first place we can distinguish functions associated with the maintenance of the body of the individual. In this category are the respiratory movements and the behavior serving to bring the fish in the proper respiratory conditions. Particular behavioral adaptations are present in the air-breathing fishes (Willmer, 1934; Leiner, 1938). Possibly phylogenetically related with some of these activities is the uptake of gas for the swim bladder occurring in physostomes (Harden Jones and Marshall, 1953). Several fish larvae (e.g., *Gasterosteus*) possess a behavioral pattern enabling them to obtain the first filling of the swim bladder at the water surface even when they are later physoclists (Wickler, 1958a). Cleaning is another body maintenance function. For this purpose most fishes use chafing (a rubbing of the body over a substrate) and fin flickering (a repeated rapid folding and spreading of the fins through which dirt can be removed). Although never investigated properly it seems likely that yawning and body-stretching movements serve primarily to facilitate certain body functions.

Feeding behavior, including the search, grasp, and swallowing of food, shows intriguing adaptations to the kind of food taken. Particularly fascinating is the behavior of the cleaner fishes and of the fishes mimicking cleaners (e.g., *Labroides dimidiatus* and *Aspidontus taeniatus*, respectively, Wickler, 1960a, 1963). The symbiosis between cleaners and their clients is warranted by mutual sensory and motoric behavioral adaptations. Cleaners advertise their nature and select their clients when the latter take up the inviting posture (Eibl-Eibesfeldt, 1955, 1959; Wickler, 1956, 1961a; Youngbluth, 1968; Casimir, 1969).

In defense against predators behavior plays a very important role. One can distinguish different kinds of defense—attacking, fleeing, hiding, and mechanisms making it difficult to pick up and swallow the fish as a prey such as (poisonous or nonpoisonous) spines (*Balistes*; three-spined stickleback, Hoogland *et al.*, 1957) or inflation (porcupine and globe fishes, Diodontidae and Tetraodontidae). Attack is usually restricted to threat, a combination of morphological structures and behavior with an intimidating effect on an opponent. Fleeing is usually a darting away, during which the fish may make itself elusive by changing color or seeking shelter. A particularly interesting behavioral adaptation is the jump and glide by which flying fishes try to escape their predators.

Hiding is often closely connected with camouflage, in which morphological structures and patterns, color changes and behavioral patterns, in combination contribute to the effect (Cott, 1940).

Schooling may be considered a protection against predators, since single fish when frightened can often be seen to join a school, while evidence exists (Eibl-Eibesfeldt, 1962) that for predators single fish in a school are difficult to catch. The behavior underlying the formation and maintenance of fish schools has been studied by Keenleyside (1955) and Shaw (1962a).

Sea anemones (Stoichactinidae) are the hiding places of *Amphiprion* and *Premnas* that live with them in symbiosis (Verwey, 1930a; Gohar, 1948). A substance in their skin protects the fish by inhibiting the discharge of the nematocysts (Davenport and Norris, 1958). The fish obtain this substance from the anemone when they approach it repeatedly, probably with a specific behavior pattern (Schlichter, 1968) which might reduce the chance of being caught by the anemone (Graefe, 1963, 1964). The fish show avoidance behavior against anemone species from which they are not protected (Eibl-Eibesfeldt, 1960).

The principles that help to hide the potential preys from their predators can also be found in predators hiding to surprise a prey.

For nearly all functions locomotion is essential; Breder (1926) has given a now classic survey of the modes of locomotion in fishes. Locomotion is perhaps most spectacular in migration. Migration may serve the escape from unfavorable abiotic conditions, the search for places where food is abundant, or the return to the spawning grounds. It is essential for long living species with eggs or larvae, which are moved away with the currents, to possess the behavioral trait to return to or near their birthplace for spawning. The mechanisms making this possible are certainly remarkable, but without them the fish would not have been able to survive.

This consideration brings us to the second group of functions: those serving the maintenance of the species, the functions of reproduction. The necessity to cooperate with one or more fellow members of the same species is obvious. This aspect was also present in one of the behavioral patterns met above, viz., schooling.

Reproduction comprises at least the behavior necessary for the external (most common in fish) or internal (e.g., poeciliid fish) fertilization. In some groups, however, there exists a more or less elaborate care for eggs and young which is always combined with the defense of an area (territory) and which usually also implies some kind of nesting behavior.

It is often said that in some schooling fish sexual behavior is restricted only to the shedding of eggs and sperm. However, in all cases that have been studied more thoroughly, special behavioral mechanisms were found to bring males and females closer together, increasing the chance that the eggs are fertilized. Whereas in a school all fish look equal and behave equally, without differences in rank, reproductive behavior is often preceded by the appearance of some inequality in the swarm. Fish may change color and behave differently, no longer observing the constant intermember distances so typical of a school.

In some species these changes make it possible for male and female to come closely together and perform a ceremony through which the sperm is brought in close contact with the eggs, as, for instance, in the pike, *Esox lucius* (Fabricius and Gustafson, 1958), and in the anabantids. In cases of internal fertilization this ceremony comprises intromission. The copulation ceremonies rarely take place within the school. This is apparently mechanically impossible, just as a predator rarely catches a prey among a school of fish. Males tend to either chase or lure a female away from the school. Baerends *et al.* (1955) have analyzed the complex behavioral sequences by which the male of the guppy, *Lebistes reticulatus*, lures the female away from the school before it attempts to copulate. Liley (1966) has shown how differences in these displays are ethological barriers preventing crossbreeding between *Lebistes* and three other sympatric species.

Very often at the onset of reproductive activity agonistic behavior may take place whereupon at least part of the school gradually changes to another type of social structure, the territorial society (Baerends, 1952), in which the distances between the members have increased and are maintained by some real fighting and a great deal of threatening.

Although females may temporarily claim and defend their own territories, it seems to be a rather general rule that the breeding territory is established by the male who is later joined by the female for a longer (substrate spawning cichlids) or shorter (mouth-breeding cichlids, sticklebacks, and anabantids) period. In the latter case the male often collects clutches from more than one female. The nest may be an open substrate (stone or plant), a hole or crevice, or a pit dug in the bottom, sometimes in a crevice, with the mouth and the pectorals (cichlids) or the tail (centrarchids). More rarely it is an elaborate structure of air bubbles (anabantids) or plant material (sticklebacks). In several cases the eggs are aerated through fanning with body and fins. When the eggs have hatched the young are often guarded for some time, in the polygamous species by the male only, but in the monogamous substrate spawning cichlids by both parents for the relatively long period of 2–4 weeks.

For pair formation, territory defense, and leading the young, communication is necessary. Consequently, one finds in fish a repertoire of different activities each with its own communicative value: the signal activities. Most of these activities are visual signals that often support or are supported by conspicuous structures or markings. Some of these activities produce acoustical signals; our knowledge about sound production in fish has rapidly increased during the last decade (Tavolga, 1960; Winn, 1964; Nelson, 1965a). Chemical signals have been found, but our knowledge about them is scant. Valone (1970) has observed differentiated electrical emissions in *Gymnotus carapo* which appeared to correspond to certain social interactions. Most signals have only intra-specific effects.

In the beginning of agonistic and sexual encounters the resident male often uses the same signals irrespective of the approach of the opponent. It then depends on the signals given by the opponent whether the encounter develops into fighting or into courting. Particularly in the polygamous species courting consists of a series of signals, with the male and female following each other in a chain of stimulus–response relations, each response in one sex being a stimulus for the next response in the other sex. Such a series of acts is a compound key–lock system by which a partner of the same species, of the right sex, and of the right physiological condition is selected. The courtship of the three-spined stickleback is a classical example of this type (Ter Pelkwijk and Tinbergen, 1937; Tinbergen, 1951); the behavioral and the morphological signals in this type are usually sexually dimorphic.

In the monogamous substrate spawning cichlids (e.g., *Aequidens portalegrensis*, Greenberg *et al.*, 1965) the chain of sexually dimorphic signals that serves the selection of the partner is relatively short, but after the pair has been formed a much longer period of mutual display follows in which male and female are using the same signals. Also, the color patterns of these fishes are either not sexually dimorphic or only slightly so. There is evidence that the main function of these signals is a reduction of the release of aggressive and flight responses between the partners. The fish become accustomed to each other's presence and probably after some time know each other individually. A real pair bond is formed and kept for at least a month; in captivity this has often been reported to continue for a much longer period and several successive spawnings. In addition the possibility exists that the performance of these activities has a stimulating effect ōn the gonadal cycle in the female (Metuzals *et al.*, 1968). Thus several mechanisms contribute in synchronizing the readiness to spawn in the partners. Barlow (1970) and Barlow and Green (1970) have tried to test experimentally in two cichlid species the likelihood that appeasement and sexual arousal are two functions of courtship, by correlating the amount of courtship with the size of each of the partners in a mating pair (size being an important factor for dominance). The data contain evidence in favor of both functions.

Because of the signal activities that first surprise the observer and then make him wonder about their causation, function, and evolution, the interest of ethologists has so far been mainly concentrated on the reproductive behavior of fish. The literature on this subject is now very extensive, and we must restrict ourselves to mentioning a few of the more important studies.

Information on several aspects of the reproductive behavior of the three-spined stickleback, *Gasterosteus aculeatus*, has been given by Leiner (1929, 1930), Wunder (1930), Ter Pelkwijk and Tinbergen (1937), Tinbergen (1951, 1953), Van Iersel (1953), Baggerman (1957), Sevenster (1961), Sevenster-Bol (1962), and Van den Assem (1967); and information on the reproductive behavior of the ten-spined stickleback has been given by Morris (1958). Cichlid fish have been studied by Breder (1934), Peters (1937, 1941), Noble and Curtis (1939), Seitz (1940, 1942, 1948), Aronson (1945, 1949), Baerends and Baerends-van Roon (1950), Ohm, (1958, 1959a,b), Wickler (1958a, 1966, 1967, 1969), Heiligenberg, (1963, 1964), Neil (1964), Myrberg (1965), Heinrich (1967), Albrecht (1968), Bergmann (1968), Blüm (1968), Apfelbach (1969), Apfelbach and Leong (1970); centrarchid fish have been studied by Breder (1936), Greenberg (1947), Carter Miller (1964), Keenleyside (1967), F. W. Clark and Keenleyside (1967); pomacentrids by Fishelson (1970); labrids have been studied by Fiedler (1964); and anabantids have been studied by Lissmann (1933), Forselius (1957), Kühme (1961), Miller (1964), Hall (1968), Hall and Miller (1968), Machemer (1970). On the reproductive behavior of *Badis badis* the work by Barlow (1961, 1962a,b, 1963, 1964) should be mentioned; on cyprinodontids that by E. Clark and Aronson (1951), E. Clark *et al.* (1954), Baerends *et al.* (1955), Liley (1966), Wickler (1967a), Franck (1964); on characids that by Nelson (1964a,b, 1965a), and on gobiids the work of Nyman (1953), Morris (1954), and Tavolga (1954). Further mention should be made here of ethological studies on the spawning behavior of *Salmo alpinus* (Fabricius, 1953), *Coregonus lavaretus* (Fabricius and Lindroth, 1954), and *Lota vulgaris* (Fabricius, 1954), and on the reproductive behavior of Syngnatidae (Fiedler, 1955). Etheostomatinae (Winn, 1958) *Blennius fluviatilis* (Wickler, 1957b), *Noemacheilus kuiperi* (Wickler, 1959) *Rasbora heteromorpha* (Wickler, 1955), *Rhodeus amarus* (Wiepkema, 1961), and the cod (Brawn, 1961a,b,c).

Studies on nonreproductive behavior in fish are still rare, but the etho-ecological study of Beukema (1968) on predation in the three-spined stickleback demonstrates the value of an analysis of less spectacular types of behavior. Beukema investigated the effect of different deprivation states, of familiarity with the environment, and of the palatability of the prey on the prey risk. Hunger was found to effect (through an increase in the searching activity) the experience with the environment and the number of prey to be encountered. Hunger increased the completeness of the response to discovered prey, but only slight hunger decreased the chance of an encountered prey to be discovered.

When studying the function of behavior, the methods and concepts of ecology have to be used. This is well illustrated by Van den Assem's study (1967) of the function of territory in the three-spined stickleback. He found a minimum territory size for successful nest building. The chance that females follow a leading male to the nest and enter it was greater in larger than in smaller territories. In larger territories there is less egg stealing, a behavior that, however, may have a positive survival value because it might help to synchronize the broods.

## III. THE ORGANIZATION OF BEHAVIOR

To make our deductions about mechanisms inside the "black box" responsible for the input–output relations observed in the intact animal, we shall work from two sides. First, we shall consider the output of the box, the motor elements, and draw conclusions about the machinery coordinating them. Second, we shall work from the receptors and attempt to reveal mechanisms dealing with the processing of the information received before this is passed on to the effectory mechanisms.

## A. Coordinating Mechanisms

The analysis of the behavioral machinery implies the distinction of component parts and of the factors activating these parts. Section II referred to the different kinds of behavior in terms of their function, i.e., their biological significance to the animal. Turning now to the study of causation we may no longer make use of functional criteria to differentiate between parts of the behavioral machinery and their relationships for it would be a mistake to start from the *a priori* assumptions that different functions must necessarily be served by different mechanisms and that there would be only one single mechanism underlying each function. Neither may we rashly assume that coordinating mechanisms for different functions would have no parts in common. On the contrary, we can expect a considerable amount of overlap between the coordinating mechanisms for functions what we consider as different, for the behavioral machinery of an animal does not originate from a design freely adapted to a program of clear-cut functional demands. It has evolved from an existing structure by trial and error, testing each novel or modified part produced by random mutation on its value for the survival of the individual. Consequently, our considerations of how the machinery works have to be based on causal criteria. Nevertheless, insight into functions will be useful in this context to find out which results of activities might act as feedback stimuli in causal chains.

## 1. THE FIXED ACTION PATTERN

To start the analysis we have to choose a behavioral unit for the occurrence of which the causes can be investigated. On the basis of our physiological knowledge the most objective choice would be the activity of a separate muscle or gland. However, so far nobody has ever used this simple basic unit for the description of behavior, undoubtedly because it was unpractical for three reasons: the impossibility for the observer to distinguish visually between discrete activities of individual muscle bundles, the impossibility to record the vast amount of activities that could be observed simultaneously, and the impossibility to deal with the masses of data that would result from observations of this kind if they were feasible. For the future, however, we should keep this possibility in mind, for electromyography, polygraphy, and automatic data processing are powerful tools to overcome these difficulties, as can be seen, for instance, in the study of respiratory behavior in fish by Ballintijn and Hughes (1965; Hughes and Ballintijn, 1965).

As long as such techniques are not available for analysis we have to refrain from starting with units that can be defined on a strictly objective basis. The next best possibility is to choose our unit on the basis of criteria of the morphology of behavior. This means that we can try to work with parts of the total behavioral sequence that to the observer appear as entities, hoping that the outcome of the causal research will show us how far the choice of these units was justified and whether they can ultimately be given a causal definition.

The capacity of our brains to lift separate patterns from a complex, i.e., to perceive "Gestalt," is our best help in this first phase of the analysis. Watching the behavior of an animal when it is hunting, fighting, courting, or cleaning itself, and trying to give a careful description of what is going on, we find ourselves distinguishing patterns of muscle activities that on repetition occur in the same form and which we are consequently inclined to call by a name. Examples are: mouth digging in *Tilapia*, fin digging in *Cichlasoma*, head butting in *Rhodeus*, zigzagging in *Gasterosteus*, shooting in *Toxotes*.

The stereotyped form of these activities makes it possible to recognize them even if they occur under abnormal circumstances in which their regular function is impeded or lost. The form is species specific; there is a variability within the species—just as in nonbehavioral characters—but this variability is usually smaller than that between similar behaviors in related species. Where investigated (in fish the shooting of *Toxotes;* Lüling, 1958; Hediger and Heusser, 1961) the muscular effort exerted was independent of the external situation. When animals grow up undisturbed in their normal environment these behavioral patterns develop in each individual of a species in the same way and, apart from the usual genotypic and phenotypic differences, with practically identical end results.

Because of all these characteristics Whitman (1898), Heinroth (1910), and particularly Lorenz (1935, 1937a,b, 1939, 1952) have emphasized that these stereotyped activities are just as typical for a species as morphological structures and, therefore, can be used as taxonomic features. Moreover, they realized that the concept of homology (Baerends, 1958; Wickler, 1967) can be applied to these activities; thus, they founded a truly comparative study of behavior. Our first task must now be to study the properties of these stereotyped activities more exactly and to see how strictly they are distinguishable from other behavioral elements.

The analysis by Lorenz and Tinbergen (1938) of the activity by which an incubating greylag goose retrieves an egg placed outside the nest bowl leads to the distinction of two components in an activity that occur simultaneously but can be separated experimentally. One component accounts for the stereotyped activity form, independent of the environment; the other serves the orientation or steering of the activity by which it is adjusted to the prevailing environmental stimuli. The egg on the nest rim was necessary to trigger the activity. But when, after the movement had started, the experimenter took the egg away, the movement was fully completed, however, without showing the corrective sideways components by which the egg is steered toward the nest. Lorenz (1937a,b) used for the stereotyped part the word *"Erbkoordination"* (fixed action pattern) and for the orienting component the term "taxis." The combination he called "instinctive activity." Because of the general discord about the interpretation of the word "instinct" the term instinctive activity has never been commonly used and should probably be abandoned. In physiology the term "taxis" is usually restricted to locomotion (see Hinde, 1970, p. 156) and then implies the orientation as well as the pattern of the movement: here we need a term suitable for only the orientation mechanisms of movements *and* postures; we prefer to speak of "orientation component" instead of using the word "taxis" in the sense, as advocated by Koehler (1950).

Unfortunately, this pioneer study by Lorenz and Tinbergen has never been extended and deepened on the same or on other patterns. Nevertheless the distinction within an activity of a centrally patterned component that is only triggered by an external stimulus, and an orientation component patterned continuously by the external situation, is a useful one. In the first place the principle is obvious for locomotory patterns. For example, the young of the mouth-breeding cichlid fish, *Tilapia mossambica*, during the first couple of days after they have left the female's mouth, will return to it in case of alarm for predators. The rapid locomotion toward the mouth is elicited by relatively strong turbulence of the water, followed by a big object moving slowly away from the young (under natural conditions this is the female withdrawing slowly after an attack on the predator). As will be seen in Section III, B, the locomotion of the young is directed by cues from the mouth of the female.

In fish the dichotomy between the fixed pattern and its components can further be demonstrated in the activity by means of which the male three-spined stickleback ventilates the eggs in its nest (Tinbergen, 1951; Van Iersel, 1953; Kristensen, 1939; Arendsen de Wolf-Exalto, 1939). Figure 1a shows the male when fanning. By bringing potassium permanganate crystals near the fish it can be shown that the pectoral fins exert a forward pressure on the water and so direct a current against and

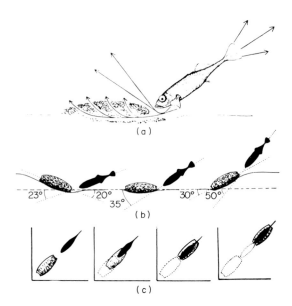

Fig. 1. Fanning in the three-spined stickleback. (a) The direction of the currents aroused (from Kristensen, 1939 and Tinbergen, 1951). (b) The influence of tilting the nest on the orientation of fanning. (c) Experiments on the relative importance of the nest and of other landmarks (i.e., the corner of the tank) on the position of the fish during fanning.

through the nest. At the same time the tail and the caudal fin exert a backward pressure that is equal and opposite to the forward pressure of the pectorals and thereby keeps the fish on the spot. This typical coordination between the pectorals and the tail is the characteristic "fixed component" of the activity. For its release, the presence of a nest with fertilized eggs is usually necessary, but sometimes it may occur in the absence of the adequate external situation (as a vacuum activity, see below). Since the coordination is unchanged the pattern is easily recognizable. In this abnormal case the longitudinal axis of the fish is held horizontal, whereas when it is fanning a nest it keeps a definite angle to the horizontal plane in which the nest usually is. When the nest was tilted, the fish tended to keep the angle between the longitudinal axis of its body and the axis of the nest constant (Fig. 1b). Meanwhile its sagittal plane remained vertical; as von Holst (1950a,b) has shown, this latter orientation is provided for by light and gravity stimuli. By displacing the nest in the horizontal plane it was proved that the exact location of the fish during fanning is determined by optical cues such as the entrance of the nest and even more by other objects in the environment, e.g., the corner of the tank (Fig. 1c).

Therefore in contrast to the constancy of the coordinations of the fixed component producing the aerating current (Sevenster, 1961) the coordinations orienting the animal in space during the action are continuously influenced by external stimuli.

Still another example is taken from a study of the courtship of the guppy, *Lebistes reticulatus*, a viviparous fish (Baerends *et al.*, 1955). In the courtship of the male two phases can be distinguished. First, in the "leading phase" the male tries to lure the female away from the swarm of guppies. If this succeeds, the "checking phase" follows in which the male stops a further movement of the female and finally attempts to copulate. Both in "leading" and "checking," a sigmoid posture, combined with a back and forth swimming movement is performed in front of the female. On closer analysis the characteristics of the sigmoid fixed pattern appear to be the same in both phases. However, in leading, this fixed pattern is combined with a set of orientation components directing the sigmoid away from the female and in checking, with a set of components orienting the activity perpendicularly toward the direction of the advancing female (Fig. 2a).

Quantification of fixed patterns is possible in different ways. First, the fact can be used that the pattern can be performed in different degrees of completeness, i.e., that not all components are always present when the activity is carried out. The order in which, with increasing completeness, the different components are added seems to be fixed.

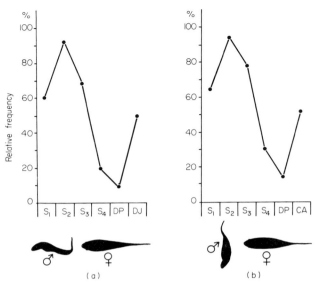

**Fig. 2.** Relative frequency of occurrence of four successive intensity stages of the sigmoid ($S_1$, $S_2$, $S_3$, and $S_4$), display posturing (DP), display jumping (DJ), and the copulation attempt (CA) in 99 leading (a) and 154 checking (b) ceremonies in the courtship of ♂ ♂ *Lebistes reticulatus*. From Baerends *et al.* (1955).

How many of its components are present depends on the probability of occurrence of the activity (the tendency of the activity), measured, for example, as the frequency of occurrence under a standard external situation. For instance, Heiligenberg (1964) found in the cichlid fish *Pelmatochromis* a positive correlation between the number of aggressive activities per time unit and the percentage of complete aggressive activities. The conclusion mentioned above that the differently oriented sigmoid postures in the two phases of the courtship of the male guppy are identical fixed patterns, although primarily based on morphological evidence, was considerably supported by the fact that the percentage in which the four different stages of the completeness of the sigmoid occurred in the courtship ceremonies was the same in both phases (Fig. 2b). Each stage appeared to correspond with a definite strength of the tendency to copulate; this tendency was found to increase in the successive stages of both phases.

Incomplete activities, sometimes called "intention" or "incipient movements," are consequently considered as activities of low intensity. This is quantal intensity in the sense of Russell *et al.* (1954), who also distinguish as quantitative intensity such measures as the speed or the tension at which an activity is executed. Tension will be difficult to quantify in practice, but speed can be measured, for instance, by the time taken for the complete movement or, when the movement is rhythmic, by the frequency of repetitions during a single performance (e.g., the number of fin beats per time unit in fanning eggs).

A second way of quantifying a fixed pattern is by measuring its occurrence in a fixed period, either by its frequency during this period (e.g., the number of fanning bouts per 15 min) or by the amount of the period it occupies (e.g., the number of seconds spent fanning per 15 min). A third method of quantification is the measurement of the latency of an activity; the time passing between the presentation of the releasing situation and the beginning of the relevant movement.

Quantitative measurements of the occurrence of fixed patterns have revealed three other characteristics.

First, the intensity or frequency of occurrence of an activity was found to correlate positively, and the latency time negatively, with the strength of the releasing stimulus. For example, in experiments by Kuenzer (1962, 1968) in the cichlid fish, *Nannacara anomala*, on the features of the parent that release the following response of the fry (see Section III, B), changes in the dummies that make their movement or darkness more similar to that of the parentally motivated adult fish made a greater percentage of the young approach, and the approach time was shorter, partly as a result of shorter latencies and partly as a result of more rapid swimming. Big females induced more copulation attempts and more high intensity sigmoids in male guppies than small ones (Baerends *et al.*, 1955). Higher concentrations of $CO_2$ in the nest entrance made stickleback males fan for more seconds per 15 min (Van Iersel, 1953).

Second, under completely constant external conditions, the occurrence and the intensity of the activity were shown to fluctuate with time. This is illustrated by Fig. 3 which shows the number of seconds a male three-spined stickleback spends in fanning during each of 24 consecutive 5-min periods when caring for its eggs (Van Iersel, 1953).

Third, in several cases the intensity of an activity released by a given stimulus was found to increase with the lapse of time since its previous occurrence. Van Iersel (1953) and Sevenster (1961) prevented the stickleback from fanning its nest for several minutes by scaring the fish mildly, and found a rise above the mean of the fanning activity after the fish had returned to the nest. The possible influence of the accumulation of carbon dioxide in the nest as an external factor was excluded from these experiments by keeping the nests completely covered with a watch glass. Heiligenberg (1965c) measured in *Pelmatochromis* a temporary in-

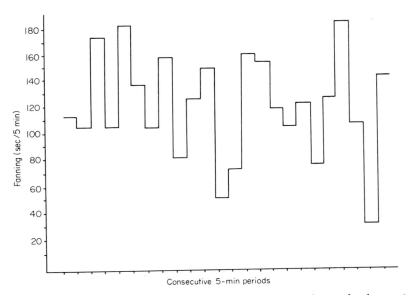

**Fig. 3.** Fluctuations with time in the fanning activity of a male three-spined stickleback. From Van Iersel (1953).

crease of aggressiveness after an interruption of aggressive behavior by the feeding activity, sifting.

Conversely, the longer the period since the activity was last performed, the weaker the minimum stimulus needed to release an activity (threshold lowering). Lüling (1958) reports that *Toxotes*, after having been prevented from shooting for some time, strongly react to minute inadequate stimuli. Baerends and Baerends-van Roon (1950), Lorenz (1950), Heiligenberg (1964), and Kuenzer (1965) have observed cichlid fish, which had failed to breed successfully, guarding a flock of *Daphnia* as if they were their own young ("overflow" of an activity; reaction to suboptimal stimuli; Bastock *et al.*, 1953). If an activity is not released for a very long time, threshold lowering may go so far that the activity is performed without any of the adequate releasing stimuli being present. This Lorenz calls "vacuum activity" (*Leerlaufreaktion*). Thus, we have several times seen single *Lebistes* males, kept in complete isolation since birth, performing the typical sigmoid display while still alone. Van Iersel (1953) mentions that at the height of the period of parental care the male three-spined stickleback, when away from the nest, may suddenly start fanning *in vacuo* (in a horizontal position) and then swim toward the nest and resume ordinary fanning.

As particularly emphasized by Lorenz (1937a,b), this variation in responsiveness indicates that besides the external releasing situation internal factors are of importance. The strength of internal factors can apparently be influenced by the frequency at which the behavior they motivate is released.

As additional evidence for the influence of internal factors Lorenz has stressed the fact that a fixed pattern is often preceded by "appetitive behavior," a searching for the stimulus that can release this pattern and that is no longer continued after the performance of this activity (consummatory act). Lorenz postulated that the occurrence of a fixed pattern and of the accessory appetitive behavior would be caused by the same factor.

This appetitive behavior can have a fixed form, but usually it is variable and often complicated. A stickleback that has left its nest for a while shows appetitive behavior when coming back to resume fanning. It uses landmarks to find its nest; its appetitive behavior may even include attempts to pull off objects which the experimenter has put on the nest (Van Iersel, 1953; Muckensturm, 1965a).

So far we have only mentioned the role of external stimuli in releasing and orienting fixed patterns. In addition we may ask how a pattern, when activated, is maintained or how it is stopped. The results of the egg-rolling experiment, in which the egg was shown to be only essential for the release but not for the continuation of the activity, led to the generalized statement that the maintenance and the completion of a fixed action pattern would be independent of stimuli resulting from its effect. The general validity of this statement has become questionable since Prechtl and Schleidt (1950, 1951) and Prechtl (1952) showed the importance of the milk stream and the contact with the areola for the continuation of suckling in mammals and since Baerends (1959, 1970) showed the importance of specific tactile and temperature feedback stimuli for the continuation of incubation in gulls. Although in fish this aspect has not yet been investigated, the notion of the effect of feedback information is of such great importance that the possibility should be mentioned here; we shall come back to it in Section VI.

Summarizing this section, we have considered evidence on the occurrence of the behavioral units commonly called "fixed action patterns" that seem to argue for a dual nature of the complex of factors causing these kinds of activities. On the one hand, external factors usually release and direct, and sometimes also maintain or stop these activities; and, on the other hand variations in threshold to the external stimuli, vacuum activity, and appetitive behavior are indications of the influence of internal factors.

The influence of internal factors has particularly been emphasized by authors (Lorenz, 1950) objecting to the statement that animals were reflex machines. The dichotomous distinction of internal and external factors originates from concepts in which the internal factors are more or less identified with subjective psychological phenomena, which then are taken to be absent in reflexes (Verwey, 1930b).

Although by now most students of behavior will consider the identification of objectively established factors working inside the animal (or inside the central nervous system) with subjectively established or presumed feelings a logical error, the dualistic approach in the study of the causation of activities has remained. It is still common in the literature and it is often used as a first practical approach to an analytical study. Consequently, we shall also follow this line when discussing in the next sections problems of causation in levels below and above that of the fixed action pattern.

## 2. Levels of Integration below the Fixed Action Pattern

A convenient way to deal with our knowledge of the mechanisms that build up the fixed action pattern is to start from the fanning activity of the three-spined stickleback (see Section III, A, 1) and to break it up in its component parts.

If we strip the fanning activity of its orientation components, the remaining core appears to consist of a forward propelling undulating movement of the tail and tail fin and backward propelling undulating and beating movements of the pectorals. Both are movements also used in locomotion, in forward and backward swimming, respectively, but here coordinated in such a way that they are acting against each other, thus fixing the fish on the spot while producing the aerating current.

With techniques that can still be called ethological, because no operation techniques were involved, von Holst (1937b, 1939) has analyzed in detail the pattern of coordination between fins and between the undulating fin rays of one fin in several species of fishes. The lowest functional unit is the fin ray that can be moved back and forth by two antagonistic muscles at its base. As Weiss (1950) pointed out, to produce the undulating movement of one fin the activity of the motor neurons of these muscles has to be coordinated at a higher level (intramember coordination). Above this level the movements of single fins must be coordinated (intermember coordination) and finally incorporated in the total coordination of the movements and posture of the body. Consequently, Weiss postulated a hierarchy of coordinating mechanisms of different order to account for the coordination of an activity. Von Holst's work mainly deals with intermember coordination. Studying the temporal relations between the undulating movement of different fins, he found two principles in which their coordinating mechanisms could interact: (1) the magnet effect, when the frequency of the rhythm of one mechanism may be enforced onto another, and (2) the superposition effect, when the amplitude of the movements is the algebraic total of the undulations of the dominating and dependent rhythms. Von Holst speaks of "absolute coordination" when the dominating rhythm gains complete control over the dependent rhythm. Since each mechanism has a stronger or weaker tendency to maintain its own frequency (*Beharrungstendenz*), the dominance is often incomplete. The dominating rhythm is then alternately more or less successful in imposing its rhythm; Von Holst has named this "relative coordination."

When a fine current is directed against part of a dominating fin the latter responds immediately with a deflection of the undulating movement but the irregularity is not induced on the dependent fins. From this kind of evidence von Holst concludes that the interacting mechanisms for intermember coordination must be situated centrally from the level of action of the peripheral reflex that helps to adapt the centrally patterned coordination to hydrodynamical changes.

Such evidence also argues against early theories (Friedländer, 1894; Philippson, 1905) that rhythmic locomotory movements are coordinated and maintained by chains of reflexes activated by peripheral proprioceptive stimuli resulting from the movements. Central patterning of coordination was also demonstrated by experiments in which reflexes were excluded over at least part of the trunk by cutting the dorsal roots (Graham Brown, 1912a,b; von Holst, 1935a; Gray, 1936; Gray and Sand, 1936; Ten Cate, 1940). In such preparations rhythmic locomotory movements continued, provided that not all connections with the periphery had been cut. Therefore, the possibility remains to consider with Lissmann (1946a,b) the entire locomotory S wave of the body as one single reflex posture, which can only be propagated along the body if a rhythmic afferent input from some part of the body is present.

With regard to the fixed pattern, its occurrence in the absence of the relevant external stimulus situation, or in response to a very incomplete situation, was considered evidence for an internal inducing factor. For an activity of lower level it is very difficult to ascertain that the adequate releasing stimuli were not present when the activity occurred in an intact or operated animal. Hence, cases of rhythmic activities occurring in the absence of a corresponding rhythmic input have been emphasized as indications that for low level activities internal factors are also important, a tonic afferent inflow being inadequate to induce a rhythmic response.

Experiments by Weiss (1941) seem to be relevant here, in which a piece of embryonic spinal cord and an embryonic leg rudiment were implanted into the connective tissue of an axolotl. Fibers grew from the implanted nerve tissue toward the implanted limb muscles, but there was no nerve contact with the spinal cord of the host. Yet, as soon as motor fibers had made contact with the muscles, the limb spontaneously performed rhythmic movements before sensory connections had been completed.

Having accepted a mechanism for central coordination it is not difficult to conceive an addition to this mechanism converting a tonic stimulus into a rhythmic pattern. Von Holst believes that the central mechanisms causing locomotory coordination produce rhythmic impulses "automatically," i.e., independently of specific afferent inflow, when the central excitatory state has reached a certain level. He compares these mechanisms with the respiratory center to which he thinks they may be closely related (von Holst, 1934b; Le Mare, 1936). Since Adrian and Buytendijk (1931) found in a completely isolated brain of a goldfish a rhythm corresponding roughly with the usual respiratory rhythm of these fishes, the existence of rhythmic electrical activity in the central nervous system has been established in many cases. Bullock (1961) has given a survey of partly hypothetical and partly established types of nerve mechanisms producing patterned movement driven by spontaneous pacemakers or by afferent inflow.

However, we will not trespass further on physiological grounds but turn back to ethological arguments for internal factors in the causation of low level motor activities.

We have seen that fixed patterns may continue after removal of the releasing situation. A comparable phenomenon has been described for reflexes by Sherrington (1948) under the name "afterdischarge."

Variation in responsiveness was another argument for the production and influence of internal factors; this variation can also be observed in simple responses. A relative change of frequency and intensity of a movement after suppression for some time (as found in a stickleback prevented from fanning) was found by von Holst (1937b) in sea horses and Le Mare (1936) in dogfish. By means of a peripheral stimulus they could inhibit the undulatory movement in the dorsal fin of the sea horse or the body of the dogfish. After removal of the stimulus the undulatory movement was resumed but for some time with an increased amplitude. The reverse also occurred: when von Holst (1934a) directed a current against the body of a goldfish in which the frontal part of the medulla was transected, the amplitude of the existing locomotory waves increased during and also for some time after the stimulation. Then, however, the amplitude decreased and was for a short period below the normal level before it reached the latter again. Similar phenomena have been described, for mammalian preparations, by Sherrington (1948) as "rebound" and "spinal contrast." Spinal contrast or rebound only occurs after a stimulus has been applied, not after a period of rest. At first sight this seems different from the changes found in the reponsiveness of fixed action patterns since these were brought about by withholding the relevant stimulus or by giving this stimulus very frequently. However, since an animal is always doing something, a period of deprivation for a certain stimulus situation can probably often be considered as a period of facilitation (disinhibition) of another activity that is incompatible with the activity that corresponds to the stimulus withheld.

From the data available at present we must conclude that the principle of a centrally patterned coordination of muscles is not restricted to the level of the fixed action pattern but can be traced down to the lowest levels of the hierarchy of coordinating systems. It is, moreover, likely that also at these levels the causation of activities is facilitated by internal factors resulting from physiological processes that act largely independently of action specific peripheral impulses.

Besides this central control of coordination, a further control from the periphery is possible through reflex mechanisms that need an input of specific stimulation from exteroceptors or proprioceptors. Such stimuli trigger special positions of the body and the fins, the character of which depends on the area stimulated. The latter control makes the adaptation of the stereotyped movements to changing external situations possible. Reflexes of these kinds have been extensively studied in fish (von Holst, 1934a,b, 1937a; von Holst and Le Mare, 1935; Eberhard et al., 1939; Le Mare, 1936; Lissmann, 1946a,b). It is outside the scope of this chapter to discuss the mechanisms of these reflexes.

Many of these reflexes are important as orientation components (for their mechanism, see von Holst, 1935b, 1950a). However, we should not forget that these reflexes are not the only orientation mechanisms of stereotyped activities. In addition, orientation can be controlled by much more complicated mechanisms, involving higher functions of the brain.

We may conclude this section with the statement that with ethological methods until now no phenomena have been found that justify, on the grounds of causation, a sharp distinction between fixed action patterns and the basic locomotory elements. We shall now turn to mechanisms above the level of the fixed pattern.

## 3. Levels of Integration above the Level of the Fixed Action Pattern

Originally Lorenz (1937a,b) believed that each fixed action pattern was an autonomous element in the behavioral repertoire of the species, not subordinated to integrative mechanisms of higher order. Thus he opposed McDougall's postulation of major instincts (superimposed on minor instincts) because—as he states—no factor coordinating different fixed patterns for the fulfillment of a functional task had ever been found.

However, this opinion was abandoned (Lorenz, 1950) when a few years later in different groups of animals such factors were actually found. On the basis of his observations on the reproductive behavior of the three-spined stickleback, Tinbergen (1942, 1950) formulated the theory of the hierarchical structure of the behavioral machinery, extending Weiss' statement for the levels below the fixed pattern (see Section III, A, 2) to the levels above it.

The basic idea is that numbers of fixed action patterns—sometimes more, sometimes less—share causal mechanisms or factors and consequently in a causal analysis appear as a group or system. When two or more systems share one or more causal factors a system of higher order can be distinguished. Tinbergen (1951) and Thorpe (1951) have proposed the term "instinct" for such a system, but this usage has never been generally accepted and cannot be recommended for the same reasons given in Section III, A, 1 for rejecting the term "instinctive activity." A further disadvantage of the term is that McDougall (1923) had already used it in a similar sense but with an emphasis on the common function of elements instead of on common causation [see however Kortlandt's (1959) defense of his method] while considering them as correlates for important subjective phenomena.

We are of the opinion that the distinction of different systems can be an important aid in the causal analysis of behavior, only, however, when the distinction is really made on the basis of causal criteria and not on criteria of common function as is still often done. We may not reason *a priori* that when an animal has a number of different behavioral elements at its disposal for the same biological function these elements must necessarily be part of one and the same causal system. Lorenz (1937a,b) has stated that in many carnivores the tendency to hunt is strongly independent of the tendency to eat. Hogan (1965) obtained evidence indicating the existence of two systems for fleeing in chicks. Moreover, we may not generalize without further investigation that particular systems found in one species are also present in others. Because this basic causal analysis is difficult and time consuming these rules have often been violated.

The analysis must start with a quantitative description of behavioral sequences. This can be subjected to a statistical analysis of the temporal relations between the elements, e.g., the frequency at which they precede or follow one another or the frequency of their occurrence in the same time span. The resulting positive and negative relationships suggest the effects of common causal factors; hypotheses can be postulated that have to be tested experimentally. Much has still to be done to improve the methods of quantitative description and to make the statistical analysis more sophisticated.

In most cases quantitative experimental tests should be made. The methods for a quantitative assessment of the occurrence of a behavioral element are a matter of concern and could probably still be considerably improved upon. One should realize that an activity can be quantified in many different ways, but that *a priori* assumptions about the quantitative relationships between these different measurements of the same activity are extremely dangerous. The problems become even more complicated when a quantitative measure for the occurrence of a group of activities is necessary. Then usually one quantitative characteristic of one member of the group is used as a measure, but this is not permitted without a preliminary study of how far this measure is representative for the behavior one wants to assess.

It is clear that the assessment could be improved by measuring a great number of characters of the same element. However, the practical difficulties in recording many characters simultaneously and in working up vast amounts of quantitative data will always compel us to compromise. The following examples of causal analytical studies undoubtedly suffer from such compromises, but they nevertheless illustrate the promising possibilities of this approach.

An ethological analysis of the reproductive behavior of the bitterling, *Rhodeus amarus*, was made by Wiepkema (1961). This fish lays its eggs in the gill cavity of freshwater mussels (*Unio* and *Anadonta*) into which the long ovipositor of the female can penetrate through the exhalent siphon. In the beginning of the reproductive period the males start defending an area around a mussel; males and unripe females are attacked and chased away, but when ripe females arrive the male tries to lead them toward the mussel. Each time the female inserts her ovipositor 1–4 eggs are laid. Before and after this the male makes skimming movements over the inhalent siphon during which he may ejaculate. The female may spawn 10–15 times a day, but between successive spawning periods there is a recovery period of at least 5 min during which she does not react positively to the leading attempts of the male and is consequently attacked and chased. For defense of the mussel and for leading the female, the male possesses several fixed action patterns that during the reproductive period occur beside some fixed patterns serving feeding and body cleaning. In his analysis, Wiepkema recorded the occurrence of 12 different fixed patterns in 13 territorial males during a total of 24 hr and in the presence of other ripe and unripe males and females. The temporal association of these 12 patterns was studied by counting the number of times each of them preceded or followed itself or each of the others and were consecutively expressed in rank correlation coefficients that were finally subjected to factor analysis. Since most of the variance among the 12 patterns could be explained by three independent factors the sequential relationships of these movements can be visualized in a three-dimensional model with three orthogonal axes. The model has to be considered as a statistical description of the chances that the activities follow (or precede) each other; the higher the correlation between activities, the smaller the angle between the vectors representing them (Fig. 4).

Three bundles of activities can be distinguished. One comprises chasing (CHS), head butting (HB), turn-beating (TU) and jerking (JK), all activities typical for agonistic encounters. A second bundle comprises quivering (QU), leading (LE), the head-down posture (HDP), and skimming (SK), all activities typical for sexual encounters. In the third bundle we find two activities with a cleaning function [chafing (CH) and fin flickering (FF)], fleeing (FL), and the feeding activity snapping (SN). The model only tells us which activities tend to occur in temporal association, but we need experimentation to find out how this correlation is brought about, or in other words, the identity of the factors represented by the orthogonal axes. Wiepkema found the arrival of the male near the mussel to be the most important external factor for the occurrence of CHS, HB, TU, and JK but the arrival of the female for the release of QU, LE, HDP, and SK. A common causal factor for both groups was the presence of a mussel. A common causal factor for CH and FF is dirt or parasites on the body of the fish. One of the causes that these two activities make relatively small angles with SN and FL is that CH, FF, SN, and FL have in common that their tendencies are all low when either those of the aggressive group or the sexual group are high (and the reverse). Another plausible cause for their occurrence in a bundle will be given in Section VI, C.

Wiepkema obtained evidence showing that when the male or the fe-

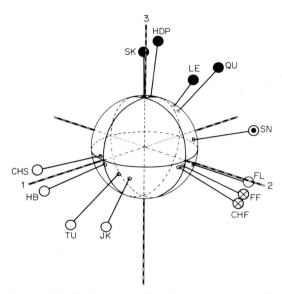

**Fig. 4.** Vector model of the temporal relations between different behavioral elements of the male bitterling in encounters with other males and females near a mussel. The positive side of the factors is indicated by the numbers 1, 2, and 3. The vectors of the 12 variables are determined by their factor loadings, (projections) on the three main axes, Axis 1 corresponds with aggressive factors, axis 3 with sexual factors, and axis 2 with nonreproductive factors. The vectors marked with black dots [skimming (SK), head-down posture (HDP), leading (LE), and quivering (QU)] represent courtship activities; those with white dots [chasing (CHS), head butting (HB), turn-beating (TU), and jerking (JK)] represent agonistic activities; the rest represent nonreproductive activities, viz., the feeding activity snapping (SN), the cleaning activities fin flickering (FF) and chafing (CHF), and fleeing (FL).

male acts as a common causal factor for the occurrence of a group of activities it does so by inducing an internal state by activating a system that makes the occurrence of these different activities (under the influence of their particular releasing stimuli) possible. For instance, the behavior of the female and the length of its ovipositor determine the frequencies in which sexual patterns occur in the male. During 5 min periods of courtship these frequencies are highly correlated among the different sexual activities; which activity is shown at a certain moment depends largely on the distance between the female and the mussel. Presentation of rival males of different sizes reveals that the proportion in which the agonistic activities are performed depends on the size of the opponent, suggesting that the attack system can be activated to different degrees and that this level of activation is an important determinant in the occurrence of CHS, HB, TU, JK, and FS.

An important argument in favor of the idea that the external stimulus activates an internal state that is maintained for some time was given by Van Iersel (1953) in his analysis of the causation of parental behavior in the three-spined stickleback. This behavior includes activities such as fanning, boring in the nest entrance, cleaning the eggs, retrieving eggs, pushing holes in the nest, nest pulling, and retrieving young. Of all these activities, fanning occurs most frequently through the whole parental phase; thus it can be used as representative of the group.

An external factor which can release the parental phase in the male stickleback (even without preceding courtship or fertilization) is the presence of fertilized eggs in the nest. Figure 5 shows how the time spent in fanning increases from day to day until the eggs are 6 days old. Then, about one day before the eggs hatch, the frequency of fanning begins to drop and falls very steeply as soon as the eggs have hatched. The slope of the curve is steeper and the peak higher, the greater the number of eggs present. In experiments when a stream of

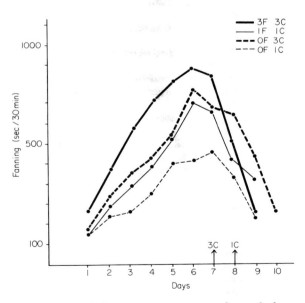

Fig. 5. The frequency of fanning on successive days of the parental cycle, measured as the number of seconds spent fanning during a 30-min test period: F indicates the number of times the male has fertilized eggs, and C indicates the number of clutches in the nest. Day 1 is the day after fertilization, the arrow indicates the average hatching day of the young. From Van Iersel (1953).

water relatively rich in carbon dioxide but poor in oxygen was siphoned
through the nest there was a rise in the frequency of parental activities
such as fanning, boring, and nosing. This suggests that normally the increasing metabolism of the developing eggs causes the gradual rise in
frequency of the parental activities. That this is not the only factor
determining the amount of fanning is shown when all the eggs in a
guarded nest are replaced after some days by an equal number of
fresh eggs. Figure 6 gives an example of what happens. The fanning
curve normally has only one peak, on the day before hatching, but here

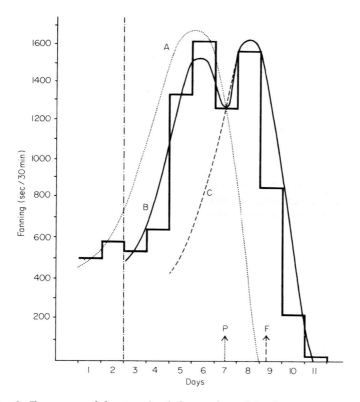

**Fig. 6.** Frequency of fanning (real data indicated by histogram; smoothed indicated by curve B) of a male three-spined stickleback which had its original clutch
replaced by a fresh one on the second day of the parental cycle. To demonstrate the
causes for bimodality, the expected frequency curves have been drawn for the
original clutch in case it had not been removed (curve A), and for the foster
clutch in case the male would not have been influenced by the clutch originally
present in the nest (curve C). Arrow P indicates hatching day of original clutch,
and arrow F indicates hatching day of foster clutch. From Van Iersel (1953).

has two peaks, one on the day before the hatching of the original clutch the other on the day before the new clutch hatches. This means that besides responding to the existing external situation the behavior of the fish continued to be influenced by stimuli that were removed 4 days earlier. These earlier stimuli must have brought about a change in some internal factor, which helps to determine the occurrence of parental activities. Apparently this internal factor can vary in strength. Once it has been activated by external stimuli it develops autonomously during the following days. Van Iersel demonstrated experimentally that the later the time when young eggs are substituted for old eggs, the smaller the influence of the new clutch. This means that the internal factor becomes stronger the longer the fish is in contact with the clutch. The eggs thus not only have a releasing effect on the parental phase but also help to build up the internal factor which, together with the external situation, determines the frequency of occurrence at any moment of the group of parental activities.

Heiligenberg (1963, 1964) found in *Pelmatochromis* a decrease of the tendency to attack after several weeks isolation of the fishes. Daily confrontations of 5 min sufficed to restore the tendency.

The words "priming" or "motivating" have been used to describe this building up effect, whereas the term "releasing" is particularly used for the action of starting the behavioral pattern. However, these different functional terms do not imply that the underlying physiological mechanisms should be necessarily different.

Beukema's work (1968) on the feeding behavior of the three-spined stickleback adds considerably to the evidence so far obtained for the existence of a common internal mechanism controlling a number of behavior elements, in this case the feeding system. He measured feeding behavior in five different ways: the number of bursts of searching-swimming, the number of feeding responses to inedible objects, the proportion of the encountered prey discovered, the proportion of the discovered prey grasped, and the proportion of the grasped prey eaten. With different deprivation times all measures appeared to run parallel. His results agree principally with those of Tugendhat (1960a) who used somewhat different measures and an entirely different environment.

The activation of a system by a causal factor not only implies the facilitation of the release of a certain group of activities but also changes in the sensitivity of the animal for certain environmental stimuli, facilitating the responsiveness to some and inhibiting that to other aspects of the external situation. For instance, a territorial male of the cichlid fish, *Tilapia mossambica*, when predominantly aggressively motivated, will react to an intruding fish that is not obviously a ripe female by fighting. But when it is highly sexually motivated it tends to lead even males toward the nest pit. Eggs or young are fanned or rescued by a fish in parental motivation but are swallowed by the same fish when it is not parentally motivated and is searching for food. Cichlid fish favor *Daphnia* as food, but several authors (see Section III, A, 1) have observed cichlids that were strongly parentally motivated guarding and leading *Daphnia* as if they were their own offspring.

In many fishes the activation of a system includes the appearance of a color pattern typical for the system. Seitz (1940, 1942, 1948), Baerends and Baerends-van Roon (1950), Neil (1964), and Blüm (1968) have studied this phenomenon in various cichlid fish, and Baerends *et al.* (1955) have studied it in the guppy, *Lebistes reticulatus*. It is particularly with this coloration that one can observe the persistence of an internal state for some time and often also make an assessment of its strength.

In the angel fish (*Pterophyllum scalare*) Bergmann (1968) observed a change in the darkness of the vertical bar pattern and of the eye spot on each of the gill covers during ritualized territorial fights, in which the fish are attacking each other mutually. Just before a fish charges the bars fade whereas the spot turns black; during the charge the bars darken while the spot disappears; just after ramming, when the attacking fish retreats, both patterns return to intermediate darkness.

We have already mentioned that besides causal factors common to more than one pattern common factors for more than one system can also be found. On this basis we can distinguish—relative to each other—systems of different order. For example, in the bitterling the possibility for agonistic and sexual behaviors only becomes available after stimulation from the mussel. In the three-spined stickleback, Baggerman (1957) found that the lengthening of the days as it occurs in spring activates the migratory behavior from the winter quarters in the sea toward the breeding places in freshwater, stimulates the development of the breeding colors in the males, and makes it possible for them to respond to the proper environmental stimuli in the breeding area while establishing a territory.

The territorial environment with green plants facilitates nestbuilding (Van Iersel, 1958; Van den Assem, 1967). Then, at the approach of a conspecific male, the repertoire of agonistic behavior patterns becomes available to it, but at the approach of a female the activities of which the courtship ceremony is built up appear (Tinbergen, 1942).

It is not surprising that activities with the same biological function often appear to belong to one causal system. Such a causal organization is likely to promote a quick responsiveness and must consequently have considerable survival value. This connection promotes the naming of systems after their function. Although this usage is practical, it paves the way to the introduction of impermissible functional criteria in a causal analysis, against which we have warned above.

The fact that behavioral elements occur in systems implies that when one system is activated the other must become inactive. Consequently, there must be interaction between systems. The evidence for these interactions will be the subject of the next section.

### 4. INTERACTION OF SYSTEMS

In the course of the reproductive season a number of different internal states become successively predominant. Figure 7 shows these changes in the three-spined stickleback. Van Iersel (1953) has studied one aspect of these changes: the mechanism by which the activation of the sexual system merges into that of the parental system. Estimating the tendency to behave sexually by counting the number of "zigzags" and the tendency for parental behavior by recording the time spent in fanning, he showed that the more frequently the male fertilized a clutch, the quicker the tendency for parental activities increased (Fig. 5) and that for courting decreased (Fig. 8). Moreover, the number of zigzags during the transition period fell more quickly when extra clutches were

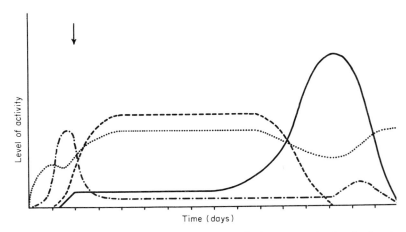

Fig. 7. Schematic diagram of the hypothetical changes in the level of activation of different behavioral systems during a breeding cycle of the three-spined stickleback: ( . . . ) level of aggressive activities, ( - · - ) level of nest-building activities, ( --- ) level of sexual activities, and (———) level of parental activities. The arrow indicates the first occurrence of creeping through (end of nest-building phase); F indicates fertilization (transition of sexual into parental phase), and Y indicates the hatching of young. From Sevenster (1961).

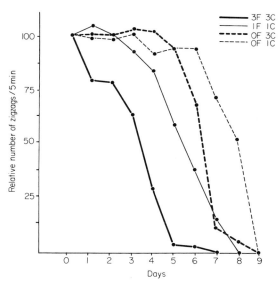

**Fig. 8.** The frequency of courting on successive days of the parental cycle, measured as number of zigzags per 5 min and expressed as percent of the value on the day of fertilization: F indicates the number of times the male has fertilized eggs, and C indicates the number of clutches in the nest. Day 1 is the day after fertilization. From Van Iersel (1953).

put into the nest by the experimenter (Fig. 8). Therefore, the change from the dominance of the sexual system to that of the parental system is facilitated by performing the final act of the former (i.e., the fertilization) and by the confrontation with the proper external situation for the latter (i.e., the eggs). The systems mutually inhibit each other; they are at higher intensities incompatible.

Incompatible systems often express themselves alternately; hence, periods of some overlap occur repeatedly. This holds in the bitterling for the systems for attack, for fleeing, and for sexual behavior. In the factor analysis model the angles between the vectors are a measure for the degree of compatibility of the activities they represent. The greater the angle between activities, the less the probability that they can occur simultaneously. Figure 4 shows that the vectors of the aggressive and the sexual patterns make very great angles with each other, while the positions of the vectors for the aggressive activities and fleeing allow for combinations. The fact that jerking and turn-beating have factor loadings on axis 1 and axis 2 suggests that those activities occur under the combined influences of the tendencies to attack and to flee. This idea was confirmed by non-factorial evidence. For instance, when rival males of different size are presented to a resident male the biggest males release a relatively great amount of fleeing, the smallest a relatively great amount of chasing and head butting, and the intermediate males mostly jerking and turn-beating, while under this latter circumstance fin spreading is maximal. Similarly, quivering and leading have both factor loadings on axis 2 and axis 3. During these behaviors the male swims away in front of the female; morphologically the movement is similar to fleeing and may in fact be induced by the female creating a fright stimulus for the male.

On the basis of his observations on *Pelmatochromis*, Heiligenberg (1963, 1964) has built up a picture of the interaction of different systems active in the behavior of this cichlid fish. His method of derivation of the different systems is open to criticism because it does not seem to be free from functional reasoning and did not emerge from an overall statistical analysis of the behavior. He has evidence for two different fleeing systems, one by forming a school, the other by hiding while assuming a camouflage pattern. Heiligenberg distinguishes a number of threat movements that for morphological reasons seem to be combinations of elements for attack and elements for fleeing. By scaring the fish he was able to induce a shift from attack, or threat behavior with only a few minor elements of fleeing, to a predominance of threat behavior containing marked fleeing components. Fleeing in a school can be induced by passing $CO_2$ through the water. If this is done during a fight the behavior shifts to schooling, but as soon as the $CO_2$ concentration is lowered aggressiveness develops again in the order: fin spreading, lateral display, tail beating, and biting.

Rasa (1969a) states that in the interterritorial fights of the coral reef fish *Pomacentrus jenkinsi* the tendency to flee is indicated by raising the dorsal fin and the tendency to attack by a darkening of the eye. During the fight in each fish both motivational factors seem to work independently of each other. Lowering of the escape motivation, not a raise of aggression, decides who wins; lowering of the aggressive motivation rather than an increase of fright determines the loser.

For the study of the interaction of systems it is very important to be able to manipulate them. Tavolga (1956a,b) found in *Bathygobius* that for some time after a fight the aggressive tendency remained at a relatively high level. The same was true for the tendency to flee after a fright response. Wiepkema used this phenomenon to study the differential effect of aggression on quivering and skimming in the bitterling. Heiligenberg (1964, 1965a,b), who in *Pelmatochromis* has made a quantitative study of the aftereffect of the presentation of an attack releasing situation on the readiness to attack in the consecutive period, found after removal of the stimulus an exponential decrease of the aggressive tendency. He found the same to hold true for fright stimuli and the tendency to flee. When at a certain level of the tendency to attack a fright stimulus is presented attack is immediately reduced, but when thereupon the tendency to flee declines exponentially the tendency to attack suddenly comes back at its previous strength. This indicates that the activity level of the attack system was not changed but that the expression of attack was temporarily inhibited.

Just as different fixed action patterns often share the use of the same elementary components, different subordinate systems often control the same fixed action patterns and different superimposed systems, the same subordinate systems. For instance, sand blowing with the pectoral fins in *Cichlasoma nigrofasciatum* is used in feeding, in uncovering a substrate for spawning, and in digging a pit for collecting the young; in several animals the same type of fighting may be used in disputes over food, over territories, and over rank order. However, the orientation of such an activity and the situation in which it is performed tends to vary between the activating systems sharing its use.

The observations on *Gasterosteus, Rhodeus, Pelmatochromis*, and similar observations on several other animals, particularly birds, show that simultaneous activation of different systems occurs frequently and, for the appearance of certain fixed action patterns, is even necessary. Simultaneous activation of different systems or patterns involves the possibility of internal conflicts, and this has important consequences. We shall come back to them in Section VI which deals with the causes underlying behavioral sequences. First, however, we shall turn to the work on sensory mechanisms processing the incoming information before the effector systems are activated.

## B. Information Processing Mechanisms

We have seen above that motor patterns of different orders of complexity can be triggered, oriented, maintained, or stopped by specific external stimulus situations. So far we have circumscribed these situations as the human observer sees them (e.g., a mussel eliciting the head-down posture in the bitterling, an intruder into the territory releasing the complex of aggressive behavior in the residential fish, the typical environment of a shallow ditch inducing nesting behavior in the three-spined stickleback). In such situations an infinite number of physical characteristics could be distinguished, corresponding to several sensory modalities. Fish have been found sensitive to visual, chemical, tactual, and acoustical stimulation and to changes in their electrostatic field. Following von Uexküll (1934) the problem of the perception of the different aspects of their environment may be considered.

When designing experiments for the analysis of the external situation causing an action, one has to realize that what at first sight might seem to be only one activity may in fact consist of a series of different responses each of which could be triggered, directed, maintained, or stopped by other aspects of the same physical situation. A hunting pike, for example, starts its activities toward a potential prey by fixating and following it with one eye, it then directs its head toward the prey, fixates it with both eyes, brings its body in line with the head, and consecutively swims slowly forward (stalking). At a distance of 5–10 cm from the prey the pike stops, curves it body in an S form, and with one powerful stroke of the tail suddenly leaps at the prey, sucks it in, and seizes it. Small prey are immediately swallowed, large prey are usually moved about in the mouth by jerky snappy movements of the head until they can be swallowed headfirst (Wunder, 1927; Hoogland et al., 1957). Up to the leaping distance all reactions of the pike are visual. Leaping is released by a combination of visual and mechanical stimuli (lateral line organ). Stimulation of tactile and taste receptors play a role after the prey has been seized. Olfactory stimuli could not be shown to be of any importance for feeding, although the pike possesses well-developed olfactory receptors.

For the experimental analysis one needs to make changes in the stimulus situation. This is usually done by replacing part of it for models in which different features can be changed at will. Since man is a visual animal, most of his attempts to analyze the effective parts of a stimulus situation have been concentrated on the visual aspects. The greatest technical problem with models is the imitation of movement.

The first study of this kind in fish was probably the work by Lissmann (1933) on the stimuli releasing fighting in males of Betta splendens carried out in von Uexküll's Institute for Umwelt research. His work was followed by the experiments by Ter Pelkwijk and Tinbergen (1937) and Tinbergen (1939, 1948, 1951) on the properties of the external situations releasing attack and courting in the three-spined stickleback. In these experiments which have now become classic, the authors used rough wax models which were attached to a wire and moved through the tank by hand. A model with red underneath elicited attack in a territorial male. Addition of a light bluish color on the back and a blue eye caused a slight increase of aggressiveness. Variations in shape and size were not effective. For the release of courtship the presence of a silvery color and a swollen abdomen on the model, and the absence of red, proved to be sufficient. Thus, the fish reacted particularly to the features most characteristic of species and sex.

Carefully prepared models with many details of shape could not be shown to be more effective than roughly shaped models, although experiments on other responses gave indications that sticklebacks are not unable to differentiate between shapes (Meesters, 1940). Consequently, the conclusion was drawn that the difference in stimulating value between components of the external situation could only result in part to limitations of the capacities of the sense organs and, in addition, had to be attributed to more centrally situated afferent mechanisms. Each activity as well as each system seemed to have its own mechanism for the evaluation of external stimuli. Ethological research on other animals, particularly birds, supported this idea.

The dominance of one or two components of an external situation for the occurrence of an activity also appears from research on chemical components. The dominating role of $CO_2$ in controlling the amount of fanning in the three-spined stickleback during the parental phase was referred to in Section III, A, 3.

An extreme case in which only one very specific part of the external situation triggers an activity was first found and studied by von Frisch (1942) in the European minnow, *Phoxinus laevis,* and later more extensively investigated by Schutz (1956) and Pfeiffer (1962). It is the responsiveness of the fleeing mechanism of Ostariophysi to a characteristic substance secreted by special epidermal cells in the skin (club cells). These secretions are released into the water when the club cells are damaged, for instance, through bites of a predator. Reactions are strongest to the alarm substance of the same species; reactions to substances of other species occur, but the effectiveness decreases with greater taxonomic distances between the species. The substance is effective at very great dilutions (e.g., 1:50,000); reception is olfactory.

Similar but less specific is the repellent effect of diluted mammalian skin extracts on migrating salmon, *Oncorrhynchus kisutch* and *O. tshawytscha,* demonstrated by Brett and MacKinnon (1954). Wrede (1932) has shown that minnows also have a species specific odor to which they respond positively, and Keenleyside (1955) and Hemmings (1966) have demonstrated the role of such specific olfactory stimuli in keeping fish schools together, particularly in the dark when the visual cues important in schooling are missing.

All experiments mentioned above have proved the existence of dominating stimuli, but the techniques used were insufficient to show that other components of the situation were not playing any role. To study the possible contribution of such components more refined and quantitative experiments, allowing for a statistical treatment, are necessary. In the first place such experiments must be quantitative because even to models differing in dominating characteristics the responses are usually only quantitatively different. For example, Sevenster (1949), when presenting models with either a slender or a swollen abdomen to males of the ten-spined stickleback, *Pygosteus pungitius*, obtained attack and courting (leading) to both models (Section VI, C), but whereas toward the slender-shaped model the frequencies of attack and leading were equal, toward the swollen model the frequency of leading was twice that of attack.

In the second place more refined studies must be quantitative to cancel out the fluctuations in the internal state of the animal (Section III, A, 1) and other possible intervening variables. This can either be done by a frequently repeated presentation in random order of the models to be compared or by simultaneous presentations of models in choice tests. A disadvantage of choice experiments is that they make it almost impossible to distinguish between factors releasing and factors directing the response.

The desire to reduce rigorously the number of intervening variables rigorously leads to standardizing the experimental environment and keeping it as constant as possible. Unfortunately this procedure promotes another intervening variable, namely, adaption (Section III, C, 2).

The experimental procedure has been satisfactory in only part of the work on fish reported here. Therefore, the data must be taken with reservation, but they do point a way to more extensive, intensive, and sophisticated research.

In fish, most studies on the influence of external stimuli on an activity have been done on the release and the orientation of the responses of young cichlid fish toward their leading and guarding parents. In the oral incubators among the cichlids the young, after having hatched and used up their yolk reserve, leave the mouth of the parent and start searching for food in the neighborhood, keeping together in a swarm. In case of danger the parent, often after having attacked the intruder, makes jerky back and forth movements. These movements and the turbulence in the water release in the young a vigorous approach of the parent that takes them up in the mouth. The stimuli directing this approach have been studied by Peters (1937) in *Haplochromis multicolor* and by Baerends and Baerends-van Roon (1950) in *Tilapia mossambica*. In young of *Haplochromis* the approach response was stronger in laterally depressed than in dorsoventrally depressed models and stronger in models with eyes than in models without eyes. The dummies used in the experiments with *Tilapia mossambica* are depicted in

Fig. 9. The young directed themselves to the lower parts of the models and to dark patches. Having reached a solid surface they pushed repeatedly and finding holes they penetrated into them. The value of the model increased when it was retreating slowly, as the female does. With the real female most of the young, through these simple responses, found the mouth into which they either entered or were picked up. Usually

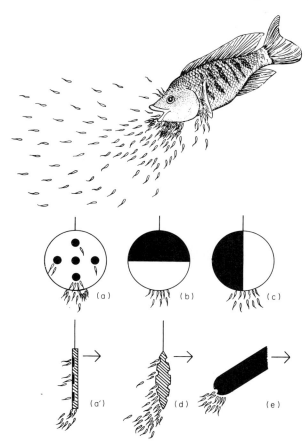

**Fig. 9.** Some model experiments on the stimulus directing the approach of young toward the female of the mouth-breeding *Tilapia mossambica*. Models (a), (b), and (c) are flat discs [cross section (a′)], (d) is a disc with pits, (e) is a black-painted test tube with an opening in the bottom. The underside of the object has a high stimulating value [(a), (b), (c), (d), and (e)], but not because it is the darkest part [(b) and (c)]. In addition, black spots attract the young [(a) and (a′)], while hollows are found by constant pushing against the surface of the model [(d) and (e)].

some young failed in finding the mouth and assembled near the black eyes or the bases of the pectoral fins.

In the substrate spawning cichlids the school of young follows the parents continuously; consequently, release and orientation cannot be separated experimentally. The stimulus situation for this following response has been studied by Noble and Curtis (1939) in *Cichlasoma bimaculatum* and *Hemichromis bimaculatus;* by Baerends and Baerends-van Roon (1950) in *Cichlasoma biocellatum, Cichlasoma meeki, Aequidens portalegrensis, Aequidens latifrons, Hemichromis bimaculatus,* and *Tilapia mossambica;* by Kuenzer (1962, 1964, 1965, 1968) and Kuenzer and Kuenzer (1962) in *Nannacara anomala, Apistogramma reitzigi,* and *Apistogramma borelli;* and by Kühme (1962) in *Cichlasoma biocellatum* and *Hemichromis bimaculatus.* In all studies simplified models of the parents were presented to the young, sometimes moved by hand, often moved mechanically. Kühme used a large arena-tank through which the models could be moved, with a central observation area.

In the substrate spawning species studied, shape or size were shown not to play a role in releasing or directing the following response. A slow, sometimes interrupted, movement was important in all species; a special analysis of movement (Kuenzer, 1962, 1968) in *Nannacara* showed a speed of 70 mm/sec to be superior to a speed of 30 mm/sec. When the intervals between successive moves of the models were 12 sec, young of *Nannacara anomala* followed less, but young of *Apistogramma reitzigi* followed more intensively than with intervals of 1.5 sec. Compared with *Nannacara* the parents of *Apistogramma reitzigi* move relatively less frequently and more slowly.

In *Etroplus maculatus* the parents "call" their young together with a characteristic flickering movement of the conspicuously black pelvic fins. Model experiments showed this movement to be of paramount importance in parental recognition (Cole and Ward, 1969, 1970).

The responsiveness to colors varied among the species roughly in relation to the colors which the parents assume when they are caring for young. In *Hemichromis,* which shows a bright red during that period, the young have a pronounced preference for red and orange over all other colors. *Apistogramma reitzigi* is chiefly yellow; the young prefer plain yellow models. But *Apistogramma borelli* has yellow combined with black, and correspondingly the young show the strongest following response to models with a pattern of yellow and black. The young of *Cichlasoma biocellatum,* both *Aequidens* species, and *Nannacara* follow green, blue, dark gray, or black about equally well and much better than yellow, orange, and red. In these species the guarding parents all have a dark pattern, in the first three species this pattern is on a dark gray or bluish background, but in *Nannacara* on a light gray background. Kuenzer (1966, 1968) demonstrated that although the *Nannacara* young are able to distinguish the checkerboard pattern of their mother, for the following response they evaluate this pattern only in terms of its overall brightness. *Cichlasoma meeki* seems particularly interesting because the young become increasingly selective in their color preference during the period they are led by the parents.

Kuenzer's study of the following response in *Nannacara* reveals another property of the information processing machinery which has been demonstrated more extensively in birds (Tinbergen and Perdeck, 1950; Baerends, 1962). When a certain component of the external situation is effective in causing a response, the order of magnitude of its contribution is roughly fixed but the exact value of the stimulus can vary within this range, in correspondence with the degree in which the component is present in the situation. Kuenzer demonstrated the importance of two characteristics of the color pattern of the leading *Nannacara* female: (1) The object should contrast darkly against a brighter background, and (2) in terms of overall reflection the object should be brighter than black. In model experiments he could realize both characteristics in different degrees by painting models and backgrounds in different shades of gray. Then several combinations could release the following response; only a few of them, however, with maximum intensity.

These results show in addition the property of the information processing mechanism in dealing with relations between different components of the external situation (*Gestalten*). Another example comes from some pilot experiments showing the importance of the position of the red patch on the male three-spined stickleback for releasing attack in another male. In a number of tests, territorial males responded with attacks and with leading to a silvery colored model, respectively, in a ratio of 1–17, to a model with red underneath in a ratio of 6–1, and to a model with a red back instead of a red belly in a ratio of 1–3.

In the ten-spined stickleback the residential male shows a black body with white ventral spines. Sevenster (1949) reports a reduction of the intimidating effect on another male of a black model with white spines after transplantation of the white spines from the rostro-ventral to the anal region.

Having filtered and evaluated the information received, the afferent mechanisms have to produce a combined output to activate the efferent machinery. This output can be built up of information from different sensory modalities. The parental behavior of *Hemichromis bimaculatus*, for instance, is directed by visual and chemical stimuli from the young (Kühme, 1963). The output increases with the completeness of the stimulus situation. For example, Lissmann (1933) and Hess (1953) have shown that the value of a model releasing fighting responses in the Siamese fighting fish, *Betta splendens,* increased in effectiveness as more visual features of the fish were added to it.

Seitz (1940) was the first to stress the fact that the intensity of a response is determined only by the total value of the releasing situation, irrespective of which of the releasing features contribute to this total. It is possible to replace a set of stimuli by another set without affecting the response. Thus, for releasing fight in the cichlid fish *Astatotilapia strigigena,* a tail-beating male in asexual dress is equivalent to a quietly posturing male in breeding dress with the median fins erected. Further, a male without the blue breeding color is equivalent to a male from which the conspicuously colored ventral fins and a patch on the dorsal fin have been removed. Another example can be taken from data on ten-spined stickleback given by Sevenster (1949). The threatening value of a black model without ventral spines is the same as that of a blue model with white spines. The deficiency in body color can thus be compensated by the addition of another important feature. Fabricius (1950) has given evidence that the same principle holds true for the complex of hydrographical factors releasing spawning in fish. Seitz called this phenomenon *Reizsummenphenomen,* a term translated by Tinbergen (1948) as "rule of heterogeneous summation."

The rule has been demonstrated in animals of different groups, but quantitative work is not yet elaborate enough to justify definite conclusions on the mathematical procedure in which the values of the features are combined. Consequently one should still be careful of taking the term "summation" too literally. Yet a study by Leong (1969) on the effect of different features in the territorial coloration of *Haplochromis burtoni* in changing the tendency to attack in an opponent argues in favor of an additive process. In the experimental tank the test fish was placed with 10–15 considerably smaller conspecifics and its attack frequency against these young recorded. A series of dummies with different color patterns was then presented behind a glass plate, each dummy for 30 sec, after which the behavior of the test fish was again recorded for 15 min. Of all the color patterns tested only two were effective in changing attack against the smaller fish: the vertical bars on the head raised the number of bites by 2.79 bites/min, the orange patch above the pectorals decreased attack by 1.77 bites/min. When both patterns were presented together, summation apparently took place, for then biting increased to 1.08 bites/min ($2.79 - 1.77 = 1.02$).

For the response specific machinery enabling the animal to react with a definite activity to a definite group of stimuli (the machinery receiving, filtering, evaluating, and combining information) Lorenz (1935) has introduced the term *Auslösendes Schema*, usually translated as "releasing mechanism." The concept is meant as a functional one; it was particularly developed for considerations on evolution where it has proved to be of great value (Section VII, A). The concept does not include any implications on the nature, localization, complexity, or way of functioning of the machinery. These can only be obtained through an anatomical and physiological follow up of this first ethological approach.

From the data presently available it appears that similar procedures as ascribed to the releasing mechanism also take place in directing a response. Therefore, it would be consistent to use in that case a term like "directing mechanism" and further to expect that future research will make it justified to distinguish in addition "maintaining" and "stopping" mechanisms.

An important consequence of the releasing mechanism is the possibility that a change of valuable properties of the releasing situation experimentally or naturally (e.g., by mutation) may cause an increase of the output of the releasing mechanism beyond the maximal output caused by the original natural situation. To such an "improved" object a stronger response would be expected than to the normal one. In this way Baerends (1962) has made models that released more intensive incubation responses in herring gulls than their real eggs; these models could, therefore, be called "supernormal." In fish as far as we know no cases have been reported of models that could successfully compete with the natural objects. However, the validity of the principle of supernormality in fish is suggested by cases in which, by exaggeration of one feature, models were made that—although deviating strongly from the normal situation—scored a stronger response than standard models carefully imitating the natural object. For example, in one of Kuenzer's experiments (1968) some rectangular models of intermediate gray shade are preferred over a fish-shaped model bearing the proper pattern. In the three-spined stickleback, supernormally big bright silvery colored models with a swollen abdomen induce more courtship in males than freshly killed dead females. Many artificial flies used for angling are likely to be supernormal compared with the usual prey.

Not always does a relatively high sensitivity for certain stimuli correspond to the presence of such stimuli in the adequate releasing situation. For instance, for directing the contact-behavior of (singly raised) young *Tilapia nilotica* and *T. tholloni* the rate of movement (70 mm/sec) found to be maximally effective far exceeded the speed by which the parents move about; neither are the colors most effective in model experiments present in the coloration of the parents (Brestowski, 1968).

Obviously in such cases the sensitivity cannot be the result of adaptation to facilitate perception of functionally important stimuli; in such cases the preference should rather be attributed to physiological characteristics primarily present in the sensory mechanism. This raises the question whether or not in a great many cases the morphology of social releasers are adaptations to such sensory "biases." Selection against such biases can only be expected in cases where they would lead to harmful effects.

## C. The Interaction between Internal and External Factors

### 1. SUMMATION

As in general internal and external factors are necessary for activating a system or releasing a fixed pattern, we must ask how differences in the quantities and in the proportions of the two kinds of stimulation affect the ensuing behavior. Experiments on the courtship of the male *Lebistes reticulatus* shed some light on this problem (Baerends *et al.*, 1955).

As already mentioned in Section III, A, 3, the different marking patterns which develop in the male during courting correspond to different values of the tendency to behave sexually. Consequently, these markings can be used as indicators for the strength of the internal factor. The external stimulus needed for courtship is the female; the bigger the female, the stronger its stimulating effect as measured by the frequency of copulation attempts. In several series of experiments a male bearing a definite marking pattern was confronted with a female of a definite size and the ensuing courting behavior was observed. The interaction of the two kinds of stimulation could thus be studied.

In Fig. 10, the marking patterns are plotted on the abscissa, which represents a scale of increasing sexual tendency. On the ordinate, the size of the female is plotted. Curves have been drawn through points representing different proportions of both kinds of stimuli that produce the same behavior. Two typical fixed patterns were studied this way: (1) following the female, a preliminary courting activity; and (2) sigmoid display, a more advanced courting performance. The lowest and the highest intensity stages of the latter are considered separately. It will be clear from the graph that the same effect can be reached with many different proportions between internal and external stimuli and that the two kinds of stimulation are interchangeable. The curve for a more advanced activity lies above that for a preliminary one (sigmoid compared with following). A high intensity performance lies above a low

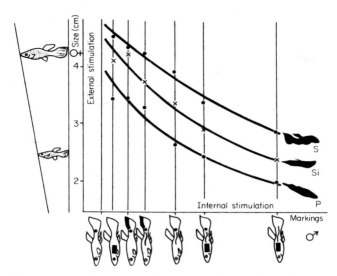

**Fig. 10.** The combined influence of internal and external stimulation on the kind and completeness of the ensuing courting behavior of the male *Lebistes reticulatus*. The different marking patterns which the male assumes during courtship are used as indicators of the tendency to perform sexual behavior (abscissa); the size of the female is a measure of the external stimulation: posturing (P), sigmoid intention movement, (Si), and fully developed sigmoid (S). From Baerends *et al.* (1955).

intensity (full sigmoid compared with sigmoid intention). Therefore, it is the total amount of information available that independently of its origin, determines the kind and the intensity stage of the behavior activated.

When Heiligenberg (1965b) presented to *Pelmatochromis* a stimulus raising the tendency to attack, the effect was additive to the existing output, its magnitude depending on the characteristics of the stimulus but not on the level of the aggressive tendency at the moment of presentation.

The above evidence indicates that the rule of heterogeneous summation also applies for the way external and internal stimuli are combined.

## 2. VARIATION IN THE EVALUATION OF THE SAME EXTERNAL STIMULI

Experiments with different kinds of animals have shown that the input–output relation of the information processing mechanism for a certain activity is not always quantitatively constant.

One cause of this variation is the degree to which different systems in the animal are activated. When Sevenster (1949) presented to several ten-spined stickleback males a series of models increasing in darkness he found that as a rule darker models release stronger aggressive responses than brighter ones. However, some individual males responded less strongly to the darkest than to somewhat brighter models. This was tentatively explained by the fact that the darkest models, like the real territory defending males they represent, not only release counter-aggressiveness but also fright responses in other males, and that the latter effect would reduce the number of attacks on fully black models more in males with at that moment a relatively high than a relatively low fleeing tendency.

Von Holst (1950b) found a comparable influence of the balance between two tendencies on the relative effectiveness of different orientation components of an activity in a quantitative study of the interaction of a geotaxis and a phototaxis in determining the upright position of a fish in the water. If the fish is illuminated from the side it takes up a position at an angle to the vertical plane. This angle is the resultant of the geotaxis that tends to keep the fish in the vertical plane and the phototaxis that in his experiments tends to turn the fish in the horizontal plane. Now this angle, and consequently the relative influence of the two taxes, depends on the degree of activation of different systems in the fish. When it is feeding, for instance, the phototaxis dominates. At night when it is at rest, geotaxis is far more important.

Another influence that can modify the input–output relations of the information processing mechanism for an activity is the repeated presentation of the same external stimulus situation: It may result in a decrement of the response. The reduced responsiveness can be of a short- or a long-term nature. The waning could theoretically result from (1) adaptation of the sensory organ, (2) a change in the input–output relation of the information processing mechanism, (3) a change in the responsiveness of the coordination mechanism of the motor pattern to the output of the corresponding information processing mechanism, and (4) muscular fatigue. The possibilities (1) and (2) may be called stimulus specific, (3) and sometimes (4) may be called response specific. All four possibilities occur, often in combination. When, after repeated presentation of the same stimulus, the response is reduced but the muscles are still able to take part with full intensity in other response patterns, the possibility of muscular fatigue is excluded. When then, provided we are dealing with a pattern that can be evoked by different stimuli, a new stimulus does not bring back full responsiveness, at least part of the decrement must result from the coordination mechanism of the pattern. If, however, the responsiveness is fully restored the cause of the decrement must be in the afferent mechanisms. If then sensory adaptation can be excluded by showing that the same receptor elements are still able to evoke another response, possibility (3) is the only one left (begging response in song birds; Prechtl, 1953). Hinde (1970) has given a survey of the work done with different kinds of animals on the waning phenomenon.

In fish response decrement has often been observed in model experiments. It is mentioned by all authors who studied the stimuli releasing and directing the following response in young cichlids (Section III, B). In these experiments one usually tries to counteract waning by moving the model in an irregular way, using the knowledge emerging from most studies of the response decrement phenomenon that recovery tends to be promoted by interruptions of a monotonous situation. It is probable that this effect prevents a harmful occurrence of waning in the young with regard to the real parents. However, Kuenzer (1968) found no differences when he compared the responses toward identical models moved mechanically with interruptions in accordance with time patterns of different complication.

Several authors have studied in males of *Betta splendens* the waning of aggressive behavior at continuous or frequently (daily) repeated presentation of either models (Laudien, 1965), their own mirror image (Baenninger, 1966; Clayton and Hinde, 1968), or conspecific males (Baenninger, 1966; Peeke and Peeke, 1970). Different aggressive activities were found to wane at different rates; activities corresponding to a higher activation level of the aggressive system seemed to wane more rapidly than less aggressive activities. The rate of waning was also dependent on the temporal pattern of simulus administration ("massed" or "distributed," Peeke and Peeke, 1970). In addition, Clayton and Hinde also found an incremental effect ("warming up"), particularly in the first few minutes after presentation of the mirror. In the experiments by Clayton and Hinde recovery to about 50% of the original frequency took two days; full recovery was either nonexistent or very slow, showing that the decrement was not due to sensory adaptation or muscular fatigue.

A decrement of aggressive responses was also observed when male three-spined sticklebacks in the sexual phase of their reproductive cycle were presented for 15 min each day for 10 days with a live nuptially colored conspecific male or with a crude wooden model (Peeke, *et al.*, 1969; Peeke, 1969). Although after 10 days the same low level of aggressiveness was reached with live fish and with models, waning proceeded more rapidly in males confronted with a live fish since fish elicit much more aggression initially than models. Orientation toward the stimulus also decreased. In the group presented with the live male stimulus the decrement of aggressiveness was correlated with an increase of the frequency of sexual responses when a gravid female was introduced. This corresponds with the idea dealt with in Section III, A, 4 that different equivalent systems inhibit each other mutually and suggests that, at least in this case, the waning process affects the system as a whole.

Several of the above-mentioned cases of response decrement can be classified as habituation, i.e., the persistent stimulus-specific waning of a response as a result of repeated stimulation which is not followed by any kind of reinforcement (Thorpe, 1963; Hinde, 1970). As a result of habituation fishes introduced into a certain environment (e.g., aquarium) become tame. This also happens with regard to the permanent environmental stimuli of the territory and while thus the tendency to flee is decreased the tendencies for the reproductive activities can increase and promote the success of the resident in its familiar environment (Braddock, 1949; Van den Assem, 1967). Van den Assem and Van der Molen (1969) have found the above-mentioned waning of aggressive responses in the three-spined stickleback to occur with respect to conspecific males in neighboring territories. Thus, in a territorial society rival males attain a hypoaggressive condition, which not only leads to a reduction of fighting but, particularly in females, also to an increase in the sexual responses towards conspecifics entering the territory. Haskins and Haskins (1949, 1950) reported that *Lebistes* males, after preliminary attempts to copulate with females of other species, finally restrict their efforts to their own species. Baerends *et al.* (1955) found a waning of the sexual behavior of *Lebistes* males toward individual females or female models with which they could not copulate. Liley (1966) has shown experimentally for *Poecilia* ( =*Lebistes*) *reticulata* that for the normal development of male sexual responsiveness to females positive reinforcement, provided by behavioral interaction with conspecific females, is necessary. This illustrates how at least some of the instances of waning are closely connected with learning and thus with the ontogeny of behavior.

## IV. THE ONTOGENY OF BEHAVIOR

In the above, when dealing with more or less complicated efferent patterns or with the sensitivity or responsiveness to more or less complicated stimulus situations, we have not touched upon the problem of the development of these behavioral characteristics in the individual. How far are the possibilities to react in a special way to certain situations and the abilities to perform special muscle coordinations encoded in the genes or acquired by appropriate information from the environment?

European ethology has emphasized the species specificity of most of the elements of behavior as well as the specific responsiveness to stimuli. In comparative ethology, the innateness of differences in responsiveness or in the form of activities has been stressed. The American psychologists in particular have criticized the ethologists for overlooking problems of ontogeny (Lehrman, 1953; Hebb, 1953). Lorenz (1965), in a reply to this criticism, has stated that the source for the information the animal needs to adapt its behavior to the environment is twofold. One source is the information laid down in the genes during evolution by the processes of trial through mutation and success in selection. The other is the acquisition of information by the individual through experience with its environment. Several ethological studies have shown that species-specific behavior can be the result of a combination of information from both sources (e.g., bird's song; see Thorpe, 1961). In such cases the genes determine what has to be learned and when. Consequently, a normal undisturbed development always leads to the same specific end result. Just as in the morphogenesis of organs this end result can only be influenced when the experimenter tampers with the developmental process or when a mutation changes its course. The development of some behavioral elements may require more bits of experience than others. It will be impossible to prove that no experience at all is incorporated in a behavioral element; this is why most modern ethologists object to calling a behavioral element innate. It is also impossible to call an element entirely learned, because all learning is programmed and has a genetic basis, sometimes a narrower, sometimes a broader one. This discussion shows that it is impossible to distinguish between innate and learned as alternative principles.

Genetic difference between homologous behavioral elements of species appear from several comparative studies of fish groups, e.g., those by Fiedler (1955) on Syngnatidae, by Fiedler (1964) on *Crenilabrus*, by Franck (1964, 1968) on *Xiphophorus*, and by Oehlert (1958) and Heinrich (1967) on cichlids. A demonstration of the genetic basis of fixed action patterns in fish can be found in the observations by Heinrich (1967) of behavior of $F_1$ hybrids between two cichlids: the mouthbreeding *Tilapia nilotica* and the substrate spawning *T. tholloni*. The hybrids showed behavioral elements of both parents in a nonfunctional arbitrary order. Quantitative features like the frequency of skimming acts per time unit showed values for the hybrids that were intermediate to those for the parents.

Franck (1970) obtained similar results when crossing different *Xiphophorus* species. In a genetic analysis of the "backing courtship" of *X. helleri* and *X. montezumae* he found the individual variability to increase in $F_2$ hybrids, while backcross hybrids showed increased similarity to the parent species involved. Some different display patterns appeared to

be inherited independently; interspecific differences between some homologous displays had a polygenic basis.

The learning capacities of fish are discussed in the chapter by Gleitman and Rozin, this volume. In this section we shall restrict ourselves
to available information on the use that is made of these capacities under
natural conditions, i.e., about the genetic programming of learning in the
course of the ontogeny. We shall distinguish between the ontogeny of
coordinations and the ontogeny of the responsiveness to specific stimulus
situations.

The ontogeny of motor patterns has been little studied in fish. Abu
Gideiri (1966, 1969) investigated the early ontogeny of locomotory
activities in different kinds of fish (*Clupea, Cyclopterus, Salmo, Tilapia,
Trachinus*) in connection with the growth of the nervous system. He
found a distinct relationship between the stage of development of nervous structures, e.g., the differentiation of intranuncial neurons, with the
appearance of simple coordinated patterns. He distinguishes four stages
in the early ontogeny: (1) a myogenic stage, at which the stimulus
causing the contraction of the myofibrils lies within the muscle, (2) a
neurogenic stage, at which the somatic motor neurons have reached the
muscle fibers and taken over the initiation of contraction, (3) a re-
flexogenic stage, at which stimulation produces responses, and (4) a
swimming stage characterized by elaborate structural development and
precise responses.

Ohm (1958) has compared the time pattern of the appearance in
ontogeny of agonistic behavioral elements in *Aequidens latifrons* and
*A. portalegrensis*. Part of the interspecies differences can be correlated
with the earlier development of territoriality in *A. portalegrensis*. Compound behavioral patterns develop gradually through superposition
of simpler elements. The most complicated activities appear later in the
ontogeny.

In some cases fish reared in isolation (*Astatotilapia strigigena* males,
Seitz, 1940; *Tilapia mossambica*, Neil, 1964; *Betta splendens*, Braddock
and Braddock, 1958, 1959; Laudien, 1965; *Gasterosteus aculeatus* males,
Cullen, 1961) developed the normal behavioral repertoire for aggression
and courtship. This means that for the development of these patterns experience with fish of the same or of other species is unnecessary. However,
such isolated fish often show exaggerated fright responses. It would be
interesting to investigate if experience with conspecifics is important
for the development of a normal equilibrium between the different
behavioral systems, as has been found in fowl (Kruijt, 1964) and monkeys
(Harlow, 1963).

In three *Mollienesia* species Parzefall (1969) found fully developed agonistic behavior already a few days after birth, whereas sexual behavior appeared considerably later, about the time sexual differentiation became generally apparent. Still, both types of behavior developed completely when contact with conspecifics was prevented.

From studies of the development of schooling in two *Menidia* species, Shaw (1960, 1961, 1962a) suggests that in accord with the idea of Schneirla (1959), by a process of withdrawal from strong stimuli (a big approaching head with eyes) and approach of mild stimuli (a silvery tail moving away), the originally random movements of the fry with regard to each other become gradually directed parallel: One fry approaches another more and more from the tail end and then stays alongside. Fry reared in isolation joined a school of conspecifics of the same age and when brought together also formed schools among themselves, but it took some time before they oriented with regard to each other. The length of this delay was inversely proportional to the length of the period they had spent in isolation. Breder and Halpern (1946) found that Brachydanio reared in isolation from the egg did not hesitate to school when confronted with an aggregation, but that in a similar experiment individuals that had spent some time in a group before isolation showed considerable hesitancy. Both findings suggest that during the early contacts a certain inhibition of approach is built up. Shaw (1962b) has shown that isolated or grouped platyfish raised in tanks with frosted glass showed as adults much less sexual behavior than platyfish that had been kept in tanks of clear glass through which they could see the surroundings.

Ward and Barlow (1967) have pointed to the interesting possibilities of a study of the ontogeny of glancing in the young of the cichlid fish, *Etroplus maculatus*. This is a skimming or bouncing movement against the side of the parent during which mucus is taken up. The rate and orientation of glancing changes with experience. Glancing is also integrated in social and sexual behavior. Bauer (1968) states that the typical contact-behavior the young of *Tilapia nilotica* show toward the mother fish develops optimally only when it has been activated during a critical period (the first ten days of swimming).

More is known in fish about the ontogeny of the responsiveness for specific stimulus situations, i.e., the ontogeny of "releasing mechanisms." This term is often preceded by the adjective "innate," although Lorenz (e.g., 1937b) stated that the extension of the innate releasing mechanism by conditioning is extremely frequent. In a paper describing the modifications which the concept has undergone since it was first formulated, Schleidt (1962) advocates a distinction between "innate releasing mechanisms," "acquired releasing mechanisms," and "innate releasing mechanisms modified by experience"; and he suggests the simple term "releasing mechanism" as long as insufficient research has been done to justify any of the other qualifications.

In analogy to Lorenz' (1935) classic studies on imprinting on the parent species in goslings following the mother greylag goose, several workers have looked for imprinting phenomena in the young or in the parents of cichlid fishes during the parental phase.

The preference for red in young of *Hemichromis bimaculatus* (Noble and Curtis, 1939; Baerends and Baerends-van Roon, 1950; Kühme, 1962) for yellow in *Apistogramma reitzigi*, for a combination of yellow and black in *A. borelli* (Kuenzer, 1962), and for certain brightness relations in *Nannacara anomala* (Kuenzer, 1968) (Section III, B) are all shown by young that from the egg stage have been reared in isolation from the parents. However, in these species improvements in the response to these colors of young kept with their parents have been noticed. Kühme (1962) succeeded in conditioning young of *Hemichromis* to follow other colors than red (e.g., blue) by rearing them with moving models of these colors. The preference for black and for colors of shorter wavelengths in young *Cichlasoma biocellatum* and *Aequidens portalegrensis* (Noble and Curtis, 1939; Baerends and Baerends-van Roon, 1950; Kühme, 1962) is not present in young reared in isolation. Such young are easily conditioned to models of another color: They change an original preference much quicker than the *Hemichromis* young change their original preference for red.

In *Cichlasoma meeki*, a species with a silver-gray body and a red throat, young reared with the parents hatch without preferences for color. This remains unchanged in young reared in isolation from the parents. In contrast young kept with the parents show from an age of about 2 weeks a decline of the response toward colors of short wavelength and toward tints of gray, and after 3 weeks also toward entirely red models. After 3½ weeks they only follow models in which red and gray occur in combination, irrespective of the pattern (Baerends and Baerends-van Roon, 1950). No evidence is as yet available whether the experience that young substrate spawners have with their parents is of importance for their discrimination of conspecifics in later life.

In *Astatotilapia strigigena*, however, Seitz (1940) found the stay of older young in a school of conspecifics to have such an influence. Males that had spent some time in a school of adolescents only courted real conspecific females and never females of other species or carefully prepared dummies. On the contrary males reared in isolation courted very simple models, such as a silvery ball, if they were moved in an adequate way. However, all males, irrespective of how they had been kept, responded to models of males with fighting behavior. Therefore, in this species, where the nonreproductive dress is very similar to the female breeding dress, it is encoded in the genes that the males acquire the knowledge about this dress during the juvenile period they spend in the schools (genetically programmed learning). However, since only ripe males assume breeding colors, and after having deserted the school, the information about this dress cannot be obtained by learning, it must more directly be encoded in the genes in a still unanalyzed way.

Cullen (1961) has found that the discrimination between the nuptial dresses of male and female in the three-spined stickleback does not depend on experience with the father or other fish.

Several authors investigated whether the discrimination which adult cichlids (*Hemichromis* and *Cichlasoma*) in the parental phase show toward young of other species is based on an imprinting process in the parents taking place in the early phase of caring for the first brood in their life. The evidence, based on experiments in which their own young were replaced by equally old young of other species, is controversial. Greenberg's data (1963) argue against the idea. The data of Noble and Curtis (1939) and part of Myrberg's data (1964), particularly those on *Hemichromis*, support it. Myrberg (1966) has tried to explain this controversy by suggesting that in part of Greenberg's tests chemical stimuli from conspecific young present elsewhere in the tank may have prevented the experienced foster parents from attacking the substituted young. This suggestion was based on Kühme's experiments (1963) showing that *Hemichromis bimaculatus* parents respond with parental behavior to water that has been in contact with conspecific fry and are even able to discriminate between water coming from their own fry and from fry of other parents of the same species and of other species.

The above-mentioned sensory learning processes might be called "imprinting" because the scanty evidence indicates that they are thus programmed that they take place in a definite stage of the life of an animal and are relatively irreversible. Not restricted to a certain life stage and definitely reversible is the learning of characteristics of the living area and the learning of food. Muckensturm (1965a,b) and Van den Assem (1967) obtained evidence that three-spined sticklebacks make use of visual landmarks for orientation in the nest area. In his study of feeding behavior in the same species, Beukema (1968) used a honeycomblike maze consisting of 18 hexagonal cells in which food could be distributed. The fishes learned to explore the maze systematically for food and to make a minimum number of turns. With experience the ratio of the number of prey encountered to the distance swum increased.

Meesters (1940) has shown that sticklebacks after having eaten a

solid prey tend to snap at models of solid objects but after eating a
worm tend to take threadlike objects. This corresponds with the idea of
"searching image formation" that Beukema could demonstrate by measuring the increase of the prey risk (i.e., the chance of an encountered
prey to be discovered) after a relatively palatable prey had been introduced for the first time in his maze. One searching image could be
changed for another one when a new type of more attractive food was
introduced.

Hoogland *et al.* (1957) showed that perch and pike, after experience
with sticklebacks which they had snapped up, became rapidly conditioned to avoid the prey on sight before they had made contact.

## V. MODELS OF THE STRUCTURE OF BEHAVIOR

### A. The Network Model

In the preceding sections we have dissected the behavioral machinery into several components and we have made some inferences on
the rules by which they operate. In summary, Fig. 11 gives a visual
representation of the fundamental characteristics of the network that in
our opinion seems to underlie the causation of behavior. The elements at
the bottom of this figure are the fixed action patterns, each of which
(as discussed in Section III, A, 2) consist of a hierarchy of subordinated
systems that coordinate motor units and cooperate with mechanisms
orienting the behavioral patterns. The latter have relatively simple in-
put-output relations working continuously and without delay as long as
the fixed pattern is being performed. Different fixed patterns share to a
large extent the use of patterns of lower order (e.g., locomotion and
reflexes).

In their turn the fixed patterns are subordinate to systems. These
systems can themselves be subordinate to systems of higher order.
Systems may share the use of fixed patterns, and different superimposed
systems may share the use of subordinate systems. Thus there is considerable overlap between the systems that therefore cannot be considered as unitary.

Systems of any order may have mutually inhibitive relations with
other systems; such systems are then called systems of the same order.
Mutually inhibitive relations also seem to occur in fixed action patterns
and other low level elements, but such inhibitions have not been depicted
in the diagram. When inhibition between systems is found it is usually

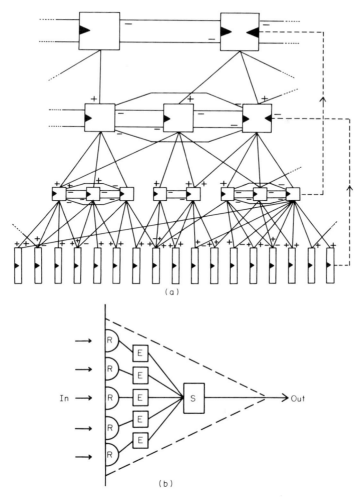

Fig. 11. (a) Hypothetic schematic representation of the hierarchical organization (functional network) of behavior. The rectangles on the bottom row represent fixed patterns; the squares represent systems of different order coordinating other systems or fixed patterns through activation ($+$) or inhibition ($-$). Systems of the same order have mutually inhibiting relations. The dotted lines represent two of the many feedback connections that are thought to be present in the network. The diagram can be extended to both sides. The black triangles in each unit represent information processing mechanisms specific for that unit. (b) The organization of the information processing mechanisms: R, receptors receiving the incoming information from the environment of the animal or from places in its body; E, mechanisms evaluating different parts of this information with regard to the behavior controlled by the unit; and S, mechanism for heterogeneous summation, producing the output of this sensory part which activates the motor part of the unit.

331

very difficult to locate with ethological methods the level where it takes place. A system of higher order may stimulate or suppress a system of lower order or fixed pattern. The activation of a system not only involves changes in the readiness to perform the various behavioral patterns but also changes in the responsiveness to different external stimuli.

Each system or pattern is released by a special stimulus situation through a response specific information processing mechanism (releasing mechanisms). Such a mechanism consists—apart from the receptors—of units for evaluating the information received with respect to the behavioral pattern concerned and of a unit for the summation of the established values. It is probable that in order to remain active several patterns and systems need feedback information which is then likely to be processed by analogous sensory mechanisms. The stimuli necessary for release or maintenance may come from receptors for external or internal stimuli. In the diagram these information processing mechanisms have been incorporated in the systems or action patterns.

The final output of these mechanisms may cause an instantaneous or a tonic response. The coordination of systems and patterns may occur centrally or through stimuli from the periphery, often by a combination of both. Systems may become and, for a while, remain active after peripheral stimulation, but they can also be activated by spontaneous internal impulse production. Different, still insufficiently analyzed, processes can produce stimulus-specific and response-specific decrements of the occurrence of activities.

The ontogeny of the systems is coded in the genes but often in the form of a program for more and less specific learning processes.

## B. Other Models

The use of models is twofold: (1) They can be used as a survey and an illustration of concepts, and (2) they can be a powerful tool to promote logical thinking.

Several different types of models and analogs are applied in biology (see Beament, 1960). Which type is chosen is much less important for purpose (1) than for purpose (2). Lorenz' hydraulic model (1950), by which he illustrated his concept of the properties of a fixed pattern, is clear and has undoubtedly stimulated ethological research. However, it has also biased ethological thinking, partly because of the energy concept inherent to it (Hinde, 1960), partly also because, like all mechanical or electronic models, it could easily interfere with seeing alternative hypothetical solutions for the functioning of the processes one wishes to understand.

This disadvantage can be avoided by a consistent use of network models in which one tries to represent all input–output relations found for the entire control structure (*Wirkungsgefüge;* von Holst and von Saint Paul, 1960) and all its components. This should be our ideal for the future.

Tinbergen's models (1950, 1951) are a beginning to keep at a distance from the mechanical analogs. The models constructed by Hayes *et al.* (1953) are a further development in the direction of more sophisticated network models. At present, however, such network models have only been designed for small parts of the structure, e.g., steering mechanisms (Mittelstaedt, 1960, 1964).

The diagram in Fig. 11 is not a true network model and is definitely biased by our personal favor for the hierarchy concept of the behavioral structure (Tinbergen, 1950). The reader should be warned here that this hypothesis—the danger of which Hinde (1956, 1959a,b, 1960, 1970) has repeatedly noted—has played an important role in the entire composition of this chapter. It is our feeling, however, that the concept is of a very great heuristic value for ethological research and that it has increased our understanding about causation and evolution of behavior considerably. Moreover, Hinde's criticism has contributed much to improving the concept.

## C. On Definitions

The use of any model involves concepts that have to be defined. Unfortunately, the terminology used by ethologists is far from uniform.

In this chapter we have often spoken of the tendency to perform a definite behavioral pattern. By this we mean the probability of occurrence of this pattern. This probability is determined by the compound output of the external and internal situation, which also determines the intensity and frequency of the occurrence. It would be practical to have a term for this total amount of actually effective information. Thorpe (1951) and Baerends *et al.* (1955) have used the term "drive" in this sense ("the complex of internal and external stimuli leading to a given behavior") but, as we shall see below "drive" is also used with other meanings.

The external situation is in principle directly observable and measurable. About the exact value and the complexity of the internal situation the ethologists can only make deductions. Only in the case of a genuine vacuum activity will the overt behavior correspond to the total internal stimulation only. Nevertheless, even if this quantity is not operationally measurable it may be useful to have a term for it. The terms "motivation" and "drive" are often used in this sense, but these terms are also used in a more restricted sense for specific components of the internal situation, and moreover may have a special meaning in the study of learning. Baerends *et al.* (1955) have used the term "specific action potential" (SAP) for total internal stimulation. This term has been suggested by Thorpe (1951) and Hinde (1954) for that state of the animal responsible for its readiness to perform the patterns of one instinct ( = system) in preference to all other behavioral patterns, a definition not excluding the influence of the releasing situation.

The total internal stimulation is likely not to have a unitary origin but to be the resultant of a number of intervening variables. Some of these originate from the different systems and are likely to interact. The terms motivation and drive are often used to denote the state of activation of a definite system (e.g., aggressive motivation or aggressive drive).

The output of a system is determined by the possible spontaneous impulse production in that system and by the input into the system from elsewhere. The effectiveness of this input is influenced by stimulus-specific changes of responsiveness (adaptation, habituation, extinction, or learning in general). Some authors exclude these variables from their concept of motivation.

Several of the concepts and the way they are being used show the influence of Lorenz' original hydraulic model (1950) in which internal and external stimuli were thought to act in a principally different way. The evidence given in Section III, C argues against this hypothesis. In general it seems much better to develop a terminology that is independent of hypothetical conceptions. We advocate the use of operational definitions, but these may prove sometimes insufficient for developing hypotheses on the way the behavioral machinery works.

This discussion will have made it clear that there is no common consensus among ethologists on the use of terms. The concept and their definitions are very much related to the model one has in mind, and they are not very stable because the theories and hypotheses change frequently in this rapidly developing young discipline. This does not matter too much as long as the definitions are properly given by the authors, actually read by the readers, and consistently applied by both.


## VI. THE CAUSATION OF BEHAVIORAL SEQUENCES

The ultimate aim of ethological analysis should be to understand any sequence of behavioral elements as it occurs in the intact animal.

In the following we shall try to see how far models of the types described above can help us to understand the appearance of behavioral elements and the order in which they succeed each other. A general survey of the means by which the units of behavioral sequences can be integrated has been given by Hinde and Stevenson (1969).

## A. Appetitive Behavior and Consummatory Act

For a first case of a behavior sequence we will return to the functional distinction, noted in Section III, A, 1, between appetitive behavior and consummatory act. This distinction, originating from Craig (1918) and introduced into ethology by Lorenz (1937a,b), emphasizes that behavioral sequences can usually be divided into a relatively variable component by which the animal searches for the situation which releases a relatively invariable part. For Lorenz, the latter part was a fixed action pattern, Holzapfel (1940) extended the idea to the attainment of suitable living conditions (e.g., right temperature, salinity or pH, and territory or familiar environment) and Baerends (1941) to the activation of systems. The latter extension implies a hierarchy of appetitive behavior because when a new system is activated this usually becomes observable through a switch to a new kind of appetitive behavior. Present evidence indicates that in some cases this appetitive behavior is of a general character and may lead to any element of the next subordinate integration level. For instance, the nipping at eggs by a cichlid parent may, when the egg is loose, lead to bringing it back to the clutch if it is alive and healthy, to devouring it if it is dead or moldy, and to transporting it to a pit if it has just hatched. In other cases, however, different kinds of appetitive behavior, each corresponding to another subordinate instinct or activity, are performed alternately. Thus a *Tilapia* male which has just established a territory may search for a place to dig a pit, for rivals to fight, or for a female to court (Baerends and Baerends-van Roon, 1950). Which of the available appetitive behaviors appears at a given moment seems to be determined chiefly by internal factors, probably by the same factors that cause changes in the threshold of the corresponding activities or systems. From the above it is clear that an appetitive behavior element leads either to the release of another such element or to a consummatory act. Questions should now be asked about the changes happening after the completion of the consummatory act.

In Lorenz' original view the performance itself of the consummatory act would remove the specific internal stimulation for this act and for the appetitive behavior belonging to it. One could, for instance imagine this to take place when somewhere in the nervous system proprioceptive feedback information coming from the muscles carrying out the consummatory act would be found to match an efference copy or corollary discharge set at the beginning of the appetitive behavior or at the release of the final act (von Holst and Mittelstaedt, 1950; Hayes *et al.*, 1953; Bastock *et al.*, 1953; Hinde, 1970). The attainment of a consummatory situation could be centrally reported in a similar way.

To check this possibility studies have been undertaken of the causes of the reduction of the tendency to behave sexually after the completion of fertilization in two fishes: in the bitterling by Wiepkema and in the three-spined stickleback by Sevenster-Bol. Wiepkema (1961) found that when, prior to spawning, the mussel of a male bitterling was exchanged for a different mussel containing freshly laid eggs the performance by the male of the aggressive activities of head butting and jerking and of the sexual consummatory act of skimming increased in frequency, whereas that of the sexual appetitive activity of quivering dropped. Wiepkema concludes that the increased activity of the aggressive system has a differential effect on quivering and on skimming, thus reducing the frequency of the appetitive behavior but not (at least not during the first few minutes after oviposition) the performance of the consummatory act. Thus, here, the behavioral sequence is changed before the consummatory act is over. Moreover, not the ejection of sperm but the smell of freshly laid eggs changed the male's behavior.

Sevenster-Bol managed to prevent the male stickleback from performing the fertilizing act by putting a wire ring in the nest entrance so that it could obtain chemical stimuli from the eggs (when carrying out its quivering movement on the tail of the female) but could not enter into the tunnel. First she (Bol, 1959) showed that this treatment led to the same reduction of the number of zigzags after the female had left the nest than as the male had been given entrance to the nest and had fertilized. Later she (Sevenster-Bol, 1962) also found an increase of the tendency to attack following fertilization, similar to that in the bitterling. Actually a reversal of the quantitative relations between the sexual and the aggressive systems took place. The author could show that this mainly resulted from chemical stimuli from the eggs but, in addition, to the performance of the quivering movements also.

This reduction of the sexual tendency by stimuli from the eggs could, by its immediate effect on the behavioral sequence and its relatively short recovery period (60 min in the presence of eggs and 20 min when the eggs were taken away), be distinguished from the long-term effect of a number of fertilizations reported by Van Iersel (1953), which lasted for days (Section III, A, 4).

It is in general true that the performance of the consummatory act or its effect strongly reduces its tendency to occur but the existing evidence, although scarce, is against the hypothesis that the appetitive behavior would not have any reducing effect on that tendency. Van Iersel (1953) compared the reduction of zigzagging after fertilization of a different number of clutches with that after the performance of courtship during different time spans. He found five fertilizations, requiring in total 30 min, to have the same influence as 120 min of courtship. One has to conclude that the appetitive behavior elements can also exert a drive reducing effect, but with less power than the consummatory act. Therefore, the difference is only relative (Hinde, 1953).

Wilz (1970) has carried out experiments on the motivational changes occurring in the male three-spined stickleback during courtship which throw some light on the complicated internal processes taking place during the courting phase. He found that pricking the courting female with the dorsal spines, an activity sometimes performed by the male after zigzagging, occurs particularly when the aggressive tendency in the male is relatively high and the sexual tendency rather low. As a result of pricking the female waits before following the male toward the nest, thus giving him the opportunity to carry out nest activities such as fanning and creeping through. Wilz showed that fanning usually precedes creeping through, and that after performance of the latter activity the sexual tendency has increased and the aggressive tendency decreased, consistent with leading the female toward the nest. When creeping through was experimentally prevented, the motivational switch remained off and pricking was repeated. It is tempting to consider here fanning and creeping through as "outlets" for aggressiveness, but this suggests a hypothetical process of motivational catharsis that—although generally accepted in human psychology—has very insufficiently been tested as to its validity. Rasa (1969b) used this principle to explain her observation that the aggressiveness of *Etroplus maculatus* within the pair is reduced when ample opportunity for fighting or threatening rivals in neighboring territories (even behind a glass wall) is present.

Although the distinction between appetitive behavior and consummatory act is sometimes useful for a preliminary classification at the beginning of an analysis, the above studies of how the completion of a sequence is brought about show that the terms may not be identified with definite causal mechanisms. The performance of the consummatory act need not be essential for the reduction of its tendency to occur, and the processes that are essential may affect the components of the chain differentially. For an understanding of the causal mechanism a functional classification does not help because evolution has developed a variety of solutions for analogous functional problems; one must study each single case.

## B. The Repetition of Behavioral Patterns in a Sequence

Changes in the behavioral sequence often primarily result from changes in the internal state. When under the influence of day lengthening the reproductive system is activated in three-spined sticklebacks wintering in the sea, they migrate until they have reached a suitable habitat for nesting in fresh water. Van den Assem (1967) found that sticklebacks, after having been introduced into a 6-meter long tank, alternated periods of long moves (migration) through the whole tank with clusters of short moves (searching for a nest area). In the beginning periods of long moves predominated, later the short moves became more numerous, indicating a gradual internal shift from the migratory to the territorial internal state. The number of short moves could be increased by external stimulation, viz., by placing plant substitutes in the tank.

Guiton (1960) has demonstrated that after the three-spined stickleback has started its nest the threshold for the different building activities gradually changes. When he removed the nest the time lapse before the fish began to dig (the start of a new nest) increased the later in the reproductive cycle the removal took place. It was concluded that the internal state did not permit digging to occur before the absence of the nest had produced a change in the internal state corresponding to a "setback" in the reproductive cycle. Other disturbing factors, such as a lowering of the temperature or strong and continued fright stimuli, can also make the behavior fall back to more preliminary phases. Then sexual behavior usually falls back to nest building, but with very strong disturbances even as far as migration (Van Iersel, 1953).

On the contrary, sudden presentation of a strong releasing stimulus for a behavioral pattern before the corresponding internal phase has properly developed is often effective. For instance, a sexually motivated male three-spined stickleback, which early in the reproductive phase is presented with a female in the nest, may respond immediately with quivering, omitting the introductory courtship.

The internal state often plays an important role in the causation of the time pattern in which a certain motor pattern is repeated within a behavioral sequence. This is illustrated in the following examples.

A relatively simple behavioral sequence, the series of responses shown by a three-spined stickleback feeding on ground-living *Tubifex* worms, was quantitatively studied by Tugendhat (1960a). She recorded the time pattern in which initiated feeding responses (fixating a possible prey) and completed feeding responses (fixating and grasping a prey) occurred with different degrees of food deprivation or satiation. The number of initiated responses per test did not change with the degree of hunger, but with longer deprivation time the number of complete responses increased. The total time spent on feeding responses was independent of hunger since with longer deprivation the duration of the complete and incomplete feeding responses was shorter. These and other quantitative effects of deprivation were reciprocal to those of satiation. When a deprived fish was fed, the intervals between successive complete feeding responses increased during the first 10 min, then decreased for another 10 min, and finally increased again, gradually to the end of the test. Tugendhat constructed a model describing the changes in feeding behavior found with increasing satiation and explaining them on the assumptions that (1) the tendency to feed, when increasing, first passes the threshold for initiated feeding responses and then has to reach a second threshold for the complete feeding response to occur, (2) after each complete feeding response the tendency to feed first drops and then builds up somewhat less steeply than for the preceding response, and (3) the extent to which the tendency drops after each feeding response decreases in accord with a hyperbolic function that reaches an asymptote at a value somewhat below the threshold for the initiated response. This means that two motivational factors may be at work, one changing back and forth after each feeding response and the other changing rapidly during the first few responses and more gradually thereafter. The rate at which the feeding responses occur is thus very much influenced by the effect of the food uptake. In which of the many possible ways this is brought about (see De Ruiter, 1963) was not further investigated.

Many courtship sequences are more complicated. In some of them (e.g., the stickleback, see Section II) the occurrence of each link in the chain is directly determined by the response of the partner, but in many species the courtship does not have such a rigid "key–lock" character. For instance, in *Lebistes* ( = *Poecilia*) (Baerends *et al.*, 1955; Liley, 1966) most copulation attempts are preceded by a number of different courtship activities that, although every activity can precede or succeed most of the others, tend to show a definite sequence correlated with an increasing (or, when in the reversed order, decreasing) tendency to copulate. If a constant test fish is presented, the male gradually passes from the preliminary activities of the courtship to more advanced ones. This is suggestive of a self-stimulatory effect of the performance of the courtship activities, although an increase of the influence of the test fish with time is not excluded. In the *Lebistes* type studied by Baerends *et al.* the change in the internal state of the male during courtship could be followed because of the simultaneously occurring changes in the pattern of black markings these males showed (see Section III, C). The sequence progressed more rapidly when the standard test fish was exchanged for a sexually responsive female; for the most advanced activities of courtship positive reactions of the female were necessary. When such a reaction was not forthcoming (thus also with a constant model), after a while the courtship of the male gradually subsided, the percentage of more advanced courting activities decreasing and those of the early courtship increasing. This decrement was at least partly stimulus specific (Section III, C, 2) but the fact that presentation of a new model did not restore the original level is suggestive of the simultaneous presence of a response specific decrement.

Simpson (1968) has studied the regularities in the occurrence of different elements in the fighting display of *Betta splendens*. The participants in a fight go repeatedly through a cycle of turning to face their opponents and turning broadside. The gill covers are raised in a fish that turns to face; they are lowered in the one that turns broadside. While broadside the fish may flicker the pelvic on the offside of its opponent and may beat or flash its tail. Although the elements may occur in any combination, in general, pelvic flickering is most likely to occur in the 1 or 2 sec, tail beating 3–5 sec, and tail flashing 4–5 sec after turning broadside. The activity of one partner influences that of the other, e.g., facing induces broadside turning and the reverse, gill cover lowering is a response to tail beating. The author obtained evidence that quantitative differences in the temporal patterning determine the result of the fight.

Comparable behavioral sequences occur in the courtship of anabantid fish. Miller and Hall (1968) distinguished in *Trichogaster leeri*, in which the male has to induce the female to spawn in the bubble nest, 15 different variations of the courtship sequences. Their classification depends on the occurrence of aggressive and of sexual behavior (in each of the partners), the occurrence of clasping and spawning, and on which of the partners took the initiative in the sequence. On the one hand, there are indications that the courtship of the male is self-stimulating; on the other hand, it is very much influenced by the courtship of the female, particularly when she has taken the initiative. In the female the kind of activity in the sequence is more dependent on the kind of the foregoing activity than in the male.

Nelson (1964a,b) subjected such behavioral sequences in some glandulocaudine fishes (characids) to statistical analysis. He defined a sequence as a series of statistically dependent actions bounded at each end by intervals separating statistically independent events. He found that the male courtship could be described and analyzed as a first-order Markov chain. Within a sequence the probability of occurrence of an action was dependent of the nature of the immediately preceding action. The data suggest a facilitatory effect of the male's courtship on his own further performance. The female responses depended on a cumulative effect of male courtship activities. These responses were necessary to bring about the final phase of courtship in the male, including the spawning act.

In all these examples there are indications of two antagonistic effects of the performance of the courting activities: an increase of the sexual tendency as well as a decrement when reinforcement is not forthcoming. Moreover, there are indications of changes in the tendencies to flee, to attack, and to behave sexually in the course of the courtship. In *Lebistes* the tendency to attack was thought to be highest in the beginning of courtship, the tendency to flee in the middle, and the tendency to copulate at the end. In *Glandulocauda* aggressive and courtship sequences could be distinguished; the former sequences became gradually shorter and less frequent in the course of the encounter. In male *Trichogaster leeri*, inactive females stimulated aggression which became particularly high when the female fled. In contrast, sexual initiatives of the female reduced aggression and flight in the male. In the female the aggressive activity of butting was used to communicate a high sexual tendency.

In substrate spawning cichlids, males and females perform the same courting activities for hours or even days, the average frequency per time unit of each of these activities passes with time through optimum curves that successively reach their maximum. Although the partner has some influence on the occurrence of the activities, the data strongly suggest that activity frequency is largely determined by internal factors (Greenberg *et al.*, 1965).

A behavioral sequence that has been intensively studied by several authors is the cyclical appearance of maxima (every 15–60 min) of zigzagging, biting, nest building, creeping through the nest, and fanning in the sexual phase of the three-spined stickleback, i.e., after the nest has been finished and before the period of parental care starts. The maxima of the different activities are clearly related to each other. Fanning is abruptly stopped after creeping through; creeping itself is followed by an increase of zigzagging. A decrease of zigzagging is correlated with an increase in aggressive activities and in nest building. Nelson (1965b) made a mathematical analysis of the special course of the cycle after presentation of a female dummy for a few minutes. Then the cycle started with creeping through, a maximum of zigzagging and a minimum of fanning, and the same situation returned after intervals that became increasingly longer in a nearly geometrical progression. The author constructed a model accounting for the time pattern of the occurrence of the activities mentioned. It consists of two variables: One (excitation) rises during presentation of the dummy, the other (threshold) rises during each occurrence of creeping through. Each variable begins to decay exponentially immediately after having reached its maximum. It is postulated that when the threshold has fallen to the momentary level of excitation creeping through occurs again and a new cycle is started. With the help of this model the occurrence of creeping through can be very satisfactorily predicted. Nelson considers this activity as a cause of the changes in motivation occurring during the cycle; he suggests four possible constructions for the network that might underlie these changes.

## C. Conflict Behavior

A different approach was followed in other studies on the causation of the behavior occurring during the sexual phase of the three-spined stickleback. If one considers this phase from the functional point of view it is understandable that an intruding fish or a model releases biting and leading to the nest (Section III, B). However, when the nest is finished the occurrence of nest-building activities as a reaction to the intruder is less easily understood and that of fanning before there are eggs in the nest seems totally out of context.

Hence questions were asked about the causation of these apparently functionally irrelevant activities (Tinbergen, 1940, 1952). It was observed that their occurrence was often temporally correlated with the simultaneous or successive performance of overt behavior of different and antagonistic systems, particularly those for escape and for attack. Moreover, on closer observations some activities, e.g., zigzagging apparently consisted of elements of two such systems (Section VII, B). Accordingly, the idea emerged that interactions between different systems might be the underlying causes for some functionally irrelevant and morphologically complicated activities.

In order to understand how a conflict between two systems could have such an effect we shall first consider what happens when two or more relatively incompatible systems are simultaneously stimulated. This, for instance, is the case in an animal approached by another before it has established the identity (enemy, mate, or parent) of the latter.

In that situation we can usually observe fragments of attack, fleeing,

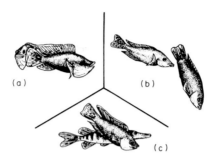

**Fig. 12.** Postures of a male *Tilapia mossambica* during (a) a fight with a con-specific male, (b) leading a female, and (c) introduction of a pike which releases behavior intermediate between (a) and (b).

and sexual behavior. Because competing tendencies exert inhibiting forces, the completeness of their patterns is reduced (Section III, A, 1). However, as long as the level of both tendencies is low, they need not exclude each other; their incomplete activities can therefore occur in close association. The incomplete patterns (intention movements) of both systems may be performed successively (as, for instance, in a slightly scared fish that makes the first head jerk of a locomotory body wave and immediately thereafter brakes the forward movement by swinging the pectorals forward) or simultaneously (as, for instance, in the behavior intermediate between threatening and leading shown by a *Tilapia* male toward intruders with characteristics of both sexes, Fig. 12).

In these examples the ambivalence referred to the stereotyped part of two fixed action patterns. Ambivalence bearing upon the orientation components of the patterns is another possibility. Successive ambivalence of this type can, for instance, often be seen in a hostile encounter at the common boundary of two *Tilapia* territories, where the opponents will alternately advance and retreat. Simultaneous ambivalence of orientation components results in the movement being carried out in a compromise direction. For instance, when flight and attack are simultaneously activated the orientation of the ensuing behavior may not be toward or away from the opponent but sideways. This principle has been called "redirection" (Bastock *et al.*, 1953). In this new orientation the animal may again direct its response to an object, which is then usually an in-adequate one. Figure 13 gives the example of a territorial male of the cichlid fish, *Cichlasoma meeki*, in which the approach of a reproductively motivated female has first activated attack, but which then, under the influence of the breeding colors and the attitude of the female, redirects its attack at a plant leaf.

**Fig. 13.** A male *Cichlasoma meeki*, instead of charging a soliciting female which has entered its territory, performs redirected aggression by biting off *Valesneria* leaves.

Heiligenberg (1965c) has shown that *Pelmatochromis*, when its aggressive tendency rises while the performance of overt attacks to an adequate object is inhibited, tends to start digging (which involves biting the substrate), provided it had recently been performing some mouth activities. It is particularly the feeding activity of sifting that under such circumstances merges into and facilitates digging. The author could show that when opponents are present the tendency to attack has increased after sifting, but not after digging or attacking. Digging can thus really be considered as an alternative for attack in reducing aggression.

A different phenomenon—the appearance of an activity of a third system—is thought to occur when the competing tendencies reach values at which the mutual inhibition becomes complete. We have seen above that in the bitterling as soon as eggs are present in the mussel the tendency to attack rises and that to lead females falls. When both tendencies are about equally high chafing, fin flickering, and snapping reach comparatively high frequencies (Wiepkema, 1961). None of these activities belongs to the aggressive or sexual systems; to the observer they seem functionally irrelevant. Other examples already noted in the three-spined stickleback are fanning and creeping through when no eggs are present in the nest, activities of the parental and of the nest building system, respectively. Temporally linked with their occurrence (Sevenster, 1961; Sevenster-Bol, 1962) are overt signs of the tendencies to attack and to behave sexually. Two other nest-building activities may occur in the sticklebacks under these circumstances: boring in the nest entrance and gluing nest material.

The theoretical importance of the appearance of a functionally irrelevant behavior when for some reason or other the expected behavior is strongly inhibited was first recognized independently by Tinbergen (1940, 1952) and Kortlandt (1940). They gave the phenomenon the name

*Uebersprung* (sparking over); later the name "displacement" became common usage in English ethological literature. The situations in which displacement has been reported are: (1) a conflict between two simultaneously activated incompatible systems, (2) a failure of the external stimulus situation for a behavior to turn up or to stay sufficiently long to evoke the behavior, and (3) deficiency in the feedback information necessary to maintain a behavior.

Several ethological studies have been concentrated on the causation of displacement (Van Iersel and Bol, 1958; Tugendhat, 1960b; Sevenster, 1961; Rowell, 1961; McFarland, 1965a,b, 1966; Baerends, 1970; for a survey, see Zeigler, 1964). Most of these are on bird behavior, but the studies of Tugendhat and Sevenster deal with fish. In her study of the feeding behavior of the three-spined stickleback, Tugendhat (1960a,b) found an increase of the comfort movements (tail-bend stretching, S-bend stretching, and chafing) when a fish, on returning from the food compartment to the living compartment of the tank, stopped moving away and turned back. This means that comfort movements occurred at a moment of conflicting tendencies. When the fish were electrically shocked before entering the food compartment or before grasping food the frequency of comfort movements rose. Near the locus of the shock the incidence of comfort movements was particularly high. Most comfort movements occurred when the fish was changing from retreating to advancing. One may in general conclude from those experiments that comfort movements tend to appear when the tendency to approach (motivated by hunger) and the tendency to avoid (motivated by the experience of the shock) are in conflict and fairly strong.

Sevenster set out to test the validity of the main two hypotheses on the causation of displacement: the "surplus hypothesis" postulated by Tinbergen and by Kortlandt and the "disinhibition hypothesis" postulated by himself and strongly supported by Van Iersel and Bol. According to the surplus hypothesis the output of the thwarted system would become available to another system. Adherents of this theory call this the "allochthonous occurrence" of an activity, in contrast to the autochthonous occurrence when the activity is directly motivated by its own system. The disinhibition hypothesis reasons that when a system is active it inhibits the occurrence of activities of other systems. Consequently, when the active system is thwarted (by any of the three causes mentioned above) other systems are disinhibited and hence get a chance to express themselves (autochthonously). This likelihood of expression increases with the level of activation of the system at that moment, a level which results from the combined effects of internal and external stimuli.

Using the ten-spined stickleback, Sevenster made numerous careful ethological measurements on the tendencies for nest building, aggression, courtship, and fanning. For the measurement of the aggressive and the sexual tendencies in a nest-owning male tests were used in which, under standard conditions, a male in breeding colors or a ripe female was presented alternatively in a glass tube. All recordings were quantitative; in particular the number of bites, the number of zigzags, and the number of leadings were used. Bites served as a measure of the tendency to behave aggressively, zigzagging and leading (that were positively correlated) as a measure of the sexual tendency. Sevenster feels that he measured in this way the absolute strengths of the tendencies. With the tests it was also possible to manipulate aggression and flight by using the aftereffect of a presentation (Section III, A, 2).

Displacement fanning was correlated with a special strength relation between the aggressive and the sexual tendencies, a conflict situation in which neither of these tendencies was dominating. This cannot be explained by separate effects of each of these tendencies on the fanning system. For in the parental phase fanning is reduced by a rise of the sexual tendency caused by a sex test and only weakly increased by a rise of the aggressive tendency by an aggression test; in contrast, displacement fanning during the sexual phase is not altered by an aggression or a sex test. This kind of evidence strongly argues for an autochthonous occurrence of fanning made possible because the mutual inhibition of aggression and sex abolishes their inhibiting forces on other systems. This reasoning is supported by further experiments, showing that the adequate external releasing stimulus for parental fanning, the $CO_2$ production of the eggs, also promotes the occurrence of displacement fanning. When $CO_2$-water is siphoned through the nest the amount for displacement fanning is increased. The same is true when for any other reason the parental system had a relatively high degree of activation. Sevenster could show that the increase resulting from this parental releasing factor is counteracted by a high sexual tendency, a strong indication that the extra amount of fanning is autochthonously produced by the fanning system. The final conclusion is that displacement fanning is caused by disinhibition of the fanning system through the conflict between the aggressive and sexual tendencies and is facilitated by an already high activity of the parental system: this may result from internal as well as external parental factors (number and age of eggs, $CO_2$ content of water).

Sevenster's data make it likely that for irrelevant nest building to occur the aggressive tendency has to be stronger than for fanning. How far absolute and how far relative levels of the conflicting tendencies are important in determining the kind of displaced activity is still not sufficiently clear. The importance, qualitatively and quantitatively, of the balance between the conflicting tendencies and of the external situation accounts for the striking phenomenon in animals that the kind of activities occurring in displacement are very often characteristic for the situation.

Rowell (1961), who objected to the idea of interacting systems, was inclined to attribute greater importance to the break in the behavioral sequence than to the disinhibition, but his analysis (of displacement grooming in chaffinches) is less penetrating than that of Sevenster. McFarland (1965a,b, 1966) has made the interesting suggestion that as an effect of thwarting the animal would become attentive to a greater number of stimuli from the environment and thus increase the chance that during the pause an unexpected behavioral pattern is elicited.

There is no consensus on the point whether displacement caused following a failure of the external situation in reaching the proper level to release or to maintain a certain activity (situation 2 on page 345), should be understood as a special type of causation of displacement or as a case of conflict. Such a conflict might arise when the insufficiency of the external situation necessary for the perpetuation of the activated system would facilitate the activation of another system, for instance, because the inadequate external situation would not suffice to keep the prevailing system sufficiently activated to maintain its inhibiting grip on other competing patterns.

Wilz (1970, see also Section VI, A) considers the results of his experiments on the function of fanning and creeping through in conflict situations as an argument in favor of the motivational catharsis function that Tinbergen and Kortlandt originally attributed to the displacement phenomenon.

In correspondence with the assumed causation, apparently irrelevant activities are often labeled ambivalent, redirected, or displacement activities. We are apprehensive about this habit because a number of cases that were analyzed more extensively have turned out to be caused by a combination of phenomena.

Thus, Sevenster-Bol (1962) found the fertilization act in the three-spined stickleback to consist of two components. One of them is creeping through, a nest-building activity for making the nest tunnel that during the sexual phase is most likely to occur when the aggressive and the sexual tendencies were found to match each other. This displacement component serves as a vehicle for depositing the sperm in the right place, but it can also occur without sperm emission. Ejaculation is considered an activity of the sexual system because its combination with creeping through could be considerably increased (from 0 to 77%) when a female was presented in a glass tube.

In the next section we shall meet more instances of activities resulting from combinations of the conflict phenomena. Therefore it is advisable to give the activities—that to the observer seem to occur out of context with the behavior going on—a more neutral name, not anticipating the results of analytical research that still has to be done. Baerends (1970) has suggested the name "interruptive behavior" to encompass all categories.

The occurrence of interruptive behavior seems to have had an enormous impact on the evolution of behavioral patterns for the purpose of communication and probably also for more direct purposes such as feeding and nest building. Therefore, it seems justified to add to the discussion on causation of behavior a short section on its evolution.

## VII. THE EVOLUTION OF BEHAVIOR

Evolution is modified ontogeny (De Beer, 1940), and ontogeny is a chain of causal processes. Consequently, knowledge of the causation of behavior may be expected to help in understanding how the different behavioral patterns have evolved (Baerends, 1958). Although this problem can be posed for any behavioral pattern, the patterns occurring in courtship ceremonies and aggressive encounters have been more frequently considered.

### A. Social Releasers

Animal ceremonies have a communicative function: Their component patterns release in another animal responses which (except for some exceptional cases discussed below) are of survival value to the animal showing the patterns. As one may expect these patterns have become specialized to fit the releasing mechanisms of the potential reactors; they are signals. This is true not only for behavioral patterns but also for structures and markings; it is usually the combined effect of both components which makes the signals conspicuous. The sensitivity of releasing mechanisms for supernormal stimuli must be considered of paramount importance for a progressive evolution of signals. Lorenz (1935) gave the communication patterns the name *Auslöser*—translated into English as "social releaser" (Tinbergen, 1948; Baerends, 1950). In addition, the terms "sign stimulus" and "key stimulus" are used. Social releasers mostly concern relationships between members of the same species, e.g., in group behavior, in fighting, in courting, and in the care of young. In group behavior, social releasers help to bring and keep the group together, to provide danger warnings, to promote the finding of food, or to maintain a rank order between the members of the group (Greenberg, 1947). In fighting behavior, they prevent the infliction of serious damage by giving the combats a ceremonial character, by limiting much of the fighting to threatening, and by giving a subordinate animal a chance to escape by displaying the signs of inferiority. Durham *et al.* (1968) have shown that in males of *Barbus stoliczkanus* threat operates through an increase of the tendency to flee in the opponent, but appeasement operates neither in this way nor through inhibiting attack but through a loss of interest in the reactor. In sexual behavior, the social releasers serve pair formation and synchronization of the partners (Section II).

In a few special cases social releasers are directed at animals of other species. This is to be expected in cases of mutualism. The cleaning fish provide an interesting example; these fish approach their customers, which are conspicuously blue or yellow colored, with a characteristic behavioral pattern. The customer seems to acquire knowledge about the features of their cleaners by experience, and this learning process is facilitated by the conspicuousness of the cleaner (Eibl-Eibesfeldt, 1955, 1959; Wickler, 1963). The example is all the more interesting because other fish (e.g., *Aspinodontus*) "parasitize" this relationship by mimicking the cleaners in behavior, color pattern, and shape when attacking fish waiting to be cleaned, to bite pieces out of their skin or fins (Eibl-Eibesfeldt, 1959; Wickler, 1960a, 1961a, 1963).

The angler fish (Antennariids) are very remarkable in that they "parasitize" the releasing mechanisms which lead other fish to their food. The lure of *Lophius*, *Antennarius*, etc. (and probably also the luminescent angles of their relatives in the deep sea), capitalize on the tendency in many predatorial fish to follow objects moving at a certain speed. Since predators attack the lure they can be led into the large mouths of these fish to be swallowed (Wilson, 1937; Wickler, 1965a).

Interspecific signals often serve to prevent attack. This holds for the eye spot patterns many fish possess: a circular black spot surrounded by a lighter ring. This pattern parasitizes on a releasing mechanism that helps protect predator fish from larger predators. The eye spots may occur on many different places: body, operculae, tail, and median fins. Often the fish can make them more or less conspicuous by changing the state of the chromatophores (*Aequidens maroni*) or by a special movement (raising of the opercula in *Cichlasoma meeki* or fluttering of the tail in *Astronotus ocellatus*). In *Lebistes reticulatus* (Baerends *et al.*, 1955) the black marking that attracts the female disappears from the tail and another appears on the flank when the orientation of the sigmoid display changes from luring to checking (Section III, C).

A very interesting intraspecific case of mimicry has been described in oral incubating *Haplochromis* species by Wickler (1962a,b). The male anal fins bear patterns mimicking eggs; when the pair is circling in the pit, just before egg laying, the male presents these dummy eggs to the female who snaps at them. The female is stimulated by this performance and moreover through the snapping movement inhales sperm that fertilizes the eggs taken up in the mouth. The pattern is differently elaborated in the various *Haplochromis* species. In *Tilapia* an interesting parallel evolution has taken place, but here appendages of the genital papilla of the male are like structures mimicking eggs (Lowe, 1956; Ruwet, 1963; Wickler, 1965b).

This mutual enhancement of structural and behavioral patterns is very common. Comparison between related species often shows the behavioral component of the signal to be more widespread than the structural pattern increasing the effect. The behavior is thus likely to be older than the marking; it can often be found to serve a noncommunicative function in the relatives of species using it in a signal. Following are some examples.

When approached many fish spread the median fins and thus increase their stability in withstanding an attack. In a number of fish, e.g., the territorial male of *Tilapia mossambica*, these fins have the same color as the body. Therefore, their erection seems to enlarge the body surface, a very important threat considering that greater body size is often a decisive factor in winning a fight. The effect of the eye patch on the tail of *Astronotus ocellatus* is enhanced by a fluttering movement of the tail which is, although less conspicuously, present in many other cichlids during aggressive encounters. The male ten-spined stickleback uses the fanning activity to show the nest entrance to the female and enhances this effect by the white ventral spines that contrast strongly against the black body.

## B. The Derivation of Social Releasers

Let us now examine the origin of these supporting movements. The causation of fin spreading and spine raising is yet insufficiently understood (Barlow, 1962a; Baerends and Blokzijl, 1963; Symons, 1965; Miller and Hall, 1968), but there are strong indications that the occurrence is promoted by the simultaneous action of two tendencies (e.g., the tendencies to approach and to avoid or to attack and to flee). The tail fluttering in the cichlids is typical for an inhibited frontal attack (incipient mouth fighting, Baerends and Baerends-van Roon, 1950). The zigzag dance of the three-spined stickleback is ambivalent behavior between leading and attack. By correlating the variability in form, orientation, and length of the "zigs" and the "zags" with overt attacking and leading, Van Iersel (1953; see also Tinbergen, 1951) showed that the zigzag dance is to some extent controlled by the sexual and the aggressive systems. Fanning in the sexual phase is considered a displacement phenomenon. Digging is used as a display in many cichlid fish, and Heiligenberg (1965c) has shown that it may be caused by redirection (Section VI, C). The three-spined stickleback defends the boundaries of its territory with an incipient digging movement; when aggression rises strongly redirected biting merges into displaced nest digging and real nest pits are made (Tinbergen and Van Iersel, 1947).

Numerous other examples show that the behavioral basis of a great many signal activities is an interruptive activity caused by a conflict between relatively incompatible tendencies. This is easily understood since, as emphasized before (Section III, B), every encounter between animals releases conflicting tendencies in all participants (tendencies to attack, to flee, to behave sexually or parentally or to beg for cover or food); it is precisely during such an encounter that communication has a high survival value.

On the basis of the hierarchy concept of the structure of behavior, Tinbergen and co-workers (see Morris, 1958) have developed the hypothesis that each of the different threat and courtship activities corresponds to a specific balance between the tendencies to attack and to escape, possibly in addition with sexual or other tendencies. Baerends and Blokzijl (1963) have used this hypothesis to explain form differences in courtship behavior between *Tilapia mossambica* and *T. nilotica* by reducing them to differences in the overall tendency to attack between the two species. They argued that in *T. nilotica* the tendency to attack is more pronounced in all displays than in those of *T. mossambica*, which results in conspicuous visual differences in homologous signals. Differences between species in the overall tendency to attack may correspond to ecological factors (e.g., amount of cover in habitat). There is no pertinent information on this in various species of *Tilapia*, but in the stickleback we know that the more aggressive *Gasterosteus aculeatus* with its efficient spines prefers open water, whereas the more timid *Pygosteus pungitius* with much smaller spines prefers a habitat in weeds (Morris, 1958). Perhaps some differences in the forms of display between males and fe-

males may be similarly explained by reducing them to differences between the sexes in the average strength of the tendency to attack. Some
cases of homosexual behavior may be explained in this way (Morris, 1952,
1955; Greenberg, 1961). Vegetative phenomena (vasomotoric reactions,
changes in the chromatophores, and contractions in the intestinal tract)
may be induced by internal conflicts. If they increase the effect of an
activity (color changes or sounds) they are likely to be incorporated in
the signal (Morris, 1955).

The change of a behavioral pattern, acquired in the course of evolution in correspondence to its new function as a social releaser, is called
"ritualization" (Tinbergen, 1940, 1952). Ritualization may thus imply
changes in the form of the movement, changes in the taxis components,
and addition of new morphological structures. Lorenz (1950, 1951) has
emphasized that ritualization must also imply a change in the underlying nervous mechanism (emancipation), making the activity independent of the systems to which it was previously subordinated. An
ambivalent activity would thus become detached from the original interfering systems and could make new contacts with superimposed ones,
or even become independent (Baerends et al., 1955).

The regular alternation of "zigs" and "zags" in the zigzag dance,
implying a lack of freedom in the occurring ambivalence, and the fact
that this pattern can be used as a measure for the sexual tendency (Section III, A, 4) indicate that a certain amount of emancipation has taken
place. The fact that correlations with attack and leading can still be found
seems to reveal that this emancipation is not complete. Morris (1957)
has noted that, in accord with the signal function, ritualized activities
often show a "typical intensity," a loss of the freedom to occur in different
intensity degrees.

## C. The Derivation of Noncommunicative Motor Patterns

Although the problem of the evolution of behavior is particularly obvious in signal behavior, phylogeny exists also for behavioral patterns
with a direct function such as spawning, nest building, feeding, and
cleaning. In Section VI, C it was mentioned that the activity in the
sticklebacks ensuring the deposit of the ejected sperm on the eggs
is a displaced nest-building movement. The digging and manipulation of plants and stones that may occur in nest-building of fish may
have been derived from attack behavior through redirection (Heiligenberg, 1965c; Baerends, 1966). The manufacturing of bubble nests in
anabantids might have originated from respiratory movements (Braddock
and Braddock, 1959). Oppenheimer and Barlow (1968) are of the

opinion that in the mouthbreeding *Tilapia melanotheron* all parental action patterns have been evolved from normal respiratory, comfort, and feeding movements. There may be considerable homology between movements for attack, feeding, locomotion, and respiration. Wickler's comparative study (1960b) on the movement of pectorals is interesting in this context, but for the most part this area of research is still underdeveloped. Those who feel attracted to this type of research should be acquainted with the comparative behavior work of the Lorenz school and with regard to fish particularly that of Wickler (1961b, 1967).

Although the activities referred to in this section are not primarily concerned with communication, they may nevertheless serve this function. Slow swimming of the parents attracts young cichlids, the fin-digging, food searching activity of *Cichlasoma nigrofasciatum* stimulates their young to concentrate around them to pick up the disturbed food particles. Keenleyside (1955) has shown experimentally that the typical head-down feeding posture in the three-spined stickleback acts as a feeding signal to nearby conspecifics.

## VIII. LINKS BETWEEN ETHOLOGICAL ANALYSIS AND PHYSIOLOGICAL RESEARCH

As explained in the Introduction it is an important task of causal ethological research to pave the way for work on the physiological mechanisms underlying behavior. The total complex of behavior has to be separated into components which can be examined by physiologists and endocrinologists. It lies beyond the scope of this chapter to review the relevant physiological literature; this will be done in other chapters. We shall restrict comment here to general remarks on the degree of understanding that has already been reached by ethologists.

This understanding can only develop satisfactorily if the physiologist uses sufficiently sophisticated ethological measurements of behavior in connection with his typical physiological techniques. In neurophysiology this has been done in only a few cases. The best example is the brain stimulation work on fowl of von Holst, so prematurely broken off by his death in 1962 (von Holst and von Saint Paul, 1960). A very promising cooperation between physiologists and ethologists has developed in studies of feeding behavior in vertebrates. Particularly De Ruiter and his co-workers (1969; De Ruiter, 1963, 1967; Beukema, 1968) are trying, with "hunger" as a subject, to convert the ethological term "system" unto physiological concepts.

Several studies deal with the relationship between brain and behavior in fish, but in only a few is the behavioral aspect examined from an ethological background. The studies of Fiedler (1966, 1967), Hale (1956a,b), Aronson (1948), Kamrin and Aronson (1954), Schönherr (1955), Segaar (1956, 1961, 1962), and Segaar and Nieuwenhuys (1963) should be mentioned here. Ablations and other lesions are the main physiological techniques used, and the effect on behavior is observed. Hale dealt with learning in a labyrinth situation and with aggressive behavior, both in *Lepomis cyanellus*. Fiedler studied the effect of forebrain ablations on aggressive behavior in *Crenilabrus* and *Diplodus*, and Aronson studied the effect of forebrain lesions on the reproduction behavior of *Tilapia* and *Xiphophorus*. Schönherr and Segaar used *Gasterosteus aculeatus*. Segaar adapted the ethological techniques of Van Iersel and his school. He concludes that the telencephalon of *Gasterosteus* males is concerned with nest-building behavior and with controlling the equilibrium between the aggressive, the sexual, and the parental tendencies. Using coagulation techniques, Segaar and Nieuwenhuys found areas for inhibition and facilitation of parental care.

The problem of the hormonal control of reproductive and parental behavior in fish has recently been reviewed by Baggerman (1968). Relatively more endocrinological than neurophysiological work has been done with the application of ethological methods. Baggerman (1957, 1959, 1960a,b, 1962) has studied day length and temperature in relation to the endocrine aspects of migration in the three-spined stickleback and Pacific salmon. Hoar (1962), Wai and Hoar (1963), Smith and Hoar (1967), and Baggerman (1965, 1966, 1968) have tried to detect the role of hormones on the occurrence of various elements in the reproductive cycle of *Gasterosteus*. This work has made it clear—although not in all details—that a combination of the effects of gonadotropic and gonadal hormones are necessary for the expression of the complete behavior of the reproductive cycle. Experimental work in the guppy (Liley, 1968) indicates that sexual behavior is under direct control of the pituitary while the gonads are exerting a regulatory influence. In the cichlids Blüm and Fiedler (1964, 1965) found indications of influences of FSH and LH on aggressive behavior and of prolactin on parental behavior in cichlids. In Symphysodon prolactin stimulated the mucous secretion of the skin by which the young are fed. Metuzāls *et al.* (1968) found histological evidence for the activity of three different regions in the proximal pars distalis of the hypophysis of the cichlid, *Aequidens portalegrensis*, during the reproductive cycle. Two areas of basophilic cells were successively active in the prespawning period, which is suggestive of the production of FSH and LH, respectively. The acidophilic cells were active after spawning during parental care, which suggests prolactin production. The coincidence between the occurrence of different prespawning behaviors and the activity in the hypophysis is such that behavior and hormone production might influence each other one way and/or the other.

It is interesting that Smith and Hoar (1967) could not induce fanning in the three-spined stickleback by injections of prolactin, but did get it in castrated males with relatively high dosages of methyl testosterone. Baggerman (1965) suggests that this may be so because in the stickleback the occurrence of fanning (and also of comfort movements) is probably indirectly influenced by hormones, namely, through activation of the sexual and aggressive systems. This example illustrates the advantage of starting from an ethological analysis while searching with physiological methods for brain areas or hormones that are involved in the occurrence of activities.

## REFERENCES

Abu Gideiri, Y. B. (1969). The development of behaviour in *Tilapia nilotica* L. *Behaviour* **34**, 17–28.

Abu Gideiri, Y. B. (1966). The behavior and neuro-anatomy of some developing teleost fishes. *J. Zool.* (*London*) **149**, 215–241.

Adrian, E. D., and Buytendijk, F. J. (1931). Potential changes in the isolated brain of the goldfish. *J. Physiol.* (*London*) **71**, 121–135.

Albrecht, H. (1968). Freiwasserbeobachtungen an Tilapien (Pisces, Cichlidae) in Ostafrika. *Z. Tierpsychol.* **25**, 377–394.

Apfelbach, R. (1969). Vergleichend quantitative Untersuchungen des Fortpflanzungsverhaltens brutpflegemono- und dimorpher Tilapien (Pisces, Cichlidae). *Z. Tierpsychol.* **26**, 692–725.

Apfelbach, R., and Leong, D. (1970). Zum Kampfverhalten in der Gattung *Tilapia* (Pisces, Cichlidae). *Z. Tierpsychol.* **27**, 98–107.

Arendsen de Wolf-Exalto, E. (1939). Unpublished report Zoology Lab., Univ. of Leiden.

Aronson, L. R. (1945). Influence of the stimuli provided by the male cichlid fish *Tilapia macrocephala* on the spawning frequency of the female. *Physiol. Zool.* **18**, 403–415.

Aronson, L. R. (1948). Problems in the behaviour and physiology of a species of African mouthbreeding fish. *Trans. N. Y. Acad. Sci.* [2] **2**, 33–42.

Aronson, L. R. (1949). An analysis of reproductive behavior in the mouthbreeding cichlid fish *Tilapia macrocephala* (Bleeker). *Zoologica* **34**, 133–158.

Baenninger, R. (1966). Waning of aggressive motivation in *Betta splendens*. *Psychonomic Science* **4**, 241–242.

Baerends, G. P. (1941). Fortpflanzungsverhalten und Orientierung der Grabwespe *Ammophila campestris* Jur. *Tijdschr. Entomol.* **84**, 68–275.

Baerends, G. P. (1950). Specializations in organs and movements with a releasing function. *Symp. Soc. Exptl. Biol.* **4**, 337–360.

Baerends, G. P. (1952). Les sociétés et les familles de poissons. *Colloq. Intern. Centre Natl. Rech. Sci.* (*Paris*) **34**, 207–219.

Baerends, G. P. (1958). Comparative methods and the concept of homology in the study of behavior. *Arch. Neerl. Zool.* **13**, 401–417.

Baerends, G. P. (1959). The ethological analysis of incubation behaviour. *Ibis* **101**, 357–368.

Baerends, G. P. (1962). La reconnaissance de l'oeuf par le Goéland argenté. *Bull. Soc. Sci. Bretagne* **37**, 193–208.

Baerends, G. P. (1966). Ueber einen möglichen Einfluss von Triebkonflikten auf die Evolution von Verhaltensweisen ohne Mitteilungsfunktion. *Z. Tierpsychol.* **23**, 385–394.

Baerends, G. P. (1970). A model of the functional organization of incubation behaviour. *In* "The Herring Gull and its Egg" (G. P. Baerends and R. H. Drent, eds.). *Behaviour Suppl.* **17**, 263–312.

Baerends, G. P., and Baerends-van Roon, J. M. (1950). An introduction to the study of the ethology of cichlid fishes. *Behaviour* Suppl. **1**, 1–242.

Baerends, G. P., and Blokzijl, G. J. (1963). Gedanken über das Entstehen von Formdivergenzen zwischen homologen Signalhandlungen verwandter Arten. *Z. Tierpsychol.* **20**, 517–528.

Baerends, G. P., Brouwer, R., and Waterbolk, H. Tj. (1955). Ethological studies on *Lebistes reticulatus* (Peters). I. An analysis of the male courting pattern. *Behaviour* **8**, 249–334.

Baggerman, B. (1957). An experimental study of the timing of breeding and migration in the three-spined stickleback (*Gasterosteus aculeatus* L.). *Arch. Neerl. Zool.* **12**, 105–318.

Baggerman, B. (1959). The role of external factors and hormones in migration of sticklebacks and juvenile salmon. *In* "Comparative Endocrinology" (A. Gorbman, ed.), pp. 24–37. Wiley, New York.

Baggerman, B. (1960a). Factors in the diadromous migrations of fish. *Zool. Soc. London* **1**, 33–60.

Baggerman, B. (1960b). Salinity preference, thyroid activity and the seaward migration of four species of Pacific Salmon (*Oncorhynchus*). *J. Fisheries Res. Board Can.* **17**, 295–322.

Baggerman, B. (1962). Some endocrine aspects of fish migration. *Gen. Comp. Endocrinol.* Suppl. **1**, 188–205.

Baggerman, B. (1965). On the endocrine control of reproductive behaviour in the male three-spined stickleback (*Gasterosteus aculeatus* L.). *Symp. Soc. Exptl. Biol.* **20**, 427–456.

Baggerman, B. (1966). On the endocrine control of reproductive behaviour in the male three-spined stickleback (*Gasterosteus aculeatus* L.). *Symp. Soc. Exptl. Biol.* **20**, 427–456.

Baggerman, B. (1968). Hormonal control of reproductive and parental behaviour in fishes. *In* "Perspectives in Endocrinology" (E. J. W. Barrington and C. Barker Jørgensen, eds.), Chapter 6, pp. 351–403. Academic Press, New York.

Ballintijn, C. M., and Hughes, G. M. (1965). The muscular basis of the respiratory pumps in the trout. *J. Exptl. Biol.* **43**, 349–362.

Barlow, G. W. (1961). Ethology of the Asian teleost *Badis badis*. I. Locomotion, maintenance, aggregation and fright. *Trans. Illinois State Acad. Sci.* **54**, 175–188.

Barlow, G. W. (1962a). Ethology of the Asian teleost *Badis badis*. III. Aggressive behavior. *Z. Tierpsychol.* **19**, 29–55.

Barlow, G. W. (1962b). Ethology of the Asian teleost *Badis badis*. IV. Sexual behavior. *Copeia* pp. 346–360.

Barlow, G. W. (1963). Ethology of the Asian teleost *Badis badis*. II. Motivation and signal value of colour patterns. *Animal Behaviour* 11, 97–105.

Barlow, G. W. (1964). Ethology of the Asian teleost *Badis badis*. V. Dynamics of fanning and other parental activities, with comments on the behavior of the larvae and postlarvae. *Z. Tierpsychol.* 21, 99–123.

Barlow, G. W. (1970). A test of appeasement and arousal hypothesis of courtship behavior in a Cichlid fish, *Etroplus maculatus*. *Z. Tierpsychol.* 27, 779–806.

Barlow, G. W., and Green, R. F. (1970). The problems of appeasement and of sexual roles in the courtship behavior of the blackchin mouthbreeder, *Tilapia melanotheron* (Pisces: Cichlidae). *Behaviour* 36, 84–115.

Bastock, M., Morris, D., and Moynihan, M. (1953). Some comments on conflict and thwarting in animals. *Behaviour* 6, 66–84.

Bauer, J. (1968). Vergleichende Untersuchungen zum Kontaktverhalten verschiedener Arten der Gattung *Tilapia* (Cichlidae, Pisces) und ihre Bastarde. *Z. Tierpsychol.* 25, 22–70.

Beament, J. W. L. (1960). Physical models in biology. *Symp. Soc. Exptl. Biol.* 14, 83–101.

Bergmann, H. H. (1968). Eine deskriptive Verhaltensanalyse des Segelflossers (*Pterophyllum scalare* Cuv. & Val.; Cichlidae, Pisces). *Z. Tierpsychol.* 25, 559–587.

Beukema, J. J. (1968). Predation by the three-spined stickleback (*Gasterosteus aculeatus* L.); the influence of hunger and experience. *Behaviour* 31, 1–126.

Blüm, V. (1968). Das Kampfverhalten des braunen Diskusfisches, *Symphysodon aequifasciata axelrodi* L. P. Schultz. (Teleostei, Cichlidae). *Z. Tierpsychol.* 25, 395–408.

Blüm, V., and Fiedler, K. (1964). Der Einfluss von Prolactin auf das Brutpflegeverhalten von *Symphysodon aequifasciata axelrodi* L. Schultz (Cichlidae, Teleostes). *Naturwissenschaften* 51, 149–150.

Blüm, V., and Fiedler, K. (1965). Hormonal control of reproductive behavior in some cichlid fish. *Gen. Comp. Endocrinol.* 5, 186–196.

Bol, A. C. A. (1959). A consummatory situation; the effect of eggs on the sexual behaviour of the male three-spined stickleback (*Gasterosteus aculeatus* L.). *Experientia* 15, 115.

Braddock, J. C. (1949). The effect of prior residence upon dominance in the fish *Platypoecilus maculatus*. *Physiol. Zool.* 22, 161–169.

Braddock, J. C., and Braddock, Z. I. (1958). Effects of isolation and social contact upon the development of aggression behaviour in the Siamese fighting fish, *Betta splendens*. *Animal Behaviour* 6, 249.

Braddock, J. C., and Braddock, Z. I. (1959). The development of nesting behaviour in the Siamese fighting fish, *Betta splendens*. *Animal Behaviour* 7, 222–232.

Brawn, V. M. (1961a). Aggressive behaviour of the cod (*Gadus callarias* L.). *Behaviour* 18, 107–147.

Brawn, V. M. (1961b). Reproductive behaviour of the cod (*Gadus callarias* L.). *Behaviour* 18, 177–198.

Brawn, V. M. (1961c). Sound production by the cod (*Gadus callarias* L.). *Behaviour* 18, 239–255.

Breder, C. M. (1926). The locomotion of fishes. *Zoologica* 4, 159–297.

Breder, C. M. (1934). An experimental study of the reproductive habits and life history of the cichlid fish *Aequidens latifrons* (Steindachner). *Zoologica* **18**, 1–42.

Breder, C. M. (1936). The reproductive habits of the North American sunfishes (fam. Centrachidae). *Zoologica* **21**, 1–50.

Breder, C. M., and Halpern, F. (1946). Innate and acquired behavior affecting the aggregation of fishes. *Physiol. Zool.* **19**, 154–190.

Brestowski, M. (1968). Vergleichende Untersuchungen zur Elternbindung von Tilapien-Jungfischen (Cichlidae, Pisces). *Z. Tierpsychol.* **25**, 761–828.

Brett, J. R., and MacKinnon, D. (1954). Some aspects of olfactory perception in migrating adult Coho and Spring Salmon. *J. Fisheries Res. Board Can.* **11**, 310–318.

Bullock, T. H. (1961). The origins of patterned nervous discharge. *Behaviour* **17**, 1–59.

Casimir, M. J. (1969). Zum Verhalten des Putzerfisches *Symphodus melanocereus* (Risso). *Z. Tierpsychol.* **26**, 225–229.

Carter Miller, H. (1964). The behavior of the Pumpkinseed sunfish *Lepomis gibbosus* (Linneaus), with notes on the behavior of other species of *Lepomis* and the Pigmy sunfish, *Elassoma evergladei. Behaviour* **22**, 88–151.

Clark, E., and Aronson, L. R. (1951). Sexual behavior in the Guppy, *Lebistes reticulatus* (Peters). *Zoologica* **36**, 49–66.

Clark, E., Aronson, L. R., and Gordon, M. (1954). Mating behavior patterns in two sympatric species of Xiphophorin fishes: Their inheritance in sexual isolation. *Bull. Am. Museum Nat. Hist.* **103**, 135–225.

Clark, F. W., and Keenleyside, M. H. A. (1967). Reproductive isolation between the sunfish *Lepomis gibbosis* and *L. macrochirus. J. Fisheries Res. Board Can.* **24**, 495–514.

Clayton, F. L., and Hinde, R. A. (1968). Habituation and recovery of aggressive display in *Betta splendens. Behaviour* **30**, 96–106.

Cole, J. E., and Ward, J. A. (1969). The communicative function of pelvic fin-flickering in *Etroplus maculatus* (Pisces, Cichlidae). *Behaviour* **35**, 179–199.

Cole, J. E., and Ward, J. A. (1970). An analysis of parental recognition by the young of the Cichlid fish, *Etroplus maculatus* (Bloch). *Z. Tierpsychol.* **27**, 156–176.

Cott, H. B. (1940). "Adaptive Coloration in Animals." Methuen, London.

Craig, W. (1918). Appetites and aversions as constituents of instincts. *Biol. Bull.* **34**, 91–107.

Cullen, E. (1961). The effect of isolation from the father on the behaviour of male three-spined sticklebacks to models. USAFRDC, Final Rept. Contr. AF 61 (052)-29, 1–23.

Davenport, D., and Norris, K. S. (1958). Observations on the symbiosis of the sea anemone *Stoichactis* and the Pomacentrid fish, *Amphiprion percula. Biol. Bull.* **115**, 397–410.

De Beer, G. R. (1940). "Embryos and Ancestors." Oxford Univ. Press (Clarendon), London and New York.

De Ruiter, L. (1963). The physiology of vertebrate feeding behaviour; towards a synthesis of the ethological and physiological approaches to problems of behaviour. *Z. Tierpsychol.* **20**, 498–516.

De Ruiter, L. (1967). Feeding behaviour of vertebrates in the natural environment. *Handb. Physiol.* **1**, 97–116.

De Ruiter, L., Wiepkema, P. R., and Reddingius, J. (1969). *Ann. N. Y. Acad. Sci.* **157**, 1204–1216.

Durham, D. W., Kortmulder, K., and Van Iersel, J. J. A. (1968). Threat and appeasement in *Barbus stoliczkanus* (Cyprinidae). *Behaviour* **30**, 15–26.

Eberhard, K., Fabricius, M., and von Holst, E. (1939). Bausteine zu einer vergleichenden Physiologie der lokomotorischen Reflexe bei Fischen. III. Z. *Vergleich. Physiol.* **26**, 467–480.

Eibl-Eibesfeldt, I. (1955). Ueber Symbiosen, Parasitismus und andere besondere zwischenartliche Beziehungen tropischer Meeresfische. *Z. Tierpsychol.* **12**, 203–219.

Eibl-Eibesfeldt, I. (1959). Der Fisch *Aspidontus taeniatus* als Nachahmer des Putzers *Labroides dimidiatus*. *Z. Tierpsychol.* **16**, 19–25.

Eibl-Eibesfeldt, I. (1960). Beobachtungen und Versuche an Anemonenfischen (*Amphiprion*) der Malediven und der Nicobaren. *Z. Tierpsychol.* **17**, 1–10.

Eibl-Eibesfeldt, I. (1962). Freiwasserbeobachtungen zur Deutung des Schwarmverhaltens verschiedener Fische. *Z. Tierpsychol.* **19**, 165–182.

Eibl-Eibesfeldt, I. (1967). "Grundriss der vergleichenden Verhaltungsforschung." Piper, Munich.

Fabricius, E. (1950). Heterogeneous stimulus summation in the release of spawning in fish. *Rep. Inst. Freshwater Res., Drottningholm* **31**, 57–99.

Fabricius, E. (1953). Aquarium observations on the spawning behaviour of the char, *Salmo alpinus*. *Rept. Inst. Freshwater Res., Drottningholm* **34**, 1–48.

Fabricius, E. (1954). Aquarium observations on the spawning behaviour of the Burbot, *Lota vulgaris* L. *Rept. Inst. Freshwater Res., Drottningholm* **35**, 51–57.

Fabricius, E., and Gustafson, K. J. (1958). Some new observations on the spawning behaviour of the Pike, Esox lucius L. *Rept. Inst. Freshwater Res., Drottningholm* **93**, 23–54.

Fabricius, E., and Lindroth, A. (1954). Experimental observations on the spawning of whitefish, *Coregonus lavaretus* L., in the stream-aquarium of the Hölle Laboratory at River Indalsälven. *Rept. Inst. Freshwater Res., Drottningholm* **35**, 105–112.

Fiedler, K. (1955). Vergleichende Verhaltensstudien an Seenadeln, Schlangennadeln und Seepferdchen (Syngnathidae). *Z. Tierpsychol.* **11**, 358–416.

Fiedler, K. (1964). Verhaltensstudien an Lippfischen der Gattung *Crenilabrus* (Labridae, Perciformes). *Z. Tierpsychol.* **21**, 521–591.

Fiedler, K. (1966). Degenerationen und Verhaltenseffekte nach Elektrokoagulationen im Gehirn von Fischen (*Diplodus*, *Crenilabrus*—Perciformes). *Verhandl. Deut. Zool. Ges., Goettingen* pp. 351–366.

Fiedler, K. (1967). Ethologische und neuroanatomische Auswirkungen von Vorderhirnexstirpationen bei Meerbrassen, (*Diplodus*) und Lippfischen (*Crenilabrus*, Perciformes, Teleostei). *J. Hirnforsch.* **9**, 482–563.

Fishelson, L. (1970). Behaviour and ecology of a population of *Abudefduf saxatilis* (Pomacentridae, Teleostei) at Eilat (Red Sea). *Animal Behaviour* **18**, 225–237.

Forselius, F. (1957). Studies of Anabantid fishes. *Zool. Bidr. Uppsala* **32**, 93–597.

Franck, D. (1964). Vergleichende Verhaltenstudien an lebendgebärenden Zahnkarpen der Gattung *Xiphophorus*. *Zool. Jahrb., Abt. Allgem. Zool. Physiol. Tiere* **71**, 117–170.

Franck, D. (1968). Weitere Untersuchungen zur vergleichenden Ethologie der Gattung Xiphophorus (Pisces). *Behaviour* **30**, 76–95.

Franck, D. (1970). Verhaltensgenetische Untersuchungen an Artbastarden der Gattung *Xiphophorus* (Pisces). *Z. Tierpsychol.* **27**, 1–34.

Friedländer, B. (1894). Beiträge zur Physiologie des Zentralnervensystems und des Bewegungsmechanismus der Regenwürmer. *Arch. Ges. Physiol.* **58**, 168–206.

Gohar, H. A. F. (1948). Commensalism between fish and anemone with a description of the eggs of *Amphipirion bicinctus* Rüppell. *Publ. Marine Biol. Sta., Ghardaqa* **6**, 35–44.

Graefe, G. (1963). Die Anemon-Fisch-Symbiose and ihre Grundlage, nach Freilanduntersuchungen bei Eilat/Rotes Meer. *Naturwissenschaften* **50**, 410.

Graefe, G. (1964). Die Anemon-Fisch-Symbiose, nach Freilanduntersuchungen bei Eilat/Rotes Meer. *Z. Tierpsychol.* **21**, 468–485.

Graham Brown, T. (1912a). The intrinsic factors in the act of progression in the mammal. *Proc. Roy. Soc.* **B84**, 308–319.

Graham Brown, T. (1912b). The factors in rhythmic activity of the nervous system. *Proc. Roy. Soc.* **B85**, 278–289.

Gray, J. (1936). Studies in animal locomotion. IV. *J. Exptl. Biol.* **13**, 170–180.

Gray, J., and Sand, A. (1936). The locomotory rhythm of the dogfish (*Scyllium canicula*). *J. Exptl. Biol.* **13**, 200–209.

Greenberg, B. (1947). Some relations between territory, social hierarchy and leadership in the green sunfish (*Lepomis cyanellus*). *Physiol. Zool.* **20**, 267–299.

Greenberg, B. (1961). Spawning and parental behavior in female pairs of the jewel fish, *Hemichromis bimaculatus* Gill. *Behaviour* **18**, 44–61.

Greenberg, B. (1963). Parental behavior and imprinting in cichlid fishes. *Behaviour* **21**, 127–144.

Greenberg, B., Zijlstra, J. J., and Baerends, G. P. (1965). A quantitative description of the behaviour changes during the reproductive cycle of the cichlid fish *Aequidens portalegrensis* Hensel. *Koninkl. Ned. Akad. Wetenschap. Proc.* **C68**, 135–149.

Guiton, P. (1960). On the control of behaviour during the reproductive cycle of *Gasterosteus aculeatus*. *Behaviour* **15**, 163–184.

Hale, E. B. (1956a). Social facilitation and forebrain function in maze performance of green sunfish *Lepomis cyanellus*. *Physiol. Zool.* **29**, 93–107.

Hale, E. B. (1956b). Effects of forebrain lesions on the aggressive behavior of green sunfish, *Lepomis cyanellus*. *Physiol. Zool.* **29**, 107–127.

Hall, D. D. (1968). A qualitative analysis of courtship and reproductive behavior in the Paradise fish, *Macropodus opcularis* (Linnaeus). *Z. Tierpsychol.* **25**, 834–842.

Hall, D. D., and Miller, R. J. (1968). A quantitative description and analysis of courtship behavior in the Pearl gourami, *Trichogaster leeri* (Bleeker). *Behaviour* **32**, 70–84.

Harden Jones, F. R., and Marshall, N. B. (1953). The structure and functions of the teleostean swimbladder. *J. Exptl. Biol.* **28**, 16–83.

Harlow, H. F. (1963). The maternal affectional system. *In* "Determinants of Infant Behavior" (B. M. Foss, ed.), Vol. II, pp. 75–88. Methuen, London: Wiley, New York.

Haskins, C. P., and Haskins, E. F. (1949). The role of sexual selection as an isolating mechanism in three species of Poecilid fishes. *Evolution* **3**, 160–169.

Haskins, C. P., and Haskins, E. F. (1950). Factors governing sexual selection as an isolating mechanism in the Poecilid fish *Lebistes reticulatus*. *Proc. Natl. Acad. Sci. U. S.* **36**, 464–476.

Hayes, J. S., Russell, W. M. S., Hayes, C., and Kohsen, A. (1953). The mechanism of an instinctive control system: A hypothesis. *Behaviour* **6**, 85–119.

Hebb, D. O. (1953). Heredity and environment in mammalian behaviour. *Brit. J. Animal Behaviour* **1**, 43–47.

Hediger, H., and Heusser, H. (1961). Zum "Schiessen" des Schützenfisches, *Toxotes jaculatrix. Natur Volk* **91**, 237–243.

Heiligenberg, W. (1963). Ursachen für das Auftreten von Instinktbewegungen bei einem Fische (*Pelmatochromis subocellatus kribensis* Boul. Cichlidae). *Z. Vergleich. Physiol.* **47**, 339–380.

Heiligenberg, W. (1964). Ein Versuch zur ganzheitsbezogenen Analyse des Instinktverhaltens eines Fisches (*Pelmatochromis subocellatus kribensis* Boul., Chichlidae). *Z. Tierpsychol.* **21**, 1–52.

Heiligenberg, W. (1965a). The suppression of behavioral activities by frightening stimuli. *Ž. Vergleich. Physiol.* **50**, 660–672.

Heiligenberg, W. (1965b). The effect of external stimuli on the attack readiness of a Cichlid fish. *Z. Vergleich. Physiol.* **49**, 459–464.

Heiligenberg, W. (1965c). A quantitative analysis of digging movements and their relationship to aggressive behaviour in cichlids. *Animal Behaviour* **13**, 163–170.

Heinrich, W. (1967). Untersuchungen zum Sexualverhalten in der Gattung *Tilapia* (Cichlidae, Teleostei) und bei Artbastarden. *Z. Tierpsychol.* **24**, 684–754.

Heinroth, O. (1910). Beiträge zur Biologie, namentlich Ethologie und Psychologie der Anatiden. *Verhandl. 5th Intern. Ornithol. Kongr. Berlin, 1910*, pp. 589–702.

Hemmings, C. C. (1966). Olfaction and vision in fish schooling. *J. Exptl. Biol.* **45**, 449–464.

Hess, E. H. (1953). Temperature as a regulator of the attack response of *Betta splendens. Z. Tierpsychol.* **9**, 379–382.

Hinde, R. A. (1953). Appetitive behaviour, consummatory act, and the hierarchical organisation of behaviour—with special reference to the Great Tit. *Behaviour* **5**, 189–224.

Hinde, R. A. (1954). Changes in responsiveness to a constant stimulus. *Brit. J. Animal Behaviour* **2**, 41–55.

Hinde, R. A. (1956). Ethological models and the concept of drive. *Brit. J. Phil. Sci.* **6**, 321–331.

Hinde, R. A. (1959a). Unitary drives. *Animal Behaviour* **7**, 130–141.

Hinde, R. A. (1959b). Some recent trends in ethology. *In* "Psychology: A Study of a Science" (S. Koch, ed.), Vol. 2, pp. 561–610. McGraw-Hill, New York.

Hinde, R. A. (1960). Energy models in motivation. *Symp. Soc. Exptl. Biol.* **14**, 199–213.

Hinde, R. A., and Stevenson, J. G. (1969). Sequences of behavior. *Adv. Study Behavior* **2**, 267–296.

Hinde, R. A. (1970). "Animal Behaviour," 2nd ed. McGraw-Hill, New York.

Hoar, W. S. (1962). Hormones and the reproductive behaviour of the male three-spined stickleback (*Gasterosteus aculeatus*). *Animal Behaviour* **10**, 247–266.

Hogan, J. A. (1965). An experimental study of conflict and fear: An analysis of behavior of young chicks toward a mealworm. Part I. The behavior of chicks which do not eat a mealworm. *Behaviour* **25**, 1–2.

Holzapfel, M. (1940). Triebbedingte Ruhezustände als Ziel von Appetenzhandlungen. *Naturwissenschaften* **28**, 273–280.

Hoogland, R., Morris, D., and Tinbergen, N. (1957). The spines of sticklebacks

(*Gasterosteus* and *Pygosteus*) as means of defense against predators (*Perca* and *Esox*). *Behaviour* 10, 205–236.

Hughes, G. M., and Ballintijn, C. M. (1965). The muscular basis of the respiratory pumps in the dogfish (*Scylliorhinus canicula*). *J. Exptl. Biol.* 43, 363–383.

Kamrin, R. P., and Aronson, L. R. (1954). The effects of forebrain lesions on mating behavior in the male platyfish, *Xiphophorus maculatus*. *Zoologica* 39, 133–140.

Keenleyside, M. H. A. (1955). Some aspects of the schooling behaviour of fish. *Behaviour* 8, 183–248.

Keenleyside, M. H. A. (1967). Behavior of male sunfishes (Genus *Lepomis*) towards females of three species. *Evolution* 21, 688–695.

Koehler, O. (1950). Die Analyse der Taxisanteile instinktartigen Verhaltens. *Symp. Soc. Exptl. Biol.* 4, 269–304.

Kortlandt, A. (1940). Wechselwirkung zwischen Instinkten. *Arch. Neerl. Zool.* 4, 443–520.

Kortlandt, A. (1959). An attempt at clarifying some controversial notions in animal psychology and ethology. *Arch. Neerl. Zool.* 13, 196–229.

Kristensen, I. (1939). Unpublished report, Zoology Lab., Univ. of Leiden.

Kruijt, J. P. (1964). Ontogeny of social behaviour in Burmese Red Junglefowl. *Behaviour* Suppl. 12, 201 pp.

Kuenzer, P. (1962). Die Auslösung der Nachfolgereaktion durch Bewegungsreize bei Jungfischen von *Nannacara anomala* Regan (Cichlidae). *Naturwissenschaften* 22, 525–526.

Kuenzer, P. (1964). Weitere Versuche zur Auslösung der Nachfolgereaktion bei Jungfischen von *Nannacara anomala* (Cichlidae). *Naturwissenschaften* 17, 419–420.

Kuenzer, P. (1965). Zur optischen Auslösung von Brutpflegehandlungen bei *Nannacara anomala* ♀ ♀ (Teleostei, Cichlidae). *Naturwissenschaften* 1, 19–20.

Kuenzer, P. (1966). Wie "erkennen" junge Buntbärsche ihre Eltern? *Umschau* 24, 795–800.

Kuenzer, P. (1968). Die Auslösung der Nachfolgereaktion bei erfahrungslosen Jungfischen von *Nannacara anomala* (Cichlidae). *Z. Tierpsychol.* 25, 257–314.

Kuenzer, P., and Kuenzer, E. (1962). Untersuchungen zur Brutpflege der Zwergcichliden *Apistogramma reitzigi* und *A. borellii*. *Z. Tierpsychol.* 19, 56–83.

Kühme, W. (1961). Verhaltensstudien am maulbrütenden (*Betta anabatoides* Bleeker) und am nestbauenden Kampffisch (*B. splendens* Regan). *Z. Tierpsychol.* 18, 33–55.

Kühme, W. (1962). Das Schwarmverhalten elterngeführter Jungcichliden (Pisces). *Z. Tierpsychol.* 19, 513–538.

Kühme, W. (1963). Chemisch ausgelöste Brutpflege- und Schwarmreaktionen bei *Hemichromis bimaculatus* (Pisces). *Z. Tierpsychol.* 20, 688–704.

Laudien, H. (1965). Untersuchungen über das Kampfverhalten der Mannchen von *Betta splendens* Regan (Anabantidae, Pisces). *Z. Wiss. Zool.* 172, 135–178.

Lehrman, D. (1953). A critique of Konrad Lorenz's theory of instinctive behavior. *Quart. Rev. Biol.* 28, 337–363.

Leiner, M. (1929). Oekologische Studien an *Gasterosteus aculeatus*. *Z. Morphol. Oekol. Tiere* 14, 360–399.

Leiner, M. (1930). Fortsetzung der oekologischen Studien an *Gasterosteus aculeatus*. *Z. Morphol. Oekol. Tiere* 16, 499–540.

Leiner, M. (1938). "Die Physiologie der Fischatmung." Akad. Verlagsges., Leipzig.

Le Mare, D. W. (1936). Reflex and rhythmical movements in the dogfish. *J. Exptl. Biol.* 13, 429–442.

Leong, C.-Y. (1969). The quantitative effect of releasers on the attack readiness of the fish *Haplochromis burtoni* (Cichlidae, Pisces). *Z. Vergleich. Physiol.* **65**, 29–50.

Liley, N. R. (1966). Ethological isolating mechanisms in four sympatric species of poeciliid fishes. *Behaviour* Suppl. **13**, 1–197.

Liley, N. R. (1968). The endocrine control of reproductive behavior in the female guppy, *Poecilia reticulata* Peters. *Animal Behaviour* **16**, 318–331.

Lissmann, H. W. (1933). Die Umwelt des Kampffisches (*Betta splendens* Regan). *Z. Vergleich. Physiol.* **18**, 65–111.

Lissmann, H. W. (1946a). The neurological basis of the locomotory rhythm in the spinal dogfish (*Scyllium canicula, Acanthias vulgaris*). I. Reflex behaviour. *J. Exptl. Biol.* **23**, 143–176.

Lissmann, H. W. (1946b). The neurological basis of the locomotory rhythm in the spinal dogfish (*Scyllium canicula, Acanthias vulgaris*). II. The effect of the de-afferentation. *J. Exptl. Biol.* **23**, 143–162.

Lorenz, K. (1935). Der Kumpan in der Umwelt des Vogels. *J. Ornithol.* **83**, 137–215, 289–413.

Lorenz, K. (1937a). Ueber die Bildung des Instinktbegriffs. *Naturwissenschaften* **25**, 289, 307, and 324.

Lorenz, K. (1937b). Ueber den Begriff der Instinkthandlung. *Folia Biotheoret. Leiden* **2**, 18–50.

Lorenz, K. (1939). Vergleichende Verhaltensforschung. *Verhandl. Deut. Zool. Ges., Zool. Anz.* **12**, Suppl., 69–102.

Lorenz, K. (1950). The comparative method in studying innate behaviour patterns. *Symp. Soc. Exptl. Biol.* **4**, 221–268.

Lorenz, K. (1951). Ueber die Entstehung auslösender "Zeremonien." *Vogelwarte* **16**, 9–13.

Lorenz, K. (1952). Die Entwicklung der vergleichende Verhaltensforschung in den letzten 12 Jahren. *Verhandl. Deut. Zool. Ges., Freiburg* pp. 36–58.

Lorenz, K. (1965). "Evolution and Modification of Behaviour." Univ. of Chicago Press, Chicago, Illinois.

Lorenz, K., and Tinbergen, N. (1938). Taxis und Instinkthandlung in der Eirollbewegung der Graugans. I. *Z. Tierpsychol.* **2**, 1–29.

Lowe, R. H. (1956). The breeding behaviour of *Tilapia* species (Pisces:Cichlidae) in natural water: Observations on *T. karoma* Poll and *T. variabilis* Boulenger. *Behaviour* **9**, 140–163.

Lüling, K. H. (1958). Morphologisch-anatomische und histologische Untersuchungen am Auge des Schützenfisches *Toxotes jaculatrix* (Pallas 1766) (Toxotidae), nebst Bemerkungen zum Spuckgehaben. *Z. Morphol. Oekol. Tiere* **47**, 529–610.

Machemer, L. (1970). Qualitative und quantitative Verhaltensbeobachtungen an Paradiesfischmännchen, *Macropodus opercularis* L. (Anabantidae, Teleostei). *Z. Tierpsychol.* **27**, 563–590.

McDougall, W. (1923). "An Outline of Psychology," 1st ed. Methuen, London.

McFarland, D. J. (1965a). The role of attention in the disinhibition of displacement activities. *Quart. J. Exptl. Psychol.* **18**, 19–30.

McFarland, D. J. (1965b). Hunger, thirst and displacement pecking in the Barbary Dove. *Animal Behaviour* **13**, 293–300.

McFarland, D. J. (1966). On the causal and functional significance of displacement activities. *Z. Tierpsychol.* **23**, 217–235.

Manning, A. (1967). "An Introduction to Animal Behaviour." Arnold, London.

Marler, P., and Hamilton, W. J. (1966). "Mechanisms of Animal Behavior." Wiley, New York.

Meesters, A. (1940). Ueber die Organisation des Gesichtsfeldes der Fische. Z. Tierpsychol. 4, 84–149.

Metuzāls, J., Ballintijn-de Vries, G., and Baerends, G. P. (1968). The correlation of cytological changes in the adenohypophysis of the cichlid fish Aequidens portalegrensis (Hensel) with behaviour changes during the reproductive cycle. Koninkl. Ned. Akad. Wetenschap., Proc. **C71**, 391–410.

Miller, R. J. (1964). Studies on the social behavior of the Blue Gourami, Trichogaster trichopterus (Pisces, Belontiidae). Copeia **3**, 469–496.

Miller, R. J., and Hall, D. D. (1968). A quantitative description and analysis of courtship and reproductive behaviour in the Anabantid fish Trichogaster leeri (Bleeker). Behaviour **32**, 85–149.

Mittelstaedt, H. (1960). The analysis of behavior in terms of control systems. 5th Conf. Group Processes, 1958, Josiah Macy, Jr. Found., New York.

Mittelstaedt, H. (1964). Basic control patterns of orientational homeostasis. Symp. Soc. Exptl. Biol. **18**, 365–386.

Morris, D. (1952). Homosexuality in the ten-spined stickleback (Pygosteus pungitius, L.). Behaviour **4**, 233–261.

Morris, D. (1954). The reproductive behaviour of the river Bullhead (Cottus gobio L.) with special reference to the fanning activity. Behaviour **7**, 1–32.

Morris, D. (1955). The causation of pseudofemale and pseudomale behavior: A further comment. Behaviour **8**, 46–56.

Morris, D. (1957). "Typical intensity" and its relation to the problem of ritualisation. Behaviour **11**, 1–12.

Morris, D. (1958). The reproductive behaviour of the ten-spined stickleback (Pygosteus pungitius L.). Behaviour Suppl. **6**, 1–154.

Muckensturm, B. (1965a). Possibilités inattendues de manipulation chez l'Epinoche (Gasterosteus aculeatus). Compt. Rend. **260**, 3183–3184.

Muckensturm, B. (1965b). Le nid et le territoire chez l'Epinoche (Gasterosteus aculeatus). Compt. Rend. **260**, 4825–4826.

Muckensturm, B. (1966). Réactions de l'Epinoche à la présence du nid et des oeufs. Compt. Rend. **262**, 2637–2639.

Myrberg, A. A., Jr. (1964). An analysis of the preferential care of eggs and young by adult cichlid fishes. Z. Tierpsychol. **21**, 53–98.

Myrberg, A. A., Jr. (1965). A descriptive analysis of the behaviour of the African cichlid fish, Pelmatochromis guentheri (Sauvage). Animal Behaviour **13**, 312–329.

Myrberg, A. A., Jr. (1966). Parental recognition of young in cichlid fishes. Animal Behaviour **14**, 565–571.

Neil, E. H. (1964). An analysis of color changes and social behavior of Tilapia mossambica. Univ. Calif. (Berkeley) Publ. Zool. **75**, 1–58.

Nelson, K. (1964a). Behavior and morphology in the Glandulocaudine fishes (Ostariophysi, Characidae). Univ. Calif. (Berkeley) Publ. Zool. **75**, 59–152.

Nelson, K. (1964b). The temporal patterning of courtship behaviour in the glandulocaudine fishes (Ostariophysi, Characidae). Behaviour **2**, 90–146.

Nelson, K. (1965a). The evolution of a pattern of sound production associated with courtship in the characid fish, Glandulocauda inequalis. Evolution **18**, 526–540.

Nelson, K. (1965b). After-effects of courtship in the male three-spined stickleback. Z. Vergleich. Physiol. **50**, 569–597.

Noble, G. K., and Curtis, B. (1939). The social behavior of the jewel fish *Hemichromis bimaculatus* Gill. *Bull. Am. Museum Nat. Hist.* **75**, 1–46.

Nyman, K. J. (1953). Observations on the behaviour of *Gobius microps*. *Acta Soc. Fauna Flora Fennica* **69**, 1–11.

Oehlert, B. (1958). Kampf und Paarbildung einiger Cichliden. *Z. Tierpsychol.* **15**, 141–174.

Ohm, D. (1958). Die ontogentische Entwicklung des Kampfverhaltens bei *Aequidens portalegrensis* Hensel und *Ae. latifrons* Steindachner (Cichlidae). *Verhandl. Deut. Zool. Ges. Frankfurt* pp. 181–194.

Ohm, D. (1958–1959a). Vergleichende Beobachtungen am Kampfverhalten von *Aequidens* (Cichlidae). *Wiss. Z. Humboldt-Univ. Berlin, Math.-Naturw. Reihe.* **8**, 1–48.

Ohm, D. (1958-1959b). Vergleichende Beobachtungen am Brutpflegeverhalten von *Aequidens* (Cichlidae). *Wiss. Z. Humboldt-Univ. Berlin, Math-Naturw. Reihe.* **8**, 590–640.

Oppenheimer, J. R., and Barlow, G. W. (1968). Dynamics of parental behavior in the Black chinned mouthbreeder, *Tilapia melanotheron* (Pisces: Cichlidae). *Z. Tierpsychol.* **25**, 889–914.

Parzefall, J. (1969). Zur vergleichende Ethologie verschiedener *Mollienesia*-Arten einschliesslich einer Höhlenform von *M. sphenops*. *Behaviour* **33**, 1–37.

Peeke, H. V. S. (1969). Habituation of conspecific aggression in the three-spined stickleback (*Gasterosteus aculeatus*). *Behaviour* **35**, 137–156.

Peeke, H. V. S. and Peeke, C. S. (1970). Habituation of conspecific responses in the Siamese fighting fish (*Betta splendens*). *Behaviour* **36**, 232–245.

Peeke, H. V. S., Wyers, E. J., and Herz, M. J. (1969). Waning of the aggressive response to male models in the three-spined stickleback (*Gasterosteus aculeatus*). *Animal Behaviour* **17**, 224–228.

Peters, H. (1937). Experimentelle Untersuchungen über die Brutpflege von *Haplochromis multicolor*, einen maulbrütenden Knochenfish. *Z. Tierpsychol.* **1**, 201–218.

Peters, H. (1941). Fortpflanzungsbiologische und Tiersoziologische Studien an Fischen. I. *Hemichromis bimaculatus* Gill. *Z. Morphol. Oekol. Tiere* **37**, 387–425.

Pfeiffer, W. (1962). The fright reaction of fish. *Biol. Rev.* **37**, 495–511.

Philippson, M. (1905). L'autonomie et la centralisation dans le système nerveux des animaux. *Trav. Lab. Physiol. Inst. Solvay* **7**, 1.

Prechtl, H. F. R. (1952). Angeborene Bewegungsweisen junger Katzen. *Experientia* **8**, 220–221.

Prechtl, H. F. R. (1953). Zur Physiologie der angeborenen auslösenden Mechanismen, *Behaviour* **5**, 32–50.

Prechtl, H. F. R., and Schleidt, W. M. (1950). Auslösende und steuernde Mechanismen des Saugaktes I. *Z. Vergleich. Physiol.* **32**, 257–262.

Prechtl, H. F. R., and Schleidt, W. M. (1951). Auslösende und steuernde Mechanismen des Saugaktes. II. *Z. Vergleich. Physiol.* **33**, 53–62.

Rasa, O. A. E. (1969a). Territoriality and the establishment of dominance by means of visual cues in *Pomacentrus jenkensi* (Pisces: Pomacentridae). *Z. Tierpsychol.* **26**, 825–845.

Rasa, O. A. E. (1969b). The effect of pair isolation on the reproductive success in *Etroplus maculatus* (Cichlidae). *Z. Tierpsychol.* **26**, 846–852.

Rowell, C. H. F. (1961). Displacement grooming in the Chaffinch. *Animal Behaviour* **9**, 38–63.

Russell, W. M. S., Mead, A. P., and Hayes, J. S. (1954). A basis for the quantitative study of the structures of behaviour. *Behaviour* 6, 154–205.

Ruwet, J. C. (1963). Observations sur le comportement sexuel de *Tilapia macrochir* Blgr. (Pisces: Cichlidae) au lac de retenue de la Lufira (Katanga). *Behaviour* 20, 242–250.

Schleidt, W. (1962). Die historische Entwicklung der Begriffe "Angeborenes auslösendes Schema" und "Angeborener Auslösemechanismus" in der Ethologie. *Z. Tierpsychol.* 19, 697–722.

Schlichter, D. (1968). Das Zusammenleben von Riffanemonen und Anemonenfischen. *Z. Tierpsychol.* 25, 933–954.

Schneirla, T. C. (1959). An evolutionary and development theory of biphasic processes underlying approach and withdrawal. *In* "Nebraska Symposium on Motivation" (M. R. Jones, ed.), 1–41. Univ. of Nebraska Press, Lincoln, Nebraska.

Schönherr, J. (1955). Ueber die Abhängigkeit der Instinkthandlungen vom Vorderhirn und Zwischenhirn (Epiphyse) bei *Gasterosteus aculeatus* L. *Zool. Jahrb., Abt. Allgem. Zool. Physiol. Tiere* 65, 357–386.

Schutz, F. (1956). Vergleichende Untersuchungen über die Schreckreaktion bei Fischen und deren Verbreitung. *Z. Vergleich. Physiol.* 38, 84–135.

Segaar, J. (1956). Brain and instinct with *Gasterosteus aculeatus*. *Koninkl. Ned. Akad. Wetenschap., Proc.* 59, 738–749.

Segaar, J. (1961). Telencephalon and behaviour in *Gasterosteus aculeatus*. *Behaviour* 18, 256–287.

Segaar, J. (1962). Die Funktion des Vorderhirns in Bezug auf das angeborene Verhalten des dreidornigen Stichlingsmännchens (*Gasterosteus aculeatus* L.)— zugleich ein Beitrag über Neuronenregeneration im Fischgehirn. *Acta Morphol. Neerl.-Scand.* 5, 49–64.

Segaar, J., and Nieuwenhuys, R. (1963). New etho-physiological experiments with male *Gasterosteus aculeatus*, with anatomical comment. *Animal Behaviour* 11, 331–344.

Seitz, A. (1940). Die Paarbildung bei einigen Cichliden. I. *Astatotilapia strigigena* Pfeffer. *Z. Tierpsychol.* 4, 40–84.

Seitz, A. (1942). Die Paarbildung bei einigen Cichliden. II. *Hemichromis bimaculatus* Gill. *Z. Tierpsychol.* 5, 74–101.

Seitz, A. (1948). Verhaltensstudien an Buntbarschen. *Z. Tierpsychol.* 6, 230–233.

Sevenster, P. (1949). Modderbaarsjes. *Levende Natuur* 52, 162–168 and 184–189.

Sevenster, P. (1961). A causal analysis of a displacement activity (fanning in Gasterosteus aculeatus L.). *Behaviour* Suppl. 9, 1–170.

Sevenster-Bol, A. C. A. (1962). On the causation of drive reduction after a consummatory act (in *Gasterosteus aculeatus* L.). *Arch. Neerl. Zool.* 15, 175–236.

Shaw, E. (1960). The development of schooling behavior in fishes. *Physiol. Zool.* 33, 79–86.

Shaw, E. (1961). The development of schooling in fishes. II. *Physiol. Zool.* 34, 263–272.

Shaw, E. (1962a). The schooling of fishes. *Sci. Am.* 206, 128–138.

Shaw, E. (1962b). Environmental conditions and the appearance of sexual behaviour in the platyfish. *In* "Roots of Behavior" (E. L. Bliss, ed.), pp. 123–141. Harper, New York.

Sherrington, C. S. (1948). "The Integrative Action of the Nervous System." Yale Univ. Press, New Haven, Connecticut.

Simpson, M. J. A. (1968). The display of the Siamese fighting fish, *Betta splendens*. *Animal Behaviour Monographs* **1**, 1–73.

Smith, R. J. F., and Hoar, W. S. (1967). The effects of prolactin and testosterone on the parental behaviour of the male stickleback *Gasterosteus aculeatus*. *Animal Behaviour* **15**, 342–352.

Symons, P. E. K. (1965). Analysis of spine-raising in the male three-spined stickleback. *Behaviour* **26**, 1–75.

Tavolga, W. N. (1954). Reproductive behavior in the gobiid fish *Bathygobius soporator*. *Bull. Am. Museum Nat. Hist.* **104**, 431–459.

Tavolga, W. N. (1956a). Pre-spawning behaviour in the gobiid fish, *Bathygobius soporator*. *Behaviour* **9**, 53–75.

Tavolga, W. N. (1956b). Visual, chemical and sound stimuli as cues in the sex discriminatory behaviour of the gobiid fish, *Bathygobius soporator*. *Zoologica* **41**, 49–65.

Tavolga, W. N. (1960). Sound production and underwater communication in fishes. *Animal Sounds and Commun. Amer. Inst. Biol. Sci. Publ.* **1**, 93–136.

Ten Cate, J. (1940). Zur Frage der rhythmischen Tätigkeit des Rückenmarks bei Haifischen. *Arch. Neerl. Physiol.* **24**, 226–249.

Ter Pelkwijk, J. J., and Tinbergen, N. (1937). Eine reizbiologische Analyse einiger Verhaltensweisen von *Gasterosteus aculeatus* L. *Z. Tierpsychol.* **1**, 193–200.

Thorpe, W. H. (1951). The definition of some terms used in animal behaviour studies. *Bull. Animal Behaviour* **9**, 34–40.

Thorpe, W. H. (1961). "Bird-song." Cambridge Univ. Press, London and New York.

Thorpe, W. H. (1963). "Learning and Instinct in Animals," 2nd ed. Methuen, London.

Tinbergen, N. (1939). On the analysis of social organisation in vertebrates, with special reference to birds. *Am. Midland Naturalist* **21**, 210–235.

Tinbergen, N. (1940). Die Uebersprungbewegung. *Z. Tierpsychol.* **4**, 1–40.

Tinbergen, N. (1942). An objectivistic study of the innate behaviour of animals. *Bibl. Biotheoret., Leiden* **1**, 39–98.

Tinbergen, N. (1948). Social releasers and the experimental method required for their study. *Wilson Bull.* **60**, 6–51.

Tinbergen, N. (1950). The hierarchical organisation of nervous mechanisms underlying instinctive behaviour. *Symp. Soc. Exptl. Biol.* **4**, 305–312.

Tinbergen, N. (1951). "The Study of Instinct." Oxford Univ. Press, London and New York.

Tinbergen, N. (1952). "Derived" activities: their causation, biological significance, origin and emancipation during evolution. *Quart. Rev. Biol.* **27**, 1–32.

Tinbergen, N. (1953). "Social Behaviour in Animals." Methuen, London.

Tinbergen, N., and Perdeck, A. C. (1950). On the stimulus situation releasing the begging response in the newly hatched herring gull chick (*Larus a. argentatus* Pont.). *Behaviour* **3**, 1–39.

Tinbergen, N., and Van Iersel, J. J. A. (1947). "Displacement reactions" in the three-spined stickleback. *Behaviour* **1**, 56–63.

Tugendhat, B. (1960a). The normal feeding behaviour of the three-spined stickleback (*Gasterosteus aculeatus* L.). *Behaviour* **15**, 284–318.

Tugendhat, B. (1960b). The disturbed feeding behaviour of the three-spined stickleback. I. Electric shock is administered in the food area. *Behaviour* **16**, 159–187.

Valone, J. A., Jr. (1970). Electrical emissions in *Gymnotus carapo* and their relation to social behavior. *Behaviour* **37**, 1–14.

Van den Assem, J. (1967). Territory in the three-spined stickleback *Gasterosteus aculeatus*. L. *Behaviour* Suppl. **16**, 1–164.

Van den Assem, J., and Van der Molen, J. N. (1969). Waning of the aggressive response in the three-spined stickleback upon constant exposure of a conspecific. I. A preliminary analysis of the phenomenon. *Behaviour* **34**, 286–324.

Van Iersel, J. J. A. (1953). An analysis of the parental behaviour of the male three-spined stickleback. *Behaviour* Suppl. **3**, 1–159.

Van Iersel, J. J. A. (1958). Some aspects of territorial behavior of the male three-spined stickleback. *Arch. Neerl. Zool.* **13**, 383–400.

Van Iersel, J. J. A., and Bol, A. C. A. (1958). Preening of two Tern species. A study on displacement activities. *Behaviour* **13**, 1–88.

Verwey, J. (1930a). Coral reef studies. I. The symbiosis between damselfishes and sea anemones in Batavia Bay. *Treubia* **12**, 305–366.

Verwey, J. (1930b). Die Paarungsbiologie des Fischreihers. *Zool. Jahrb., Abt. Allgem. Zool. Physiol. Tiere* **68**, 1–120.

von Frisch, K. (1942). Ueber einen Schreckstoff der Fischhaut und seine biologische Bedeutung. *Z. Vergleich. Physiol.* **29**, 46–145.

von Holst, E. (1934a). Studien über die Reflexe und Rhythmen beim Goldfisch (*Carassius auratus*). *Z. Vergleich. Physiol.* **20**, 582–599.

von Holst, E. (1934b). Weitere Reflexstudien an spinalen Fischen. *Z. Vergleich. Physiol.* **21**, 658–679.

von Holst, E. (1935a). Erregungsbildung und Erregungsleitung im Fischrückenmark. *Arch. Ges. Physiol.* **235**, 345–359.

von Holst, E. (1935b). Ueber den Lichtrückenreflex bei Fischen. *Publ. Staz. Zool. Napoli* **15**, 143–158.

von Holst, E. (1937a). Bausteine zu einer vergleichenden Physiologie der lokomotorischen Reflexe bei Fischen. II. *Z. Vergleich. Physiol.* **24**, 532–562.

von Holst, E. (1937b). Von Wesen der Ordnung im Zentralnervensystem. *Naturwissenschaften* **25**, 625–631 and 641–647.

von Holst, E. (1939). Entwurf eines Systems der lokomotorischen Periodenbildungen bei Fischen. *Z. Vergleich. Physiol.* **26**, 481–528.

von Holst, E. (1950a). Die Arbeitsweise des Statolithenapparates bei Fischen. *Z. Vergleich. Physiol.* **32**, 60–120.

von Holst, E. (1950b). Quantitative Messung von Stimmungen in Verhalten der Fische. *Symp. Soc. Exptl. Biol.* **4**, 143–172.

von Holst, E., and Le Mare, D. W. (1935). Bausteine zu einer vergleichenden Physiologie der lokomotorischen Reflexe bei Fischen. I. *Z. Vergleich. Physiol.* **23**, 223–236.

von Holst, E., and Mittelstaedt, H. (1950). Das Reafferenzprinzip. *Naturwissenschaften* **37**, 464–476.

von Holst, E., and von Saint Paul, U. (1960). Vom Wirkungsgefüge der Triebe. *Naturwissenschaften* **37**, 464–476; English translation: On the functional organisation of drives. *Animal Behaviour* **11**, 1–20 (1963).

von Uexküll, J. B., and Kriszat, G. (1934). "Streifzüge durch die Umwelten von Tieren und Menschen" [English translation: "Instinctive Behaviour" (C. H. Schiller, ed.) Methuen, London, 1957].

Wai, E. H., and Hoar, W. S. (1963). The secondary sex characters and reproductive behavior of gonadectomized sticklebacks treated with methyl testosterone. *Can. J. Zool.* **41,** 611–628.

Ward, J. A., and Barlow, G. W. (1967). The maturation and regulation of glancing off the parents by young orange chromides (*Etroplus maculatus*: Pisces-Cichlidae). *Behaviour* **29,** 1–56.

Weiss, P. (1941). Autonomous versus reflexogenous activity of the central nervous system. *Proc. Am. Phil. Soc.* **84,** 53–64.

Weiss, P. (1950). Experimental analysis of coordination by the disarrangement of central-peripheral relations. *Symp. Soc. Exptl. Biol.* **4,** 92–109.

Whitman, C. O. (1898). "Animal Behaviour," Biol. Lect. Marine Biol. Lab., Woods Hole, Massachusetts.

Wickler, W. (1955). Das Fortpflanzungsverhalten der Keilflackbarbe, *Rasbora heteromorpha* Duncker. *Z. Tierpsychol.* **12,** 220–228.

Wickler, W. (1956). Eine Putzsymbiose zwischen *Corydoras* und *Trichogaster*. *Z. Tierpsychol.* **13,** 46–49.

Wickler, W. (1957a). Das Verhalten von *Xiphophorus maculatus* var. Wagtail und verwandten Arten. *Z. Tierpsychol.* **14,** 324–346.

Wickler, W. (1957b). Vergleichende Verhaltensstudien an Grundfischen. I. Beiträge zur Biologie, besonders zur Ethologie von *Blennius fluviatilis* Asso im Vergleich zu einigen anderen Bodenfischen. *Z. Tierpsychol.* **14,** 393–428.

Wickler, W. (1958a). Ueber die erste Schwimmblasenfüllung bei Cichliden (Pisces Acanthopterygii). *Naturwissenschaften* **46,** 94–95.

Wickler, W. (1958b). Vergleichende Verhaltensstudien an Grundfischen. II. Die Spezialisierung des *Steatocranus*. *Z. Tierpsychol.* **15,** 427–446.

Wickler, W. (1959). Vergleichende Verhaltensstudien an Grundfischen. III. Die Umspezialisierung von *Noemacheilus kuiperi* De Beaufort. *Z. Tierpsychol.* **16,** 410–423.

Wickler, W. (1960a). Aquarienbeobachtungen an *Aspidontus*, einem ektoparasitischen Fisch. *Z. Tierpsychol.* **17,** 277–292.

Wickler, W. (1960b). Die Stammesgeschichte typischer Bewegungsformen der Fischbrustflosse. *Z. Tierpsychol.* **17,** 31–66.

Wickler, W. (1961a). Ueber das Verhalten der Bleniiden *Runula* und *Aspidontus* (Pisces, Blenniidae). *Z. Tierpsychol.* **18,** 421–440.

Wickler, W. (1961b). Oekologie und Stammesgeschichte von Verhaltensweisen. *Fortschr. Zool.* **13,** 304–365.

Wickler, W. (1962a). Ei-Attrappen und Maulbrüten bei afrikanischen Cichliden. *Z. Tierpsychol.* **19,** 129–164.

Wickler, W. (1962b). "Egg-dummies" as natural releasers in mouth-breeding cichlids. *Nature* **194,** 1092–1093.

Wickler, W. (1963). Zum Problem der Signalbildung, am Beispiel der Verhaltens-Mimikry zwischen *Aspidontus* und *Labroides* (Pisces, Acanthopterygii). *Z. Tierpsychol.* **20,** 657–679.

Wickler, W. (1965a). Specialization of organs having a signal function in some marine fish. *Intern. Conf. Tropical Oceanogr., Miami, Florida,* pp. 539–548.

Wickler, W. (1965b). Signal value of the genital tassel in the male *Tilapia macrochir* Blgr. (Pisces: Cichlidae) *Nature* **208,** 595–596.

Wickler, W. (1966). Sexualdimorphismus, Paarbildung und Versteckbrüten bei Cichliden. *Zool. Jb. Syst.* **93,** 127–138.

Wickler, W. (1967a). Vergleichende Verhaltensforschung und Phylogenetik. *In* "Die Evolution der Organismen" (G. Heberer, ed.), 3rd ed., Vol. I, pp. 420–508.

Wickler, W. (1967b). Vergleich des Ablaichverhaltens einiger paarbildender sowie nicht-paarbildender Pomacentriden und Cichliden. *Z. Tierpsychol.* **24**, 457–470.

Wickler, W. (1969). Zur Sociologie des Brabantbuntbarsches, *Tropheus moorei* (Pisces, Cichlidae). *Z. Tierpsychol.* **26**, 967–987.

Wiepkema, P. R. (1961). An ethological analysis of the reproductive behaviour of the Bitterling (*Rhodeus amarus* Bloch). *Arch. Neerl. Zool.* **14**, 103–199.

Willmer, E. N. (1934). Observations on the respiration of certain tropical freshwater fishes. *J. Exptl. Biol.* **11**, 283–306.

Wilson, D. P. (1937). The habits of the angler fish (*Lophius piscatorius* L.) in the Plymouth aquarium. *J. Marine Biol. Assoc. U. K.* **21**, 477–496.

Wilz, K. J. (1970). Causal and functional analysis of dorsal pricking and nest activity in the courtship of the three-spined stickleback *Gasterosteus aculeatus*. *Animal Behaviour* **18**, 115–124.

Winn, H. E. (1958). Comparative reproductive behavior and ecology of fourteen species of darters (Pisces-Percidae). *Ecol. Monographs* **28**, 155–191.

Winn, H. E. (1964). The biological significance of fish sounds. *Marine Bio-Acustics, Proc. Symp. Bimini, 1963*, pp. 213–231.

Wrede, W. L. (1932). Versuche über den Artduft der Elritzen. *Z. Vergleich. Physiol.* **17**, 510–519.

Wunder, W. (1927). Sinnesphysiologische Untersuchungen über die Nahrungsaufnahme bei verschiedenen Knochenfischarten. *Z. Vergleich. Physiol.* **6**, 67–98.

Wunder, W. (1930). Experimentelle Untersuchungen am dreistachlichen Stichling (*Gasterosteus aculeatus* L.) während der Laichzeit. *Z. Morphol. Oekol. Tiere* **16**, 453–498.

Youngbluth, J. (1968). Aspects of the ecology and ethology of the cleaning fish, *Labroides phthirophagus* Randall. *Z. Tierpsychol.* **35**, 915–932.

Zeigler, H. P. (1964). Displacement activity and motivation theory: A case study in the history of ethology. *Psychol. Bull.* **61**, 362–376.

# 6

# BIOLOGICAL RHYTHMS

## HORST O. SCHWASSMANN

## I. INTRODUCTION

A chapter dealing exclusively with biological rhythms is a new feature in a book on the physiology of fish, and it attests to the importance currently conceded these physiological rhythms as adaptations to our periodic environment. During the last two decades, substantial revival of interest in the study of the physiological mechanisms by which organisms adapt to the temporal conditions of their surroundings has resulted in a strong body of evidence supporting the thesis that these overt rhythms are expressions of a biological time-measuring system. Although investigations of endogenous periodicities in plants, initially concerning leaf movements, began more than 200 years ago, only early in the present century was the endogenous nature of these periodic phenomena clearly recognized and demonstrated. Several symposia have been held

during the last 10 years on the subject of rhythms and several textbooks
are available (Cloudsley-Thompson, 1961; Harker, 1964; Bünning, 1958,
1967) of which "The Physiological Clock" by Bünning probably offers the
broadest and most updated treatment.

This chapter cannot provide a thorough review of rhythmic phenomena in plants and animals; instead, it will summarize the evidence
concerning physiological and behavioral rhythms in fish and discuss these
in their appropriate context. Such a review appears to be necessary and
timely especially since one easily gains the impression that little experimental evidence for endogenous rhythms in fish exists. Bünning's text
(1967), for example, includes only nine references to data obtained
on fish, four of them concerning the time sense involved in sun orientation; and only very few references are found in two other monographs
(Cloudsley-Thompson, 1961; Harker, 1964). There are plausible reasons
for the lack of reference to work involving fish. Until very recently,
it has been difficult to obtain long-term records of activity in fish. On
the other hand, much work was done with a few selected organisms resulting in extensive data which provide the detail necessary for analytic
evaluation. It is a useful, although perhaps doubtful proposition that
the basic mechanisms underlying the time-measuring ability are essentially
alike in all organisms. Thus, there would be no obvious necessity for
extensive duplication. This "unified" approach toward an understanding
of the biological clock has been most rewarding. As can be expected,
there are several striking exceptions from established generalizations in
certain animals; and it is not always clear if these are artifacts of experimental procedure, specific peculiarities of some species, indicating
differing evolutionary lines in the development of physiological time
measuring, or if they are sufficiently compelling to make us reassess our
theories.

Until very recently the evidence for or against the endogenous nature
of daily rhythms of activity in fish has been somewhat contradictory.
In addition, ichthyologists have been very conservative and cautious
in accepting the idea of endogenous persistent rhythms in fish in spite
of the overwhelming evidence obtained with other organisms. The
older concept that the environmental changes in light level, temperature, etc., actually trigger certain activities still pervades even recent
texts in ichthyology. Although mentioning the possibility of "innate timing
mechanisms" and listing a "physiological clock" as one of the biological
factors influencing migrations, Lagler *et al.* (1962) continue to emphasize
the trigger concept. The slow acceptance of the idea of innate timing
mechanisms may also result from incorrectly placed emphasis concerning their ecological usefulness as mere daily regulators restricting

certain activities to specific and appropriate times of day or night in the presence of strong periodic factors of the environment which could be sufficient to directly cause, or properly time, these activities. A more important role of the daily rhythms lies in their basic involvement in photoperiodic induction phenomena, photoperiodism.

Therefore, it seems expedient not only to summarize the available experimental evidence concerning rhythms in fish but to relate these data to recently developed ideas and theories about biological rhythms in general. The currently prevailing concepts concerning biological time measuring will be reviewed briefly in the following section.

## II. DEVELOPMENT OF CONCEPTS AND GENERALIZATIONS

As might be expected in any rapidly growing branch of biology, there was considerable conflict of ideas until our present knowledge of the basic principles underlying the time-measuring capability of organisms was reached. Perhaps the greatest achievement attained in the field of biological rhythms was the substitution of principles of living organization, amenable to experimental analysis, for the previously rather mystical and simply descriptive concepts of cycles or rhythms.

### A. Certain Important Principles

#### 1. EARLY EVIDENCE FOR THE ENDOGENOUS NATURE OF RHYTHMS

The study of physiological rhythms is a relatively recent interest, at least when one considers the time since the experimental approach to this problem began. The earliest laboratory studies concerned the diurnal rhythm of leaf movements, and a full account of this early work can be found elsewhere (Bünning, 1960a). According to Bünning, it was the astronomer De Mairan who reported on the persistence of diurnally periodic leaf movements in constant darkness in 1727. During the following years, his observations were confirmed by several other workers. About 100 years later, De Candolle (1835) demonstrated that the period length in the rhythm of leaf movements in *Mimosa pudica,* when maintained in constant darkness, was not exactly 24 hr but ranged from 22 to 23 hr. In nature, the daily light–dark cycle apparently enforced the 24-hr period.

The hereditary nature of the time-measuring principle underlying those persistent periodic movements was already postulated by some early investigators; however, this view was not generally accepted at that time. Instead, the persistence of periodic changes in constant conditions was mostly interpreted as an aftereffect of a previous light–dark cycle. Pfeffer (1915) added further evidence for the non-24-hr period of the rhythm in constant conditions, and many more examples became known. Today, the term "diurnal" is replaced by the more appropriate "circadian" (Halberg *et al.*, 1959).

## 2. Response Curve as Basis of Entrainment

The history of rhythm study is full of examples of discoveries made at an early date when their importance and general validity was not fully recognized. For example, Kleinhoonte (1929) noticed that brief breaks in the dark period could delay or accelerate the phase of the leaf-movement rhythm depending on the time when these signals occurred. Other demonstrations of this phenomenon followed (Bünsow, 1953; Webb *et al.*, 1953; Pittendrigh, 1954); however, it was apparently Rawson (1956) who formulated the theory that these differences in responsiveness to light, dependent on the time when it acted, could be the basis for entrainment of circadian rhythms by a light–dark cycle. Complete "response curves" were subsequently obtained by Pittendrigh and Bruce (1957) for the *Drosophila* eclosion rhythm and for the activity rhythm of the flying squirrel by DeCoursey (1959). Such response curves are now known for a variety of plants and animals, and their fundamental importance for circadian rhythms has been generally recognized (Aschoff, 1965b).

## 3. Range of Entrainment and Zeitgeber

Another early discovery by Kleinhoonte (1932) was that the period of the leaf-movement rhythm in constant conditions was not affected by a preceding treatment with light–dark cycles, the period of which differed greatly from 24 hr. Today, one of the generalizations about circadian rhythms is that there is a limited range near 24 hr within which the period of the endogenous oscillation can be stretched by an external cycle. Periodic changes in light intensity and also temperature can affect the period. The term "zeitgeber" (Aschoff, 1951, 1954) is now generally used to describe the external synchronizing oscillation. If the zeitgeber has a period outside of the range of entrainment, the rhythm will not be coupled to the zeitgeber periodicity but will "free-run." There are interesting relationships between the range of entrainment and the previously mentioned light response curve (Enright, 1965). When no zeitgeber is acting, the frequency of the rhythm depends on the intensity of constant illumination.

## 4. Aschoff's Generalizations

Early investigators of endogenous rhythms in animals were concerned mostly with recording of locomotor activity in small rodents (Richter, 1922; Johnson, 1926). Hemmingsen and Krarup (1937) noted that in constant light the period of the activity rhythm in the rat was longer than 24 hr. Confirmation of these observations came from Johnson (1939) who found that the period increased with increasing intensity of continuous light. Aschoff (1952, 1958, 1959) recognized a correlation between diurnal and nocturnal habits and the observed direction in change of period length of the rhythm at different intensities of constant light. "Aschoff's Rule" (Pittendrigh, 1960), which has been reaffirmed for a large number of species, states that the spontaneous period in constant conditions increases with increasing illumination in dark-active animals, whereas the opposite effect is noted in light-active animals. Later, Aschoff (1960) formulated the "Circadian Rule" according to which the spontaneous frequency, the ratio of activity time to rest time within one period, and the amount of activity should increase with increasing intensity of constant illumination in light-active animals, while these three parameters should decrease with increasing light intensity in dark-active animals. Several possible exceptions to this generalization are discussed by Hoffmann (1965) and Lohmann (1967).

## 5. The Physiological Clock Concept

More than a century ago, Sachs (1857) already expressed the idea that the observed periodicity of leaf movements must be controlled by a whole complex of growth processes. This distinction between an underlying self-sustained periodicity, the "clock," and the many overt rhythms, its "hands," which are only indirect indicators of the function of the basic timing system, has been currently adopted by many. The measurable periodic changes may exhibit different degrees of coupling to the driving periodicity.

The current clock concept, implying a self-sustained timing system which is oscillating continuously, has replaced the older "hourglass" analogy which assumed that some environmental stimulus, e.g., sunrise or sunset, would initiate certain physiological changes to run off for a predetermined time. A new stimulus would be required to start a new cycle. The discovery of the time sense of honey bees (Beling, 1929; Wahl, 1932; Kalmus, 1934) and finally the demonstration of time-compensated sun orientation in bees (von Frisch, 1950) and birds (Kramer, 1950, 1951; Kramer and von Saint Paul, 1950) revolutionized our concepts about biological time measuring and made it imperative to regard the underlying physiological systems as continuously running clocks. An interesting theory about sun navigation in birds (G. V. T. Matthews, 1955) assumed a time sense of great accuracy. The demonstrated ability of diverse animals to utilize the continuously changing position of the sun for orientation indicated a clock mechanism which was running during the day and night. The latter was easily shown by resetting experiments, shifting the light–dark cycle out of phase with the natural day–night cycle, which permitted assaying the rhythm during the animal's subjective night (Hoffmann, 1953; Birukow and Busch, 1957; Pardi and Grassi, 1955; Braemer, 1960).

## 6. The Oscillator Analogy, Coupling and Phase Control

The Cold Spring Harbor Symposium entitled "Biological Clocks" (Chovnick, 1960) and the recent Feldafing Summer School entitled "Circadian Clocks" (Aschoff, 1965a) can be considered highlights in the modern period of formulation and testing of theories. In view of the greatly increasing body of observations about rhythmic phenomena in many organisms, Pittendrigh (1960) advanced the plea for a concerted effort to recognize and analyze the general principles. Increased emphasis was placed on functional analysis and conceptualization. Of the proposed models which compared the circadian timer with a self-sustained oscillation, the two-oscillator analogy of Pittendrigh and Bruce (1957; also see Pittendrigh, 1958, 1960, 1965; Pittendrigh and Bruce, 1959; Pittendrigh et al., 1958) and the model based on Aschoff's Circadian Rule by Wever (1960, 1962, 1964a,b, 1965; Aschoff and Wever, 1962) should be mentioned. These models treat biological rhythms in mathematical terms and have been most helpful in illustrating and clarifying many of the functional aspects. The basic clock, as self-sustained oscillation, is the inherent feature of the organism. Its period is remarkably circadian and can be stretched within a narrow range (range of entrainment) around its natural endogenous frequency by the zeitgeber frequency. The zeitgeber is understood as an exogenous oscillation to which the endogenous self-sustained oscillation becomes coupled, or entrained, with a specific phase angle difference. This phase difference depends on such zeitgeber parameters as the period of the entraining cycle, the ratio of light time to dark time (photofraction), the relative intensities of the light and dark phases, and the rate of transition between the light and dark levels (twilight duration). The phase response curve expresses the differing sensitivity to light during the circadian period and illustrates the mechanism by which entrainment is effected and phase relations are determined. However, its general application in the system-analysis approach is severely limited (Wever, 1965).

## 7. Effects of Temperature

A physiological clock which marked time at greatly different rates depending on tissue temperature would be of limited value, especially when one considers its critical role in sun orientation and navigation. Temperature compensation, therefore, is an important issue in studies of biological time measuring. Experimental evidence pertaining to this subject has been reviewed by Sweeney and Hastings (1960) and Wilkins (1965). It was generally found that the endogenous period of the clock is only very slightly dependent on temperature. But entrainment of a free-running rhythm by temperature cycles was possible, and resetting of the phase of the rhythm as well as a pronounced effect on amplitude seems to be common. Hoffmann (1968) demonstrated synchronization by sinusoidal temperature cycles of small amplitude for the locomotor activity rhythm of lizards in constant light. In general, poikilotherms are more easily entrainable by temperature cycles than are homoiothermous animals.

## B. Recording Methods and Choice of Reference Points

The functional aspects of the circadian clock can be investigated only indirectly by recording one or several periodic processes, and it is always uncertain to what degree these overt periodicities are coupled to the basic oscillator. The indicator processes chosen for recording cover a broad spectrum. In plants, leaf movements, flower opening, sporulation in algae and fungi, fluctuations in photosynthetic capacity, luminescence, cell division, and many other processes have been studied. The most frequently used methods for animals are the recording of locomotor activity, pupal eclosion in metamorphosing insects, egg deposition, oxygen consumption, heart rate and temperature changes, pigment migration in crustacea, fluctuations in amount and composition of excretory products, time sense in honey bees, and directional preference indirectly measuring this time sense. For a time, it was hoped that experiments on sun-compass orientation could provide the most direct and detailed measures of the clock since there was no limitation of measurable points within one period, as, for example, the onset or midpoint of activity in locomotor studies. However, the great variance of data obtained in sun-orientation experiments combined with the necessary handling of the animals during these tests rendered this method less useful. Especially for experiments under strictly controlled conditions, those procedures of recording are considered most objective and relevant which involve the least disturbance of the animals. One example of a recently developed recording method is the continuous monitoring of discharge frequency in certain electric fish (Lissmann and Schwassmann, 1965). Merely recording any periodic fluctuations is usually not sufficient to obtain records which are suitable for analysis. Especially for determinations of free-running period and phase differences, the records must contain at least one reference point which can be recognized and measured with accuracy throughout many periods. The onset of running activity of small rodents has been a most useful and precise criterion, although midpoint or maximum amplitude might be more relevant criteria (Aschoff, 1965c).

## C. Circadian Clock and Photoperiodism

An important role of the circadian clock in the timing of annual events was assumed by Bünning who postulated that the physiological basis of photoperiodic induction rested within the endogenous daily rhythm (Bünning, 1936). Photoperiodic control of annual cycles like flower formation, reproductive events, and insect diapause has been demonstrated in a large number of organisms. Early pioneering studies are those of Garner and Allard (1920), Marcovitch (1924), and Rowan (1926). However, Bünning's original idea about its underlying mechanism has only very recently been more favorably considered after several studies demonstrated that the circadian rhythm is indeed involved in the timing of photoperiodic responses (Bünsow, 1953, 1960; K. C. Hamner and Takimoto, 1964; W. M. Hamner, 1963, 1964, 1965).

Two main lines of evidence can be distinguished in the experimental support of Bünning's theory. A frequently followed approach was to subject the organism to light–dark cycles, in which the light duration was insufficient in inducing a known photoperiodically controlled reaction, and to scan the dark period with brief light breaks (interrupted-night experiments, skeleton photoperiods); the time of maximum sensitivity to light exposure could thus be demonstrated. Or groups of organisms were maintained in light–dark cycles consisting of, for example, a uniformly brief light period coupled to dark periods of varying duration (ahemeral cycle experiments). The interpretation of these experiments is that a circadian rhythm of differential sensitivity to light continues to oscillate during the long dark period and that the next following brief light pulse will either induce or fail to induce the critical physiological response depending on the time it acts in these light–dark cycles of differing period. The other approach, even more convincing, involved concurrent monitoring of one or more circadian periodicities in order to demonstrate the actual working and the phase relations of the time-measuring system of the same organism assayed for the photoperiodic response. For a full discussion of the problem recent summaries must be consulted (Bünning, 1960b, 1967; Pittendrigh, 1966; Pittendrigh and Minis, 1964; K. C. Hamner and Takimoto, 1964).

Several aspects of Bünning's original hypothesis have been more clearly defined. The "scotophil" portion of the oscillation which is sensitive to light, originally comprising about half of the total period, is now replaced by "photoperiodically inducible phase" (Pittendrigh, 1966). This inducible phase is the restricted time within the circadian oscillation which is sensitive to photoperiodic induction by light. It is not to be confused with the response curve which is a measure for the effect of light on the phase of the circadian oscillation. Although these two parameters are not independent of each other, for the response curve illustrates the mechanism of phase control and thus also determines when the inducible phase will occur, the sensitivity to light of the circadian entraining system covers the entire period and has negative and positive values, whereas the inducible phase may be restricted to a very narrow portion of the period. These two control systems probably involve different pigments mediating the different responses to light, and the work of Hendricks, Borthwick, and others indicates that a phytochrome "red–far-red" pigment system is the mediator for photoperiodic induction in plants (Hendricks, 1960; Borthwick, 1964; Borthwick et al., 1948).

So far, the evidence for the involvement of the circadian timing system in photoperiodism is based mainly on functional analysis and our understanding of the mechanism rests on formal analogy to models. The concrete biochemical processes involved still need to be investigated. Perhaps the main importance of Bünning's far-sighted hypothesis is the recognition that the adaptive significance of the circadian clock lies not only in its functional role as a built-in daily timer but that living organisms depend on it as the control mechanism for orientation to the proper time of the year.

## III. CIRCADIAN RHYTHMS IN FISH

### A. Introductory Remarks

There is now sufficient evidence available to permit the generalization that virtually all organisms, with the possible exception of bacteria and some algae, exhibit endogenous circadian oscillations in physiological functions. These periodic changes of approximately daily frequency manifest themselves in certain overt rhythms of the organism's behavior which can be measured in the field and also in the laboratory. In certain favorable artificial conditions, these overt rhythms will persist in the absence of environmental periodic fluctuations, mainly light and temperature; however, the period under these "constant" conditions is almost always significantly different from 24 hr (Aschoff, 1960). The recorded periodic phenomena are to be understood as indirect expressions of an underlying time-measuring oscillation to which they are coupled. If the overt rhythm should fail to persist under adverse artificial conditions, it could mean nothing more than the loss of a suitable criterion for measuring the basic oscillation. In the past, overt rhythms which did not seem to continue in constant laboratory conditions have sometimes been classified as "exogenous," versus "endogenous" rhythms which were observed to persist. This descriptive distinction appears to have lost its meaning in the light of our current knowledge about circadian rhythmicity, especially since it involves the danger of confusing the method of recording periodic phenomena with the underlying physiological system which we attempt to investigate.

Most of the literature dealing with recorded daily periodic activities in fish contains no information about the endogenous nature of these rhythms. Most of the work was not concerned with this problem, and some observations appear to demonstrate a lack of significant periodicities in the activity of certain species. It has been only during the last few years that sufficient experimental evidence in favor of the endogenous nature of circadian rhythms of activity in several species of fish became available. The older literature concerning rhythms in fish is included in this section of circadian rhythms with the assumption that we are dealing with one and the same principle of living organization, although in many instances its experimental demonstration is still lacking.

## B. Rhythms of Activity

### 1. Field Observations and Early Laboratory Work

It is common knowledge that fish in fresh water as well as in the ocean show a cyclic pattern in daily activity. Most species are more active at certain times of the day than at others. Like terrestrial vertebrates, fish can also be classified into diurnally active species, relying predominantly on cone vision, and nocturnal species which rely more on tactile, chemical, or electrical senses. Two examples of observations on daily cyclic movements are those of Hasler and Villemonte (1953) on the freshwater perch, and the extensive work of Barlow (1958) on the desert pupfish. The periodic movements of these fish are related to periodically changing physical characteristics such as light intensity in the first study and predominantly temperature in the second. Simple catch statistics, when correlated with time of day, are also indicative of movement or activity patterns in fish, as are the examinations of stomach content and observation of feeding activity. Hart (1931) noted on the basis of gill-net catches that different species have active periods at different times of day or night. Mužinić (1931) found that herring have two main feeding periods, one from afternoon until nightfall and one lesser one in the early morning. No feeding was noted during the hours of bright daylight. Blaxter (1965) reviewed the evidence concerning the feeding pattern in herring larvae in relation to time of day and to light intensity. Other studies which show a daily activity pattern with increased activity at dawn are those of Sushkina (1939) on herring larvae and of Oliphan (1951) on a species of grayling and other freshwater species. Analysis of gill-net catches (Carlander and Cleary, 1949) indicates that some species such as sauger, yellow pike perch, and others are more active at night, whereas perch and northern pike show their main activity during daytime. Spoor and Schloemer (1939) had already proposed that the high catch of freshwater rock bass during morning and evening hours could result from a rhythmic activity pattern. Von Seydlitz (1962) reported on higher catches of *Sebastes marinus* during daytime than at night.

Schooling behavior also shows a daily pattern. For example, Steven (1959) noted that schools of *Hepsitia stipes* and *Bathystoma rimator* disperse at night and that these fish show an activity increase with decreasing illumination. Working on the schooling behavior of the jack mackerel, Hunter (1966) maintained groups of six fish in constant illumination and observed that these schools became more compact at a time which corresponded to the first hours of darkness of the preceding light regimen. While working on the schooling behavior of *Rasbora*, Thines and Vandenbussche (1966) reported evidence suggestive of the presence of diurnal fluctuations in the readiness to school in response to external stimuli.

The daily vertical movements of fish schools have received considerable attention. Balls (1951) recorded an upward migration in herring schools during the night and a movement to deeper waters at daybreak. Richardson (1952) suggested that these vertical movements may be a direct response to changes in light intensity. Brawn (1960) reported seasonal variations in the depth distribution pattern of herring schools during day and night. Similar diurnal migrations of planktonic organisms are known to exist. The study by McNaught and Hasler (1961) on schooling and feeding behavior of white bass demonstrates that maximum feeding activity of these fish coincides with high concentrations of their planktonic food organisms at the surface during morning and evening. Thus, it could be that the vertical movements of the plankton-feeding fish might be in response to the diurnal migration of their principal food source. Many authors consider the change in light intensity as a stimulus for vertical migration of zooplankton organism (e.g., Clarke, 1930, 1933). However, some authors concluded from experimental evidence of their studies that an internal rhythm must be involved at least in certain species (Esterley, 1917, 1919; Harris, 1963). Enright and Hamner (1967) found evidence for endogenous control in several species of planktonic crustaceans by demonstrating the persistence of vertical movements in constant dim light.

Experiments in the laboratory under controlled conditions allow investigation of the effect of environmental factors. A pronounced daily rhythm in feeding activity of goldfish was reported by Hirata and Kobayashi (1956) and Hirata (1957). Such diurnal variations in the pattern of feeding and locomotor activity are quite common and were documented in controlled laboratory experiments on salmonids by Hoar (1942, 1958). Spencer (1939) obtained records for several species of freshwater fish and concluded that some species were diurnally active, others were nocturnal, and the carp and many others were probably arhythmic. A good correlation of oxygen consumption with the degree of activity was demonstrated by Spoor (1946) whose mechanical–electrical recording method resulted in excellent records of the activity pattern. Earlier work by Clausen (1936) demonstrated a daily periodicity with morning and late afternoon maxima in oxygen consumption for the largemouth bass and low activity during daytime alternating with increased activity at night in the black bullhead. Clausen, however, could not observe rhythmic fluctuation in fish from rapidly flowing waters. Many other studies recording oxygen consumption of fish were not concerned with the demonstration of a periodically changing pattern (Schuett, 1933, 1934). A diurnal rhythm in phototactic behavior was reported by Kawamoto and Konishi (1955) for *Girella punctata*, but it was not noticeable in *Mugil cephalus* and two species of eels. Swift (1962, 1964) could obtain good records on the diurnal and annual activity pattern of brown trout when the fish were confined in cages in their natural streambed. Recording the locomotor activity of *Phoxinus* in the laboratory, Jones (1956) reported that these minnows were active during daylight hours but that this pattern reversed if their tank contained a hollow brick where they could hide from bright light. When cover was provided, they were very active around sunrise and sunset. Jones found no evidence for an inherent daily rhythm of locomotor activity under his experimental conditions. Recent experiments with the same species, utilizing a different recording method (Müller and Schreiber, 1967), demonstrate convincingly a free-running circadian periodicity in swimming activity (Müller, 1968). The free-running period for two fish, recorded over 7 days under continuous natural light during the summer near the polar circle, was surprisingly long (27 and 30.5 hr) but seems to be paralleled by similar long free-running periods found in the laboratory under conditions of constant light and temperature (Müller, 1968). According to recent work by Müller (1969), *Salmo trutta* found near the polar circle are day-active during winter and night-active in summer. Especially during the times of reversal, and in midsummer, nonsynchronized states occur with free-running periods of 29 hr. A similar change from light activity in winter to summer dark activity is reported for *Cottus poecilopus*, also from the Arctic (Andreasson, 1969).

Other studies concerning daily periodicities in the activity of fish are those of Harder and Hempel (1954) on sole and flounder, of Wikgren (1955) on the burbot, Kruuk (1963) on the sole, and Davis (1963) on bluegill and largemouth bass and some other species (Davis and Bardach, 1965). An earlier investigation by Jones (1955) failed to show evidence in favor of an inherited rhythm of locomotor activity in ammocoete larvae of a brook lamprey, *Lampetra planeri;* but a more recent report describing the use of photoelectric sensing system (Kleerekoper *et al.*, 1961) demonstrated a persistent endogenous activity rhythm in transforming ammocoete larvae and adult *Petromyzon marinus*. Although the rhythm gradually declined in amplitude, it continued for a sufficiently long time to allow a free-running period of 22 hr and 58 min to be measured (Kleerekoper *et al.*, 1961).

## 2. ON THE IMPORTANCE OF SUITABLE RECORDING METHODS

Considering the often contradictory evidence frequently involving the same species of fish, one cannot avoid concluding that the particular experimental conditions, and especially the suitability and sensitivity of the chosen recording method, must be decisive factors which affect the results of studies on locomotor activity. Workers in this field have realized the importance of sensitive recording methods. One of the earliest methods, the ichthyometer (Spencer, 1929), consisted of thin threads tied to the tail of the fish which activated a kymograph scriber. Other examples of recording devices for aquatic animals are those of Szymanski

(1914), Kalmus (1939), Spoor (1941, 1946), and DeGroot and Schuyf (1967). An interesting approach utilizing the Doppler effect for recording the swimming speed of fish has been tried recently (Cummings, 1963; Muir et al., 1965; Meffert, 1968). Another method makes use of the heat loss near a thermoregulator caused by increased swimming activity of fish (Beamish and Mookherjii, 1964). A very sensitive recording device employing the same principle has been successfully used in activity measurements with small amphipods (Heusner and Enright, 1966). It consists principally of a pair of matched thermistors incorporated in the arms of a Wheatstone bridge circuit. One thermistor is shielded from water movements, the other freely exposed. Any slight disturbance of the water will lower the temperature of the exposed thermistor and increase its electrical resistance, resulting in voltage variations across the bridge which in turn activate a capacitor-relay circuit.

## 3. CIRCADIAN ACTIVITY PATTERNS IN GYMNOTID ELECTRIC FISH

Gymnotid fish have recently been found suitable for studies of circadian activity rhythms (Lissmann and Schwassmann, 1965). Since many recent results are unpublished, these are summarized in this section.

*a. General Notes on Gymnotid Behavior and Electric Discharge.* The gymnotids are a family of South and Central American freshwater fishes, perhaps best known for being one of the few groups which have evolved an electrosensory system for orientation. Although they have recently become one of the favored objects for studies in sensory physiology, our knowledge about their ecology and behavior is very rudimentary and their taxonomy is direly in need of revision. All gymnotids emit continuous low voltage discharges, the frequency and shape of the electrical pulses being species-characteristic. One group of these fishes emits an approximately sinusoidal discharge at relatively stable and usually high frequency (Type I of Lissmann, 1961). The other members of the family produce brief and polyphasic pulses at relatively low frequency, ranging from 2 Hz in *Hypopomus* sp. to slightly more than 100 Hz in other species (Type II of Lissmann, 1961). The discharge frequency of the latter group is variable, in contrast to the former group with a sinusoidal discharge pattern. All gymnotids are nocturnal in habit and show marked differences in their activities between day and night, an observation documented by Lissmann's field studies (1961). During daytime, these fish are hiding in vegetation, under rocks or in crevices, or even buried in the sand, while soon after sunset they start roaming about their nearby territory. Thus, they show a clear pattern of activity and rest which can easily be followed in their natural habitat by recording and listening to their electrical discharge pattern (Lissmann, 1961; Lissmann and Schwassmann, 1965). In most species of Type II with a variable frequency of brief polyphasic pulses, the discharge pattern is related to the phase of activity or rest. For example, a more uniform discharge rate is observed while resting, while many high frequency bursts, correlated with swimming and other activities, are noted during the active phase in *Gymnotus carapo*. In addition to the high frequency bursting, a generally higher rate of discharge frequency is characteristic of the active phase in several species of *Hypopomus*.

*b. Earlier Studies with Gymnorhamphichthys hypostomus.* In *Gymnorhamphichthys hypostomus* which "sleeps" in the sand during the daytime, the differences in discharge frequency between the two states of activity are very pronounced. A low rate of 10–15 Hz, characteristic for the resting phase while the fish are in sand, shows a slight but significant increase to 20–30 Hz during the 2 hr prior to actual emergence, at which time it changes instantaneously to a stable high level of 50–100 Hz (Lissmann and Schwassmann, 1965). Different individual fish under apparently identical conditions discharge at different frequencies, predominantly in the 60–90 Hz range, and each individual maintains its specific rate during the active phase with very little variation. Occasional increases of up to 20% were observed during feeding and also during

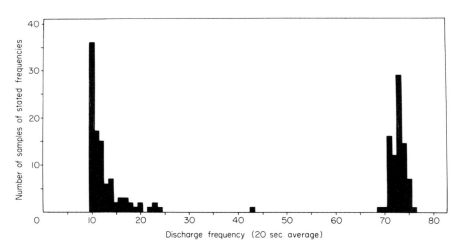

Fig. 1. Frequency spectrum of electric organ discharge of an undisturbed *Gymnorhamphichthys* recorded for 24 hr at 8 min intervals. A low rate of discharge (10 Hz+) is noted when the fish rests in the sand, a high rate (around 73 Hz) when the fish is active (Lissmann and Schwassmann, 1965).

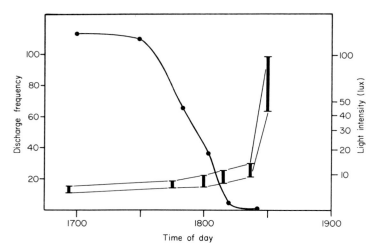

**Fig. 2.** Recorded changes in discharge frequency of *Gymnorhamphichthys* in their natural habitat during the hours prior to emergence from the sand. Vertical bars indicate the total range in frequency at the different times of sampling. All fish left the sand between 1820 and 1830 hr; (●) recorded light level (Lissmann and Schwassmann, 1965).

mechanical disturbance of the fish, but these bursts are usually less than 1 sec in duration. The distinct levels of discharge frequency and their precise correlation with the states of activity and rest made the rate of electric organ discharge a suitable criterion for long-term recording of the activity rhythm in *Gymnorhamphichthys*. The bimodal distribution of frequencies over a 24-hr period in this species can be seen in Fig. 1. A study of a population of these fish in their natural habitat revealed that they emerge from their daytime rest in the sand, precisely synchronized with each other, about a half-hour after sunset. The frequency ranges of about 20 *Gymnorhamphichthys*, recorded at different times before and after emergence from their resting places in the sand, are shown in Fig. 2. All the observed fish were found swimming actively about at 1830 hr when the total frequency range suddenly attained a high level. They reentered the sand before daybreak. By recording the discharge pattern of individual isolated fish, it could be demonstrated that the light–dark cycle entrained the rhythm which persisted in conditions of constant dim light with a free-running period significantly different from 24 hr. The consecutive activity onsets, measured by the sudden rise in discharge frequency, showed very little variation from the straight line indicating the slope of the free-running rhythm (Fig. 3).

    *c. Recent Results with Gymnorhamphichthys and Hypopomus.* Subse-

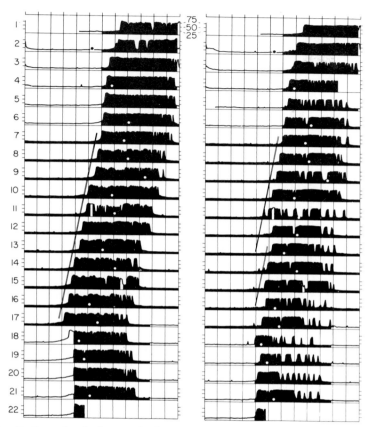

Fig. 3. Records of electric discharge frequency for two *Gymnorhamphichthys* (right and left block) over 22 days. Successive 24-hr days are mounted in vertical order. Recorded frequencies are indicated on day 1 between the blocks. Records are left open during light exposure and are solid black during dark. Light–dark cycle until day 6 is replaced by constant dark until day 17 and by a final light–dark cycle to the end of recordings. A 15-min light exposure on day 11 results in a delay of activity onset (steep rise in discharge frequency) in the fish on the right (Lissmann and Schwassmann, 1965).

quent unpublished experiments with the same and various other species of gymnotids have resulted in additional information. The period of the free-running circadian rhythm is a function of the intensity of constant illumination. As predicted for nocturnal animals by Aschoff's Rule (Aschoff, 1952, 1958, 1960), the period becomes longer with increasing light intensity. This effect of the light level on the circadian period can be seen in the graph (Fig. 4) which summarizes results from four *Hy-*

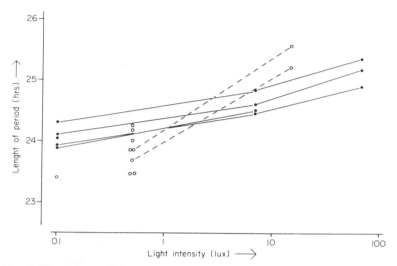

**Fig. 4.** Dependence of free-running (spontaneous) period on intensity of constant light: (○) the average period of each of nine *Gymnorhamphichthys*, and (●) the average period of four *Hypopomus*. Lines connect data from the same individual at different light intensities.

*popomus* and nine *Gymnorhamphichthys*. In evaluating the original records, the onset of activity was used for *Gymnorhamphichthys*, whereas the midpoints of the active phase proved more suitable and also showed less variance in the *Hypopomus* data. The examples of original records in Fig. 5 illustrate a certain limitation of our method of recording discharge rate for evaluating parameters other than the free-running period. One of the generalizations contained in the Circadian Rule (Aschoff, 1960) is the negative correlation between period length and the ratio of activity to rest time in their dependence on light intensity. If this generalization is applied to our nocturnal gymnotids, the activity time should decrease and the resting time should increase with increasing illumination. All of the results clearly substantiate this generalization. However, as already demonstrated in a previous report (Lissmann and Schwassmann, 1965), light seems to have an additional strong direct influence on the activity pattern. As shown in Fig. 5a, high light intensity apparently delays the emergence of the fish from the sand by more than 2 hr (days 11–18). This delaying effect is clearly seen when compared with another measurable point, that particular frequency during the "preemergence rise" in discharge rate at which the same fish used to leave the sand while under the preceding low light intensity treatment. This phenomenon in the activity records of *Gymnorhamphichthys* is interpreted as a negative masking effect (Lohmann, 1967) and is similar to the masking effect often observed under the influence of a strong zeitgeber (Aschoff, 1960; Wever, 1965). Since the two levels in discharge frequency are precisely correlated with the two distinct phases of activity in these gymnotid fish, it must be concluded that the activity rhythm cannot be considered a true measure of the functioning of the underlying circadian system because of this masking influence of light on the beginning, and possibly also on the cessation, of the active phase. Estimates of the free-running period, however, seem to be unaffected. Another case of negative masking is illustrated in the records of another fish of the same species in Fig. 5b. These data show the locomotor rhythm in different L:D ratios when activity apparently remained phase-locked to the light–dark transition, even when a brief dark pulse of less than 10 min was used (day 18). Negative masking of activity is apparent on days 17–18 when the animal enters the sand; however, a significantly increased level of discharge frequency continues with the same phase and is replaced by locomotor activity during the last days of free-run in constant dark.

## C. Other Functions of Circadian Period

### 1. MELANOPHORE AND OVIPOSITION RHYTHMS

Locomotor activity may not always be a suitable parameter for assaying the circadian time sense in fishes. Other overt rhythms which have been recorded are, for example, the melanophore movements in the brook lamprey larva (Young, 1935). Young not only found that the melanophore rhythm continued in constant dark and constant temperature but he also demonstrated a controlling influence of the pituitary on the melanophore movement. A cyprinodont, the medaka (*Oryzias latipes*), exhibits a daily rhythm of oviposition. The female lays eggs daily just before dawn, and the studies of Robinson and Rugh (1943) and Egami (1954) demonstrate that the light–dark cycle is the zeitgeber in this oviposition rhythm. Inverting the cycle of artificial light, the oviposition time shifted in a few days to the new time of dawn, 180° out of phase with the egg-laying time prior to the shift. In continuous illumination, the spawning rhythm became irregular. Marshall (1967) reported on two anabantoid species where spawning occurs during the last three to four light hours under a cycle of natural or artificial illumination. Shifting the light cycle by 12 hr, these fish almost immediately shifted their spawning to the end of the new light phase. In continuous light, spawning became less frequent and was not confined to specific

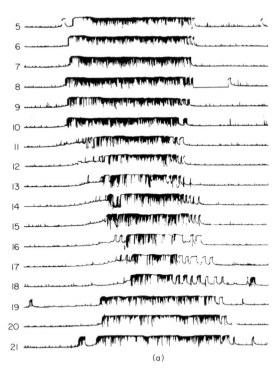

**Fig. 5.** Excerpts of activity recordings of two *Gymnorhamphichthys*. (a) Effect of intensity of constant light on period and amount of activity. Days 5–10 and 19–21 free-run in 1 lux, days 11–18 in 15 lux. (b) Activity pattern in varied light–dark ratios until day 18; light 50 lux; the duration of darkness (0.1 lux) is indicated by horizontal bars. Free-run in continuous 0.1 lux from day 19. See text for discussion of the records. (From data obtained in collaboration with D. Kuffler and D. G. Flaming.)

times. Many other species are known to restrict spawning activity to a certain time of day or night during the spawning season. For example, Gamulin and Hure (1956) reported that in the sardine spawning takes place in the evening.

## 2. Retinomotor Rhythm

Adaptation of the fish eye to vision in different light levels is effected by photomechanical movements. In the light-adapted state, the cones are aligned along the external limiting membrane, freely exposed to light, whereas the rods with greatly elongated myoids are enveloped by pigment granules inside pigment cell processes. In the dark, the cone

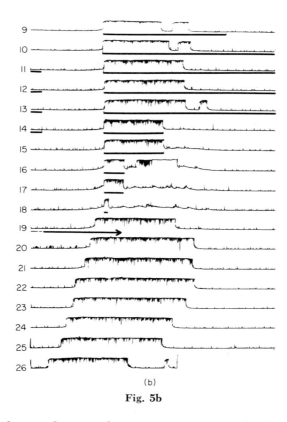

(b)

Fig. 5b

myoids lengthen and move the cones away from the limitans externa while the rods migrate out of the pigment toward the limiting membrane by contraction of their myoids. The pigment granules, usually fuscin, also move out of the pigmented extensions, enveloping the receptor cells, and concentrate inside the more peripherally located cell bodies. Welsh and Osborn (1937) demonstrated the persistence of photomechanical changes in the eye of *Ameiurus* over several days in constant darkness. The amplitude of this rhythm of rod and cone movements diminished in constant conditions. This observation was confirmed by Arey and Mundt (1941) and Wigger (1941) who also found a persistent rhythm in the pigment movement in the eye of goldfish. A very interesting report by Davis (1962) must be mentioned in connection with the rhythm in photomechanical changes in the fish eye. While working with bluegill, Davis observed that during the night, or when kept in the dark, the fish would react to a sudden light exposure by sinking to the bottom and would take several minutes to recover from this light shock. Davis demonstrated a pronounced rhythm in this recovery time which depended greatly on the time during the dark phase when the light was turned on. Recovery time was longer in the early dark phase than later in the night. John and Haut (1964) correlated the retinomotor rhythm with schooling behavior in *Astyanax*. When groups of these fish were kept in the dark and a light was turned on suddenly during the time which corresponded to their previous day, they needed only about 15 sec to school. But when the light was turned on during the time of their subjective night, about 6 min elapsed before schooling took place. A retinomotor rhythm which persisted for 6 days was demonstrated in the eye of *Astyanax* (John and Kaminester, 1969).

A wealth of interesting rhythmical habits has been brought to our attention by observations of the natural behavior of fishes. Winn (1955) reported that several species of parrot fish secrete a mucous envelope around themselves during the night. This envelope is provided with one intake near the mouth and one outlet at the caudal fin to allow respiratory water to pass through. Another peculiar habit which apparently was noticed only recently is found in the señorita, *Oxyjulis californica*, a member of the labrid family. These fish enter the sand during evening hours and apparently remain there all night. In the aquarium under artificial light conditions, they go into the sand before the light turns off, indicating that this behavior is an endogenous rhythm and not merely a response to changes in illumination (Wilkie, 1966). A similar behavior of entering the sand when disturbed has been reported for *Crystallodytes cookei* (Gosline and Brock, 1960) and might also occur in other fish. The habit of the señorita is of special interest since it seems to parallel that of the electric gymnotid, *Gymnorhamphichthys hypostomus*, reported in the preceding pages. The latter is nocturnally active and spends the resting phase in the sand during the daytime, whereas the diurnally active señorita rests in the sand at night.

## 3. TIME SENSE IN SUN ORIENTATION

Sun-compass behavior has been documented in a great variety of animals, including many species of fishes. These animals can find compass directions throughout the day by utilizing the momentary position of the sun as a reference and by making appropriate compensations for its apparent movement during the day. This ability of the orienting animal to alter the direction of its movement with respect to the sun progressively in time depends on a circadian clock. Several details concerning sun orientation in fish have been investigated and have resulted in certain conclusions which help us understand how this orientation mechanism can operate, not only at a particular latitude and during a certain season but also north or south of the equator and throughout the seasons. Many of these findings were reviewed recently by Hasler (1966); however, those data which illustrate the circadian timing involved in sun orientation will be discussed briefly here.

Sun-compass orientation in fish is clearly the expression of an endogenous circadian rhythm. For these orientation experiments the fish either have been trained to swim into a certain direction at a particular time of day or they display a natural directional tendency. The orientation rhythm is recorded as angles to the left or right of the sun's horizontal projection, its azimuth. The change of this angle with time is an indirect measure of the circadian time sense. The advantage of this method lies in the possibility of measuring the phase of the rhythm at any time, whereas many other methods provide only one measuring point per period, as, for example, the beginning of locomotor activity. A distinct disadvantage is that there is little precision in the sun-orientation method; in fish as well as in other animals, the scatter around the mean direction of repeated tests and among several animals is large. The method also requires considerable handling of the animal and exposure to environment factors which might influence the time sense and also the conditioned behavioral response.

In sun-orientation behavior every specific angle to the sun is indicated repeatedly at 24-hr intervals. The light–dark cycle synchronizes the rhythm with its 24-hr daily period. Six green sunfish, *Lepomis cyanellus*, tested over six full periods under zeitgeber control showed orientation into the trained compass direction with an average day-to-day accuracy of ±6°. The precision in indicating the trained angle, measured as standard deviation of the mean directions of all scores, decreased from 26° on the first day to 35° on the last day of the experiments (Schwassmann, 1962). The rhythmic orientation behavior persisted in constant light for 4–6 days in four trained centrarchid fish; after that time, the conditioned directional response deteriorated. The free-running period in constant light was approximately 23 hr in three fish, but it was longer than 24 hr in one animal which was subjected to testing at 25.5 hr intervals, thus supporting the possibility that the exposure to the bright sun during the repeated tests was affecting the phase and thereby the period of the rhythm (Schwassmann, 1960). Earlier experiments by Braemer and Schwassmann (reported in Schwassmann, 1962, 1967) demonstrated rhythmic sun orientation in young centrarchid fish which were raised from the fertilized egg in constant light. Further evidence for the entraining action of the light–dark cycle resulted from phase-shifting experiments. Resetting the times of onset and termination of artificial light resulted in a predictable shift of the previously trained direction which was complete in about 4 days (Braemer, 1960). It was also possible to read the phase of the clock involved in this compass orientation during the subjective night of the fish by recording the directional behavior in animals which had been subjected to an artificial light–dark cycle sufficiently out of phase with the natural day so that a large part of the day coincided with the fish's subjective night (Braemer, 1960; Braemer and Schwassmann, 1963). In addition, an influence of the photoperiod on the amplitude of the rhythm of angular change could be demonstrated. Groups of fish subjected to "long" days changed the orientation angle to the sun during a comparable time interval around noon faster than other fish which were subjected to "short" days (Schwassmann and Braemer, 1961).

Summarizing the available evidence concerning sun orientation in fish, we find that sun-compass behavior is a circadian rhythm which is influenced by daily and seasonal changes of light and dark. Its free-running period in constant light is different from 24 hr. A photoperiodic effect on the circadian timer seems to adjust the rhythm to quantitative changes in the sun's azimuth movement during the different seasons at one and the same latitude, whereas other experiments indicate that the sun's daily changes in altitude have a significant role in adjusting the angular change in orientation to the local latitude (Schwassmann and Hasler, 1964; Braemer and Schwassmann, 1963).

## D. Ecological Significance of Circadian Rhythms

The circadian organization of living systems is certainly an inherited and historically very ancient feature. Its selective advantage appears to consist in the unique manner in which it causes periodic changes in the organism's physiological state, regulating its activities which are often best performed at specific times of the daily cycle. Although most organisms in nature are sufficiently exposed to the direct action of strong environmental fluctuations of 24-hr periodicity, especially the day–night changes in light intensity, these external factors are not immediately causing periodic changes in the organism's physiology and behavior, but are merely acting as zeitgeber, imposing their phase on the endogenous system which is periodic in itself and comparable to a self-sustained oscillation.

Certain reservations must be exercised when we consider the adaptive value of those overt rhythms in physiological and behavioral functions which can be recorded. It was demonstrated and emphasized that the measurable functions are the only available means by which we are able to investigate functional properties of the underlying time-measuring system; they are not the clock itself but only its hands. Especially under experimental conditions deviating drastically from the normal environment, these indicators may become uncoupled from the basic oscillator or show otherwise abnormal behavior. In addition, overt rhythms must not be confused with the endogenous timing system itself; they are merely proof of its existence and useful as indicators for its response characteristics. Therefore, an adaptive significance of all the separate manifestations of the circadian clock may not always be obvious or may be difficult to recognize.

Many circadian functions clearly show an ecological usefulness. The adaptive value may be a restriction of certain activities to the most favorable time of the environmental daily cycle, either determined by abiotic factors such as temperature or humidity or by biological factors such as the availability of food organisms. Most desert reptiles exhibit a behavioral rhythm which protects them from extremes in the prevailing temperature changes. Time of insect eclosion is frequently restricted to times of high relative humidity; the same coincidence is known for the main flight activity of certain insects. Adaptive synchronization with biological rhythms of other organisms is often noted. Many flowers open only for restricted times of day or night, and for many the flow of nectar and pollen production is greatest during certain limited periods. The collecting activity of many insects, nectar- and pollen-feeding bats, and hummingbirds shows a periodic pattern which is synchronized with the times of food availability. The crepuscular flight activity of insectivorous bats is correlated with maximum numbers of flying insects. Of course, the most impressive example for adaptive significance is the physiological clock as the basis for time-memory and for time-compensated orientation to the sun.

A functional significance of circadian rhythms in the aquatic environment may not always be readily recognized; however, this may partially result from our incomplete knowledge concerning natural history and behavior of aquatic organisms. A correlation between main feeding activity of many fish species with the times of concentration of plankton near the surface, or the catching of low flying insects in the evening by trout and many tropical species, is easily seen. The strikingly nocturnal activity of the gymnotids, for example, is thought to be of adaptive value since it renders them less vulnerable to the predominantly visually oriented predators in their native waters, the piranhas and the trahira. In almost all cases where the observed rhythms were investigated, it could be demonstrated that those physical or biological factors of the environment to which the overt rhythmic activity was temporally adapted were not causally responsible for the synchronization. Light and, to a lesser degree, temperature cycles were shown to be the effective zeitgeber.

The general opinion today is that in the past too much emphasis was placed on finding some adaptive value in any and all of the overt rhythmic activities of organisms. These are to be looked upon as manifestations of a circadian organization, and the many overt periodicities by themselves may or may not be of adaptive significance. The most powerful argument in favor of this modern viewpoint is the evidence concerning the causal action of the circadian system in photoperiodic induction (see Section II, C). One could, then, assume that the important role of the circadian rhythmicity in photoperiodic control would constitute the most meaningful functional adaptation in an ecological and evolutionary sense.

## IV. RHYTHMS OF OTHER THAN CIRCADIAN PERIOD

### A. Control of Annual Breeding

1. INTRODUCTORY REMARKS

Successful reproduction is an essential factor for species survival, and it is not surprising to find that breeding periods of animals are adjusted in time to that particular phase of the seasonal cycle which is suitable for rearing of the offspring. In many fish the actual process of spawning, involving release and fertilization of ova, is limited to a relatively brief time span; however, gonadal development is a complicated physiological process of long duration. This internal physiological process makes the animal ready for actual breeding behavior to occur at the most appropriate time. Since, in many instances, extensive migrations to special breeding grounds occur preparatory to spawning, one must assume that it is not the sudden incidence of directly acting stimuli of favorable environmental conditions which trigger reproductive behavior but the entire annual sexual cycle is subject to synchronization by external factors.

Of all the environmental fluctuations correlated with the annual cycle of seasonal change, only the systematically changing length of day seems to provide a reliable time marker which could account for the often spectacular accuracy in occurrence of annual breeding behavior. The effect of day length in controlling seasonal flowering in plants was first demonstrated by Garner and Allard (1920), and its role in the timing of reproductive rhythms in animals was shown by Rowan (1926). Since then a large amount of work on plants and animals has established the widespread occurrence of photoperiodic control of annual cycles. There is still insufficient evidence to assume that photoperiodism is the only means of external synchronization to the seasons which has been utilized in the evolution of organisms nor that it must apply to all existing species. It is, however, the one principle which was found to be effective in plants and animals of middle and higher latitudes where seasonal environmental changes are severe, and it is a theory which has been tested experimentally.

Concerning our understanding of photoperiodism in terms of functional analysis, an important theory was advanced by Bünning (1936) who proposed that the endogenous daily rhythm is involved in photoperiodic induction. There is now significant evidence from experiments on plants, insects, and birds attesting to the validity of Bünning's original hypothesis (see Section II, C).

A survey of the evidence concerning the timing of annual breeding rhythms in fish poses difficulties. Although there is adequate demonstration for photoperiodic control in many species, there are also many reports which appear to argue against accepting photoperiodism as the principal seasonal timing agent. A unified approach, which has been so fruitful in the case of circadian rhythmicity, does not seem justifiable in the case of annual reproductive periodicities.

A great deal of information exists concerning the role and interaction of endocrine glands and gonads in the reproductive cycle in fish. This aspect is treated in detail in earlier volumes of this treatise, and the present section will be restricted to a discussion of the role of external factors as initiators or synchronizers of the physiological reproductive cycle.

## 2. Survey of Experimental Evidence

An extensive review of the literature on this subject was published by Atz (1957) with an addition in 1964 (Atz and Pickford, 1964). Other pertinent reviews are those of Hoar (1955) and Harrington (1959a). The present survey will be restricted to the experimental approach which involves manipulation of environmental factors and their effect on time of actual breeding or gonadal development. It will also be assumed that annual periodicity and restriction in time of breeding is the rule rather than the exception and needs no further documentation. Information about breeding times in fish is contained in Breder and Rosen (1966).

Early recognition of a close correlation of reproductive cycles with annual seasonal changes led to an investigation of morphological changes of the gonads during the reproductive cycle; the classic study by Turner on the freshwater perch (Turner, 1919) is such an example. The earlier prevailing opinion that seasonal changes in temperature, light intensity, rainfall, etc., were responsible for causing the appropriate adjustment in reproductive cycles was subjected to a critical reevaluation following Rowan's pioneering work with migratory birds (Rowan, 1926, 1929). The experimental approach in evaluating the role of external factors employed primarily the manipulation of day length and temperature; other factors, for example, salinity changes, were rarely tested. Craig-Bennett (1931), working with *Gasterosteus*, found that the male reached potential maturity before the female and that spermatogenesis was completed several months in advance of the next breeding season. His conclusion that day length was not affecting gonadal development was based on the appearance of secondary sex characteristics and has been contradicted by several later workers (van den Eeckhoudt, 1946; Kazanskii, 1952; Baggerman, 1957; Schneider, 1969). Van den Eeckhoudt (1946) employed gradual increases in photoperiod, reaching 16 hr of light per day, from February to April, and found spermatogenesis to be continuing normally but observed no nuptial color development. Those males which were exposed to consistently short days, under otherwise identical conditions, seemed to revert to earlier stages of spermatogenesis. Increasing day length caused progressing oogenesis which became arrested in the short-day group. A significantly higher rate of egg maturation was noted when, instead of whole and increasing photoperiods, light was administered in doses of 20 min, gradually increasing to 40 min for every hour, a technique which resembles modern experiments for testing the involvement of the circadian system in photoperiodism which utilize this method of "breaking" the dark phase. Van den Eeckhoudt noted complete maturation and nestbuilding in those animals which had been treated with continuous light in March and April. It is unfortunate that the later phase of experimental treatment practically coincided with the onset of breeding in nature, thus making interpretation of the results difficult.

From the extensive study by Baggerman (1957) on the three-spined stickleback, *Gasterosteus aculeatus*, a reasonably clear account of the complex interactions between an internal rhythm and day length–temperature combinations as external factors seems to result. Nestbuilding in males and oviposition in females were the criteria employed for inferring positive or negative responses owing to the different treatments as well as estimating the physiological state of the fish. Although a great deal of detailed information is contained in Baggerman's work, the principal conclusions appear to be the following: Four successive physiological response phases can be distinguished in the annual breeding cycle of the three-spined stickleback, of which phase 0 probably ought not to be considered part of the breeding cycle since it is shown to occur in immature fish only; those fish which have spawned at least once never pass through this stage again. Fish in the following phases, 1a and 1b, seem to require exposure to long day length and high temperature in order to reach maturity rapidly. Short day length, even when combined with high temperature, inhibits maturation at these stages, but a sequence of short day–low temperature followed by short day–high temperature abolishes this inhibition in phase 1a only. Phase 2 is characteristic for the condition just prior to maturity when the physiological rhythm is so far advanced that manipulation of day length and temperature is of little influence. In the last two stages gonadal development even proceeds at 4°C when day length is long, and nestbuilding by males occurs at temperatures as low as 6°C. At long day length the rate of gonadal development increases with increasing temperature during the last stages. In the last two stages, sudden increases in length of day have a stronger accelerating effect on gonadal maturation than do gradual increases. In constant conditions of long days and high temperature, males and females go through cycles of alternating reproductive and nonreproductive behavior with a period of approximately 200 days/cycle. Baggerman employed only two contrasting day lengths in her experiments, 8 and 16 hr, respectively; therefore, no information is available about any possibly critical length of daylight.

McInerney and Evans (1970) could demonstrate a broad action spectrum of photoperiodically effective light in *Gasterosteus*. About 5 lux at 420 m$\mu$ was found to be sufficient for inducing normal seasonal changes.

A study by Merriman and Schedl (1941) on the four-spined stickleback *Apeltes* showed no difference in spermato- and ovogenesis for groups of fish exposed from October to November–December either to a gradually increasing length of day (to 13.25 hr), or to continuous 11.75-hr days, or to only a few minutes of light per day. A significant increase in rate of oocyte growth was observed in fish after 2 months in continuous illumination.

There are a few observations about the effect of light on breeding in the live-bearing *Gambusia* and the guppy which showed no difference in number of broods produced, or maturation attained, under different light conditions (Dildine, 1936; Medlen, 1951); but one report indicates a reduction of the intervals between successive broods in the guppy as a result of continuous light (Scrimshaw, 1944). The sexual development of *Astyanax mexicanus* was found to be delayed when these fish were placed in continuous darkness (Rasquin, 1949; Rasquin and Rosenbloom, 1954). Control of light was essential for successful spawning in zebra fish (Legault, 1958), and short days inhibited egg laying while long days advanced the reproductive season in the Japanese killifish, *Oryzias latipes* (Yoshioka, 1962). Continuous light during the winter resulted in gonadal weight increase in *Ameiurus nebulosus* (Buser and Blanc, 1949) and in several other species (Buser and Lahaye, 1953).

More detailed information exists for cyprinid fish. S. A. Matthews (1939) detected no difference from controls in weight or in microscopic appearance of the testes in *Fundulus* maintained either in darkness in December or in continuous light in spring. Similarly, elaborate treatment with decreasing day length in late summer or increasing light in winter did not produce a difference in microscopic appearance of the testes in the same species, but high temperature in spring caused active spermatogenesis (Burger, 1939a,b, 1940). Day length was found to be of influence on spermatogenesis in *Phoxinus* (Bullough, 1939). The presence of an inherent rhythm of gonadal maturation is indicated in experiments by the same author who noted merely a delay of maturation when the fish were exposed to short days in spring (Bullough, 1940), perhaps comparable to the ineffectiveness of environmental factors in inhibiting final maturation during the last phases in the stickleback (Baggerman, 1957). Barrington and Matty (1955) found that increasing artificial day length during fall activated the testes of *Phoxinus*. Working with the bitterling, *Rhodeus amarus,* Verhoeven and van Oordt (1955) found a lengthening of the ovipositor owing to long-day treatment 6 to 3 months before the spawning season. A different effect of the same treatment, depending on the time during the seasonal cycle, was claimed by the authors when the response failed to occur later in the season. However, Harrington (1959a) believed that postspawning refractoriness could have been the cause. Harrington's early work on the bridled shiner, *Notropis bifrenatus* (Harrington, 1950), showed that spawning was advanced 3 months by treatment with 17 hr of light per day during January and February. Harrington's earlier and later studies on this species (Harrington, 1947, 1957) demonstrated that the annual breeding cycle consisted of a postspawning refractory period from July to November, leading into a prespawning period during which treatment by long days could advance the following spawning period by up to·5 months. Similar conditions seem to prevail in a centrarchid, *Enneacanthus*, which could be made to spawn in November, 45 days after treatment with 15-hr day length was begun (Harrington, 1956).

By subjecting *Fundulus confluentus*, a species with a prolonged breeding season to four combinations of long and short days and low and high temperature, Harrington (1959b) investigated the effects of these factors on oogenesis. High temperatures accelerated later maturation stages but retarded earlier ones, whereas low temperatures tended to have the opposite effect. The retardation of early stages by high temperature seemed to be reinforced by long days.

The interplay of temperature and photofraction on gametogenesis and reproductive behavior in *Cymatogaster aggregata*, one of the live-bearing sea perches (Embiotocidae), was investigated by Wiebe (1968). Increasing or long photoperiods from late winter to early summer induce spermatogenesis and reproductive behavior, especially at warm temperature, whereas low temperature and short photoperiod, normally occurring in winter, enhance restitution of gonadal tissues and spermatogonia growth.

Manipulation of day length was successfully employed in hatchery rearing of brook trout as early as 1936 in the eastern United States. *Salvelinus fontinalis*, as many other salmonids, normally spawns from October to December when the days are shortening. By first increasing and later decreasing the light duration much in advance of the natural photofraction change, the spawning season could be advanced by about 4 months to August (Hoover, 1937; Hoover and Hubbard, 1937). Similar results were reported by Hazard and Eddy (1951) who also found that decreasing the day length by 1 hr each week for 9 weeks in August–September alone caused a spawning advancement of about 1 month. The advancing effect seemed greater when the addition and subtraction of light was begun earlier. By adding 4 hr of light to the normal day length between September–December, a delay in spawning resulted. Also working with the brook trout, Allison (1951) reported similar results. The method of manipulating day length, originally introduced by Hoover (1937), seems to have found widespread application for controlling spawning time (Corson, 1955). Henderson (1963) observed that an acceleration of the seasonal change in photofraction, though effective with adult fish, did not advance spawning in fish maturing for the first time. A responsiveness to photoperiod was also noted in other salmonids. For example, spawning could be delayed by lengthening the light duration and advanced by shortened days in *Oncorhynchus nerka* (Combs *et al.*, 1959); and Hoar (1953) assumed that the changing photoperiod, by acting on the pituitary, would initiate the transformation to the smolt stage in salmon and that a similar mechanism would bring about prespawning and spawning behavior in adult salmon.

Such a decisive effect of day length cannot always be found. Hubbs and Strawn (1957) stated that the reproductive rate in a darter, *Etheostoma lepidum*, was controlled by temperature and the condition of the fish and that duration of the light period was unimportant. In most temperate zone species, interaction of a day length–temperature complex with an internal physiological rhythm seems to be involved. This has been recognized in most of the more extensive studies. Another recent example is provided by the work of Ahsan (1966) on the spermatogenetic cycle of the lake chub, *Conesius plumbeus*. Temperature was found to be a major factor controlling spermatogenesis, but a clear effect of day length was present at low temperatures during the later part of the cycle, and an endogenous rhythm seemed partially responsible for the timing of testicular changes.

A strong effect of temperature is indicated in the report by Lagler and Hubbs (1943), who noted a second spawning season of the mud pickerel in mid-November after abnormally warm weather had prevailed during the preceding October. This species usually has a brief spawning period in spring. John (1963) described two spawning periods within one year for the speckled dace in Arizona. The major peak of this "bimodal rhythm" in April–May is correlated with flooding from melting snow. A lesser peak in spawning, in late July–August, coincides with flooding by rainwater. If the late summer freshets occurred before late July, no spawning was observed.

A clear effect of seasonally changing day length was demonstrated by the experiments of Turner (1957) with a poecilid, *Jenynsia lineata*. This live-bearer is limited to southern South America where spawning occurs during spring and summer, October to January–February. After having been transported to the northern hemisphere in September, already gravid females initially produced broods at a time corresponding to the summer of their native habitat. Under the influence of a light–dark cycle, shifted by 180° from that in the southern hemisphere, they went through another breeding period corresponding to the northern summer, presumably stimulated by the increasing day length.

The frequently inferred role of an internal or endogenous breeding rhythm seems to be in need of conceptual clarification. It is not always clear if simply the physiological changes, mainly endocrine–gonadal interactions, are meant or if it ought to be understood in functional terms as a feature of living organization similar to the useful oscillator analogy of circadian systems. Meske *et al.* (1967) reported that carp could be made to spawn every 5 months by pituitary administration when they were maintained in warm running water in aquaria. How this observation, however, could be interpreted as showing a "lack of an endogenous sexual rhythm" is not clear. It seems to demonstrate clearly an internal reproductive cycle which can be completed within 5 months when external conditions are very favorable and when spawning is induced artifically. Tropical characins in northeast Brazil have only one breeding period but can be induced to spawn three times a year by pituitary application in pond culture (Fontenele *et al.*, 1946). There are examples to show that alternation of breeding and nonbreeding periods continues under favorable conditions in the absence of changing day length and temperature with a period which is less than a full year. Baggerman (1957) found a period of 200 days for the stickleback in continuously long days and at high temperature. Gonadal maturation in minnows was merely delayed by a few months when these fish were exposed to short days from late winter until summer (Bullough, 1940). Another interesting observation is that of Henderson (1963) who reported that acceleration of the seasonal day length change, usually advancing the spawning time in trout, was ineffective in young fish which had not spawned previously.

A further example of gonadal maturation in previously immature fish, proceeding under conditions of approximately constant temperature and day length, must be mentioned here. Many centrarchids are known to go through an extended annual breeding period which consists of several successive spawning cycles (Breder, 1936). In a study of the spawning behavior in natural populations of the green sunfish, *Lepomis cyanellus*, in southern Wisconsin, Hunter (1963) could document the occurrence of 8–10 nesting cycles between late May and early August. When the fish of one pond were breeding as a single colony, the successive nesting periods were synchronized and showed a periodicity with intervals of 8–9 days (Hunter, 1963). These fish normally reach maturity within one year, and small yearling males participated in spawning later in the breeding season than the older males. Twenty immature fish were maintained in the laboratory from early September 1959 in a regimen of artificial light with a 12-hr-long light period. In October, they were transported to Belém, Brazil, of 1°30' southern latitude. At this equatorial location they lived in a 100-meter² shallow pond under continuously high temperature (27–29°C) and unchanging 12-hr day length. The fish grew rapidly, and samples taken at regular intervals indicated that testicular growth began in late December. By February 1960, sampling had reduced their numbers to four males and five females. Nestbuilding, spawning, and hatching of young began in late February and continued into early August. Two further spawnings were observed in mid-September. The data are shown in Fig. 6 in a summarized form, where the incidence of nest construction by initially two, and later by all four, males indicates periodic intervals of 7–11 days, resembling the

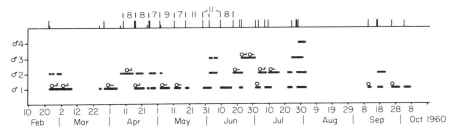

Fig. 6. Spawning of green sunfish, *Lepomis cyanellus*, at the equator. Nest-building and occupancy are indicated for the individual males by horizontal bars, successful spawning with eggs present by a circle above the bar. Prolonged presence of eggs is shown as horizontal dash after the circle and successful hatching by the upward dash. A synchronized pattern of nest construction developed in April 1960 and the upper row shows the intervals between the times of new nest occupancy.

periodicity of intermittent nesting demonstrated for the parent population in Wisconsin (Hunter, 1963). Since immature fish were used in
our experiment at the equator, comparison with the results obtained by
Harrington (1956) on adult *Enneacanthus*, another centrarchid, does
not seem possible. It appears that gonadal maturation in immature green
sunfish does not require stimulation by changing day length as long as
temperature is high. However, once they have passed through at least
one phase of postspawning refractoriness, they require this environmental
factor. A comparable condition seems to exist in the trout (Henderson,
1963). Our experiment also indicates that 12 hr of daylight are sufficient
for gonadal maturation to proceed in previously immature green sunfish. It is also possible that day length plays only a synchronizing role
and that, in their natural habitat in the temperate zone, days of less
than 12-hr light prevent renewed breeding in fall at still high temperatures, whereas, spawning activity during the following year cannot begin
before sufficiently high temperatures are present.

There is substantial evidence for different spawning seasons in certain
races of the same species which makes it difficult to evaluate the effect
of changing day length and temperature in timing the reproductive cycle.
Einsele (1965) reported that some species and races of coregonid fishes
spawn in autumn, most of the others spawn during winter, and a few
spawn even in the spring. Hempel (1965), reviewing the ecology of
herring, *Clupea harengus*, in the North Sea and in the Baltic Sea,
described different tribes, or races, which appear to have distinctly different times of spawning. For example, the Shetland Islands' population
begins spawning in July and continues through August–September along
the Scottish coast. Off the North Sea coast of England and in the Dog-

gerbank region, the reproductive period lasts through September–October. The Downs herring near the English Channel spawn in November–December. Herring in the Baltic spawn mainly in spring and summer. It appears from Hempel's report that each stock of herring has a specific spawning area and also a specific spawning season.

## 3. THE PROBLEM OF THE TROPICS

It is not surprising to find an effect of changing photoperiod on the timing of breeding rhythms in animals living in temperate zones where the annual cycle of changing day length is of substantial magnitude, thereby providing a reliable indication of progressing season. In contrast, temperature fluctuations are often of great variability except, perhaps, in the ocean, large lakes, and rivers. In the tropics climatic conditions do not vary as much as they do at higher latitudes. Most areas near the equator are continuously humid and, concerning the aquatic environment, annual temperature fluctuations are exceedingly small. The time of the daily light period from sunrise to sunset remains practically constant at 12 hr throughout the year. Even at 10° northern or southern latitude the total amplitude of annual fluctuation in day length is less than 1 hr. In spite of a certain apparent uniformity in climatic conditions, there are many tropical areas where rather distinct seasons are discernible because of an alternation of a rainy with a drier period. To anyone having spent several years in such regions, the decisive difference between these periodically changing seasons has been obvious; and the apparently coinciding synchronization of reproductive periods of many animal and plant species seems impressive. In the Amazon Basin the conventional terms "summer" and "winter" are applied to the drier and wetter season, respectively. Although the so-called dry period in regions of tropical rain forest is never really dry, the difference in precipitation between the two seasons is considerable. Occurring predominantly at the beginning of the wet season, heavy rains cause a profound change in the aquatic environment. One instructive example is the island of Marajó in the Amazon estuary situated just south of the equator. From January until May, the interior of this largest estuarine island is covered with water of 2-meter average depth and travel on the island during this period is possible only by boat. When the rains subside after May, the island slowly begins to dry up, partially by limited runoff through a few small rivers but mostly by evaporation. The water, including all aquatic life, retreats into a few shallow lakes and interconnecting rivers. Toward the end of the dry season, the largest lake on the island resembles a highly concentrated brine solution (Egler and Schwassmann, 1962). The reproductive period of most fish species in this area begins after the first heavy rains cause substantial flooding. Conditions in other areas nearby, especially along the river courses, show similar changes under the influence of the seasonal rainfall, but the observed changes are less extreme than on Marajó. The mixing zone of river with seawater experiences an annual displacement of about 200 km. In the middle and upper Amazon, the enormous rainfall causes a large annual difference in discharge volume and water level, bringing about flooding of adjacent areas.

In view of these incisive annual changes which are known to occur also in other regions of the tropics, for example, the flooding caused by the monsoon rains in Asia, the assertion by Bünning (1967) that information about seasonal changes would be entirely superfluous in tropical areas which are continuously moist must be considered with caution. Several reports indicate that spawning activity in many species of carp coincides with seasonal flooding by periodic rains. The observations by Khanna (1958), extending over 8 years, are quite impressive and indicate that failure to spawn in 3 out of the 8 years was obviously related to insufficient flooding. Other studies confirming the coincidence of flooding and spawning in carp are those of David (1959), also in India, and of Tang (1963) in Taiwan. It is necessary, however, to realize that these observations about a precise onset of breeding activity under favorable environmental conditions leave us in ignorance about the existence of possible zeitgeber which could account for anticipatory timing of the annual sexual cycle and which could result in attainment of a gonadal phase of readiness so that spawning can be accomplished almost immediately when the proper environmental situation is realized. There is, of course, the possibility that long-term phasing by external factors is absent in tropical species and that the internal physiological rhythm proceeds on its own and maintains a prespawning condition for a rather long time. The few existing reports concerning tropical fish would not contradict this possibility. A study by Lake (1967) suggests the presence of a soil substance which, when leached out by flooding, could initiate spawning.

The success of the method to induce spawning in fish by pituitary treatment possibly results from a prolonged period of prespawning readiness in tropical species. The method was developed in Brazil in the early 1930's and was applied to fish in pond culture where natural reproduction appeared difficult to achieve (Cardoso, 1934; von Ihering and Wright, 1935; de Menezes, 1944; Fontenele, 1955). It was practiced extensively in Ceará, a northeastern state of Brazil, which is known for occasional years of severe drought conditions resulting from failure of annual rains. Von Ihering and Wright (1935) reported that the characins used in these pond cultures showed well-developed gonads several months before conditions in nature were favorable for reproduction. Spawning in nature is supposed to occur on the day following a good rain. Fontenele *et al.* (1946) were able to obtain three successive spawnings with pituitary injection, in February, May, and August, in a species of *Prochilodus.* The widespread utilization of pituitary hormones in fish culture was reviewed by Atz and Pickford (1959).

Insufficient knowledge concerning spawning habits of tropical fish in their natural habitat renders difficult a discussion of timing of breeding cycles in the tropics. It would be important to know whether two annual breeding periods existed in those regions where two wet seasons occur in one year. For example, Miller (1959a) finds that one of the bird species investigated by him in Colombia showed two annual breeding cycles correlated in time with the late parts of each of the two wetter periods. This species, *Zonotrichia capensis,* was also found to have a latent light-response mechanism in its testicular maturation cycle and showed no postbreeding refractoriness (Miller, 1959b), a condition which may also exist in tropical fish.

The extensive Amazon drainage basin provides unique features of diverse environmental conditions. Although most areas have a typical sequence of a very wet and a drier season, the upper Rio Negro and the upper Solimões areas experience a more uniform distribution of annual precipitation. Furthermore, the upper Rio Branco region has a rainy and a drier season which are occurring at times opposite to those elsewhere in Amazonia. There must be many species of fish which are common to these areas, and knowledge of the time of reproduction would be valuable. There also exist several species which have a rather wide geographical distribution, north as well as south of the equator, for example, some species of *Astyanax* and some gymnotids. Information concerning the time of spawning for extreme northern and southern populations could indicate whether photoperiodism is involved and, if so, how close to the equator it may be effective.

There are many fish which do not confine their breeding activity to a short time but which go through many successive spawning periods, extending perhaps over the greater part of the year. A good example are the cichlids, in which this extended reproductive period might be considered to be an adaptation to an environment where precise timing of seasonal breeding is difficult. On the other hand, we also find a long reproductive period with many successive spawning sessions in temperate zone species such as most centrarchids (Breder, 1936) or the California grunion, *Leuresthes tenius* (Walker, 1952). Our observations indicate that reproduction in several cichlids, mainly *Cichlasoma severum* and *C. festivum*, takes place between December and April and that spawning occurs predominantly in shallow areas near lake shores which are filled with water only during the rainy season. Another species, *Astronotus ocellatus*, reproduces from October to April in pond culture, according to an 8-year study of Fontenele (1954), although occasional spawnings were observed in the remaining period. The osteoglossid, *Arapaima gigas*, an important food fish, spawns during the rainy season between late December and May (Fontenele, 1953).

In contrast to these reports, Aronson (1951a, 1957) observed that the African cichlid, *Tilapia macrocephala*, has a major peak of spawning activity in March and a lesser one in October and that a decline in breeding activity, sometimes correlated with gonadal regression, occurred during periods of heavy rains.

Entrainment of reproductive cycles by periodic environmental changes is not only advantageous for spawning to occur during external conditions favorable to the offspring but also would assure that both sexes attain maturity at the same time. This latter factor would seem more important for species which have a brief spawning period and might not be as crucial for those with prolonged breeding activity like the cichlids. Mutual synchronization seems to occur in fish with elaborate mating and nestbuilding behavior. Such mutual synchronization in the males of a breeding colony has been demonstrated for the green sunfish (Hunter, 1963).

### 4. Conclusions

Most species of fishes show an annual periodicity in their reproductive behavior. An internal physiological rhythm of gonadal maturation, involving predominantly pituitary–gonadal interactions, is adjusted in time by the action of environmental variables. This ensures that breeding will occur at a time when environmental conditions are most favorable for survival of the offspring. The duration of daylight, changing systematically throughout the year in higher latitudes, is considered to be the most reliable external factor for long-term timing of breeding cycles, and photoperiodic control has been demonstrated for the annual rhythm of reproduction and related events in several fish of the temperate zone. Temperature has also been shown to affect the rhythm of gonadal maturation and spawning; aside from directly influencing physiological processes, there seems to be some interaction with day length. In some instances, the breeding cycle was found to proceed when day length and temperature were kept uniform. Under long days at high temperature, the cycle was completed in less than one year in the stickleback, whereas short days delayed gonadal maturation in *Phoxinus*. The experiments by van den Eeckhoudt (1946) concerning the effect of short light pulses on gonadal maturation in the stickleback suggest that in fish the circadian system is also involved in the mechanism of photoperiodic control. Although the external variable under natural conditions is an uninterrupted light period, the effective agent in photoperiodism is only that light which acts during a sensitive phase. The time when this phase occurs in the daily cycle is determined by the circadian rhythm (see Section II, C). Photoperiodic control of annual breeding may not be the only mechanism by which fish and other organisms orient to the appropriate season; however, its effectiveness and its mode of action could be experimentally demonstrated. Concerning sun-compass orientation of centrarchid fish, changing photoperiod was found to be effective in adjusting the daily compensation of the sun's movement to the different seasons (Schwassmann and Braemer, 1961).

Control by means of changing day length could not possibly account for many examples of strictly seasonal reproduction in regions near the equator where the amplitude of annual photoperiodic change is very small. There are laboratory results which point toward the possible role in photoperiodism of the slightly changing duration of twilight in the tropics (Wever, 1967; Aschoff and Wever, 1965). Reproductive activity in many tropical fishes often coincides with extensive flooding during the rainy season. Since spawning is frequently reported to occur almost immediately after heavy rains, this factor cannot be held responsible for long-term anticipatory timing of the sexual cycle. It could be possible that the internal physiological rhythm of gonadal maturation remains arrested at a prespawning phase for a considerable length of time and that the consummatory breeding behavior is triggered by the same environmental situation to which the reproductive rhythm is ultimately timed. The relative ease of inducing spawning by pituitary administration in tropical fish seems to support this hypothesis. It would, however, be difficult to account for other seasonal events such as the prespawning migration of several species in the Amazon as well as migratory and breeding behavior of oceanic tropical species. At present, the most urgent need seems to be for information concerning the natural times of reproduction of tropical fish in relation to local meteorological conditions.

## B. Lunar and Tidal Rhythmicity

Many marine organisms show a rather precise correlation in the timing of certain developmental processes with the lunar cycle or with the incidence of spring and neap tides which are determined by the phases of the moon. The swarming of the Palolo worm (Caspers, 1961) and the spectacular periodic breeding activity of the California grunion (Thompson, 1919; Clark, 1925; Walker, 1949, 1952) are some of the better known examples. In addition to the semilunar or lunar periodicities, several intertidal species are known for synchronization of behavioral and physiological functions with the approximately twice daily fluctuations in water level, the tides.

Several studies have been reported which demonstrate that the overt tidal rhythm persists for a few periods in constant laboratory conditions; some rhythms could be recorded for a time sufficient to observe a drifting out of phase with the tidal schedule. One such example is the work of Enright (1963) on locomotor activity of an intertidal amphipod; another is that of Naylor (1958) and B. G Williams and Naylor (1967) on *Carcinus maenas*. Rao (1954) examined the tidal-periodic filtering rate in a mussel *Mytilus*, which persisted for several weeks in the laboratory without losing its phase relation with the tidal schedule. Many

Fig. 7. Grunion on the Scripps Beach, La Jolla, California (photograph by author).

other apparently precise rhythms of lunar and tidal periodicity were shown in a large variety of organisms and were reviewed by Webb and Brown (1959). These periodicities, however, are mostly unlike those of overt rhythms and could be detected only after statistical treatment of the data; they will not be discussed here.

Concerning tidal rhythms of activity in fish, G. C. Williams (1957), who studied the behavior of tide pool animals, noted that many of these species show a strict correlation with the periodic flooding of the tide pools, especially the opaleye, *Girella nigricans*, and the wooly sculpin, *Clinocottus analis*. A tidal rhythm also seems indicated for *Bathygobius soporator* according to Aronson's study (1951b). Fishelson (1963) reported similar tide-related behavior in a blenny. In a laboratory study on another blenny, *Blennius pholis*, Gibson (1965, 1967) found that a tidal rhythm of activity persisted for up to 5 days and that the period increased under conditions of continuous light, or darkness, in the laboratory. Other species which Gibson tested failed to exhibit such persistent overt rhythm in his experiments.

The breeding habits and the precise timing of successive semilunar spawning runs of the grunion, *Leuresthes tenuis*, are a spectacular phenomenon on the beaches of Southern California. The grunion is a member of the silverside family and has an extended breeding period from March to August or September with successive spawning runs spaced 14–16 days apart. The spawning runs occur only at night when the fish come out of the water on the beaches to bury and fertilize their eggs in the sand (Figs. 7–9). The extensive work of Walker (1949) has resulted in valuable information about this peculiar behavior and about the predictability of the spawning runs. Highest tides occur with full and new moon, and the tides at night are higher than those during the day in spring and summer along the California coast. The grunion spawn during a receding tide series, when the high water levels are getting less and less each night, and from 1 to 4 nights are utilized for spawning. The fertilized eggs will not be washed out of the sand until 2 weeks later when a new series of high tides occurs. Upon agitation at this time the larvae hatch immediately; if they are not washed out of the sand during a high tide, the eggs remain viable for another semilunar cycle. By evaluating extensive records from several years, Walker was able to reach certain conclusions concerning the timing pattern of the spawning runs. The incidence of spawning runs does not appear to be directly caused by the tides, for when in certain years the sequence of moon phases exhibited alternating long and short intervals, the series of grunion runs showed a similar long and short interval pattern but with a one-period delay. The midpoints of spawning run series were

**Fig. 8.** Grunion on the Scripps Beach, La Jolla, California (photograph by author).

correlated not to the immediately preceding full or new moon but to the one prior to the preceding moon phase. The time of spawning seems to be determined by the interaction of a physiological rhythm of gonadal maturation, showing a period of about 2 weeks, with some factor related to the second preceding full or new moon. The external factors responsible for the precise timing of the runs are not known. Another closely related species, *Hubbsiella sardina*, inhabits the northern part of the Gulf of California and is reported to have similar spawning habits. According to Walker (1952), this species runs on the same dates as *Leuresthes tenuis* but is also known to spawn during daylight hours and has an earlier spawning season. Walker cited reports concerning similar semilunar spawning periodicities in *Galaxias attenuatus*, a teleost from New Zealand (Hefford, 1931), and *Enchelyopus cimbrus*, a member of the gadid family from the Canadian Atlantic coast (Battle, 1930).

Behavioral and physiological rhythms of tidal periodicity are quite common in fish occupying the intertidal zone. The behavior and movements shown by populations of *Anableps microlepis*, one of the three species of four-eyed fish, provide another example (Schwassmann, 1967).

**Fig. 9.** Grunion on the Scripps Beach, La Jolla, California (photograph by author).

The tidal movements near the Amazon estuary occur in a rather regular pattern with two high and two low tides a day on which a semilunar monthly amplitude change is superimposed involving water level differences of more than 4 meters at full and new moon and intervening low amplitudes of a little more than 1 meter. *Anableps microlepis* shows a pronounced tidal rhythm of moving up on the beaches with every rising tide. At a location which supported a large population of these fish, extensive brackish water lagoons were regularly flooded by the high spring tides in March. During these high tides mature *Anableps* entered the lagoons and left again before the water level receded. Reproductive activity was frequently observed while the fish were in these sheltered waters. The females of this live-bearing cyprinodont must also give birth to their young inside the lagoons since large schools of young were present. The strong urge of the older fish to enter the lagoons during rising water level, day or night, literally stranding themselves on the sand with every incoming wave, was most spectacular, and it closely resembles the behavior of the grunion during their spawning runs.

A study by Lang (1967) on the guppy gives evidence for periodic fluctuations of lunar periodicity in the sensitivity to light of different spectral ranges. By utilizing the dorsal light response as a measure of sensitivity to light, these fish (*Lebistes reticulatus*) were found to be most sensitive to yellow light at full moon and less sensitive during new moon, whereas violet and red light showed the reversed relation.

Several instances of overt tidal and lunar rhythmic behavior, persisting in constant conditions, suggest the participation of some endogenous component, especially when a free-running period can be demonstrated. However, the paucity of data demonstrating an endogenous nature of tidal rhythms is paralleled by a lack of knowledge about the external factors which adjust these rhythms in nature. An interesting observation is that of Hauenschild (1960) who showed that the lunar breeding rhythm in a polychaete *Platynereis*, persisting for more than two periods in the laboratory, could be influenced by very dim nighttime illumination of moonlight intensity. Entrainment of the hatching rhythm in *Clunio marinus*, a tidal chironomid, was demonstrated by Neumann (1966). Bünning and Müller (1961) reported that the semilunar rhythm of egg liberation in a brown alga was sensitive to phase control by very dim light at night, even when this "moonlight" was acting for one night only. The 16-day period of this rhythm in laboratory conditions could be reduced by employing shortened days of 23.5 hr. The same authors also suggested that lunar and semilunar rhythms could possibly be the result of the interaction of a circadian periodicity with an endogenous tidal component in such a way that the difference in period of the two would cause a reinforcement by phase coincidence at 14–15- or 29-day intervals. A "multiple clock" hypothesis had already been suggested by Naylor (1960) and was supported by the work of Blume et al. (1962) on the same species, *Carcinus maenas*. Webb and Brown (1965) also inferred two interacting rhythms from their studies on locomotor activity in *Uca*. Neumann (1969) discussed data which depend on a combination of endogenous rhythms of differing periods.

Some evidence is available for the time sense in the seaward orientation of an amphipod, *Talitrus saltator* (Papi, 1960). By utilizing the sun's position during daytime and the moon at night, often involving opposite positions of the two stellar bodies, these animals appear to switch from a circadian clock to a lunar one at dusk. Both clocks must be supposed to run continuously in order to allow for the changing phase difference between the two cycles.

In view of the limited knowledge concerning tidal rhythms, the possibility of an underlying single rhythm should also be considered. One could assume that the present tidal rhythms are derived from historically older circadian rhythms which gradually lost their sensitivity to phasing by light; different factors present in the intertidal environment could then have become effective as zeitgeber. Two amplitude peaks within one period are sometimes known to occur in the circadian pattern (Aschoff, 1957) and could account for the period of the tidal regime which is about one-half that of the circadian period. Environmental factors of the local tidal conditions could modify the pattern so that one or the other peak of the bimodal pattern is enhanced. However, it appears difficult to explain a persistence of the often strongly modified pattern under constant laboratory conditions as was demonstrated by Enright (1963) for a beach amphipod. An interesting observation is that of B. G. Williams and Naylor (1967), who reported that *Carcinus maenas*, when reared from the egg in the laboratory in a light–dark cycle, exhibited a circadian locomotor pattern.

## V. SYNOPSIS AND PROSPECTUS

Biological rhythms of daily, tidal, lunar, and annual periodicity, which are an inherent feature of organismic organization, are recognized as adaptations to our periodically changing environment. Considering the experimental evidence, it is obvious that rhythmic phenomena in many species of fish are in no way different from those known in other organisms regarding the endogenous nature and the control of phase and period by periodic environmental variables. Therefore, certain established generalities resulting from studies on different organisms must also be valid for biological rhythms in fish.

Experimental evidence is available for circadian rhythms in several species of fish; most of it, however, is limited to demonstrating a persistence of overt periodic functions in constant conditions. Concerning annual rhythms, several studies investigated mainly change of day length and temperature for their effect in timing annual reproductive cycles in about 10 teleost species. Rhythms of tidal, semilunar, and lunar periodicity in fish of the intertidal zones are known from a few rather spectacular examples, but with one exception they have not been investigated in the laboratory.

Most progress has been made recently in the field of functional analysis of circadian rhymicity. Circadian organization seems to be the phylogenetically oldest feature and might well be of common origin, whereas the many diverse overt functions could be considered secondary consequences of the circadian system. A major role of circadian **organization** appears to lie in its involvement in the mechanism of photoperiodic control as an adjustment to the temporal order of annual environmental cycles. In photoperiodism, the circadian oscillation makes possible the sensitivity to the length of the daily light period. The ecologically significant effect of photoperiodic control, especially evident from studies on annual breeding in fish, appears to be in adjusting the temporal sequence of a physiological rhythm of gonadal maturation rather than to actually trigger certain specific physiological events. Photoperiodic control cannot account for the timing of reproduction and preceding migratory movements of species living in the tropics, where a coincidence of spawning activity with the onset of the rainy season appears to be a fairly common phenomenon.

Most experimental studies involved animals of temperate zones which may have led to the current emphasis of photoperiodic control mechanisms. Experimental work concerning possible timing mechanisms of breeding cycles in tropical fish seems to be hampered by the scarcity of information about their natural behavior and the times of reproductive activity in natural habitats of different meteorological conditions. In spite of the great progress achieved by laboratory studies of rhythmic phenomena in diverse organisms, including fish, essentially in terms of functional systems analysis, it is this writer's opinion that further achievements will depend on field studies which not only provide the basis for any experimental analysis in the laboratory but also test present generalizations and theories.

## REFERENCES

Ahsan, S. N. (1966). Some effects of temperature and light on the cyclical changes in the spermatogenic activity of the lake chub, *Conesius plumbeus* (Agassiz). *Can. J. Zool.* **44**, 161–171.

Allison, L. N. (1951). Delay of spawning in eastern brook trout by means of artificially prolonged light intervals. *Progr. Fish Culturist* **13**, 111–116.

Andreasson, S. (1969). Locomotory activity patterns of *Cottus poecilopus* Heckel and *C. gobio* L. (Pisces). *Oikos* **20**, 78–94.

Arey, L. B., and Mundt, G. H. (1941). A persistent diurnal rhythm in visual cones. *Anat. Record* **79**, 5–11.

Aronson, L. R. (1951a). Factors influencing the spawning frequency in the female cichlid fish *Tilapia macrocephala*. *Am. Museum Novitates* **1484**, 1–26.

Aronson, L. R. (1951b). Orientation and jumping behavior in the gobiid fish *Bathygobius soporator*. *Am. Museum Novitates* **1486**, 1–22.

Aronson, L. R. (1957). Reproductive and parental behavior. *In* "The Physiology of Fishes" (M. E. Brown, ed.), Vol. 2, pp. 271–304. Academic Press, New York.

Aschoff, J. (1951). Die 24-Stunden-Periodik der Maus unter konstanten Umweltbedingungen. *Naturwissenschaften* **38**, 506–507.

Aschoff, J. (1952). Frequenzänderung der Aktivitätsperiodik bei Mäusen in Dauerdunkel und Dauerlicht. *Arch. Ges. Physiol.* **255**, 197–203.

Aschoff, J. (1954). Zeitgeber der tierischen Tagesperiodik. *Naturwissenschaften* **41**, 49–56.

Aschoff, J. (1957). Aktivitätsmuster der Tagesperiodik. *Naturwissenschaften* **44**, 361–367.

Aschoff, J. (1958). Tierische Periodik unter dem Einfluss von Zeitgebern. *Z. Tierpsychol.* **15**, 1–30.

Aschoff, J. (1959). Periodik licht- und dunkelaktiver Tiere unter konstanten Umgebungsbedingungen. *Arch. Ges. Physiol.* **270**, 9.

Aschoff, J. (1960). Exogenous and endogenous components in circadian rhythms. *Cold Spring Harbor Symp. Quant. Biol.* **25**, 11–28.

Aschoff, J., ed. (1965a). "Circadian Clocks." North-Holland Publ., Amsterdam.

Aschoff, J. (1965b). Response curves in circadian periodicity. *In* "Circadian Clocks" (J. Aschoff, ed.), pp. 95–111. North-Holland Publ., Amsterdam.

Aschoff, J. (1965c). The phase-angle difference in circadian periodicity. *In* "Circadian Clocks" (J. Aschoff, ed.), pp. 262–276. North Holland Publ., Amsterdam.

Aschoff, J., and Wever, R. (1962). Aktivitätsmenge und $\alpha$: $\rho$-Verhältnis als Messgrössen der Tagesperiodik. *Z. Vergleich. Physiol.* **46**, 88–101.

Aschoff, J., and Wever, R. (1965). Circadian rhythms of finches in light-dark cycles with interposed twilights. *Comp. Biochem. Physiol.* **16**, 507–514.

Atz, J. W. (1957). The relation of the pituitary to reproduction in fishes. *In* "The Physiology of the Pituitary Gland of Fishes" (G. E. Pickford and J. W. Atz, eds.), pp. 178–269. N. Y. Zool. Soc., New York.

Atz, J. W., and Pickford, G. E. (1959). The use of pituitary hormones in fish culture. *Endeavour* **18**, 125–129.

Atz, J. W., and Pickford, G. E. (1964). The pituitary gland and its relation to the reproduction of fishes in nature and in captivity. An annotated bibliography for the years 1956–1963. *FAO Fish. Biol. Tech. Paper* **37**, 1–61.

Baggerman, B. (1957). An experimental study of the timing of breeding and migration in the three-spined stickleback (*Gasterosteus aculeatus* L.). *Arch. Neerl. Zool.* **12**, 105–317.

Balls, R. (1951). Environmental changes in herring behavior. A theory of light avoidance, as suggested by echo-sounding observation in the North Sea. *J. Conseil, Conseil Perm. Intern. Exploration Mer.* **17**, 274–298.

Barlow, G. W. (1958). Daily movements of the desert pupfish *Cyprinodon macularis*, in shore pools of the Salton Sea, California. *Ecology* **39**, 580–587.

Barrington, E. J. W., and Matty, A. J. (1955). The identification of thyrotrophin-secreting cells in the pituitary gland of the minnow (*Phoxinus phoxinus*). *Quart. J. Microscop. Sci.* **96**, 193–201.

Battle, H. I. (1930). Spawning periodicity and embryonic death rate of *Enchelyopus cimbrius* (L.) in Passamaquoddy Bay. *Contrib. Can. Biol. Fisheries* [N.S.] **5**, 363–380.

Beamish, F. W. H., and Mookherjii, P. S. (1964). Respiration of fishes with special emphasis on standard oxygen consumption. I. Influence of weight and temperature on respiration of goldfish, *Carassius auratus* L. *Can. J. Zool.* **42**, 161–175.

Beling, I. (1929). Über das Zeitgedächtnis der Bienen. *Z. Vergleich. Physiol.* **9**, 259–338.

Birukow, G., and Busch, E. (1957). Lichtkompassorientierung beim Wasserläufer *Velia currens* F. (Heteroptera). Orientierungsrhythmik in verschiedenen Lichtbedingungen. *Z. Tierpsychol.* **14**, 184–203.

Blaxter, J. H. S. (1965). The feeding of herring larvae and their ecology in relation to feeding. *Calif. Coop. Oceanic Fish. Invest. Rept.* **10**, 79–88.

Blume, J., Bünning, E., and Müller, D. (1962). Periodenanalyse von Aktivitätsrhythmen bei *Carcinus maenas*. *Biol. Zentr.* **81**, 569–573.

Borthwick, H. A. (1964). Phytochrome action and its time displays. *Am. Naturalist* **98**, 347–355.

Borthwick, H. A., Hendricks, S. B., and Parker, M. W. (1948). Action spectrum for floral initiation of a long-day plant, Wintex barley (*Hordeum vulgare*). *Botan. Gaz.* **110**, 103–118.

Braemer, W. (1960). Versuche zu der im Richtungsgehen der Fische enthaltenen Zeitschätzung. *Verhandl. Deut. Zool. Ges. Muenster* pp. 276–288.

Braemer, W., and Schwassmann, H. O. (1963). Vom Rhythmus der Sonnenorientierung am Äquator (bei Fischen). *Ergeb. Biol.* **26**, 182–201.

Brawn, V. M. (1960). Seasonal and diurnal vertical distribution of herring (*Clupea harengus* L.) in Passamaquoddy Bay. *J. Fisheries Res. Board Can.* **17**, 699–711.

Breder, C. M. (1936). The reproductive habits of the North American sunfishes (Family Centrarchidae). *Zoologica* **21**, 1–48.

Breder, C. M., Jr., and Rosen, D. E. (1966). "Modes of Reproduction in Fishes." *Proc. Zool. Nat. Hist. Press, Garden City, New York.

Bullough, W. S. (1939). A study of the reproductive cycle of the minnow in relation to the environment. *Proc. Zool. Soc. London* **109**, 79–102.

Bullough, W. S. (1940). The effect of the reduction of light in spring on the breeding season of the minnow (*Phoxinus laevis* Linn.). *Proc. Zool. Soc. London* **110**, 149–157.

Bünning, E. (1936). Die endonome Tagesrhythmik als Grundlage der photoperiodischen Reaktion. *Ber. Deut. Botan. Ges.* **54**, 590–607.

Bünning, E. (1958). "Die Physiologische Uhr." Springer, Berlin.

Bünning, E. (1960a). Opening address: Biological clocks. *Cold Spring Harbor Symp. Quant. Biol.* **25**, 1–9.

Bünning, E. (1960b). Circadian rhythms and the time measurement in photoperiodism. *Cold Spring Harbor Symp. Quant. Biol.* **25**, 249–256.

Bünning, E. (1967). "The Physiological Clock." 2nd rev. ed. Springer, Berlin.

Bünning, E., and Müller, D. (1961). Wie messen Organismen lunare Zyklen? *Z. Naturforsch.* **16b**, 391–395.

Bünsow, R. C. (1953). Endogene Tagesrhythmik und Photoperiodismus bei *Kalanchoë blossfeldiana*. *Planta* **42**, 220–252.

Bünsow, R. C. (1960). The circadian rhythm of photoperiodic responsiveness in *Kalanchoë*. *Cold Spring Harbor Symp. Quant. Biol.* **25**, 257–260.

Burger, J. W. (1939a). Some preliminary experiments on the relation of the sexual cycle of *Fundulus heteroclitus* to periods of increased daily illumination. *Bull. Mt. Desert Isl. Biol. Lab.* pp. 39–40.

Burger, J. W. (1939b). Some experiments on the relation of the external environment to the spermatogenic cycle of *Fundulus heteroclitus* (L.). *Biol. Bull.* **77**, 96–103.

Burger, J. W. (1940). Some further experiments on the relation of the external environment to the spermatogenic cycle of *Fundulus heteroclitus*. *Bull. Mt. Desert Isl. Biol. Lab.* pp. 20–21.

Buser, J., and Blanc, M. (1949). Action de la lumière sur l'ostéogénèse réparatrice chez le poisson-chat. *Bull. Soc. Zool. France* **74**, 170–172.

Buser, J., and Lahaye, J. (1953). Etude expérimentale du déterminisme de la régénération des nageoires chez les poissons téléostéens. *Ann. Inst. Oceanog. Monaco* **28**, 1–61.

Cardoso, D. M. (1934). Relação gênito-hipofisária e reprodução nos peixes. *Arquiv. Inst. Biol. (São Paulo)* **5**, 132–136.

Carlander, K. D., and Cleary, R. E. (1949). The daily activity patterns of some freshwater fishes. *Am. Midland Naturalist* **41**, 447–452.

Caspers, H. (1961). Beobachtungen über Lebensraum und Schwärmperiodizität des Palolowurmes *Eunice viridis. Intern. Rev. Ges. Hydrobiol.* **46**, 175–183.

Chovnick, A., ed. (1960). Biological clocks. *Cold Spring Harbor Symp. Quant. Biol.* Vol. 25.

Clark, F. N. (1925). The life history of *Leuresthes tenuis*, an atherine fish with tide controlled spawning habits. *Calif. Div. Fish Game, Fisheries Bull.* **10**, 1–51.

Clarke, G. L. (1930). Changes of phototropic and geotropic signs in *Daphnia* induced by changes of light intensity. *J. Exptl. Biol.* **7**, 109–131.

Clarke, G. L. (1933). Diurnal migration of plankton in the Gulf of Maine and its connection with changes in submarine irradiation. *Biol. Bull.* **65**, 402–436.

Clausen, R. G. (1936). Oxygen consumption in fresh water fishes. *Ecology* **17**, 216–226.

Cloudsley-Thompson, J. L. (1961). "Rhythmic Activity in Animal Physiology and Behaviour." Academic Press, New York.

Combs, B. D., Burrows, R. E., and Bigej, R. G. (1959). The effect of controlled light on the maturation of adult blueback salmon. *Progr. Fish Culturist* **21**, 63–69.

Corson, B. W. (1955). Four years' progress in the use of artificially controlled light to induce early spawning of brook trout. *Progr. Fish Culturist* **17**, 99–102.

Craig-Bennett, A. (1931). The reproductive cycle of the three-spined stickleback, *Gasterosteus aculeatus* L. *Phil. Trans. Roy. Soc. London* **B219**, 197–279.

Cummings, W. C. (1963). Using the Doppler effect to detect movements of captive fish in behavior studies. *Trans. Am. Fisheries Soc.* **92**, 178–180.

David, A. (1959). Observations on some spawning grounds of the Gangetic major carps with a note on carp seed resources in India. *Indian J. Fisheries* **6**, 327–341.

Davis, R. E. (1962). Daily rhythm in the reaction of fish to light. *Science* **137**, 430–432.

Davis, R. E. (1963). Daily "predawn" peak of locomotion in bluegill and largemouth bass. *Animal Behaviour* **12**, 272–283.

Davis, R. E., and Bardach, J. E. (1965). Time-co-ordinated prefeeding activity in fish. *Animal Behaviour* **13**, 154–162.

de Candolle, A. P. (1832). "Physiologie Végétale." Paris, J. Röper, transl., Stuttgart, Tübingen, 1835; (quoted from Bünning, 1960a).

DeCoursey, P. J. (1959). Daily activity rhythms in the flying squirrel, *Glaucomys volans*. Ph.D. Thesis, University of Wisconsin.

DeGroot, S. J., and Schuyf, A. (1967). A new method for recording the swimming activity in flatfishes. *Experientia* **23**, 574–575.

de Menezes, R. S. (1944). Nota sôbre a hipofisação de peixes do Rio Mogi-Guaçu com extrato glicerinado de hipofises de peixes. *Bol. Inst. Animal (São Paulo)* **7**, 36–44.

Dildine, G. C. (1936). The effect of light and temperature on the gonads of *Lebistes. Anat. Record* **67**, Suppl. 1, 61.

Egami, N. (1954). Effect of artificial photoperiodicity on time of oviposition in the fish, *Oryzias latipes*. *Annotationes Zool. Japon.* **27**, 57–62.

Eg'er, W. A., and Schwassmann, H. O. (1962). Limnological studies in the Amazon estuary. *Bol. Museo Paraense E. Goeldi* [N.S.] **1**, 2–25.

Einsele, W. (1965). Problems of fish larvae survival in nature and their rearing of economically important middle European freshwater fishes. *Calif. Coop. Oceanic Fisheries Invest. Rept.* **10**, 24–30.

Enright, J. T. (1963). The tidal rhythm of a sand beach amphipod. *Z. Vergleich. Physiol.* **46**, 276–313.

Enright, J. T. (1965). Synchronization and ranges of entrainment. *In* "Circadian Clocks" (J. Aschoff, ed.), pp. 112–124. North-Holland Publ., Amsterdam.

Enright, J. T., and Hamner, W. M. (1967). Vertical diurnal migration and endogenous rhythmicity. *Science* **157**, 937–941.

Esterley, C. O. (1917). The occurrence of a rhythm in the geotropism of two species of plankton copepods when certain recurring external conditions are absent. *Univ. Calif. (Berkeley), Publ. Zool.* **16**, 393–400.

Esterley, C. O. (1919). Reactions of various plankton animals with reference to their diurnal migrations. *Univ. Calif. (Berkeley), Publ. Zool.* **19**, 1–83.

Fishelson, L. (1963). Observations on the littoral fishes of Israel. I. Behaviour of *Blennius pavo* Risso (Teleostei, Blenniidae). *Israel. J. Zool.* **12**, 67–80.

Fontenele, O. (1953). Hábitos de desova do Pirarucu, *Arapaima gigas* (Cuvier) (Pisces: Isospondyli, Arapaimidae), e evolução de sua larva. *Dept. Nacl. Obras Sêcas, Serv. Piscicult. Publ.* **153**, Ser. I-C, 1–22.

Fontenele, O. (1954). Contribuição para o conhecimento da biologia do Apaiarí, *Astronotus ocellatus* (Spix), (Pisces, Cichlidae), em cativeiro. Aparelho de reprodução, hábitos de desova e prolificidade. *Dept. Nacl. Obras Sêcas, Serv. Piscicult. Publ.* **154**, Ser. I-C, 1–29.

Fontenele, O. (1955). Injecting pituitary (hypophyseal) hormones into fish to induce spawning. *Progr. Fish Culturist* **17**, 71–75.

Fontenele, O., Camacho, E. C., and de Menezes, R. S. (1946). Obtenção de três desovas anuais de curimata comum, *Prochilodus* sp. (Pisces: Characidae, Prochilodinae), pelo método de hipofisção (Nota previa). *Bol. Museu Nacl. (Rio de Janeiro), Zool.* **53**, 1–9.

Gamulin, T., and Hure, J. (1956). Spawning of the sardine at a definite time of day. *Nature* **177**, 193–194.

Garner, W. W., and Allard, H. A. (1920). Effect of relative length of day and night and other factors of the environment on growth and reproduction in plants. *J. Agr. Res.* **18**, 553–606.

Gibson, R. N. (1965). Rhythmic activity in littoral fish. *Nature* **207**, 544–555.

Gibson, R. N. (1967). Experiments on the tidal rhythm of *Blennius pholis*. *J. Marine Biol. Assoc. U. K.* **47**, 97–111.

Gosline, W. A., and Brock, V. E. (1960). "Handbook of Hawaiian Fishes." Univ. of Hawaii Press, Honolulu, Hawaii.

Halberg, F., Halberg, E., Barnum, C. P., and Bittner, J. J. (1959). Physiologic 24-hour periodicity in human beings and mice, the lighting regimen and daily routine. *In* "Photoperiodism and Related Phenomena in Plants and Animals," Publ. No. 55, pp. 803–878. Am. Assoc. Advance. Sci., Washington, D. C.

Hamner, K. C., and Takimoto, A. (1964). Circadian rhythms and plant photoperiodism. *Am. Naturalist* **98**, 295–322.

Hamner, W. M. (1963). Diurnal rhythm and photoperiodism in testicular recrudescence of the house finch. *Science* **142**, 1294–1295.

Hamner, W. M. (1964). Circadian control of photoperiodism in the house finch demonstrated by interrupted-night experiments. *Nature* **203**, 1400–1401.

Hamner, W. M. (1965). Avian photoperiodic response-rhythms: Evidence and inference. In "Circadian Clocks" (J. Aschoff, ed.), pp. 379–384. North-Holland Publ., Amsterdam.

Harder, W., and Hempel, G. (1954). Studien zur Tagesperiodik der Aktivität von Fischen. I. Versuche an Plattfischen. *Kurze Mitt. Inst. Fischerei Biol. Univ. Hamburg* **5**, 22–31.

Harker, J. E. (1964). "The Physiology of Diurnal Rhythms." Cambridge Univ. Press, London and New York.

Harrington, R. W. (1947). The breeding behavior of the bridled shiner, *Notropis bifrenatus* (Cope). *Copeia* pp. 186–192.

Harrington, R. W. (1950). Preseasonal breeding by the bridled shiner *Notropis bifrenatus*, induced under light-temperature control. *Copeia* pp. 304–311.

Harrington, R. W. (1956). An experiment on the effects of contrasting daily photoperiods on gametogenesis and reproduction in the centrarchid fish, *Enneacanthus obesus* (Girard). *J. Exptl. Zool.* **131**, 203–223.

Harrington, R. W. (1957). Sexual photoperiodicity of the cyprinid fish, *Notropis bifrenatus* (Cope), in relation to the phase of its annual reproductive cycle. *J. Exptl. Zool.* **135**, 1–47.

Harrington, R. W. (1959a). Photoperiodism in fishes in relation to the annual sexual cycle. In "Photoperiodism and Related Phenomena in Plants and Animals," Publ. No. 55, pp. 651–667. Am. Assoc. Advance. Sci., Washington, D. C.

Harrington, R. W. (1959b). Effects of four combinations of temperature and daylength on the ovogenetic cycle of a low-latitude fish, *Fundulus confluentus* Gooden and Bean. *Zoologica* **44**, 149–168.

Harris, J. E. (1963). The role of endogenous rhythms in vertical migration. *J. Marine Biol. Assoc. U. K.* **43**, 153–166.

Hart, J. L. (1931). On the daily movements of the coregonine fishes. *Can. Field Naturalist* **45**, 8–9.

Hasler, A. D. (1966). "Underwater Guideposts." Univ. of Wisconsin Press, Madison, Wisconsin.

Hasler, A. D., and Villemonte, J. R. (1953). Observations on the daily movements of fishes. *Science* **118**, 321–322.

Hauenschild, C. (1960). Lunar periodicity. *Cold Spring Harbor Symp. Quant. Biol.* **25**, 491–498.

Hazard, T. P., and Eddy, R. E. (1951). Modification of the sexual cycle in brook trout (*Salvelinus fontinalis*) by control of light. *Trans. Am. Fisheries Soc.* **80**, 158–162.

Hefford, A. E. (1931). Report on fisheries for the year ended 31st March, 1941. *New Zealand Marine Dept.* pp. 1–20.

Hemmingsen, A. M., and Krarup, N. B. (1937). Rhythmic diurnal variations in the oestrus phenomena of the rat and their susceptibility to light and dark. *Kgl. Danske Videnskab. Selskab, Biol. Medd.* **13**, 1–61.

Hempel, G. (1965). On the importance of larval survival for the population dynamics of marine food fish. *Calif. Coop. Oceanic Fish. Invest. Rept.* **10**, 13–23.

Henderson, N. E. (1963). Influence of light and temperature on the reproductive cycle of the eastern brook trout, *Salvelinus fontinalis* (Mitchill). *J. Fisheries Res. Board Can.* **20**, 859–897.

Hendricks, S. B. (1960). Rates of phytochrome as an essential factor determining photoperiodism in plants. *Cold Spring Harbor Symp. Quant. Biol.* **25**, 245–248.

Heusner, A. A., and Enright, J. T. (1966). Long-term activity recording in small aquatic animals. *Science* **154**, 532–533.

Hirata, H. (1957). Diurnal rhythm of the feeding activity of goldfish in winter and early spring. *Bull. Fac. Fisheries, Hokkaido Univ.* **8**, 96–107.

Hirata, H., and Kobayashi, S. (1956). Diurnal rhythm of the feeding activity of goldfish in autumn and early winter. *Bull. Fac. Fisheries, Hokkaido Univ.* **7**, 72–84.

Hoar, W. S. (1942). Diurnal variations in feeding of young salmon and trout. *J. Fisheries Res. Board Can.* **6**, 90–101.

Hoar, W. S. (1953) Control and timing of fish, migration. *Biol. Rev.* **28**, 437–452.

Hoar, W. S. (1955). Reproduction in teleost fish. *Mem. Soc. Endocrinol.* **4**, 5–24.

Hoar, W. S. (1958). The evolution of migratory behavior among juvenile salmon of the genus *Oncorhynchus*. *J. Fisheries Res. Board Can.* **15**, 391–428.

Hoffmann, K. (1953). Experimentelle Änderung des Richtungsfindens beim Star durch Beeinflussung der "inneren" Uhr. *Naturwissenschaften* **40**, 608–609.

Hoffmann, K. (1965). Overt circadian frequencies and circadian rule. In "Circadian Clocks" (J. Aschoff, ed.), pp. 87–94. North-Holland Publ., Amsterdam.

Hoffmann, K. (1968). Synchronisation der circadianen Aktivitätsperiodik von Eidechsen durch Temperaturcyclen verschiedener Amplitude. *Z. Vergleich. Physiol.* **58**, 225–228.

Hoover, E. E. (1937). Experimental modification of the sexual cycle in trout by control of light. *Science* **86**, 425–426.

Hoover, E. E., and Hubbard, H. F. (1937). Modification of the sexual cycle of trout by control of light. *Copeia* pp. 206–210.

Hubbs, C., and Strawn, K. (1957). The effects of light and temperature on the fecundity of the greenthroat darter, *Etheostoma lepidum*. *Ecology* **38**, 596–602.

Hunter, J. R. (1963). The reproductive behavior of the green sunfish, *Lepomis cyanellus*. *Zoologica* **48**, 13–24.

Hunter, J. R. (1966). Procedure for analysis of schooling behavior. *J. Fisheries Res. Board Can.* **23**, 547–562.

John, K. R. (1963). The effect of torrential rains on the reproductive cycle of *Rhinichthys osculus* in the Chiricahua mountains, Arizona. *Copeia* pp. 286–291.

John, K. R., and Haut, M. (1964). Retinomotor cycles and correlated behavior in the teleost *Astyanax mexicanus* (Fillipi). *J. Fisheries Res. Board Can.* **21**, 591–595.

John, K. R., and Kaminester, L. H. (1969). Further studies on retinomotor rhythms in the teleost *Astyanax mexicanus*. *Physiol. Zool.* **42**, 60–70.

Johnson, M. S. (1926). Activity and distribution of certain wild mice in relation to biotic communities. *J. Mammal.* **7**, 245–277.

Johnson, M. S. (1939). Effect of continuous light on periodic spontaneous activity of white-footed mice (*Peromyscus*). *J. Exptl. Zool.* **82**, 315–328.

Jones, F. R. H. (1955). Photo-kinesis in the ammocoete larvae of the brook lamprey. *J. Exptl. Biol.* **32**, 492–503.

Jones, F. R. H. (1956). The behaviour of minnows in relation to light intensity. *J. Exptl. Biol.* **33**, 271–281.

Kalmus, H. (1934). Über die Natur des Zeitgedächtnisses der Bienen. *Z. Vergleich. Physiol.* **20**, 405–419.

Kalmus, H. (1939). Das Aktogram des Flusskrebses. *Z. Vergleich. Physiol.* **25**, 798–802.

Kawamoto, N. Y., and Konishi, J. (1955). Diurnal rhythm in phototaxis of fish. *Rept. Fac. Fisheries, Prefect. Univ. Mie* **2**, 7–17.

Kazanskii, B. N. (1952). Experimental analysis of intermittent spawning in fish. *Zool. Zh.* **31**, 883–896 (in Russian, quoted from Atz, 1957).

Khanna, D. V. (1958). Observations on the spawning of the major carps at a fish farm in the Punjab. *Indian J. Fisheries* **5**, 282–290.

Kleerekoper, H., Taylor, G., and Wilson, R. (1961). Diurnal periodicity in the activity of *Petromyzon marinus* and the effects of chemical stimulation. *Trans. Am. Fisheries Soc.* **90**, 73–78.

Kleinhoonte, A. (1929). Über die durch das Licht regulierten autonomen Bewegungen der *Canavalia*-Blätter. *Arch. Neerl. Sci., Ser. IIIB* **5**, 1–110.

Kleinhoonte, A. (1932). Untersuchungen über die autonomen Bewegungen der Primärblätter von *Canavalia ensiformis*. *Jahrb. Wiss. Botan.* **75**, 679–725.

Kramer, G. (1950). Weitere Analyse der Faktoren, welche die Zugaktivität des gekäfigten Vogels orientieren. *Naturwissenschaften* **37**, 377–378.

Kramer, G. (1951). Eine neue Methode zur Erforschung der Zugorientierung und die bisher damit erzielten Ergebnisse. *Proc. 10th Intern. Ornithol. Congr., Uppsala, 1950* pp. 271–280.

Kramer, G., and von Saint Paul, U. (1950). Stare (*Sturnus vulgaris* L.) lassen sich auf Himmelsrichtungen dressieren. *Naturwissenschaften* **37**, 526–527.

Kruuk, H. (1963). Diurnal periodicity in the activity of the common sole, *Solea vulgaris* Quensel. *Neth. J. Sea Res.* **2**, 1–28.

Lagler, K. F., and Hubbs, C. (1943). Fall spawning of the mud pickerel, *Esox vermiculatus* Lesueur. *Copeia* p. 131.

Lagler, K. F., Bardach, J. E., and Miller, R. R. (1962). "Ichthyology." Wiley, New York.

Lake, J. S. (1967). Rearing experiments with five species of Australian freshwater fishes. I. Inducement to spawning. *Australian J. Marine Freshwater Res.* **18**, 137–153.

Lang, H. J. (1967). Über das Lichtrückenverhalten des Guppy (*Lebistes reticulatus*) in farbigen und farblosen Lichtern. *Z. Vergleich. Physiol.* **56**, 296–340.

Legault, R. (1958). A technique for controlling the time of daily spawning ánd collecting of eggs of the zebra fish *Brachydanio rerio* (Hamilton-Buchanan). *Copeia* pp. 328–330.

Lissmann, H. W. (1961). Ecological studies on gymnotids. Bioelectrogenesis. *Proc. Symp. Comp. Bioelectrogenesis,* pp. 215–226.

Lissmann, H. W., and Schwassmann, H. O. (1965). Activity rhythm of an electric fish, *Gymnorhamphichthys hypostomus. Z. Vergleich. Physiol.* **51**, 153–171.

Lohmann, M. (1967). Zur Bedeutung der lokomotorischen Aktivität in circadianen Systemen. *Z. Vergleich. Physiol.* **55**, 307–332.

McInerney, J. E., and Evans, D. O. (1970). Action spectrum of the photoperiod mechanism controlling sexual maturation in the threespine stickleback, *Gasterosteus aculeatus. J. Fisheries Res. Board Can.* **27**, 749–763.

McNaught, D. C., and Hasler, A. D. (1961). Surface schooling and feeding behavior in the white bass, *Roccus chrysops* (Rafinesque), in Lake Mendota. *Limnol. Oceanog.* **6**, 53–60.

Marcovitch, S. (1924). The migration of the aphidae and the appearance of the sexual forms as affected by the relative length of daily light exposure. *J. Agr. Res.* **27**, 513–522.

Marshall, J. A. (1967). Effect of artificial photoperiodicity on the time of spawning in *Trichopsis vittatus* and *T. pumilis* (Pisces, Belontiidae). *Animal Behaviour* **15**, 510–513.

Matthews, G. V. T. (1955). "Bird Navigation." Cambridge Univ. Press, London and New York.

Matthews, S. A. (1939). The effects of light and temperature on the male sexual cycle in *Fundulus*. *Biol. Bull.* **77**, 92–95.

Medlen, A. B. (1951). Preliminary observations on the effects of temperature and light upon reproduction in *Gambusia affinis*. *Copeia* pp. 148–152.

Meffert, P. (1968). Ultrasonic recorder for locomotor activity studies. *Trans. Am. Fisheries Soc.* **97**, 12–17.

Merriman, D., and Schedl, H. P. (1941). The effects of light and temperature on gametogenesis in the four-spined stickleback, *Apeltes quadracus* (Mitchill). *J. Exptl. Zool.* **88**, 413–419.

Meske, C., Luhr, B., and Szablewski, W. (1967). Fehlender Sexualrythmus bei Karpfen in Warmwasserhaltung. *Naturwissenschaften* **54**, 291.

Miller, A. H. (1959a). Reproductive cycles in an equatorial sparrow. *Proc. Natl. Acad. Sci. U. S.* **45**, 1095–1100.

Miller, A. H. (1959b). Response to experimental light increments by Andean sparrows from an equatorial area. *Condor* **61**, 344–347.

Muir, B. S., Nelson, G. J., and Bridges, K. W. (1965). A method for measuring swimming speed in oxygen consumption studies on the Aholehole *Kuhlia sandricensis*. *Trans. Am. Fisheries Soc.* **94**, 378–382.

Müller, K. (1968). Freilaufende circadiane Periodik von Ellritzen am Polarkreis. *Naturwissenschaften* **55**, 140.

Müller, K. (1969). Jahreszeitlicher Wechsel der 24 h Periodik bei der Bachforelle (*Salmo trutta* L.) am Polarkreis. *Oikos* **20**, 166–170.

Müller, K., and Schreiber, K. (1967). Eine Methode zur Messung der lokomotorischen Aktivität von Süsswasserfischen. *Oikos* **18**, 135–136.

Mužinić, S. (1931). Der Rhythmus der Nahrungsaufnahme beim Hering. *Ber. Deut. Komm. Meeresforsch.* **6**, 62–64.

Naylor, E. (1958). Tidal and diurnal rhythms of locomotor activity in *Carcinus maenas* (L.). *J. Exptl. Biol.* **35**, 602–610.

Naylor, E. (1960). Locomotor rhythms in *Carcinus maenas* (L.) from non-tidal conditions. *J. Exptl. Biol.* **37**, 481–488.

Neumann, D. (1966). Die lunare und tägliche Schlüpfperiodik der Mücke *Clunio*. Steuerung und Abstimmung auf die Gezeitenperiodik. *Z. Vergleich. Physiol.* **53**, 1–61.

Neumann, D. (1969). Die Kombination verschiedener endogener Rhythmen bei der zeitlichen Programmierung von Entwicklung und Verhalten. *Oecologia* **3**, 166–183.

Oliphan, V. A. (1951). Daily feeding rhythms in the fry of the Baikal grayling. *Dokl.—Biol. Sci. Sect.* (*English Transl.*) **114**, 591–593.

Papi, F. (1960). Orientation by night: The moon. *Cold Spring Harbor Symp. Quant. Biol.* **25**, 475–480.

Pardi, L., and Grassi, M. (1955). Experimental modification of direction-finding in *Talitrus saltator* (Montagu) and *Talorchestia deshayesei* (And.) (Crustacea-Amphipoda). *Experientia* **11**, 202–211.

Pfeffer, W. (1915). Beiträge zur Kenntnis der Entstehung der Schlafbewegungen. *Abhandl. Math. Phys. Kl. Kgl. Sachs. Ges. Wiss.* **34**, 1–154.

Pittendrigh, C. S. (1954). On temperature independence in the clock-system controlling emergence time in *Drosophila*. *Proc. Natl. Acad. Sci. U. S.* **40**, 1018–1029.

Pittendrigh, C. S. (1958). Perspectives in the study of biological clocks. *Perspectives*

*Marine Biol., Symp. Scripps Inst. Oceanog., 1956* pp. 239–268. Univ. of California Press, Berkeley, California.

Pittendrigh, C. S. (1960). Circadian rhythms and the circadian organization of living systems. *Cold Spring Harbor Symp. Quant. Biol.* **25**, 159–184.

Pittendrigh, C. S. (1965). On the mechanism of the entrainment of a circadian rhythm by light cycles. In "Circadian Clocks" (J. Aschoff, ed.), pp. 277–297. North-Holland Publ., Amsterdam.

Pittendrigh, C. S. (1966). The circadian oscillation in *Drosophila pseudoobscura* pupae: A model for the photoperiodic clock. *Z. Pflanzenphysiol.* **54**, 275–307.

Pittendrigh, C. S., and Bruce, V. G. (1957). An oscillator model for biological clocks. In "Rhythmic and Synthetic Processes in Growth" (D. Rudnick, ed.), pp. 75–109. Princeton Univ. Press, Princeton, New Jersey.

Pittendrigh, C. S., and Bruce, V. G. (1959). Daily rhythms as coupled oscillator systems and their relation to thermoperiodism. In "Photoperiodism and Related Phenomena in Plants and Animals," Publ. No. 55, pp. 475–505. Am. Assoc. Advance. Sci., Washington, D. C.

Pittendrigh, C. S., and Minis, D. H. (1964). The entrainment of circadian oscillations by light and their role as photoperiodic clocks. *Am. Naturalist* **98**, 261–294.

Pittendrigh, C. S., Bruce, V. G., and Kaus, P. (1958). On the significance of transients in daily rhythms. *Proc. Natl. Acad. Sci. U. S.* **44**, 965–973.

Rao, K. P. (1954). Tidal rhythmicity of rate of water propulsion in *Mytilus* and its modifiability by transplantation. *Biol. Bull.* **106**, 353–359.

Rasquin, P. (1949). The influence of light and darkness on thyroid and pituitary activity of the characin, *Astyanax mexicanus*, and its cave derivatives. *Bull. Am. Museum Nat. Hist.* **93**, 497–532.

Rasquin, P., and Rosenbloom, L. (1954). Endocrine imbalance and tissue hyperplasia in teleosts maintained in darkness. *Bull. Am. Museum Nat. Hist.* **104**, 359–426.

Rawson, K. S. (1956). Homing behavior and endogenous activity rhythms. Ph.D. Thesis, Harvard University.

Richardson, I. D. (1952). Some reactions of pelagic fish to light as recorded by echo-sounding. *Fishery Invest., London* **18**, 1–19.

Richter, C. P. (1922). A behavioristic study of the activity of the rat. *Comp. Psychol. Monogr.* **1**, 1–55.

Robinson, E. J., and Rugh, R. (1943). The reproductive processes of the fish, *Oryzias latipes. Biol. Bull.* **84**, 115–125.

Rowan, W. (1926). On photoperiodism, reproductive periodicity and the annual migrations of birds and certain fishes. *Proc. Boston Soc. Nat. Hist.* **38**, 147–189.

Rowan, W. (1929). Experiments in bird migration. I. Manipulation of the reproductive cycle: Seasonal histological changes in the gonads. *Proc. Boston Soc. Nat. Hist.* **39**, 151–208.

Sachs, J. (1857). Über das Bewegungsorgan und die periodischen Bewegungen der Blätter von *Phaseolus* und *Oxalis. Botan. Ztg.* **15**, 809–815.

Schneider, L. (1969). Experimentelle Untersuchungen über den Einfluss von Tageslänge und Temperatur auf die Gonadenreifung beim Dreistachligen Stichling (*Gasterosteus aculeatus*). *Œcologia* **3**, 249–265.

Schuett, F. (1933). Studies in mass physiology: The effect of numbers upon the oxygen consumption of fishes. *Ecology* **14**, 106–122.

Schuett, F. (1934). Studies in mass physiology: The activity of goldfishes under different conditions of aggregation. *Ecology* **15**, 258–262.

Schwassmann, H. O. (1960). Environmental cues in the orientation rhythm of fish. *Cold Spring Harbor Symp. Quant. Biol.* **25**, 443–449.

Schwassmann, H. O. (1962). Experiments on sun orientation in some freshwater fish. Ph.D. Thesis, University of Wisconsin.

Schwassmann, H. O. (1967). Orientation of Amazonian fishes to the equatorial sun. *In* "Atas do Simpósio sôbre a Biota Amazônica" (H. Lent, ed.), Vol. 3, pp. 201–220.

Schwassmann, H. O., and Braemer, W. (1961). The effect of experimentally changed photoperiod on the sun-orientation rhythm of fish. *Physiol. Zool.* **34**, 273–286.

Schwassmann, H. O., and Hasler, A. D. (1964). The role of the sun's altitude in the sun orientation of fish. *Physiol. Zool.* **37**, 163–178.

Scrimshaw, N. S. (1944). Superfetation in poeciliid fishes. *Copeia* pp. 180–183.

Spencer, W. P. (1929). An ichthyometer. *Science* **70**, 557–558.

Spencer, W. P. (1939). Diurnal activity rhythms in freshwater fishes. *Ohio J. Sci.* **39**, 119–132.

Spoor, W. A. (1941). A method of measuring the activity of fishes. *Ecology* **22**, 329–331.

Spoor, W. A. (1946). A quantitative study of the relationship between the activity and oxygen consumption of the goldfish and its implications to the measurement of respiratory metabolism in fishes. *Biol. Bull.* **91**, 312–326.

Spoor, W. A., and Schloemer, C. L. (1939). Diurnal activity of the common sucker, *Catastomus commersonii* (Lacepède), and the rock bass, *Ambloplites rupestris* (Rafinesque), in Muskellunge Lake. *Trans. Am. Fisheries Soc.* **68**, 211–220.

Steven, D. M. (1959). Studies on the shoaling behavior of fishes. I. Responses of two species to changes in illumination and to olfactory stimuli. *J. Exptl. Biol.* **36**, 261–280.

Sushkina, A. P. (1939). The nutrition of the Caspian migratory herring larvae during the river period of their life. *Zool. Zh.* **182**, 221–230.

Sweeney, B., and Hastings, J. W. (1960). Effects of temperature upon diurnal rhythms. *Cold Spring Harbor Symp. Quant. Biol.* **25**, 87–104.

Swift, D. R. (1962). Activity cycles in the brown trout (*Salmo trutta* L.). 1. Fish feeding naturally. *Hydrobiologia* **20**, 241–247.

Swift, D. R. (1964). Activity cycles in the brown trout (*Salmo trutta* L.). 2. Fish artificially fed. *J. Fisheries Res. Board Can.* **21**, 133–138.

Szymanski, J. S. (1914). Eine Methode zur Untersuchung der Ruhe und Aktivitäts-perioden bei Tieren. *Arch. Ges. Physiol.* **158**, 343–385.

Tang, Y.-A. (1963). The testicular development of the silver carp, *Hypophthalmichthys molitrix* (C. and V.), in captivity in relation to the repressive effects of wastes from fishes. *Japan. J. Ichthyol.* **10**, 24–27.

Thines, G., and Vandenbussche, E. (1966). The effects of alarm substance on the schooling behaviour of *Rasbora heteromorpha* Duncker in day and night conditions. *Animal Behaviour* **14**, 296–302.

Thompson, W. F. (1919). The spawning of the grunion (*Leuresthes tenuis*). *Calif. Fish Game Comm., Fish Bull.* **3**, 1–27.

Turner, C. L. (1919). The seasonal cycle in the spermary of the perch. *J. Morphol.* **32**, 681–711.

Turner, C. L. (1937). Reproduction cycles and superfetation in poeciliid fishes. *Biol. Bull.* **72**, 145–164.

Turner, C. L. (1957). The breeding cycle of the South American fish, *Jenynsia lineata*, in the northern hemisphere. *Copeia* pp. 195–203.

van den Eeckhoudt, J. P. (1946). Recherches sur l'influence de la lumière sur le cycle sexuel de l'épinoche (*Gasterosteus aculeatus*). *Ann. Soc. Zool. Belg.* **77**, 83–89.

Verhoeven, B., and van Oordt, G. J. (1955). The influence of light and temperature on the sexual cycle of the bitterling. *Rhodeus amarus*. *Konnkl. Ned. Akad. Wetenschap., Proc.* **C58**, 628–634.

von Frisch, K. (1950). Die Sonne als Kompass in Leben der Bienen. *Experientia* **6**, 210–221.

von Ihering, R., and Wright, S. (1935). Fisheries investigations in northeast Brazil. *Trans. Am. Fisheries Soc.* **65**, 267–271.

von Seydlitz, H. (1962). Untersuchungen über die Tagesperiodizität des Rotbarsches, *Sebastes marinus*, auf Grund von Fanganalysen. *Kurze Mitt. Inst. Fischerei Biol. Univ. Hamburg* **12**, 27–35.

Wahl, O. (1932). Neue Untersuchungen über Zeitgedächtnis der Bienen. *Z. Vergleich. Physiol.* **16**, 529–589.

Walker, B. W. (1949). Periodicity of spawning in the grunion, *Leuresthes tenuis*. Ph.D. Thesis, University of California, Los Angeles, California.

Walker, B. W. (1952). A guide to the grunion. *Calif. Fish Game* **38**, 409–420.

Webb, H. M., and Brown, F. A., Jr. (1959). Timing long-cycle physiological rhythms. *Physiol. Rev.* **39**, 127–161.

Webb, H. M., and Brown, F. A., Jr. (1965). Interactions of diurnal and tidal rhythms in the fiddler crab, *Uca pugnax*. *Biol. Bull.* **129**, 582–591.

Webb, H. M., Bennett, M. F., Graves, R. C., and Stephens, G. C. (1953). Relationship between time of day and inhibiting influence of low temperature on the diurnal chromatophore rhythm of *Uca*. *Biol. Bull.* **105**, 386–387.

Welsh, J. H., and Osborn, C. M. (1937). Diurnal changes in the retina of the catfish, *Ameiurus nebulosus*. *J. Comp. Neurol.* **66**, 349–359.

Wever, R. (1960). Possibilities of phase control, demonstrated by an electronic model. *Cold Spring Harbor Symp. Quant. Biol.* **25**, 197–206.

Wever, R. (1962). Zum Mechanismus der biologischen 24-Stunden-Periodik I. *Kybernetik* **1**, 139–154.

Wever, R. (1964a). Ein mathematisches Modell für biologische Schwingungen. *Z. Tierpsychol.* **21**, 359–372.

Wever, R. (1964b). Zum Mechanismus der biologischen 24-Stunden-Periodik III. *Kybernetik* **2**, 127–144.

Wever, R. (1965). A mathematical model for circadian rhythms. *In* "Circadian Clocks" (J. Aschoff, ed.), pp. 47–63. North-Holland Publ., Amsterdam.

Wever, R. (1967). Zum Einfluss der Dämmerung auf die circadiane Periodik. *Z. Vergleich. Physiol.* **55**, 255–277.

Wiebe, J. P. (1968). The effects of temperature and daylength on the reproductive physiology of the viviparous seaperch, *Cymatogaster aggregata* Gibbons. *Can. J. Zool.* **46**, 1207–1219.

Wigger, H. (1941). Diskontinuität und Tagesrhythmus in der Dunkelwanderung retinaler Elemente. *Z. Vergleich. Physiol.* **28**, 421–427.

Wikgren, Bo-J. (1955). Daily activity pattern of the burbot. *Mem. Soc. Fauna Flora Fennica* **31**, 91–95.

Wilkie, D. W. (1966). Personal communication.

Wilkins, M. B. (1965). The influence of temperature and temperature changes on biological clocks. *In* "Circadian Clocks" (J. Aschoff, ed.), pp. 146–163. North-Holland Publ., Amsterdam.

Williams, B. G., and Naylor, E. (1967). Spontaneously induced rhythm of tidal periodicity in laboratory reared *Carcinus J. Exptl. Biol.* **47**, 229–234.

Williams, G. C. (1957). Homing behavior of California rocky shore fishes. *Univ. Calif. (Berkeley), Publ. Zool.* **59**, 249–284.

Winn, H. E. (1955). Formation of a mucous envelope at night by parrot fishes. *Zoologica* **40**, 145–147.

Yoshioka, H. (1962). On the effects of environmental factors upon the reproduction of fishes. I. The effect of day length on the reproduction of the Japanese killifish, *Oryzias latipes. Bull. Fac. Fisheries, Hokkaido Univ.* **13**, 123–136.

Young, J. Z. (1935). The photoreceptors of lampreys. II. The function of the pineal complex. *J. Exptl. Biol.* **12**, 254–270.

# 7

# ORIENTATION AND FISH MIGRATION

*ARTHUR D. HASLER*

## I. INTRODUCTION

A fisheries biologist once remarked that among the many riddles of nature not the least mysterious is the migration of fishes (Scheuring, 1930). The homing of salmon is a particularly dramatic example. The Chinook salmon of the northwestern United States is born in a small stream, migrates downriver to the Pacific Ocean as a young smolt, and, after living in the sea for as long as 5 years, swims back apparently unerringly to the stream of its birth to spawn. Its determination to return to its birthplace is legendary. No one who has seen a 20-kg salmon fling itself into the air (Fig. 1) again and again until it is exhausted in a vain effort to surmount a waterfall can fail to marvel at the strength of the instinct that draws the salmon upriver to the stream where it was born.

How do salmon remember their birthplace, and how do they find their way back, sometimes from thousands of miles away? This enigma,

**Fig. 1.** Salmon flinging themselves against a waterfall as they strive to reach their home stream and spawning ground.

which has fascinated naturalists for many years, is the subject of this chapter.

Man's knowledge of and interest in the salmon have been confined largely to the runs of fish from the sea and up the rivers to the spawning beds. However, cooperative studies by Japanese, Canadian, and American scientists have recently shown that salmon are incredibly intermingled during their sea phase over thousands of miles of the northern Pacific, only to sort themselves out neatly and precisely as spawning time approaches and to head for their stream of origin, be it in Asia or North America.

## Homing Salmon: The Kinds and Their Migrations

There are seven species of salmon belonging to two genera: the Atlantic salmon, *Salmo salar* Linn., belongs to the genus *Salmo*, while six

Pacific species belong to the genus *Oncorhyncus*—the Chinook or king, *O. tschawytscha* (Walbaum); sockeye, *O. nerka* (Walbaum); coho or silver, *O. kisutch* (Walbaum); chum, *O. keta* (Walbaum); pink, *O. gorbuscha* (Walbaum); and masu, *O. masu* (Brevoort)—and sometimes the steelhead trout, *S. gairdneri* Richardson, is counted among the Pacific salmon. Some of the Pacific species range as far from their home shores as 4000 km at sea, live there for 1–7 years, and return to distant rivers such as the Yukon and Columbia to spawn. Once at the mouth of these rivers, they swim upstream to find the rivulet of their birth, in some cases 4000 km from the river mouth.

While the Atlantic salmon do not travel as far within a stream system as do the Pacific salmon, some are known to migrate over 2500 km. According to Carlin (1962), one specimen of *S. salar* was recorded which accomplished a sea journey of 4000 km from southwestern Sweden to the western coast of Greenland. American Atlantic salmon from the Narraguagus River in Maine have been recovered (in oceanic commercial catches as far removed as 30 miles) above the Arctic Circle on the west coast of Greenland. In addition to *S. salar*, *S. gairdneri* and *S. trutta* Linn. (brown trout) undertake homing migrations.

The salmon runs vary, as to time of migration and river system ascended, according to species, and also according to particular groups or races within a species. In fact, in some rivers there are early and late runs of the same species during a single season; even these temporally divergent intraspecies runs give rise to distinct races.

Once at the ancestral spawning ground, the fully mature fish seek beds of gravel, from which the females excavate the redds. Into these the females shed their eggs while the males provide the milt. All representatives of the genus *Oncorhynchus* (Pacific salmon) spawn once in a lifetime and die after spawning. They are born and they perish in the same river. In the genus *Salmo*, adults may go back out to sea after spawning and return in subsequent times to the same place to breed. During the winter, the fertilized eggs develop slowly and hatch out toward spring.

The small salmon which emerges and wriggles up through the pebbles of the stream where the egg was laid and fertilized still retains a large mass of yolk from the egg. This larval stage is known as the "alevin." While the yolk remains, searching for food and feeding are largely unneccesary, and the fishes tend to remain on the bottom, although they are not entirely inactive. After the yolk is absorbed and depleted, the fry, now 3–5 cm long, feeds on insects and small aquatic animals for several weeks. In some species each fry roams within a limited radius of its redd, establishing a more or less definite territory for itself. As the fry grows, it is often dubbed with the unspecific, if descriptive, title of "fingerling."

The fry or fingerling of some salmon species remain in the river or stream for one or more years until a certain maturity, known as the "smolt" stage, is reached and the journey toward the sea begins. Sømme (1941) has reported smolts of Atlantic salmon, S. salar, entering the sea in Norway as old as 7 years, although the majority of them migrate in their third or fourth year of life. At the other extreme, the pink salmon of the Pacific start migrating toward the sea immediately after hatching.

## II. STREAM PHASE OF SALMON HOMING

### A. Hypotheses

Homing in migrating fishes, such as salmon, is a complex, physiologically dictated behavioral pattern: The animal spends its early life in one locality and, after undertaking migratory journeys of long or short duration to areas where the environment is radically different, ultimately returns to the original locality. This remarkable behavioral pattern is not limited to ocean-going fishes alone, for fishes inhabiting smaller bodies of water, such as a stream-pool, pond, or lake, show similar abilities to return to their home territories when displaced (Hasler and Wisby, 1958; Gunning, 1959). However, no example of fish homing is more impressive than that of salmon, and probably for this reason a great body of both speculative and experimental literature on aquatic migrations concerns the salmon.

### 1. Previous Hypotheses and Their Limitations

Among the various hypotheses that have been advanced to explain the mechanism by which salmon detect their home stream, two which have been predominantly discussed are based on physiochemical characteristics of the water and the salmon's presumed ability to follow gradients of these characteristics. Ward (1921a,b, 1939a,b) proposed that salmon always swim in the direction of the coolest water and that temperature gradients underlie the salmon's selection. Powers (1939), Powers and Clark (1943), and Collins (1952) have clearly shown that relatively high carbon dioxide tensions can repel migrating fishes, and on that basis these workers have suggested that salmon might be guided by carbon dioxide gradients.

Both of these hypotheses, as Scheer (1939) and Lissmann (1954) have pointed out, are subject to the same difficulty. Neither hypothesis accounts for the stream selectivity shown by salmon: Why does one salmon choose one tributary while another salmon chooses a second tributary which may have presumably less conducive properties? If there were a single general attractant, such as cooler temperature or lower carbon dioxide tension, then it would be more reasonable to expect all salmon to ascend the one most favorable stream. It is apparent that a more specific attractant than the temperature or $CO_2$ of the water must be involved. Even if we were to assume that fishes may become conditioned to a specific water temperature, there are still sufficient streams with the same characteristics that stream specificity would be negligible; moreover, these factors are too inconstant even over a brief period to be uniquely identified by a fish.

Gradient theories in general have a further drawback since, for the gradient to be operative, the change in physiochemical property must occur often enough that the animal does not adapt to the stimulus and cease to respond. Thus, the rapid physiological adaptation of olfactory and similar sensors is such that the migrating fish would have to be subjected to relatively steep gradients of temperature and carbon dioxide in order to receive continuous information relative to these differences in the environment. It seems unlikely that fish, in their migration, experience gradients sufficiently steep to enable their sensory systems to function in this manner. Alternatively the fish might pass in and out of the gradient-carrying current to prevent adaptation, but such movement would necessitate a continual recall of the latest level in the reference current. This seems improbable; furthermore, if the waters outside the reference current had lower values of these properties, the fish would be disposed toward them and would, in effect, choose incorrectly away from the current.

Buckland (1880), Kyle (1926), Craigie (1926), and Scheuring (1930) have postulated that homing in fishes might be ascribed to scent perception. Of these, Craigie alone conducted a preliminary experiment with 500 homing sockeye salmon, of which half had had the olfactory nerves severed, when they were still in an oceanic bay a considerable distance from their presumed home river, to determine whether scent might be a guiding factor. Craigie found that, of 65 normal recaptures, 56 (about 86% of the normal recaptures) were recovered in some part of the river system (Fraser River) toward which the salmon were believed, on initial capture, to be headed; of the 42 operated recaptures, 19 (about 45% of the operated recaptures) were found within the Fraser River system. However, of the 42 operated recaptures, only 28 had left the original capture site, and the remaining 14 appeared to be "sulking"

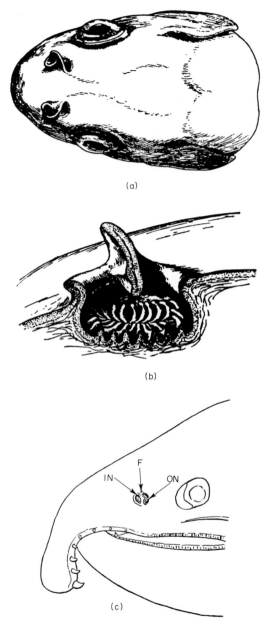

(a)

(b)

(c)

Fig. 2a–c

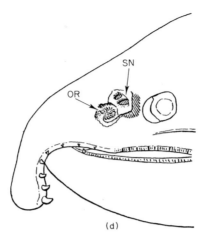

(d)

**Fig. 2.** Teleost olfactory capsules: (a) head of typical bony fish showing intake and outlet vents of the nasal capsule; (b) same head showing a cross section of the nasal capsule to illustrate the shunting of water over the sensory tissues; (c) head of adult male coho (silver) salmon showing nares, inflow opening (IN), outflow opening (ON), and flap (F) for directing water flow; and (d) same head with skin of nares (SN) folded back to reveal olfactory rosette (OR). Diagrams (a) and (b) from von Frisch (1941).

from the operative trauma. Thus, the operated recaptures in the Fraser River system constituted about 68% of those operated salmon which had continued their migratory journey. Craigie's findings apparently were not sufficiently convincing to be pursued by other workers, but perhaps the greatest deficiency of his proposal was its failure to explain what stream factor the salmon's olfactory system was detecting.

In about 1945, Hasler revived some of the olfactory theories and attempted to delve more deeply into the several aspects of olfaction as a migratory guide. His findings suggest that the odor of the natal stream is imprinted in the salmon when they are fry or fingerling, and further, that the odor may be organic in origin, possibly derived from the unique plant community of the stream's drainage basin and the flora within that stream. In short, the fish "smells" its way home from the coastline of the sea, tracking a familiar scent as would a fox hound. This theory we have termed the "odor hypothesis"; the experimental evidence for this hypothesis shall be the main emphasis of this section.

Figure 2 illustrates the position and anatomical structure of the nasal type common to most bony fishes, including the salmon. As the fish moves forward, a flap of skin shunts the water into the nasal sac, over the sensory olfactory epithelium, and out the posterior opening or naris. The fish, in a sense, continuously sniffs the chemical environment through which it moves (for further details, see Hasler, 1954, 1957).

The fish's paleocortex, which receives the nerve impulses from the olfactory tissue, is the dominant portion of the fish brain (Fig. 3), whereas the paleocortex of the human brain, and even that of other mammals which have more acute olfaction than human beings, is anatomically much less significant. Therefore, to consider odors and odor responses anthropomorphically is to vastly underplay their importance in the behavior of these lower forms in which pure stimulus–response behavior governs most activity without modification by capabilities mediated at higher levels of the brain. We are dealing with an acuity which certainly matches any attainment of terrestrial animals.

## 2. Properties of Olfaction as a Basis of the Odor Hypothesis

In the nasal passages of the human being and other land vertebrates, substances can be detected only if they are soluble. Thus, because a substance is not smelled until it passes into solution or diffuses into the mucus film of the nasal passage, smell may be described as fundamentally aquatic. For fishes, of course, the odors are already in solution in their watery environment.

It seems evident from our research (Walker and Hasler, 1949) that plants are capable of imparting their individual aromatic properties to

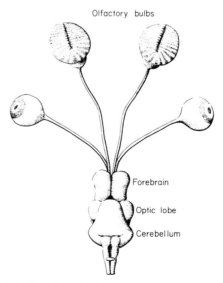

Fig. 3. Brain of bullhead. Redrafted from Adrian and Ludwig (1938).

water and that these properties can be detected and discriminated by fish. It is possible, therefore, that plants play a more important part in the existence of freshwater fishes than those obvious roles of food and shelter. For example, plant odors may guide fishes to feeding grounds when visibility is poor, as in muddy water or at night. The attraction of the odor may deter young fish from straying from protective cover. Furthermore, trained fishes could discern and selectively avoid several phenols in concentrations far below the threshold of man's perception (Hasler and Wisby, 1950). These data strongly suggest that olfaction may be an effective determinant in fish behavior in general.

## B. Laboratory Tests of the Odor Hypothesis

This section will be devoted largely to description, results, and interpretation of experiments, both field and laboratory, conducted to test the odor hypothesis. The series of tests presented in detail in this chapter was designed, first, to ascertain whether fish can, in fact, detect and discriminate between the waters from two different streams; second, to determine whether the detection and discrimination are made by olfaction, implying that the distinctive factor is an odor; third, to demonstrate whether odor imprinting and long-term retention can occur in fish; and finally, to indicate into what general chemical classification the distinguishing odor factor falls.

### 1. DISCRIMINATION BETWEEN WATERS

The general experimental plan for this first series of tests was to condition bluntnose minnows, by reward (food) or punishment (electrical shock), to react positively or negatively to one of two waters and, when the reinforcement was withdrawn, to determine whether the fish continued to react appropriately to the different waters. If the reactions remained correctly associated, it would be clear that the fish were discriminating between the waters.

*a. Experimental Method.* The experimental water samples for this series of training tests were obtained from two creeks which drained watersheds of different (soil and topographic) conditions. Otter Creek heads in an area composed of about 90% quartzite rock, the remainder being mainly sandstone and dolomite. Honey Creek, on the other hand, runs over moraines composed principally of sandstone (95%), cemented with dolomite, with lesser amounts of quartzite (Wanemacher *et al.*, 1934). Quite different plant associations, which would be expected to contribute divergent organic components, grow on these soils as well as within the streams (Fassett, 1960). The collecting procedures and containers used were identical for both streams (Hasler and Wisby, 1951). The filled containers were placed in a deep freezer (*ca.* $-13°C$) for storage to preserve the quality of the water. Before a test was initiated, the frozen water was thawed to room temperature.

Several specially equipped 7-gal aquaria were used (Fig. 4). At both ends of each aquarium a siphon-airlift circulation system was installed (Fig. 5). Water was siphoned from the aquarium, returned by air pressure, and discharged into a 6-inch funnel which was suspended above the tank. The funnel was connected to a glass tube which lay across the end of the aquarium. Perforations in the tube directed the incoming water across the bottom of the aquarium. Water from the jet on one side flowed only about halfway across, because there it met the stream from the other end, and both were deflected upward. This produced two currents or convection cells, each of which involved one-half of the tank. Each experimental sample was introduced in a measured amount from a separatory funnel. The separatory funnels—one at each end— were connected in turn to the siphon tubes beyond the tank edge so

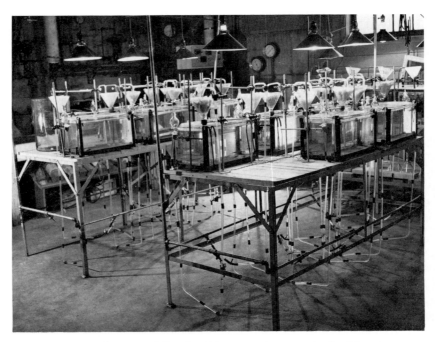

Fig. 4. Special equipped 7-gal training aquaria set up in the laboratory.

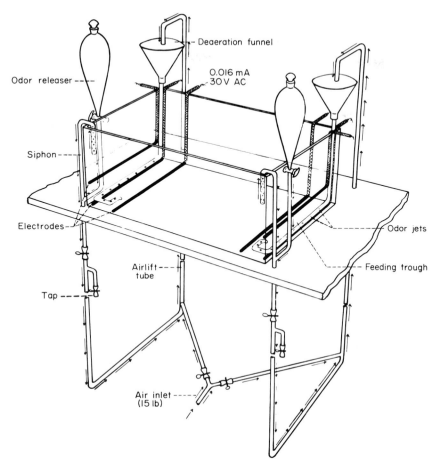

**Fig. 5.** Experimental tank built in the University of Wisconsin Laboratory of Limnology to train fish to discriminate between two odors. In the isometric drawing the vessel at the left above the tank contains water of one odor. The vessel at the right contains water of another odor. When the valve below one of the vessels is opened, the water in it is mixed with water siphoned out of the tank. The mixed water is then pumped into the tank by air. When the fish (minnows or salmon) move toward one of the odors, they are rewarded with food. When they move toward the other odor, they are punished with a mild electric shock from the electrodes mounted inside the tank. Each of the fish is blinded to eliminate associations of reward and punishment with movements of the experimenters.

that samples entered the tank unobtrusively in the same manner as the ordinary circulating water.

Three electrodes were placed at each end of the tank (see Hasler and Wisby, 1951) in such a way that it was possible to punish, without injuring, any fish entering an end region ($5 \times 5 \times 15$ cm) below the level of the upper electrode and above the floor of the aquarium where the other two electrodes were placed. The punishment, an electric shock of 2.3 V and 20 mA, was applied whenever a fish entered the end region into which a water sample intended to yield a negative reaction was flowing; higher voltages could be used without adversely affecting the fish, thereby impressing the training to the negatively reinforced sample if desired. The 60 cm$^2$ training region was designated the "end zone."

The end zone of positive reinforcement or reward was defined by the same limits and dimensions as the region of punishment but, of course, the demarcating electrodes were not used. The fish were rewarded by placement of food, pressed onto perforated celluloid strips, into the end zone. Since, in this method of training, hunger is the principal motivating force, tests were timed to coincide with the periods of greatest hunger. An attempt was made to test at different times each day and to feed no more than was necessary for the well being of the experimental animals. The fish were fed very heavily every sixth day and tests omitted on the seventh.

Because a set of electrodes and a sample outlet had been placed at each end of the aquarium, it was possible to randomize the presentation of the positive and negative samples. A table of random numbers (Snedecor, 1946) was used to establish for each day which odor was to be presented first and from which end of the aquarium. In order to preclude possible cueing to the operator, the fish were blinded by chemical cautery (injection of phemerol into the posterior chamber of the eye).

Fishes of two aquaria received positive training to the water of Otter Creek (that is, the fish were presented food in the appropriate end zone immediately after the water sample was added) and negative training to the water of Honey Creek (that is, the fish were punished by a mild electrical shock if they entered the respective end zone as the water sample was flowing in). The fishes of two other aquaria were trained to respond inversely to the two water samples: Honey Creek was the positive water; Otter Creek, the negative. If the training was successful and if the fishes were able to distinguish between the two waters, they would learn to associate reward with the positive water and thus enter the correct end zone promptly and with little hesitation when the positive sample was flowing in. Conversely, they would associate the negative water with punishment and would avoid the end zone into which the negative sample was flowing or would hesitate at length if they entered at all.

*b. Results.* Positive and negative tests were given daily, and all scores were recorded and evaluated. At the outset the fishes demonstrated natural, unconditioned orientative responses for the water of either creek; that is, they tended to enter the end zone more often during the introduction of a new water, whether positive or negative, than they did during the pretest. Only after a moderate amount of training did they clearly discriminate between the waters.

Table I gives the records of activity during the tests. It is evident that the minnows learned equally well regardless of which creek was positive or negative. A noticeable degree of discrimination was achieved in a month of training. After the discrimination level shown in Table I was reached, training was continued for two more months in order to

**Table I**

Mean Training Scores in Seconds for Fishes Reacting to Odor Stimuli[a]

| Month of training | Odor tested | Pretest | Test | Hesitation | Pretest | Test | Hesitation |
|---|---|---|---|---|---|---|---|
| | | | Tank 7 | | | Tank 8 | |
| First[b] | Positive | 63 | 97 | 69 | 39 | 142 | 43 |
| | Negative | 66 | 46 | 104 | 35 | 61 | 82 |
| Second[b] | Positive | 85 | 181 | 55 | 57 | 160 | 47 |
| | Negative | 73 | 20 | 200 | 52 | 18 | 196 |
| Third[b] | Positive | 68 | 194 | 46 | 43 | 176 | 31 |
| | Negative | 47 | 21 | 217 | 40 | 19 | 209 |
| Sixth[b] | Positive | 47 | 200 | 6 | 39 | 242 | 9 |
| | Negative | 39 | 15 | 241 | 40 | 13 | 226 |
| | | | Tank 2 | | | Tank 3 | |
| First[c] | Positive | 25 | 76 | 76 | 40 | 113 | 34 |
| | Negative | 58 | 52 | 132 | 68 | 43 | 146 |
| Second[c] | Positive | 47 | 123 | 56 | 87 | 156 | 35 |
| | Negative | 42 | 40 | 180 | 76 | 44 | 137 |
| Third[c] | Positive | 43 | 146 | 47 | 64 | 182 | 23 |
| | Negative | 57 | 27 | 203 | 39 | 19 | 129 |
| Sixth[c] | Positive | 46 | 343 | 6 | 34 | 336 | 0 |
| | Negative | 31 | 13 | 285 | 40 | 17 | 273 |

[a] Data presented are averages for training tests at the end of the month indicated. See text for explanation.

[b] Positive odor, Otter Creek; negative odor, Honey Creek.

[c] Positive odor, Honey Creek; negative odor, Otter Creek.

attain the maximum level of discrimination, evident by a plateau in the learning curve.

## 2. Seasonal Influence on the Distinctive Stream Factor

To test the possibility of seasonal changes in the water, samples were collected at other seasons and presented to fish that had been trained to water collected from the same streams in summer. The fishes' responses were unaltered, indicating that the characteristics recognized by the fishes did not lose their identity in either stream with the change in season.

## 3. Evidence of Olfactory Detection of Stream Odors

To test our hypothesis and prove that the fish's nasal tissues are the perceptive mechanism, we anesthetized trained fish and destroyed their olfactory capsules by heat cautery. After the tissues had healed, these fishes were again tested with the same experimental waters. Their test scores corresponded to their preoperative random movement scores. The operated fishes neither reacted appropriately to the odors of their training nor showed even the orientative but unconditioned response to "new" waters, as had been apparent in their pretraining days. They were impervious to retraining. Thus, it is clear that the reaction to the factor is dependent on the olfactory system, and it seems evident that odor is the principal factor.

## 4. Retention of Learning

Thus far, we have demonstrated that fishes are able to discriminate effectively between natural waters and that the distinguishing characteristic in each water is perceived by the olfactory organs, indicating that the factor is an odor. If these odors serve to influence the orientation of salmon migrations, the salmon's early associations to the odor must be retained during the 4-year sojourn at sea. The final and crucial test of the odor hypothesis, then, is to show whether fishes are capable of "remembering" imprinted odors over a period of time.

To determine the length of retention of discrimination in trained bluntnose minnows, daily training was stopped and odors were presented weekly without reward or punishment. After 6 weeks the fishes were confusing the two odors so completely that it was obvious they had lost the ability to discriminate (Table II).

That this method of testing does not provide a definitive measure of true retention is well known. It is because an animal which has been

**Table II.** Extinction Tests. Scores in Seconds for Fishes Reacting to Odor Stimuli of Training Odors[a]

Old fish

| Days after training | Water tested | Tank 2[b] Pretest | Test | Hesitation | Tank 3[b] Pretest | Test | Hesitation | Tank 8[c] Pretest | Test | Hesitation |
|---|---|---|---|---|---|---|---|---|---|---|
| 7 | Honey Creek | 50 | 320 | 0 | 50 | 325 | 10 | 40 | 15 | 150 |
|  | Otter Creek | 20 | 40 | 220 | 50 | 15 | 220 | 35 | 235 | 0 |
| 17 | Honey Creek | 60 | 190 | 0 | 45 | 325 | 10 | 65 | 70 | 180 |
|  | Otter Creek | 60 | 50 | 125 | 70 | 60 | 115 | 80 | 270 | 0 |
| 29 | Honey Creek | 50 | 150 | 25 | 60 | 175 | 60 | 70 | 95 | 125 |
|  | Otter Creek | 70 | 145 | 60 | 40 | 100 | 85 | 65 | 90 | 90 |
| 39 | Honey Creek | 110 | 115 | 80 | 45 | 75 | 75 | 115 | 125 | 130 |
|  | Otter Creek | 120 | 130 | 120 | 25 | 75 | 130 | 120 | 140 | 115 |

Young fish

| Days after training | Water tested | Tank 2[c] Test | Hesitation | Tank 5[b] Test | Hesitation |
|---|---|---|---|---|---|
| 0 | Otter Creek | 322 | 8 | 22 | 297 |
|  | Honey Creek | 21 | 304 | 315 | 17 |
| 52 | Otter Creek | 276 | 15 | 102 | 254 |
|  | Honey Creek | 53 | 272 | 304 | 24 |
| 66 | Otter Creek | 261 | 34 | 43 | 266 |
|  | Honey Creek | 29 | 256 | 205 | 51 |
| 95 | Otter Creek | 93 | 97 | 102 | 119 |
|  | Honey Creek | 71 | 134 | 182 | 96 |
| 105 | Otter Creek | 74 | 114 | 107 | 76 |
|  | Honey Creek | 82 | 127 | 121 | 90 |

[a] Data are for tests made on indicated days after complete cessation of training.
[b] Honey Creek positive; Otter Creek negative.
[c] Otter Creek positive; Honey Creek negative.

trained to associate food with an odor will be subjected to the reverse of this training process if fed without prior introduction of the odor. Thus, during these retention tests the minnows were being detrained, and at most the results of the tests can be considered only an absolute minimum indication of true retention.

## 5. CHEMICAL NATURE OF THE ODOR

A complete set of experiments proved that the fish did not associate the residue of either creek with either of their training odors since their scores were within the range of the control tests with distilled water. Thus, the inorganic fraction was eliminated as a possible factor, and it seemed increasingly likely that the odorous stimulus was either in the organic fraction or possibly in an organic–inorganic complex. Further experiments on the organic fraction have borne out this supposition.

In this research, we have explored one of the major unknowns in our understanding of movements directed toward the parent stream, that is, whether or not there is a specific stream factor by which fish actually do discern their home streams. We showed that at least some streams have odor characteristics which can elicit persisting differential responses in certain fishes and, further, that the odors of streams are probably aromatic substances present in the volatile organic fraction of stream water. However, our evidence for olfactory discrimination of stream water by fishes does not constitute proof that it is the only controlling factor in oriented movement toward the parent stream. Certainly other factors might contribute alternately or concurrently, or might outweigh the odor factor in importance to the fish.

Any factor which is to serve as a signal for returning salmon must fulfill certain rigid qualifications, which are defined as follows: First, it must remain relatively constant in any one stream over a period of years because an interval of 3–5 years may elapse between the initial imprinting of the salmon and the ultimate reactivation of that imprinting at the time of the spawning migration to the home stream. Changes in the factor must not take place with more rapidity than evolution of the species if the homing and attendant reproductive success of the salmon are not to be disrupted. Second, since the salmon's age of return to a given stream for any 1 year-class may vary from 3 to 5 years and since runs may occur at different times within a single year, the factor cannot be cyclical but must be present in the same form continuously through the many seasons. Third, the factor must have significance only to those returning salmon which originated from that given stream, and it must be neutral to all other salmon; any factor which is attractive or repellent in general would induce all salmon to enter or reject a stream or tributary, whether or not they were native to that stream. Fourth, the factor must remain detectable despite upheavals in the chemical and physical characteristics of the stream, through seasonal and meteorological changes, floods, and pollution, which may have occurred during the salmon's period at sea.

Our experiments indicate that odorous substances, probably carried into a stream by runoff from the vegetation and soils of a drainage basin, and combined with the bouquet of the aquatic flora and fauna, lend to the stream a distinctive scent which can be perceived, learned, and recognized again by fishes after a protracted period of nonexposure. We have also shown above that the characteristic odor remains present in the water throughout the changing seasons. Since the natural plant community and soil composition of a drainage basin do remain constant and comparatively balanced over long periods of time, it is likely that the characteristic odor, even in the presence of other odors, would still be detectable.

## C. Field Tests of the Odor Hypothesis

The results of the laboratory tests on minnows and on salmon fry, as described in the foregoing section, seemed sufficiently promising to warrant further experimentation into the olfactory powers of the migrating salmon. We established that fishes have the ability to detect and select one specific stream, evidently through its characteristic odor, to which they have been conditioned. However, we have not yet proved that the adult, sexually mature salmon actually needs or uses its nose to find its way upstream. Thus, the experiments which follow were designed to test the laboratory results and to determine whether such olfaction is used by the fish in their natural environment.

### 1. IMPORTANCE OF OLFACTORY ORIENTATION TO THE ADULT SALMON

A salmon returning to its parent stream is confronted with a problem of "correct" selection each time a new tributary joins the main stream which the salmon is ascending. If, at each juncture, the salmon chooses the stream leading ultimately to its natal tributary, it will finally attain its birthplace. If, on the other hand, a nonparent tributary is entered at any one of these confluences, or if the home stream itself is passed inadvertently, the salmon must retrace its path to the intersection where the error was made and resume its journey. That such backtracking does occur has been documented for sockeye salmon by Ricker and Robertson (1935) and for pink salmon by Wickett (1958).

The general experimental plan originally called for elimination of several of the sensory systems, including visual, common chemical, and olfactory. Scarcity of manpower and migrating salmon at the proposed site militated against such an ambitious program. Therefore, on the strength of laboratory investigations, it was elected to confine manipulations to occlusion of the sense of smell, which would then at least either affirm or negate the laboratory experiments and the odor hypothesis.

*a. Experimental Methods.* The site selected for the experiment was a point of confluence of two streams, Issaquah Creek and its East Fork, near Issaquah, Washington (Fig. 6). Each stream supports a natural run of coho (silver) salmon, and on the larger stream, the Issaquah, the

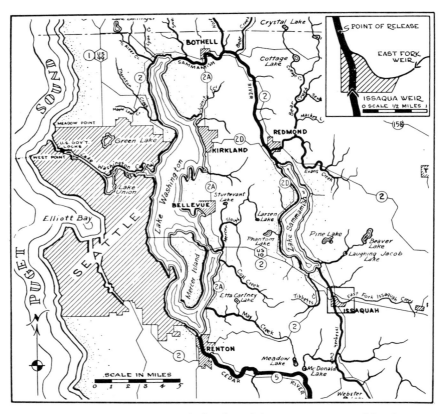

Fig. 6. Lake Washington watershed, adapted from a map prepared by the State of Washington Department of Fisheries.

State of Washington Department of Fisheries maintains a salmon hatchery and fish trap about 1 mile above the junction. On the smaller tributary about 1½ miles upstream a second fish trap was constructed. Thus, we were able to capture from either stream those salmon which, presumably, had entered their natal stream.

Each fish was tagged according to a code system (Wisby and Hasler, 1954) which identified whether the fish was experimental or control and where it was caught. The tagged experimental and control fishes were subsequently displaced three-fourths of a mile downstream from the junction of the streams. Recapture of the fishes at the two traps revealed whether they had been able to retrace their original route from the fork.

The sense of olfaction was obstructed by cotton plugs or, in some cases, cotton plugs coated with Vaseline or benzocaine ointment, inserted into the olfactory pits in such a way (Fig. 7) that water was prevented from flowing over the rosette of olfactory tissue.

In addition, the olfactory nerves of some salmon were severed by a surgical incision behind the olfactory pits. An equal number of controls were traumatized by an incision comparable in size and depth but made anterior to the olfactory pits. Unfortunately, because only very few neurotomized fishes and traumatized controls were recaptured, their recapture location was meaningless, since chance alone could account for their distribution. Therefore, they were omitted from the analysis. All those recaptured were initially Issaquah captures, a fact which is consistent with the considerably larger representation of fish from the Issaquah. All were recaptured at the Issaquah weir, and this is accountable on purely rheotactic grounds, for the Issaquah River is the main stream and salmon would be inclined to follow the dominant current unless "summoned" away from it by the odor of the home stream. Probably the severity of the operation completely thwarted the upstream

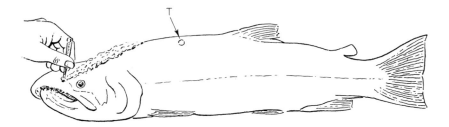

Fig. 7. Method of inserting a cotton wad into the olfactory pit of a tagged, anesthetized coho salmon (approximate length, 26 in.); T indicates tag.

drive, both physically and motivationally, of those salmon which did not resume the migration.

*b. Edaphic Characteristics of the Streams.* In view of the different hypotheses set forth regarding homing mechanisms, it is pertinent to review here some of the physical characteristics of these two streams. Issaquah Creek, after it is joined by its East Fork, empties into Lake Sammamish, which, in turn, is connected with Lake Washington through the Sammamish River. Lake Washington Canal establishes the link with the Pacific Ocean. A fish returning to Issaquah Creek must, therefore, travel some 40 stream miles, bypassing 15 or 20 tributary streams on the way. Even if the fish reaches its spawning ground with no back-tracking, it will nonetheless have moved in almost all compass directions, through both streams and lakes.

*c. Results.* Of a total of 302 fishes captured and displaced downstream, 149 were controls and 153 had plugged olfactory pits; 226 had been captured at the Issaquah trap and 76 at the East Fork trap. All the control fishes originally from Issaquah which were recaptured at the traps had returned to that stream on their second attempt, while 71% of the recaptured East Fork controls had returned to the East Fork (Table III). In contrast, of the experimental salmon initially captured at Issaquah and ultimately recaptured, 23% had entered the East Fork trap on their second attempt and, of the experimental salmon originally captured at the East Fork trap and subsequently recaptured, only 16% had returned accurately to the East Fork (Table IV).

To assess whether the occlusion procedure grossly affected the salmon's behavior as did the surgical incision, the total number of control fish recaptured was compared with the total number of experimental fish recaptured (Table V). No significant differences existed between the two. The nasal obstruction affected only their ability to choose correctly at the juncture and did not, as had been feared, deter them unduly from their migration.

### Table III

Distribution of Recaptured Control Silver Salmon ($\chi^2 = 43.72$, $P < 0.001$)

| Capture site | Recapture site | |
|---|---|---|
| | Issaquah | East Fork |
| Issaquah | 100% | |
| (46 fishes) | (46) | (0) |
| East Fork | 29% | 71% |
| (27 fishes) | (8) | (19) |

**Table IV**

Distribution of Recaptured Plugged Silver Salmon ($\chi^2 = 0.488$, $P = 0.49$)

| Capture site | Recapture site | |
| --- | --- | --- |
| | Issaquah | East Fork |
| Issaquah | 77% | 23% |
| (51 fishes) | (39) | (12) |
| East Fork | 84% | 16% |
| (19 fishes) | (16) | (3) |

*d. Interpretation and Significance of Data.* The data indicate that the normal fish were readily able to repeat their original choice at the stream juncture, thus furnishing additional support for the home-stream theory. Those with olfactory occlusion, however, were unable to select accurately. Interference with olfaction seriously disrupted their orientation and reduced their ability to retrace their original route. These experimental findings are consistent with the results which would be expected if the fishes were relying on their sense of smell to differentiate and select between streams. It is, of course, possible that these fishes, having just ascended one of the two streams, learned that route by other cues and therefore were able to retrace their path up the same stream after downstream displacement. If this belated learning were the explanation for the "correct" selection the second time, it would suggest that the initial ascent was motivated not by a specific "homing" drive but rather only by the general tendency for all salmon to swim upstream. It is, however, unrealistic to suppose that this rheotactic tendency could account for the accuracy of homing since any such unspecific response would probably direct all the migrating salmon in a river system toward one main stream; but on the contrary, not all salmon do select the same stream. Moreover, adult fish have been demonstrated to have slower learning ability and lesser retention of learning than young fish so that such rapid learning in these waning adults seems unlikely (Hasler and Wisby, 1951). In this experiment at least, orientation appears to have been accomplished by olfaction.

**Table V**

Effect of Olfactory Occlusion on Recapture of Tagged Silver Salmon, Compared with No Treatment ($\chi^2 = 0.30$, $P = 0.60$)

| Fishes | Total tagged | Recaptured | Not captured |
| --- | --- | --- | --- |
| Control | 149 | 73 | 76 |
| Treated | 153 | 70 | 83 |

## 2. FURTHER BASIC EXPERIMENTS ON OLFACTORY ORIENTATION IN FISHES

Stuart (1957) designed a series of comparable experiments on olfaction in the homing of brown trout. The brown trout, inhabiting a reservoir, were displaced from their home-spawning streams to streams on the other side of the reservoir. The nares of half had been occluded; those fishes were found, upon recapture, to have strayed in almost random fashion with reference to the home stream, but the control fish returned with great accuracy. While this evidence suggests that olfaction is essential for trout to identify their home stream, Stuart wisely deferred final judgments on the significance of olfaction until a greater experimental sample could be tested. Similarly, Gunning (1959) reported that displaced sunfish, *Lepomis magalotis* Rafinesque, with occluded nares but normal vision did not return to a home-pool territory, while both normal fish and some blinded fish with normal noses did home accurately. In this instance, the migratory distance was not great, but the sense of smell was evidently required for ascertainment of the home territory.

Hartman and Raleigh (1964) also provided an interesting study of displacement–return phenomena. They displaced adult sockeye salmon from the mouths of their home streams off Brooks and Karluk Lakes in Alaska to the mouths of other streams entering the lakes and attempted to induce the displaced salmon to enter these other tributaries. Despite the presence of other stocks of adult salmon at the alternative stream mouths and the apparent attractiveness of the streams to salmon, the displaced fishes were clearly predisposed to spawn exclusively in a particular tributary and could not be conditioned to enter any alternative stream.

Sato *et al.* (1966) have displaced spawning chum salmon from a Japanese stream into the sea and noted a failure of salmon to return to the home site if the olfactory system were occluded.

With regard to the stream factor or odor, Fagerlund *et al.* (1963) have supplied impressive findings of the ability of migrating sockeye salmon to discern their home-stream water in the laboratory, far removed from the other possible cues of temperature differentials, currents, and visual landmarks. These salmon were removed from their migration toward Great Central Lake, Vancouver Island, British Columbia, and taken to the biological laboratory of the Fisheries Research Board at Nanaimo, British Columbia, where they were presented water from Cultus Lake and outlet water from Great Central Lake. The salmon responded positively to the water from Great Central Lake but responded only weakly to the Cultus Lake Water. Furthermore, they responded only to the volatile fraction of the Great Central Lake water, confirming our hypothesis that the olfactory stimulant is not only unique but also organic in nature.

Schäffer (1919) made observations from simple displacement experiments on terrestrial migrations of eels and maintained that their ability to "smell the sea" on a light wind over several kilometers directed their overland trek. While his observations are elementary and the experiments themselves would never permit any conclusions regarding the sense or senses involved, they are nonetheless suggestive and challenging for interested researchers to replicate with eels having one sensory organ or another excluded from function.

### 3. THE EVIDENCE FOR ODOR IMPRINTING

With respect to the olfactory hypothesis of the homing mechanism in salmon, we have thus far shown evidence, from both the field and the laboratory, that each stream has a unique scent; that salmon can differentiate selectively among the myriads of individual, characteristic odors; that adult salmon deprived of their nasal function cease to discriminate among streams; that fish, including salmon, appear to retain experimental conditioning to odors; and that adult salmon in their migrations are responding to the odor factor of a given stream, through either an inherited or a learned reaction. The question then arises whether olfactory detection of the ancestral home stream occurs through some specific genetic character, "selected for" from generations of isolation and inbreeding, or through environmental imprinting of the olfactory cortex during youthful stages, synaptically maintained throughout adult life.

### 4. ARTIFICIAL IMPRINTING WITH DECOY ODORS

Still untested in the field is an experiment which would confirm or negate the odor-imprinting hypothesis. This would involve diverting a portion of a stream through a hatchery in order to condition a stock of sockeye fingerlings to that stream and subsequently marking and releasing them. In the year in which they are expected to return this stream could be diverted into a downstream tributary in the system.

An earlier suggestion of Hasler and Wisby (1951) was to use an artificial odorous chemical which was cheap, chemically stable, perceptible in low concentrations, and a repellent nonattractant. Wisby (1952) suggested morpholine for this test. Hartman (personal communication) has made field tests in Alaska in which he was unable to decoy conditioned fish with morpholine.

5. NATURAL IMPRINTING TO SCENT OF STREAM

*a. Electrophysiological Evidence.* Hara *et al.* (1965) have obtained electrophysiological data which indicate that waters from different sources produce noticeably different brain-wave patterns in Chinook and coho salmon, suggesting that a home-stream scent may in fact produce stimuli which are meaningful. This result needs to be confirmed with marked fingerling salmon which have been conditioned to a specific stream and tested upon return to the river and contrasted with controls. More exacting electrophysiological evaluations are also needed.

On the assumption that odor might play a role in returning to an ancestral homing site in a lake fish such as the white bass (see Section III, B, 1), the author and Horrall (Horrall, 1961) displaced white bass with noses occluded with cotton (322 controls and 328 experimentals). The results were somewhat ambivalent but suggested inferior homing in the treated fish.

One of the most convincing studies in support of a conditioned-response concept (that is, that imprinting of fry and fingerling salmon by the home-stream water is the determinant in spawning-stream choice) has been contributed by Donaldson and Allen (1957). These workers removed the stock of coho salmon eggs from an ancestral stream and transported them to a different stream in which the eggs hatched and the fry developed, and from which the year-old, marked smolts migrated to the sea. Some years thereafter when these adult salmon returned to freshwater to spawn, they ascended the adopted stream of their youth rather than the parent or ancestral stream. This study clearly favors the view that the odor characteristics of the stream are learned and identified through an imprinting process (i.e., a learning of environmental factors) rather than through any hereditary mechanism. It is interesting that, using the methods of their experiment, Donaldson has built up a run of salmon to the University of Washington's hatchery at Lake Union.

In an earlier study (International Pacific Salmon Fisheries Commission, 1949) salmon eggs, taken from the Horsefly River in British Columbia, were hatched and reared in a hatchery on a tributary of the Horsefly, called the Little Horsefly. At the smolt stage, they were flown a considerable distance and released into the main Horsefly River, from which they migrated to the sea. Three years later, 13 of these salmon had returned to their rearing place in the Little Horsefly to spawn, having ignored their ancestral home en route.

Still more recently, Carlin (1963) has transplanted fingerlings of Atlantic salmon from their ancestral home streams, now obstructed by a hydroelectric dam, to the hatcheries of the Swedish power companies which are on tributaries of the same river system as the ancestral streams but downstream from the dam and very near the Baltic Coast. Thus far, 700,000 salmon have been raised (Lindroth, 1963) to the smolt stage, marked and released into tributaries of the Baltic Sea. After reaching maturity, these marked salmon have returned to the streams of their youth, and among the 65,000 recaptured salmon, very little straying has been observed. Carlin believes that only a few weeks of conditioning are necessary for imprinting to take place, because when some of these smolts, which had been raised in the northern Baltic hatcheries, were transferred to streams in southern Sweden for imprinting and release, their subsequent adult returns were to the adopted southern Baltic sites rather than to the northern Baltic hatchery in which they were raised. Carlin considers that imprinting can occur when the smolts are as long as 20 cm and in their second year of hatchery life.

The diverse field tests on stream homing conducted thus far have significantly borne out the several aspects of the olfactory hypothesis. However, the tests have covered too small a sample of salmon for reliable generalizations; there is much that still needs testing, to rule out other possible factors, to detect contributory factors, and to provide unequivocal evidence for the hypothesis. Future investigations should be focused on methods of inducing salmon to enter streams other than their natal stream in an effort both to test the hypothesis and to contribute a practical answer to the problem of impending extinction of some runs of salmon by pollution and hydroelectric power dams.

## 6. THE SIGNIFICANCE OF OLFACTION FROM THE RIVER MOUTH TO THE NATAL STREAM

The full significance of olfaction in migration is probably not yet completely appreciated nor are its mechanisms thoroughly revealed. Among the problems remaining are what part olfaction and other senses play in determination of the home-river system from the shoreline of the sea and precisely how the odor track is followed throughout the entire journey upstream. At the present time, we can only speculate on the answers.

a. *Sensing the River from the Sea.* Although many sensory impressions and physical characteristics are complexly intermingled throughout the salmon's journey, sensory detection of the home-river system at a convoluted coastline must surely present one of the most confusing orientation choices for the salmon. Among species of salmon which spawn in large rivers near their outlets to the sea, the odor hypothesis could account adequately for the selection of the home river. In this instance the spawning ground is close enough to the coastline that odor-carrying currents could reasonably be expected to reach the sea. On the other hand, when the homing run involves a long inland journey, it seems unlikely that the odor of the natal tributary would be recognizable at the great dilutions which it would undergo by the time it reached the sea.

How, then, can the mouth of the river system be recognized at sea by salmon whose home tributary is far inland? Recently, Heath (1960) has drawn attention to assemblages of salmon at the coastal outlets of two blocked creeks in Oregon. Sand bars which appear during the dry summer completely obstruct surface flow from the outlets into the sea, and behind these bars fresh or slightly brackish water backs up. The dates in autumn on which the bars break down permanently may vary over several months. Heath has suggested that the salmon which gather in large numbers at the stream mouths prior to breach of the barriers must be capable of sensing the water from these two creeks as it seeps through the sand bars. Therefore, he investigated the physical character of the bars to ascertain the presence of seepage and the quality of the ocean water off the sand bars to determine whether chemical differences existed sufficient to cue the salmon to the presence of the stream beyond.

His conclusions were that even before the very first breach of the bars, "there must certainly be plenty of chemical influence in the adjoining sea such that the concentration of fish in preparation for the opening is explainable on this basis."

Columbia River water has been identified as far as 115 km from its mouth by isotopes originating at the Hanford Atomic Energy Plant (Gross et al., 1965); hence, it may be within the capabilities of the sense of smell to perceive molecules of odors from the home river at this distance.

From this evidence one might postulate that the "chemical influence" could be an odor, and although, as we have already suggested, it seems improbable that the odor of a distant spawning ground would reach the salmon in the sea. Still it is possible that the salmon have also become conditioned to a second odor—the combination of odors at the river mouth. Since salmon smolt tend to linger in estuarine waters at the ocean–river junction for several weeks (Manzer, 1956; Manzer and Shepard, 1962; McInerney, 1964), there is ample time for imprinting to occur. It is difficult to design a field experiment to test this hypothesis directly; however, some of Carlin's (1955) releases of smolt directly into tidal zones of the ocean just beyond the rivermouth (so that the young salmon were not allowed visual or topographical cues of the river mouth itself but were within the influence of the river's waters) and their subsequent return into the main river supply some foundation for this theory.

The study of eels has furnished evidence that at least some fish species can detect the organic properties of inland waters from the tidal zones of the ocean. Teichman (1957) demonstrated the incredible acuity of the eel's olfactory perception of pure chemicals. Concentrations of $3 \times 10^{-18}$ ml of $\beta$-phenylethyl alcohol in water were detected by young eels conditioned by training to this chemical. This dilution of the compound corresponds in magnitude to 1 ml of the aromatic alcohol dissolved in a lake of a volume 58 times as great as that of the Lake of Constance (Bodensee). Teichmann computed that at such a dilution as few as two or three molecules would be in the eel's olfactory sac at any one time. Creutzberg (1959) applied the findings of Teichmann's study to natural phenomena, suggesting that elvers of the eel use this sensitive olfactory sense to discriminate between the waters of ebb and flow tide. Through this distinction, the elvers are able to take advantage of the motion of the flow tide to transport them from the sea to freshwater. In laboratory tests, Creutzberg found that elvers were totally unable to discriminate between ebb and flow waters after the water had been filtered through charcoal, although they had made the distinction before the filtering. The fact that the distinctive characteristic was adsorbed by charcoal indicates once again that the detected odor is organic in nature. Evidently, then, an identifiable odor must flow into the tidal zone of the ocean, and the concentration which need reach the animal to be perceived can be very small indeed.

Wright (1964) calculated dilution factors of the water with its odor-bearing constituents from small salmon tributaries of the Fraser River and attempted to determine the relative concentration of any individual home odor at the river mouth. He concluded that a comparatively modest addition of scent from a home stream could put its mark on the whole downstream system. These statistics also appear to support the suggestion that detectable quantities of scent from the natal stream might be present at the river's mouth.

Yet another suggestion (Hasler, 1956a) of how the home-river system might be detected is that each sea–river junction has a unique conformation and hence the tactile and sound vibrations, arising from the movement of the shallow water through the unique topography, may provide a characteristic signal which the fish recognizes. Recently, Stuart (1962) conducted brilliant experiments which demonstrated the significance of such stimuli to migrating salmon. He showed that by sensing currents and sound vibrations to locate the standing wave at the base of a waterfall a salmon is able to place itself in the position most favorable to a successful leap.

Each of these divergent theories makes use of an established sensory capability of the salmon, and each may be valid. But infinitely more evidence is needed to assess the relative merits and importance of each. Perhaps the salmon uses only one of them and experimentation may some day reveal which; or, perhaps each sense contributes at a particular point. Certainly future workers have a fertile experimental field.

### III. OCEANIC PHASE OF SALMON HOMING

### A. Open-Sea Migration

Prior to World War II, little attention was paid to the open-sea movements of salmon because it was generally believed that salmon stayed on the continental shelves of the oceans after they had emigrated from their home streams. This belief, furthermore, fit in rather comfortably with the home-stream theory, for under the presumption that the fish remained within the influence of the home-river system, no new

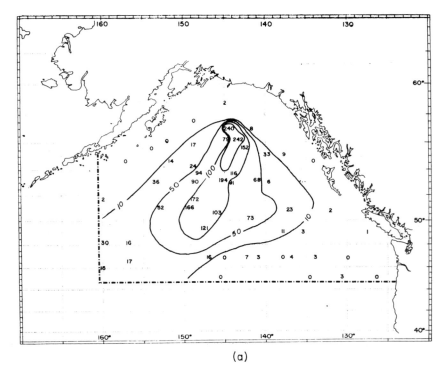

(a)

Fig. 8a

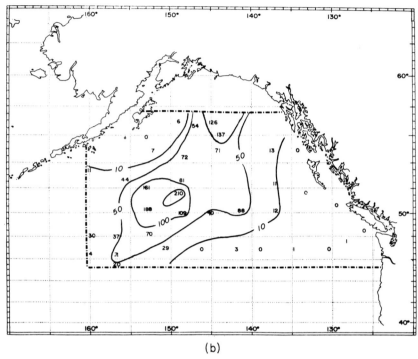

(b)

Fig. 8. Relative numbers of sockeye salmon in long line catches: (a) April 9 to May 6, 1962; (b) May 10–24, 1962. From Neave *et al.* (1962).

theories were necessary to account for the salmon's correct entry of the river system leading to the natal tributary. Since 1955, however, a great deal has been learned of the oceanic movements of salmon, the most noteworthy disclosure being that Pacific salmon are found across the entire northern Pacific Ocean and the Bering Sea during at least part of the year. Atlantic salmon, also, have subsequently been shown to migrate as far as Greenland from the North American and Scandinavian coastlines. Here, however, we will concentrate on the Pacific salmon since the scope of the studies, as well as the international commercial value, is greater.

## 1. THE OCEANIC LIFE OF SALMON

*a. Intermingling of Salmon in the Ocean.* Because of their importance to the international fisheries, the sockeye salmon has been studied more extensively than other species of salmon and provides the best statistical example of midoceanic intermingling of salmon stocks. An indication of their far-ranging distribution, reported by Hartt (1962), was revealed by

recaptures in a Kamchatka stream (160°E) of sockeye tagged in the
Aleutians at 180°.

Chartings by Neave *et al.* (1962) showed the seasonal distribution
of North American sockeye. Figure 8 illustrates the great concentration
of sockeye in the northern part of the Gulf of Alaska from early April
until late June. Upon maturation, the ripening fish dissociated themselves from the immature group, by early July the center of concentration of mature fish shifted toward the coast of British Columbia, and
by late July (Fig. 9) there were no sexually mature salmon of this
species in the Gulf of Alaska; they had already begun to enter their
home rivers.

This same survey (Neave *et al.*, 1962) furthermore provided con-
siderable evidence not only for the intermingling of sockeye in the
Gulf of Alaska but also for their sorting out as spawning time approached.
When the catches of salmon which were tagged and released in the
Gulf were labeled according to freshwater recapture area and plotted

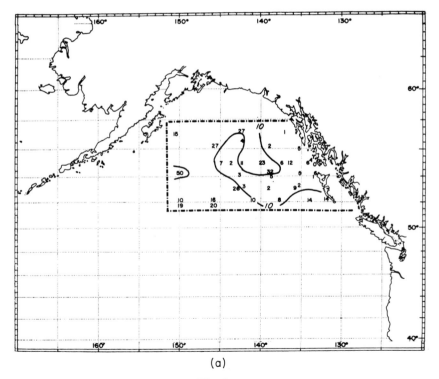

(a)

Fig. 9a

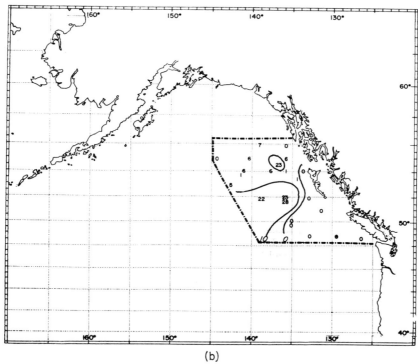

(b)

Fig. 9. Relative numbers of sockeye salmon in long line catches: (a) July 3–14, 1962; (b) July 15–26, 1962. From Neave *et al.* (1962).

to show their geographical distribution in the Gulf, it was clear (Figs. 10, 11, 12 and 13) that sockeyes from the Columbia River, Fraser River, Smith Inlets, Chignik Bay, and Bristol Bay matured together in the Gulf of Alaska, only to order themselves neatly into these diverse regions hundreds of miles apart. The significance of these data can be realized from the impressive number of salmon tagged at sea and the unusually high percentages of salmon later recaptured in freshwater (Table VI). It is also interesting to note that salmon returning to a single stream may often converge from widely separated oceanic areas (Neave, 1964; Margolis *et al.*, 1966).

## 2. OCEANIC DISTRIBUTION OF SALMON CORRELATED WITH HOME-STREAM LOCATION

Hartt (1962) recorded recaptures of fish which had originally been caught by gill nets and purse seines and tagged in the Aleutian Islands,

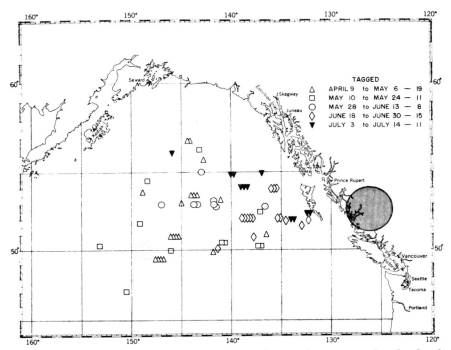

**Fig. 10.** Recoveries of sockeye salmon, tagged at sea between April and July of 1962, in the Rivers and Smith Inlets area as reported by October 10, 1962. Identified according to tagging period and location. From Neave *et al.* (1962).

but which had migrated considerable distances in easterly, northerly, and westerly directions to reach their home streams (Fig. 14).

A few highlights of the oceanic marking and home-stream recapturing reports may be mentioned here to indicate some of the remarkable migrations that take place. Pink salmon tagged south of Kamchatka and

**Table VI**

Canadian High Seas Tag and Recapture Data for 1962[a]

| Fish | Tagged | Recaptured | Percent |
|------|--------|------------|---------|
| Sockeye | 6260 | 668 | 10.7 |
| Pink | 7929 | 547 | 6.9 |
| Chum | 2640 | 106 | 4.0 |
| Coho | 409 | 35 | 8.6 |
| Chinook | 23 | — | — |
| Steelhead | 246 | — | — |

[a] Data from Neave (1962).

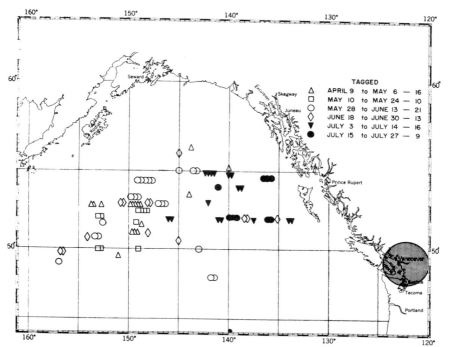

**Fig. 11.** Recoveries of sockeye salmon, tagged at sea between April and July of 1962, in the Fraser River area as reported by October 10, 1962. Identified according to tagging period and location. From Neave *et al.* (1962).

in the Sea of Okhotsk return to Kamchatka, and it would appear that most of the pinks returning from south of this peninsula are not captured far east of Kamchatka (Birman, 1958). However, of the fish tagged in the Gulf of Alaska by the Canadian teams (Neave *et al.*, 1962), six specimens of pink salmon were recovered in Asia; and one even swam to the Sea of Japan, opposite Korea, a distance of 3500 miles (7800 km) from the point of tagging. Bristol Bay and the Yukon River supply pink salmon to the Aleutians, and pinks that are tagged throughout the Gulf of Alaska return to Kodiak Island, where 15,000,000 pinks were caught in 1962. The Gulf of Alaska appears also to be the feeding ground for the pinks of Cook Inlet, Prince William Sound, southeast Alaska, and the Nass and Skeena Rivers in British Columbia. Chums from home streams in British Columbia are also found throughout the Gulf of Alaska. Cleaver (1964) has recently provided an excellent review of the detailed data from oceanic distribution studies of sockeye salmon.

A pertinent fact to our consideration of migration is that all five

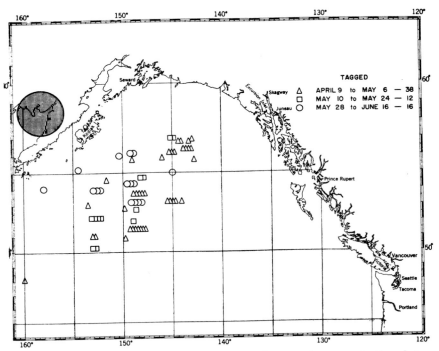

Fig. 12. Recoveries of sockeye salmon, tagged at sea between April and June of 1962, in Bristol Bay as reported by October 10, 1962. Identified according to tagging period and location. From Neave *et al.* (1962).

species of Pacific salmon ascend the Yukon River (the Chinook traveling inland through that river system for at least 1000 miles or 1600 km). These Yukon River fish all migrate to the Gulf of Alaska to spend their adult lives, and in making this migration, they must travel around and through the Aleutian Islands, a rather remarkable feat of orientation.

In only a few instances have biologists obtained evidence of the total migration from river to sea and back to river. Huntsman (1942), Pritchard (1943), and Blair (1956) have each reported single examples in which the salmon was marked in the home stream, caught and re-marked at sea, and ultimately recaptured in the home stream. Nonetheless, if one considers the tremendous mortality at sea as well as in the spawning run and the relatively small numbers of taggings that have been done, these examples are impressive.

More recently the Oregon State Game Commission (DeLacy, 1967) released a report of a remarkable series of recaptures:

*April, 1958*—Steelhead fingerlings (probably about 150–200 mm long)

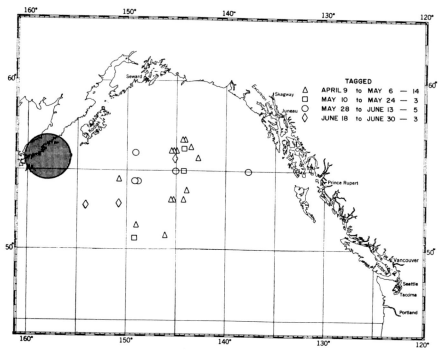

**Fig. 13.** Recoveries of sockeye salmon tagged at sea between April and June of 1962, in Chignik Bay as reported by October 10, 1962. Identified according to tagging period and location. From Neave *et al.* (1962).

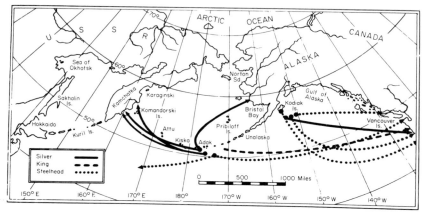

**Fig. 14.** Generalized distribution of recaptures of Chinook (king) and coho (silver) salmon and steelhead trout tagged at sea, 1956–1960. The routes are the shortest distances between marking and recapture points. From Hartt (1962).

were marked by fin clipping and released from the Alsea River hatchery on the central Oregon coast.

*September 5, 1958*—One of these original fingerlings was captured 75 miles southeast of Geese Island (southwest of Kodiak Island, Alaska) and marked with a spaghetti tag before release. The salmon was then 365 mm long.

*February 5, 1960*—This same fish was recovered at the Alsea River hatchery, measuring 558 mm in length.

## 3. ORIENTATION IN THE OPEN SEA

Acknowledging the paucity of information on which to base judgments, olfaction appears to be important to salmon migrations within a stream system but far less important to migrations within the open sea. The factors which guide the salmon to the oceanic feeding ground and which govern its oriented return through the sea are indeed complex.

In the ocean, odor might play the role principally of a sign-stimulus for home recognition. If a salmon were swimming within a water mass, that fish would have no sense of displacement as the mass moves (similar to the experiences of balloonists in a cloud) unless there were fixed visual or tactile features in the environment; on the other hand, the surface of contact between two water masses might have perceptible differences in salinity, dissolved gases, and odor (Hasler, 1954, 1957). Unpublished data from the laboratory satisfy the author that minnows can smell the difference between the water from the Georges Bank and samples from the Sargasso Sea. However, this sensing of salinity, gases, or odors at sea would seem to be more meaningful as appropriate cues for a salmon's recognition of, for example, an oceanic spawning site, once the fish had arrived there, than as cues for directional orientation.

Two quantitative findings that have emerged from the international surveys might have some relevance to the mechanism of salmon migrations. The first, obtained during the Canadian survey (Neave *et al.*, 1962), was the minimum swimming velocity of the fish, estimated from the recapture interval. Sockeye traveled an average of 30 miles/day; pink salmon, 24 miles/day; and chum salmon, 16 miles/day. The relatively low rate of travel of the chums may perhaps be ascribed to their late runs. In all of these cases, the velocity exceeded that of the ocean currents and the movement was often against the currents. Neave (1964) suggested that, since salmon migrate actively during only about 8 hr a day, the swimming velocity must be in excess of 3–3½ miles/hr.

The second important measurement evolved from the depth distribution studies of Japanese scientists (International North Pacific Fisheries

Commission, 1963). Salmon were caught in their gill nets principally in the upper 20 meters. Only early in the year, when the temperatures were fairly uniform (5°C) throughout all depths, were the fish found in depths as great as 40 meters. Manzer (1964) essentially confirmed these findings. After stratification of the water, the salmon hovered above the thermocline, usually within the 10-meter layer, in water warmer than 5°C. Cleaver (1964) and Neave (1964) report that the salmon were near the surface at night and at sunrise but tended to swim in deeper waters (to 61–62 meters in the absence of a thermocline) during the day. That the fish were in a shallow vertical distribution at sunrise is particularly interesting in the light of Braemer's finding (1959) in the laboratory that fish appear to adjust their orientation through internal biological clocks for latitude (season) and longitude (time of day) according to the appearance of first light. This disclosure strongly suggests that the salmon's open-water migrations could be oriented by a sun-compass mechanism, possibly directed by the changing diurnal azimuth of the sun, the altitude of the sun, or a combination of the two.

## B. A Model of Oceanic Orientation

Supported by significant data establishing the existence of a sun-compass mechanism—an ability in some animals to use the sun's position in determining a particular directional choice—in other vertebrates (Kramer, 1950, 1952; and summarized by Griffin in 1964) and invertebrate animals (von Frisch, 1949, 1950a,b; Pardi and Papi, 1952), it is proposed that fish, including salmon, are also oriented in part by a sun-compass mechanism in open-water migrations, and, indeed, the vertical distribution studies just discussed would seem to substantiate this possibility. With such a guiding factor, the fish could set a course in a given compass direction from as far as 150 km away and thus arrive in a near-shore zone within reasonable proximity to its home. Once in that zone, perception and recognition of visual, auditory and tactile, and olfactory features could direct the fish more precisely. That fishes can depend upon olfaction in the most precise aspects of orientation has already been shown and previous speculation made on the auditory and tactile cues of the waters at the sea–river junction. That visual cues are also sometimes important is documented by Hasler and Wisby (1958), who found that largemouth bass, *Micropterus salmoides* Lac., and green sunfish, *Lepomis cyanellus* Raf., can use visual references in locating food, nests, or home territories. In the laboratory, as well, a common European minnow, *Phoxinus laevis* Linn., appears to orient itself to a feeding site by local visual landmarks (Hasler, 1956b). Thus, the sun-compass proposal could account for a considerable amount of the movement from open water where, except in direct currents, the odor of the home-river system could not penetrate and where specific landmarks would be absent; yet this proposal is consistent with observed data on open-water migrations and with the odor hypothesis for stream homing.

Initial demonstration of the probability of the sun-compass hypothesis in open-water migrations had, as its subjects, the white bass, *Roccus chrysops* (Raf.), of Lake Mendota. It was the author's objective to study homing behavior in situations less complex than the high seas, where the fish would move shorter distances and be available for experimentation at the appropriate times in their life cycle. Yet, despite the differences in species, space, and time, positive proof that white bass utilize a sun compass in homing orientation would strongly suggest that salmon might well, also. The white bass must locate its specific spawning ground from an open-lake area in which it has spent most of its adult life, just as the salmon must migrate from open water back to its juvenile environment, and in both migrations there is a lack of visual or olfactory cues.

## 1. Accuracy of Homing after Displacement

Lake Mendota is a very productive but comparatively small lake with an area of 39.4 km² and a shoreline of 32.3 km. Although we have studied the natural history of the white bass in this lake for many years, we have been able to locate only two major spawning sites in the entire lake, and these two are of very limited area. This economy of space assured us of a simple yet efficient model of open-water migration, which could be studied with relative ease. Both spawning grounds, Maple Bluff and Governor's Island, are situated on the north shore of the lake but separated by a distance, measured across the water, of 1.6 km (Fig. 15). The white bass congregate on these grounds at spawning time in late May and early June, when the water temperatures range from 16° to 24°C.

Throughout three spawning seasons (1955, 1956, and 1957) white bass were captured from each spawning ground with fyke nets marked with numbered Petersen tags and transported in open tanks to a release point (release station 1 in Fig. 15) in the open lake, 2.4 km away from each spawning ground, for daytime release. It was obvious, even before computations were made, that a large percentage of displaced fish returned to the nets. This is particularly remarkable since a fish returning to the area of the net might not be caught immediately because of the

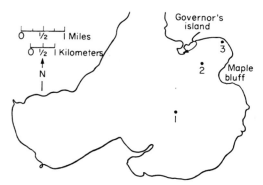

Fig. 15. Summary of white bass tagging and recapture data, Maple Bluff and Governor's Island spawning runs, 1955–1959, Lake Mendota, Wisconsin.

relative inefficiency of the trap. Moreover, as the observations accumulated and capture–recapture correlations were made, we were impressed with the precision of return of the spawners to the home ground where they were originally captured (Table VII).

It is clear that white bass do home to a specific spawning ground with high accuracy and, furthermore, as Table VIII shows, that they return to the same home ground year after year to spawn. Therefore, the fact is established that white bass, like salmon, migrate predictably from open water to specific remote areas to spawn. However, whether it is an oriented migration cannot be determined from these data. The method of recapture was such that one cannot eliminate random searching as a possible mechanism of return; the rapidity of return and the directness of return course were not ascertainable by the crude net recaptures. The observation that the fish could distinguish accurately between the two spawning grounds implied that selective orientation must be involved at least to some extent, and further experimentation seemed warranted.

## 2. TRACKING OF HOMING MIGRATIONS AFTER DISPLACEMENT

To determine if the return was prompt and reasonably direct, it was necessary to know the direction and speed of movement from the release site. A small, yellow fisherman's bobber (64 mm in diameter), which could be plainly seen at the water surface, was attached to a 2–5-meter length of fine monofilament nylon line to serve as the indicator of the take-off direction while the fish swam below surface.

Sexually ripe fish were removed from the fyke nets and transported to a predetermined point in the center of the lake. Just before release

**Table VII**

Summary of White Bass Tagging and Recapture Data, Maple Bluff and Governor's Island Spawning Runs, 1955–1959, Lake Mendota

| Fish | Number tagged | Number recaptured | % recaptured | Number recaptured at other site | % correct return |
|---|---|---|---|---|---|
| | | 1955[a] | | | |
| Males | 181 | 12 | 6.6 | Unknown | — |
| Females | 14 | 0 | 0.0 | Unknown | — |
| | | 1956 | | | |
| Males | 1082 | 47 | 4.3 | 6 | 87.2 |
| Females | 121 | 0 | 0.0 | 0 | — |
| Re-releases | | | | | |
| Males | 37 | 2 | 5.4 | 0 | 100.0 |
| | | 1957 | | | |
| Males | 1810 | 269 | 14.9 | 32 | 88.1 |
| Females | 291 | 6 | 2.1 | 1 | 83.3 |
| Re-releases | | | | | |
| Males | 288 | 42 | 14.6 | 6 | 85.7 |
| | | 1958 | | | |
| Males | 1389 | 261 | 18.8 | 30 | 88.5 |
| Females | 303 | 61 | 20.1 | 5 | 91.8 |
| Re-releases | | | | | |
| Males | 365 | 127 | 34.8 | 13 | 89.8 |
| Females | 76 | 36 | 47.4 | 8 | 77.8 |
| | | 1959[b] | | | |
| Males | 2531 | 466 | 18.4 | 50 | 89.3 |
| Females | 583 | 55 | 9.4 | 4 | 94.4 |
| | Totals 1955–1959 | | | | |
| Males | 6993 | 1055 | 15.1 | 118 | 88.8 |
| Females | 1312 | 122 | 9.3 | 10 | 91.8 |
| Re-releases | | | | | |
| Males | 690 | 171 | 24.8 | 19 | 88.9 |
| Females | 76 | 36 | 47.4 | 8 | 77.8 |
| Totals | 9071 | 1384 | 15.3 | 155 | 88.8 |

[a] Fish tagged and recaptured from Maple Bluff only.

[b] Only year when all releases were made without displacement. There were 521 re-releases; however, recaptures were not remarked, and any multiple recaptures are included with original recaptures.

of each fish, the line with its attached bobber was fastened to the dorsal flesh, posterior to the dorsal fin, with a small fish hook. The fish were released individually, about 3 min apart, until four or five had been liberated. After 1 hr had elapsed, each fish was located and its position charted. Figure 16 shows that the fish which were liberated on clear days in midlake (release station 1 in Fig. 15), even when corrected for

Table VIII

Summary of White Bass Recapture Data 1 or 2 Years after Tagging, Maple Bluff and Governor's Island Spawning Runs, 1957–1960, Lake Mendota

| Year of tagging | Year of recapture | Fish[a] | Number recaptured | Number recaptured at other side | % correct return |
|---|---|---|---|---|---|
| 1956–1958 | 1957–1960 | Male (O) | 124 | 14 | 88.7 |
| | | Male (D) | 58 | 4 | 93.1 |
| | | Female (O) | 7 | 4 | 42.9 |
| | | Female (D) | 3 | 0 | 100.0 |
| 1959 | 1960 | Male (O) | 307 | 62 | 79.8 |
| | | Male (D) | 107 | 17 | 84.1 |
| | | Female (O) | 12 | 2 | 83.3 |
| | | Female (D) | 3 | 2 | 33.3 |
| | | Totals | | | |
| 1956–1959 | 1957–1960 | Male (O) | 431 | 76 | 82.4 |
| | | Male (D) | 165 | 21 | 87.3 |
| | | Female (O) | 19 | 6 | 68.4 |
| | | Female (D) | 6 | 2 | 66.7 |
| | | Totals | 621 | 105 | 83.1 |

[a] O indicates original recaptures and D indicates double or more recaptures.

current, moved generally north, toward the spawning grounds. The final position of each of these fish was corrected for drift displacement (Fig. 16), because the water currents measured at the time and depth of the fish's swimming from midlake to the spawning ground areas were

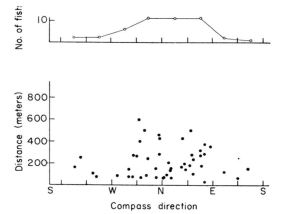

Fig. 16. Direction take-off and distance traveled, after 1 hr, by white bass from a midlake release point (station 1 in Fig. 15). Corrected for drift by currents.

fairly homogeneous and directional and could, therefore, favor or oppose the fish's movement. Swimming velocities were not calculated since the bobber appeared to exert a substantial drag force on the fish.

On the basis of the movements of bass released at the end of the bay (release station 3 in Fig. 15), it would appear that white bass, once near the shore, follow the shoreline and find their spawning area through visual or olfactory recognition of the local environment. Studies of white bass with occluded olfactory sacs certainly suggest that olfaction is essential in discernment of the correct spawning ground; nonetheless, the exact mechanism of orientation of the white bass near the spawning grounds needs more intensive study.

One of the most significant findings of this study, in terms of the sun-compass proposal, was that white bass did appear to use the sun in their open-water migrations. Fishes set free on cloudy days and fishes provided with opaque plastic eye-caps (Wisby, 1958; Gunning, 1959) were usually found to be randomly distributed in all compass directions from the release point (Fig. 17). An appraisal of these experiments suggested that the white bass do possess a sun-compass mechanism which is used for orientation in open water where the shore is not seen nor the home ground smelled, and olfactory experiments seem to imply further that once the white bass reach odor-bearing currents by means of the sun compass, they use their olfactory sense to locate the specific homing area. The white bass, migrating from midlake, were able to maintain a relatively constant compass direction, regardless of the time of day or the seasonal relationship of the day within the spawning

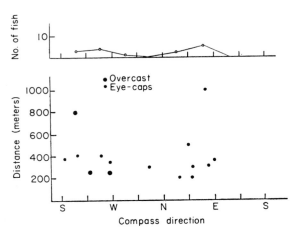

Fig. 17. Direction tendency of blinded fish (with eye-caps) and of normal fish on heavily overcast days, released in midlake (station 1 in Fig. 15).

period, evidently by using the sun as a point of reference. Because they were clearly able to compensate for the changing solar azimuth, these fish must also possess a "biological clock."

## 3. THE SUN AND THE ANIMAL

To facilitate an understanding of sun-compass orientation, let us first review some of the properties of incident light from a point source that appears to move both in a short cycle and a long cycle. The point source to the animal in its natural environment is the sun, and the two cyclic periods are the day and the year.

In the northern hemisphere in middle latitudes above 23.5°N the sun always appears to move diurnally from east to west but the deflection is toward the south. In the southern hemisphere in middle latitudes below 23.5°S the path again is from east to west but the deflection is toward the north. At the equator the deflection is toward the north during the summer and toward the south during the winter, with zero deflection at the equinoxes when the sun culminates in the zenith. At 23.5°N the deflection is toward the south except at the summer solstice when the deflection is zero, the sun culminating in the zenith; the converse is true of 23.5°S where the sun culminates in the zenith at the winter solstice. Between the equator and 23.5°N the sun's deflection is toward the south for the greater portion of the year (the number of days becoming greater with increasing distance from the equator), but the deflection is toward the north during part of the spring–summer season. Again, the deflections are the opposite between the equator and 23.5°S. At the poles the sun is hidden for half the year, is visible on the horizon throughout 24 hr at the equinoxes, and parallels the horizon throughout 24 hr (at an elevation increasing until June 21 and decreasing thereafter) for the other half of the year.

At any time in the sun's path, an imaginary circle can be constructed with the observer as center, which passes through the sun and is perpendicular to the horizontal plane of observation. This great circle will intersect the horizontal plane in a line (the intersection of two planes always being a line); the line of intersection may be called the sun's "horizontal component." So that the horizontal component will be meaningful, it must be described relative to a fixed point, and for that purpose the north point is the most frequently used reference in navigation. The relationship between a north line and the horizontal component is the sun's "azimuth," and it is measured clockwise as the angle of the arc on the horizon from the north point to the sun's horizontal component.

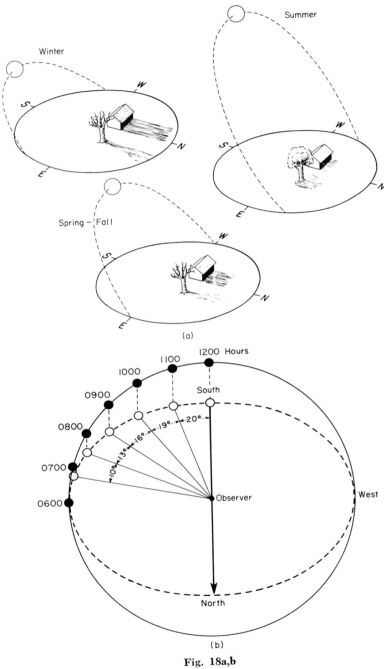

Winter

Summer

Spring - Fall

(a)

1200 Hours
1100
1000
0900
0800
0700
0600

South
20°
19°
16°
13°
10°

Observer

West

North

(b)

Fig. 18a,b

472

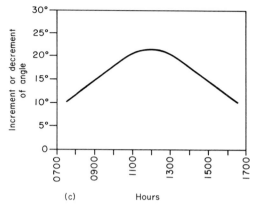

Fig. 18. Pattern of sun's movement at northern middle latitudes during the equinox: (a) deviation of the sun's path from the zenith (summer and winter solstices as well as vernal and autumnal equinoxes); (b) projected image of the sum on the horizontal plane; and (c) locus of the diurnal change of the sun's azimuth, hour of day as the abscissa and increment or decrement of angle as the ordinate.

Obviously, as the sun moves across the sky during the day, its azimuthal angle will continually change; however, the rate of change varies with latitude and with the longer, yearly cycle. Let us first consider the variation with latitude and set as a constant the time of the year, which for our present purposes we will choose to be the equinoxes. In middle latitudes the rate of angular change of the sun's azimuth is not constant. Because the sun follows a vertically arched path that is skewed from the zenith (Fig. 18a), its image hypothetically projected onto the horizontal plane describes an elliptical curve (Fig. 18b). Since the sun's curvilinear movement across the sky is at a constant velocity, the velocity of its distorted projection will change with the amount of distortion from a true circle on the horizontal plane. Thus, the angular velocity of the projection must increase as the ellipsoidal curve flattens out toward noon and must decrease as the curvature increases toward evening. Since this angular change is identical with the azimuthal angular change, we may say that the rate of change of the sun's azimuth is not constant. If we then plot the increment (or decrement) in the azimuthal angles (taking noon and midnight as maxima and sunrise and sunset as intercepts on the abscissa) against the hour of day, we obtain a periodic curve (Fig. 18c). It should be noted that the night curve is simply the day curve for the same latitude in the opposite hemisphere. At the equator, the vertical arch of the sun culminates in the zenith, and therefore the projection of the sun on the earth appears as a straight

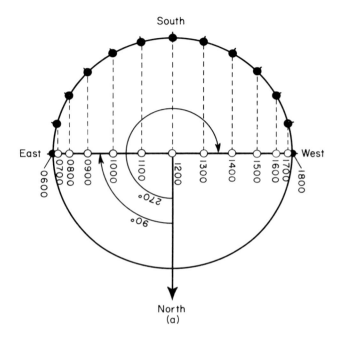

South

East

West

0600
0700
0800
0900
1000
1100
1200
1300
1400
1500
1600
1700
1800

270°

90°

North
(a)

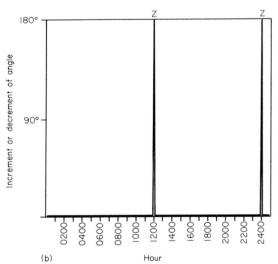

180°

Increment or decrement of angle

90°

Z                    Z

0200
0400
0600
0800
1000
1200
1400
1600
1800
2000
2200
2400

(b)                                    Hour

Fig. 19a,b

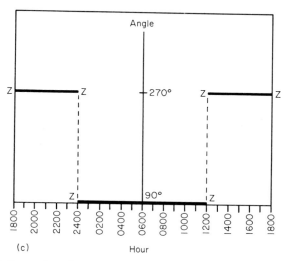

(c)

**Fig. 19.** Pattern of sun's movement at the equator during the equinox: (a) projected image of the sun on the horizontal plane; (b) locus of the diurnal change of the sun's azimuth, hour of day as the abscissa and increment or decrement of angle as the ordinate (Z indicates time of culmination is actually instantaneous and is shown on the graph only in concept); and (c) locus of the sun's azimuth, hour of day as the abscissa and azimuth angle as the ordinate.

line (Fig. 19a). The locus (or graph) of azimuthal angular change is noteworthy; at noon and at midnight, the azimuth of the sun does not exist, while between 1 min before noon and 1 min after noon the azimuthal angle changes by 180° (Figs. 19b and 19c) from 90° to 270°. Finally, at the poles, the projected image as well as the true image of the sun follows the horizon in a perfect circle (Fig. 20a). The linear distances traveled by the sun for each hour are identical, and because the projection has no distortion, the angles of each arc are identical. Since we know the linear movement of the sun to be constant, then its angular velocity must also be constant, and its azimuth at the poles changes at a constant rate of 15°/hr (Fig. 20b).

Before we turn to the larger, yearly cycle, let us contemplate the orienting animal's use of the simple diurnal azimuth of the sun. Assume that the orienting animal is motivated to swim southeast. The animal perceives the sun's position, responds to some positional characteristic, such as the horizontal component, and moves at an angle to it. But the animal must continually change its angle to that component throughout the day if it wishes to maintain a constant single compass direction, because the angle of the sun's horizontal component to the

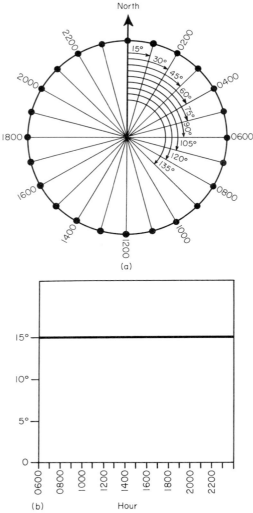

Fig. 20. Pattern of sun's movement at the poles during the equinox: (a) projected image of the sun on the horizontal plane and (b) locus of the diurnal change of the sun's azimuth, hour of day as the abscissa and increment or decrement of angle as the ordinate.

compass points (or more specifically the sun's azimuth, referred to the north point) is continually changing. Therefore, to make use of the azimuthal angle, the animal must have a mechanism which either provides the north reference point or tells the time of day and gives an awareness of what the azimuth should be for that particular hour; further-more, the animal must be able to compensate for the varying velocity of the azimuthal change. The orienting animal, then, moves at a constant angle from the north point, forming its own azimuth. The difference be-tween the sun's changing azimuth and the animal's constant azimuth is the animal's "bearing" (Fig. 21). For convenience, we will always meas-ure the smaller angle of difference from the sun's azimuth to the animal's azimuth and indicate whether the measurement was clockwise or counter-clockwise, to avoid negative angles in our description of the animal's bearing.

The long period of the sun's azimuthal change is the annual cycle. Clearly, in middle latitudes, the sun will appear to be higher in the

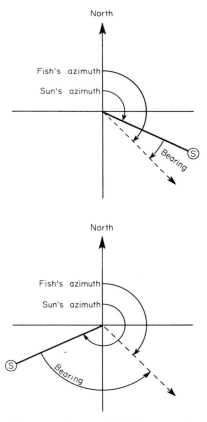

Fig. 21. Diagrams relating sun's azimuth, fish's azimuth, and the fish's bearing from the sun—the angle of difference measured from the sun's azimuth to the fish's azimuth.

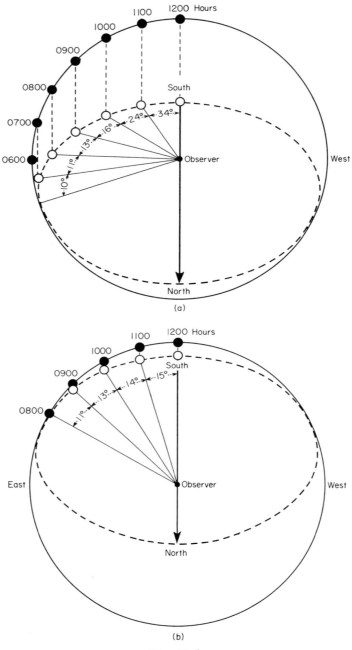

Fig. 22a,b

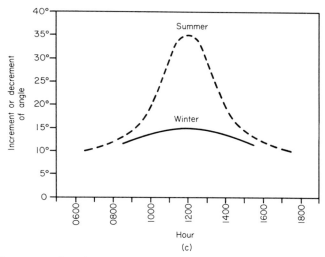

**Fig. 22.** Pattern of sun's movement at northermn middle latitudes in (a) summer and (b) winter; (c) comparison of loci of azimuthal change between winter (solid line) and summer (broken line). The night curve is the day curve for the same latitude in the opposite hemisphere.

sky during summer than during winter. The projected diurnal image of the sun on the earth in summer at the middle latitudes approaches the straight line which we saw at the equator during the equinoxes (and which, incidentally, also occurs at the summer solstice at 23.5°N—this latitude being the angle of maximum inclination of the earth toward the sun), while in winter it approaches the circle which we saw at the poles during the equinoxes. Therefore, the rate of angular change during a day in summer will increase much more rapidly toward noon than it does during a day in winter (Fig. 22). At the equator, between the two equinoxes, the sun deviates north in summer and south in winter; consequently, the equinoctial straight-line projection takes on a curved shape and the angular change becomes slightly less abrupt (Fig. 23) than it was at the equinoxes. One can readily interpolate the seasonal changes that occur in tropic latitudes to 23.5° north and south of the equator.

When we speak of the sun's being "higher in the sky," we immediately imply a second component of the sun's position—*vertical component* which is more simply known as the "altitude." The altitude of the sun is the angular elevation of the sun above the horizontal plane of observation, and it is conveniently measured as the angle from the horizontal plane to a sighting of the sun by an observer (Fig. 24).

The orienting animal, taking its bearing from the sun and com-

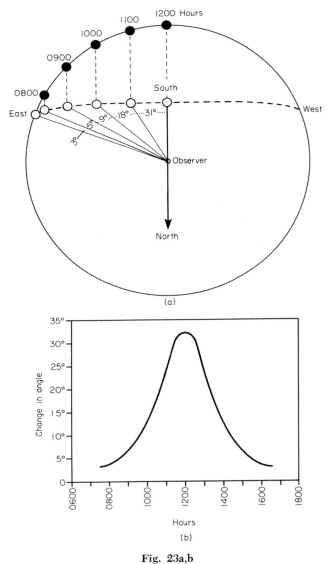

Fig. 23a,b

pensating for the inconstant rate of azimuthal change, must know at what rate to compensate in order to maintain a constant compass direction. If the animal compensates at a constant rate of 15°/hr, it will be perfectly oriented at the poles during the equinoxes but elsewhere would

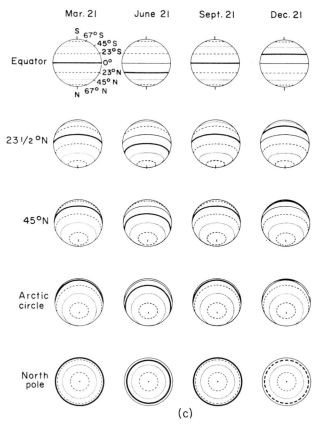

Fig. 23. Pattern of sun's movement at the equator during winter: (a) projected image of the sun on the sun's azimuth, (b) hour of day as the abscissa and increment or decrement of angle as the ordinate, (c) projection on the earth of the sun's movement as seen from selected latitudes during the various seasons. Note that as one views the sun from successively higher latitudes, the sun's arc rotates from an ellipse to a circle. Therefore, the rate of hourly change in the azimuthal angle decreases at noon and increases morning and evening until, at the poles, the hourly change in azimuthal angle is an equal rate of 15°/hr.

find itself off course. If the animal compensates at a gradually increasing rate (for example, if it compensates 5° the first hour, 6° the second, and 8° the third), it might be perfectly oriented at some middle latitude but only there. If the animal does not compensate at all but takes its bearing at a constant angle from the sun's azimuth in the morning and instantly changes its bearing by 180° at the culmination, it will be perfectly oriented at the equator during the equinoxes. To determine

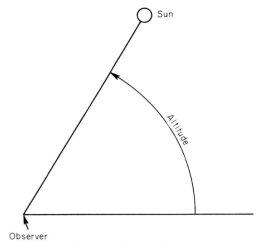

Fig. 24. Altitude of sun.

which rate of compensation to use, the animal must possess a biological clock to know the diurnal sun time and the season, and it must perceive the altitude of the sun to know the altitude. From this information it can "calculate" the bearing that must be taken to the sun, much as a human navigator employs his sextant. That an animal is actually able innately to accomplish something for which men require instruments, charts, and tables is quite incredible, but through experimental observations we know that this does, indeed, happen, and moreover, this ability has provided some of the strongest evidence for the existence of the biological clock.

In addition to the rate of compensation, the animal must apply a direction of compensation, for the deflection of the sun's path may be south or north, depending upon latitude and, in the tropics, on season, as noted previously. When the sun's path is deflected south, the sun appears to move clockwise; when the sun's path is deflected north, the sun appears to move counterclockwise. Therefore, sun-compass animals which live within the tropics or which during their migrations cross the equator or the tropics must have the ability to change their bearing to the sun's azimuth by 180° at the appropriate moment. How this is accomplished no one has discerned since the only unequivocal reference point for the shift is the culmination of the sun in the zenith, and if the animal crossed that particular latitude at an hour other than 1200, the reference point would be unseen.

A final unsolved puzzle which arises in sun-compass orientation by an animal in an aquatic environment is that the refraction of light by the greater density of the medium distorts the sun's altitude and compresses the 180° of the sun's arch through the sky into only 97.6° (Fig. 25). Accordingly, a rising sun, which appears on the horizon to the terrestrial animal, is at an altitude just above 41.2° for the aquatic animal. Similarly, throughout the day, except in the latitude and hour at which the sun culminates at the zenith, the sun's altitude seems greater to the aquatic animal than to the land animal. One could theorize that the animal is accustomed to seeing nothing other than the compressed circle

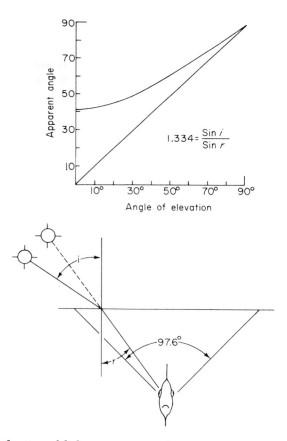

**Fig. 25.** Refraction of light rays entering the water surface. Because of the refraction, the view out of the water is compressed for a water animal into a visual angle of 97.6°. The apparent sun altitude is always higher than the actual. From Braemer (1960).

of light, but this is immediately refuted by the ability of fish to compensate for the refraction when they leap for flying insects above the surface of the water, or, as another example, the ability of the archer fish to knock down an insect from an overhanging branch with a projectile of water which the fish "spits" from the surface. Therefore, this altitudinal distortion should somehow be taken into account in the ultimate understanding of sun-compass orientation. Distortion of the sun's image by surface waves on the water could also present some difficulty to the orienting fishes. However, Leibowitz (1967, see also 1965), an eminent experimental psychologist of optical phenomena, contends that such diffuse signals are resolved by the nervous system into a fairly accurate mean position. Moreover, as discussed later, there is some evidence that dispersion of the rays by waves may actually make the sun's position be perceived more easily (Henderson, 1963).

Some workers have assumed that the compass animal utilizes only the azimuth of the sun's position, and furthermore, because the earth rotates through 360° in 24 hr, they have assumed that the animal can compensate at a constant rate of 15°/hr for the changing azimuth of the sun. On the other hand, the horizontal and vertical components of the sun's light are interdependent, as we have seen, and it would seem impossible to isolate a single factor, such as the azimuth of the sun, by which the animal would orient itself. Nor does it seem reasonable to assume that the animal's bearing from the sun's azimuth changes at a constant rate, for the animal might soon be far off course, the amount of error depending upon latitude and season. It should be mentioned, however, that at the latitudes of salmon migration, compensation of about 15°/hr might be sufficient for the rough orientation which could occur in much oceanic migration. On the other hand, Neave (1964) suggests that the orientation of salmon to the river mouth is often very precise and direct, and when coastal swimming occurs, it appears to be of an oriented, nonrandom nature.

In working with fish, it was first assumed that the animal has a biological clock and some ability to perceive small differences in both azimuthal and altitudinal angles of the sun. It was then theorized that both the sun's azimuth and the sun's altitude contributed to the animal's determination of its bearing and attempted experimentally to distinguish the relative roles which these two factors might play in directional orientation. At this point in our knowledge, we must still ignore the means by which the animal can distinguish between apparent clockwise and counterclockwise movement of the sun and by which it can compensate for the different refractive indices of water and air. Furthermore, there is no evidence of any unique property of the fish eye which permits any remarkable ability of light perception or compensation of refraction (Polyak, 1957; Walls, 1952).

## C. Experiments to Assess the Role of the Sun's Azimuth in Sun-Compass Orientation

To investigate the role of the sun's azimuth in sun-compass orientation, the sun's altitude was held as constant as possible by conducting all experiments in the same latitude and within as brief a portion of season as was compatible with training and experimentation. The general experimental plan was to determine if a fish, under rigorous laboratory conditioning, could be trained to respond predictably in a single given compass direction at two different times of day with substantially different solar azimuths. However, two times were selected, one in the forenoon and the other in the afternoon, for which the angles of deviation of the sun from its culmination were nearly identical in opposite directions (e.g., 45°E and 45°W of the noon position) and therefore for which the animal's bearing measured clockwise from the forenoon azimuth would be identical to its bearing measured counterclockwise from the afternoon azimuth. Thus, a "constant angle" choice would cause the fish to err by about 90°, a significant and readily apparent deviation.

In these studies, fish were trained, with the sun as the only point of reference, first to take food at a feeding disk oriented in a precise compass direction but subsequently, when food was found not always to provide adequate motivation, to seek cover in the one particular enclosure, out of 16 identical enclosures, which was placed in a given compass direction. The test tank (Figs. 26a and b) was designed to be rotated in order to eliminate orientation by any consistently positioned visual landmarks on the tank walls, and it was supplied with water from Lake Mendota at summer and early fall temperatures (a range of 16°–28°C). The tank was situated at the end of a pier on the south side of the lake, 26 meters from shore. The experimental fish, conditioned to seek refuge, was an immature bluegill, *L. macrochirus* Raf., 70 mm in length.

The bluegill lived throughout the experimental period in the test tank (Fig. 26a) under the open sky, and its behavior was scored according to an escape or cover-seeking response.

After the bluegill was released from the center of the circular tank, through a remote-control device to lower the plastic cage into a recessed position, the fish was free to swim toward the margin of the tank and enter one of 16 compartments. However, all the compartments, except

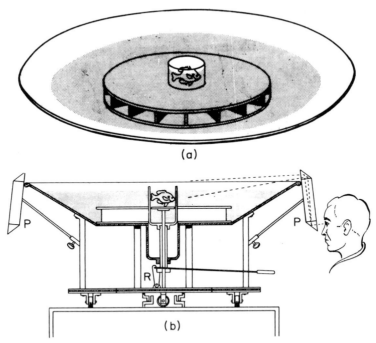

**Fig. 26.** Tank for training fish to a compass direction: (a) as seen from above showing hiding boxes; (b) side view showing periscopes (P) for indirect observation and the release lever (R) to permit the cage to be recessed by remote control when fish is released to score.

one in a predetermined compass direction, were closed by a metal band. The small containers were arranged beneath the release point so that the fish could not see any of them initially. To preclude other visual cues, the tank was randomly rotated between tests and a different chamber left open but always in the same compass direction (north, in this experiment). In addition, so that he would not be seen, the experimenter viewed the fish's behavior through a periscope (Hasler *et al.*, 1958; Hasler and Schwassmann, 1960).

Training tests were conducted at frequent intervals. Upon depression of the plastic cage, the fish was subjected to a weak electric shock to coerce it into seeking cover. Because the only cover available was the single open compartment at the north, the fish became conditioned always to swim north to escape. After training was complete, trials were conducted with all 16 of the boxes open and available to the fish. Each correct entry into the north compartment was counted as one point, and when the scores indicated that the fish had learned to select the north box consistently, the critical tests were begun.

Tests were made with all compartments open, both at 0800–0900 hr and at 1500–1600 hr, CST. The sun's azimuthal angle in the morning was about 90° to the right of the north line in which the fish must swim to reach the north box, while in the afternoon the azimuthal angle was about 90° to the left of the fish's northward swimming direction; or, according to our earlier definition, the fish's bearing in the morning was about 90° measured counterclockwise from the sun's azimuth while in the afternoon it was about 90° measured clockwise from the sun's azimuth. Despite the fact that 16 choices were available, the fish usually chose one of the three north-lying boxes (Figs. 27a and b).

On the other hand, as Fig. 27c demonstrates, when the sky was so heavily overcast that the experimenter could not detect the presence of the sun, test results showed completely unoriented responses. This clearly proves that the sun's azimuth (constant altitude again being assumed)

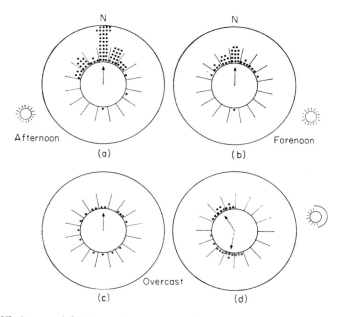

Fig. 27. Scores of fish trained to cover in the north compartment: (a) tested in the afternoon with 16 possible choices; (b) tested in the forenoon with 16 possible choices; (c) scores of fish B tested under completely overcast sky on two different days; (d) scores of fish B using an artificial light, where the altitude was the same as that of the sun. (●) Scores of fish trained to north and tested in the forenoon; (○) scores of same fish tested in the afternoon.

was the point of reference and that the fish compensated for the changing azimuth at different times of day, always escaping in the same compass direction.

A final crucial and definitive test was then conducted by substitution of an artificial "sun" indoors for the natural sun. It is obvious from Fig. 27d that the bluegill chose the hiding box which lay in the same angle from the artificial sun as that of the north box from the natural sun at that time of day (see Kramer, 1950).

Although these experiments used centrarchid fish as the subjects, subsequent investigations have shown that white bass can also be trained to swim in a given compass direction. Their responses affirmed that the sun's azimuth provided the orienting reference. Furthermore, the fact that two such different species, of different families, responded similarly provides some basis for a generalization of this mode of orientation to other fish species.

## 1. DISPLACEMENT OF TRAINED FISH TO THE EQUATOR

To test whether a fish trained at one latitude maintains the same degree of compensation at a different latitude, trained fish were displaced to an area where the rate of the sun's azimuthal change was radically different. A green sunfish was trained in Madison, Wisconsin, at 43°N until it was well oriented to a single compass direction. This fish was then transported to Belém, Brazil, at 1°S, where the change of the diurnal azimuth of the sun proceeds at a very different rate. (For example, in low latitudes at the equinox, one minute's time change from 1200 hr results in an azimuthal change of nearly 180°, while in middle latitudes at the equinox one hour's time change from 1200 hr results in an azimuthal change of only a little more than 15°; see Figs. 18b and 19a.) This fish, which had compensated correctly for the movement of the sun in Madison, Wisconsin, was quite disoriented when it applied the same degree of compensation to its movements in the tropics (Fig. 28). Thus clearly the sun's azimuth is a prime factor in fish orientation, and their ability to compensate for change is adjusted to the latitude in which they have lived. This, however, raises the problems of what factor determines this conditioned rate of compensation and what factor permits fish migrating over long distances to readjust their rate of compensation.

## D. Experiments to Assess the Role of the Sun's Altitude in Sun-Compass Orientation

Experiments in Madison, Wisconsin, showed that at middle latitudes, over time periods of sufficiently short duration, that the altitude of the

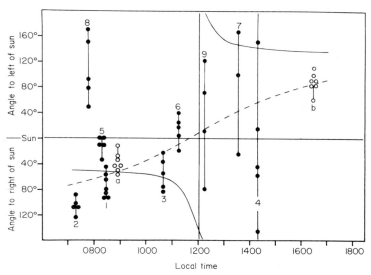

**Fig. 28.** Green sunfish after displacement from Madison to Belém. Solid line: change in sun's azimuth at Belém; broken line: same during tests at Madison.

sun remained essentially constant, a uniform rate of angular compensation per hour was sufficient for an orienting animal to maintain an approximately consistent course in one compass direction. In the tropics, however, an animal that orients itself by this method is misled considerably from its goal because a locus of the diurnal change in the sun's azimuth is very distorted (see Figs. 19 and 23). The slope of the locus changes abruptly twice during the daylight hours, in contrast to the gradually changing slope of the locus at higher latitudes and the complete linearity of the locus at the poles. Furthermore, the daily progression of the sun's azimuth in the tropics changes profoundly over the year, with a full 180° shift for a given hour taking place between summer and winter. Therefore, orienting animals living in or migrating through the tropic zones must have some mechanism that corrects for these radical changes. Clearly, as the fish displaced to Brazil proved, it is impossible that all orienting animals at all latitudes compensate at the same constant rate of about 15°/hr.

Braemer (1959) showed that sunrise and sunset were the references by which the fish synchronized its orientation rhythm to the sun's movement at a particular longitude. When natural sunrise and sunset were replaced by artificial onset and termination of electric light and these two events were delayed or advanced suddenly, a trained fish shifted its conditioned compass orientation to a new direction that was consistent with the specious "new latitude" or "new season" which the fish ascertained from the changed light rhythm. Moreover, Schwassmann and Braemer (1961) demonstrated that centrarchid fishes, trained and tested at 43°N, changed their response from their trained direction in a manner quantitatively correlated with their compensation for the contrived change in day length.

Thus, it seemed likely that the altitude of the sun functioned as a correction factor in orientation to the azimuth of the sun by indicating to the fish the seasonal or latitudinal rate of compensation necessary. Therefore, several experiments were designed with both the natural sun and artificial light to determine the importance of altitude and the interaction of altitude and azimuth in sun-compass orientation. Answers were sought to the following questions:

(1) If a fish swims toward its trained direction at any time of the day, does it do so by compensating for the sun's movement at a constant rate of 15°/hr or does its correction for the changing azimuth of the sun approximate the varying velocity of the azimuthal change? To obtain the answer to this question, fish were trained and tested near the equator where, to swim in a single compass direction, the fish would obviously have to depart from a constant rate of compensation.

(2) Does the fish learn the direction and rate of movement of the sun, or is the pattern of daily change in orientation to the sun's azimuth evident in the initial tests? Is this a conditioned phenomenon or an innate characteristic?

(3) Does the altitude of the sun affect the orientation of the fish to the sun's azimuth? Or will the fish continue to show an unchanged pattern in its bearing from the sun's azimuth when it is tested with the sun's altitude higher than that to which the fish is accustomed at the particular time of day? If the sun's altitude has no influence on the fish's bearing from the sun's azimuth, then the "azimuth hypothesis" alone may be considered to account for directed orientation in open water. This would mean that the orienting animal determines its position relative only to a horizontal component of the sun rather than to a combination of horizontal and vertical components.

In the following experiments on the effect of altitude, we trained green sunfish, *Lepomis cyanellus* Raf., bluegill, *Lepomis macrochirus* Raf., and two South American cichlids, *Cichlaurus severus* Heckel and *Uaru amphiacanthoides* Heckel. All fish were less than 1 year old and measured from 6 to 10 cm in length. In the first group of experiments, the fish lived under natural conditions in the latitude of their training. In the second group of experiments, the fish were trained for brief periods either indoors, under an electric light which substituted for the sun, or to the natural sun outdoors for not more than 5 min. Otherwise, they saw neither the sun's cycle nor daylight but were kept in artificially lighted tanks with the position of the lights fixed and with the light period regulated.

### 1. VARIABILITY OF COMPENSATION BY LATITUDE

Six green sunfish, living under natural conditions, were trained at 43°N latitude to swim in a single compass direction. When training was complete, these fish were tested in August at all times of day, and a continuous curve was plotted from the mean bearings which the fish took from the sun's azimuth at these times. These results are recorded in Fig. 29. The S-shaped curve is the locus of the change in the sun's azimuth, the circles are the means of the angles taken by the fish, and the vertical lines indicate the deviations. The fishes, then, are reasonably accurate in duplicating the actual rate of change of the sun's azimuth at

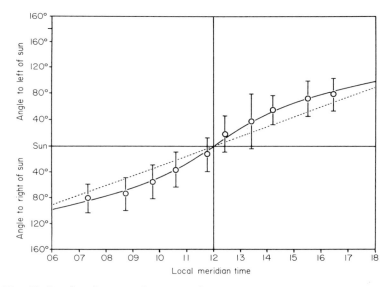

Fig. 29. Results of tests to determine if compensation is constant or quantitative, showing daily pattern of angular change which six trained green sunfish made to the sun's azimuth at 43°N. Data are shown as mean angles with their standard deviations computed for 10-hr intervals from 59 tests and 347 scores. The dotted curve is the compass direction which the fish would take relative to the sun's azimuth (when the sun's position is arbitrarily considered fixed) if the sun's azimuth changed at a constant rate of 15°/hr. The solid curve is the compass direction, as computed from the actual change of the sun's azimuth, which quantitatively compensating fish would take. Training and testing were conducted in Madison, Wisconsin, at 43°N. Training took place at different times of day, but principally in the forenoon.

this latitude. Their compensation for the change is not at a regular rate of 15°/hr; it varies as the sun's movement varies.

A comparable experiment was conducted with five cichlid fishes trained and tested at 1°S. These fishes also compensated in May for the true rate of the sun's azimuthal movement, the angle of their bearing from the sun's changing at a variable rate rather than a constant rate (Fig. 30). The curve of the fishes' angular change of bearing from the sun's azimuth closely approximates the curve of the angular change of the sun's azimuth itself.

Both experiments illustrate how accurately the trained fishes conform to the sun's movement at the latitude and during the season of training and testing. As the rate of change in the azimuthal angle increases, the rate of change in the fishes' angle of bearing increases by the same degree; as soon as the sun's azimuth and the fishes' azimuth are identical or at a straight angle, then the changes become reversed, the rate of change of bearing decreasing with decreasing rate of change of azimuth. The direction of the angles of bearing from the sun, as defined earlier, also becomes reversed. Because of this adaptation to a specific rate of change, it is evident that some factor other than an innate ability to compensate uniformly (i.e., 15°/hr) must influence the fishes' pattern of changing bearing from the sun's azimuth.

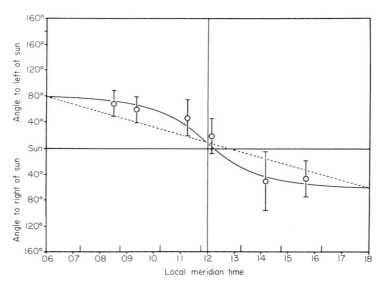

**Fig. 30.** Similar to Fig. 29, but data are from cichlid fishes trained and tested at 1°S, computed from 34 tests and 226 scores for six equal time intervals. Time of testing was 2 hr during the forenoon.

## 2. THE INFLUENCE OF CHANGING ALTITUDE OF THE SUN ON SUN-COMPASS ORIENTATION

The experiment which followed the tests just described provided an indication that the altitude of the sun is that factor. A Port cichlid, *Aequidens portalegrensis*, was trained to swim south and tested in Madison, Wisconsin, at 43°N in August. In October, the fish was retrained in Miami, Florida, at 26°N, when the altitude and the azimuth of the sun were comparable to those in August at Madison. Tests showed that the fish oriented well in both Madison and Miami (Fig. 31, open circles). Then the fish was flown to Solemar, Brazil, at 24°S and tested there in late November (Fig. 31, solid circles), where the sun's altitudes were always greater than those at the training latitudes at corresponding times. Moreover, the sun's azimuthal change (the sun's movement

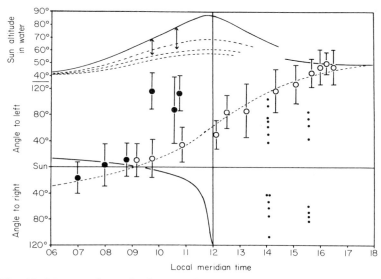

Fig. 31. Mean angles with their standard deviations of a trained Port cichlid. Each symbol is a mean of at least five individual scores: (○) tests at Madison (43°N) and Miami (26°N); (●) tests at Solemar (24°S); and (·) individual scores for tests in the afternoon on two successive days, showing a disruption of the orientation very similar to the equator in 1959 (see Hasler and Schwassmann, 1960). Broken s-shaped curve: average compass direction for Madison and Miami. Solid curve (intersecting broken curve at 0900 hr): compass direction as determined by the sun's movement at Solemar. Again the sun's position is arbitrarily considered fixed. Upper left: altitude curves of the sun at Madison and Miami, dotted, and at Solemar, solid. This demonstrated the effect of changed solar altitude on orientation by the azimuth of the sun. Fish were displaced in latitude, and training was conducted throughout the day.

around the horizon) occurs in the opposite direction from that in northern latitudes, i.e., clockwise in the northern hemisphere and counterclockwise in the southern hemisphere. Until 0900 hr, the Port cichlid changed its bearing from the sun's azimuth in the same amount and direction as it had done at the northern latitudes, the rate of change of its bearing measured clockwise from the sun increasing from sunrise until 0900 hr and the actual angle of bearing becoming smaller. After 0900 hr, as all three tests indicated, the fish's heading had shifted counterclockwise from the sun, as would have been appropriate in the early afternoon in the latitudes of training. This finding suggested that the change in altitude of the sun between 0900 and 1000 hr was sufficiently distinct that the fish was cued to a new season or latitude and began to change its rate of compensation accordingly, whereas in the early hours of the morning, the altitudinal difference was not enough to be perceived. The perceptible altitudinal difference, as measured between the altitude at the training latitudes at 1000 hr and that at the testing latitude at 1000 hr, was 10°. The true perceptible difference might have occurred between 0900 and 1000 hr and consequently be smaller than 10°.

The importance of altitude in determining rate of compensation was further confirmed by a study in which the rate of the sun's azimuthal change was kept nearly the same from one test to another and the altitude was artificially varied (Braemer and Schwassmann, 1963b). The fish used in this experiment was reared from the egg in a daily cycle of electric light and was tested in a tank which was always shaded from

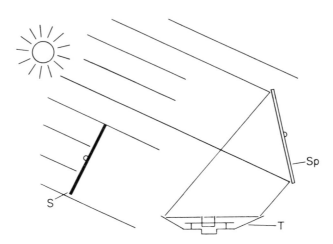

Fig. 32. Sketch of experimental tank (T) as in Fig. 26, with a shadow-throwing shield, S, and a mirror, Sp. From Braemer and Schwassmann (1963a).

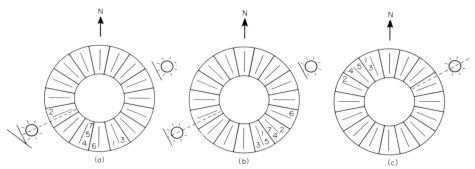

**Fig. 33.** Example of an experiment in which the sun is displaced in altitude by a reflecting mirror. Diagrams give the date and time of the experiment and position of the sun. (a) Azimuth is displaced by 180°; the sun is 26° higher. (b) Azimuth displaced 180°; height of sun unchanged. (c) Unchanged sun. The numbers are the individual scores of a trained fish given in sequence of choice. Redrafted from Braemer and Schwassmann (1963a).

direct sunlight by a large screen. A large mirror (Fig. 32) was used to reflect the sun into the testing tank at any altitudinal angle. When the mirror was in a vertical position, the light was reflected at the true altitude, but when the mirror was tilted downward, the reflected sunlight was at a greater altitude (Figs. 33 and 34). Thus, the altitude could be changed abruptly and the angles in which the fish swam relative to a single compass direction could be immediately determined. Since the sun's azimuthal angle in the brief time span remained virtually constant, any change in swimming direction was necessarily dependent upon the change in altitude. A significant deviation of the mean direction of swimming was observed under the experimentally elevated sun. This deviation

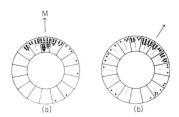

**Fig. 34.** Summary of mirror experiment. Each point represents a single choice. (a) The 11 directions of choice (M) with the sun's position transposed with a mirror but unchanged in altitude; all choices are in the trained direction, north. (b) Azimuth of sun changed by 180° and altitude of sun 22.5° higher. The difference between (a) and (b) in the fish's choice of direction is 34°. From Braemer and Schwassmann (1963a).

was shown by all of the fishes tested, and it was in the same direction as that chosen by our fish at Solemar.

Subsequent studies (Schwassmann and Hasler, 1964) have shown that it is the rate of change of the sun's altitude which determines the pattern of angular compensation. In addition, it was found that "inexperienced" fish, unaccustomed to any particular latitude of the sun, adapt more readily to a different rate of compensation than do "experienced" fish; this reaffirms that the rate of compensation is a factor conditioned by the latitudinal reaches of the fish and the solar altitudes there. However, Braemer (1960) and Braemer and Schwassmann (1963b) have shown that the ability to use the sun and to allow for some daily movement is present in the initial tests of completely inexperienced fish. The sun compass is innate; its application is conditioned.

## 3. SUMMARY

An animal which is orienting to the sun needs a fairly accurate sense of sun time and of season (this sense usually being described as a biological clock), and, by perceiving the sun's position in the sky, the animal "calculates" appropriate horizontal angles to the sun at different times of day in order to swim in a single compass direction. Since the compass orientation of animals necessarily occurs in the horizontal plane, the bearing from the sun is related to the sun's horizontal coordinate, its azimuth, and in fact, the rate of angular change of the animal's bearing from the sun is of the same magnitude as the rate of angular change of the sun's azimuth. In addition, the orienting animal must possess stereotaxis—a sense of its position in the environment—supplemented when possible by a visible horizon, so that the animal can relate its movements on the horizontal to the position of the sun as the altitude of the sun varies. The ability to change bearing from the sun's azimuth as shown in laboratory tests seems to be an intrinsic property of many migrating animals, while the rate of that change seems to be a learned trait.

Let us now speculate on how this sun-compass orientation might be used by the salmon in the North Pacific several hundred miles from shore. Obviously, a sun-azimuth mechanism, while it might aid in directing the salmon toward the correct continent (i.e., a Fraser River salmon might direct its course generally southeast and a Kamchatka salmon might direct its course generally west), is a very crude direction finder indeed. The prevailing currents alone, not to mention a storm, would certainly drive a fish migrating from the Gulf of Alaska to the Fraser River off its course, so that it might be carried many miles north of its goal. Since field observations indicate that the salmon are more accurate in locating the mouth of their natal river systems than orientation to the sun's azimuth alone would permit, we naturally search for other factors which might be perceived as a reference. The most relevant additional factor, according to experiments, is the sun's altitude, by which the fish can relate the azimuth it perceives to a particular latitude, providing a piscine counterpart of the sailor's sextant. Thus, in speculation, it may be able to change its bearing when the altitude indicates that its course has been shifted, and so arrive in reasonable proximity to the natal river system.

Of course, we have no understanding of how a fish can "compute" these functions to take advantage of the information they provide. Furthermore, we cannot say how the fish evaluates features of the sun when the view of the sun's arch across the sky is compressed from 180° to about 97°. Finally, we must bear in mind that waves on the rough water act as lenses and individually refract the sun's light. Thus, the sun, viewed vertically, appears to the fish as a cluster of suns, distorted continually in both shape and position.

## 4. MIGRATION AT NIGHT

Adult salmon are known to travel during the night in the sea; it is to take advantage of this activity that gill-net fishermen set their nets at night. Recently, moreover, it has been observed that the salmon's nocturnal movements appear to be directed. Professor Clifford Barnes of the University of Washington (1961) noted a school of large salmon migrating at right angles to his oceanographic research vessel which was on a course at night in the northeastern Pacific. Because of a luminescent sea, the fish were clearly visible and were seen to maintain a fairly straight course until out of sight. This observation would seem to justify a full-scale study of nocturnal movements of salmon and the mechanism of orientation at night. White bass, when displaced at night and tracked by means of a miniature transmitter (see following section), follow fairly straight courses (Hasler et al., 1969).

## 5. RANDOM MOVEMENT IN MIGRATION

Recently, Professor Saul Saila and R. A. Shappy (1963) constructed a mathematical model of homing in the salmon. They used a modern computer and a Monte Carlo system of analysis of randomness. In addition, they provided in their formula for some known characteristics of salmon during homing such as distance and speed of travel per day. Then by programming these various constants and variables into a high-speed digital computer, probability statistics were obtained for the accurate homing of salmon by random movement. This approach is very thought-provoking indeed, because their hypothesis contends that almost negligible orientation is required by a salmon to return near enough to the mouth of its home river system (i.e., 64 km north or south of its entrance into the ocean) that it could pick up the scent of the natal stream.

From the results of extensive trials, it became clear that return probability increased significantly with small changes in A, the coefficient of "directivity" or orientation. From these trials we found that an A value of 0.3 gave a return probability of 0.37 from a series of 100 hypothetical fish. This considerably exceeds the observed return of mature salmon tagged on the high seas, which range from less than 10 to 22%. Of course, these tag returns give a low estimate of actual returns because of mortality from handling and tagging as well as from incomplete tag reporting. There is no objective way at present to evaluate these losses. A value of A as high as 0.3, however, certainly does not suggest that precise orientation on the part of the fish is necessary.

One of the first arguments against the paramount role of random movement stipulated by Saila's hypothesis is that, if the salmon were migrating largely at random, the sea fishery would continue to catch large numbers of adult, sexually mature salmon even after the presumably successfully migrating adults had arrived at their home streams. This is not the case, as can be seen from the extensive experimental fishing efforts of the Canadians under the leadership of Neave (see Figs. 9a and 9b). The salmon destined to spawn in a given season move out of the Gulf of Alaska and distribute themselves to the coastal areas near their respective home streams in Alaska, Canada, the northwestern United States, and Asia.

A second, and biologically powerful, argument is that, if there were a great deal of random distribution, the concentration of the genes at the spawning ground could not continue, and the evolution of races with distinctive meristic characters and different spawning seasons would not proceed as it has in the past.

Moreover, Saila has assumed some small amount of sun orientation to account for his A value of 0.3. However, if there were only one clear day out of ten to give the fish sun orientation—and, indeed, in the North Pacific such continually cloudy conditions may be common—this would be insufficient to account for even so small an A. Purely random movement would not take the salmon in the direction of home, yet they do reach home. Incidentally, this point again emphasizes the need for exploration of possible cues other than the sun which might explain salmon migrations.

Data of salmon migrations suffer from a lack of details on the actual fish uses its sun-orienting capabilities. Because previous tracking methods have been proposed. It is hoped they will provide evidence of the salmon's type of movement, whether directed or random, and suggest how the fish uses its sun-orienting capabilities. Because previous tracking methods have been to little avail, we have had no accurate data on the consistency of a salmon's course.

As a result of dissatisfaction with the Ping-Pong ball float used in earlier tracking experiments on open-water orientation of displaced white bass, an ultrasonic transmitter was developed that was small enough to fit into the body cavity of salmon or other migrating fish such as the white bass. Preliminary tests on the ultrasonic transmitters were made during the summers of 1963 and 1964 using the white bass of Lake Mendota (Henderson *et al.*, 1966). Subsequently, during the white bass spawning seasons of 1965 (Hasler *et al.*, 1969) and 1966, a large series of ultrasonic tracking experiments were conducted with fish that had been displaced from their spawning grounds. The results of these tracks essentially confirmed the earlier tracking study with Ping-Pong ball floats (bobber floats), namely, that the white bass return to their spawning ground in a directed and well-oriented fashion (Fig. 35). Future tracking experiments on the white bass will include fish that have had their light cycles shifted, fish that have been blinded, and fish with their olfactory sense destroyed. Some tracks will be conducted during the night and some tracks will start on the spawning grounds, with displacement, to determine the pattern and timing of movement on and off the spawning ground. An ultrasonic tracking project involving the salmon commenced in the summer and fall of 1967 under the auspices of the National Science Foundation and Office of Naval Research.

## 6. OTHER CUES

While gravity, magnetic fields, oceanic waves, and flotsam have been imputed as possible cues for migrating fishes, no one has designed and carried out critical experiments to evaluate them. The new ultrasonic tracking transmitter system provides a tool for such studies in the future.

## E. Migration from the Stream to the Sea

The migratory journey of the young salmon smolt seaward is undoubtedly a more passive type of movement than the subsequent return migration upstream because the downstream currents favor and supple-

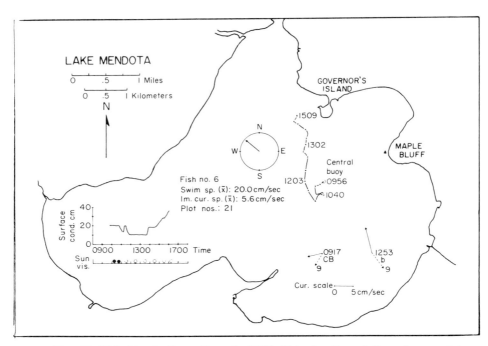

Fig. 35. Fish No. 6. Date: June 3, 1965. Note location of the release point (labeled Central buoy) relative to Maple Bluff and Governor's Island. The black triangle near Maple Bluff indicates the position of the fyke net used to capture all the fish tracked during the spawning season. The mean wind vector is shown in a unit circle; the length of the vector is an indication only of angular dispersion about the mean and is not indicative of wind speed. The 1-meter water current vectors are shown by a solid line, whereas the deeper currents are shown by a broken line with the specific depth indicated. The length of the current vectors represent current speed; the scale for vector length is given by Cur. scale in the figure. The letters be neath each series of current vectors indicate the locations of the measurements on the lake; for example, the letters CB indicate the measurements were taken at Central buoy, and the letter b indicates the measurements were taken at position b which is shown on the map of the track itself. The mean swimming speed is given along with the mean 1-meter current speed (labeled Swim sp. ($\bar{x}$) and 1 m. cur. sp. ($\bar{x}$), respectively). The actual number of position determinations of the fish is given by Plot nos. The graph labeled Surface cond. shows the height of surface waves in centimeters estimated at each position determination.

ment the smolt's swimming. Nevertheless, much evidence indicates that this movement, too, is oriented and not merely random drifting.

Hoar (1953, 1954) pointed out the importance of age and physiological condition of salmon as determinants of the manner and time of their response to environmental factors. He has shown that young sockeye salmon are able to maintain their position in the stream, despite the strong downstream currents, by day and by night at an early age (larval to early fry stages), but that upon attainment of a certain physiological state, usually dictated by age, the young sockeye fry cease to respond to the restraining stimulus at night. They then allow themselves to be nocturnally transported downstream, holding their positions only during the day, until they reach the "nursery" lake. The salmon's movements in the lake become guided by a new set of stimuli, which are later abandoned for yet another set at the smolt stage when the fishes head toward sea.

Neave (1955) also stressed the importance of light in the fry stages of chum and pink salmon. In the larval and very early fry stages the fish remains buried in the gravel of its natal stream except for brief feeding forays. However, as the fry matures and begins its migration, it rises from the bottom at night and returns at daybreak to bury itself, remaining under cover until dark. In short streams the migratory journey to the ocean may be accomplished overnight. In longer streams there is a progressive of nocturnal drifting and diurnal hiding from the sun over a period of several days. This pattern of seaward migration occurs among those species—chum and pink—that migrate when only a few weeks old, in the fry stage.

Those species, including sockeye, coho, Chinook, and Atlantic salmon, that migrate to the ocean as smolts after several seasons in the natal waters participate more actively in the seaward journey. It is in these species that complex oriented movement particularly appears. In searching for directive cues, one is naturally inclined first to test those references which the salmon appears to use on the other portions of its migratory cycle, namely, odor and sun-compass orientation. While coho salmon fry could not be trained to respond to different odors, it seems unlikely that odor would play a role in the downstream migration. The direction of the water currents would certainly preclude their carrying the odor of the sea, and it is also improbable that airborne odors of the sea could be a factor in the seaward migration.

Johnson and Groot (1963) have examined the importance of sun-compass orientation among smolts undertaking their seaward migration, through a study of the movements of sockeye salmon smolts in Babine Lake, their juvenile home, during their migration around the lake to the outlet (Fig. 36). Their observations indicated that as the migratory date progressed, the young salmon changed their directional headings to enable them to reach the lake's outlet; for example, smolts from the Morrison Arm moved south down that arm, made the appropriate westerly movement when they came abreast of the arm of Babine Lake (North Arm) which leads to the outlet, and then pursued a generally

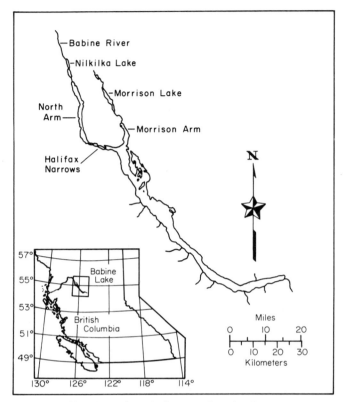

Fig. 36. Babine Lake, British Columbia.

northward course. These observations regarding the natural movement of smolt in the lake were confirmed by Johnson and Groot, who removed the smolts periodically during this migratory season and placed them into a vertically walled tank with a transparent bottom. By recording the average headings of the school during the noon period when there were no shadows in the tank, Johnson and Groot again found that the bearings changed as the season progressed, hence suggesting that the sun was being used for orientation and that the salmon smolts' response to it changed during the summer of their migration enabling them to reach the outlet. The fish took random positions on cloudy days. Groot (1965) suggests from preliminary experiments that polarized light may serve as a directional cue.

It would appear that laboratory studies on the fishes' use of sun orientation, combined with a biological clock mechanism, find a field application, at least in this phase of the migratory cycle. Still more recent evidence that sun-compass orientation in fishes occurs in the natural habitat has been presented by Winn *et al.* (1964); these workers found that parrot fish on the shoals of Bermuda use the sun's position to find the direction from their feeding grounds to their caves. Moreover, Braemer (1959), working in the University of Wisconsin laboratory, has shown that coho salmon smolt, of the age at which the stream to sea migration takes place, can be readily trained in the testing tank (see Fig. 26) to swim in a specific compass direction and that they like the sunfish, respond accurately at any time of day.

TRAVERSING THE ESTUARY

Once the smolt has left the river mouth, it must pass through an estuarine environment where the fish gradually adapts itself from a freshwater to a marine habitat. The salinity gradients in the estuaries of salmon rivers present a physiological stress to the migrating fish, but Black (1957) has found that the salmon's mechanism for regulating water balance under changing salt concentrations adjusts fairly rapidly. McInerney (1964) has found that smolts prefer increasingly saline waters as their age increases. Working in the laboratory, he discovered that the different salinity levels preferred as time progressed corresponded to the different salinity concentrations in a typical estuary with progression toward the sea. McInerney theorized, then, that the smolts are physiologically held within the limits of increasingly saline thresholds, until gradually the tolerance for seawater is established.

These experiments of smolt migration through estuaries should not be considered to demonstrate any orientative mechanism. Rather they test the fishes' preference for differing salinities, and these preferences serve to increase the searching behavior. By increased random movement plus appropriate responses to still unknown directive stimuli, the salmon traverses the estuary and reaches the seawater to which it has finally become physiologically tolerant.

IV. SUMMARY

The life cycle and homing migrations of salmon are defined within three distinct phases: (1) A sojourn in the inland stream beginning with the fertilized egg, through at least the larval and early fry stages, terminating with a short downstream migration of fry of pink or chum salmon or with a long downstream migration of smolts of sockeye, coho, Chinook, or Atlantic salmon to the sea; (2) a period of rapid growth and far-ranging movement in the open sea, concluding at sexual maturity with an extensive homing migration of several hundred kilometers to the mouth of the home-river system; and (3) a strenuous upstream drive over barriers and waterfalls with rejection of tributary after tributary until the stream of origin is reached, where spawning takes place and the enervated salmon's life cycle is ended. The fertilized eggs in the gravel-covered redds remain to perpetuate the cycle.

In this chapter stress has been made on the sensory mechanisms which the salmon possess and the environmental cues which they detect in order to find their way out of the river, around in the sea, and back to the river again with such remarkable accuracy. I have reviewed and discussed some of the more credible theories which attempt to explain the salmon's return to its home river, and while extensive use has been made of the work of others, generally more attention has been paid to the details of experiments which were conducted in and from our laboratory.

Because the natal stream is both the beginning and the conclusion of the salmon's journey, the complete hypothesis must start and end with the characteristics of those waters. It is proposed that each home stream acquires, from springs, soils, and plant communities of its drainage basin and in its bed, a unique organic quality which young salmon learn in the early weeks of life, and recognition of which remains imprinted throughout their adult lives. That a unique stream factor exists and that fish may become conditioned to respond to that factor have been abundantly substantiated in the laboratory.

When the young salmon has attained a certain, still-undefined physiological state that causes it to abandon the stimuli which have heretofore kept it in the natal pools and riffles, the fry or smolt begins its downstream drift toward the sea. Among some salmon species which begin their sea life when only a few weeks old, this downstream migration may be a completely passive movement, while among those species which remain in a stream or lake for 1 or 2 years, the migration may be more actively directed, perhaps by a sun-compass mechanism. Here, as with other phases of the salmon's migratory movements, much remains to be discovered.

Once in the ocean, the Pacific salmon's movements carry them far over the North Pacific Ocean in search of productive feeding grounds. There, and perhaps with the greatest concentration in the Gulf of Alaska, the fish remain for 2 or 3 years until full growth and maturity are attained. Similarly, the Atlantic salmon range considerably distances in the North Atlantic, usually for 2 years or more, until sexual maturity impels them to leave the sea. Then the ripe salmon begin one of the

most remarkable journeys to be found in the animal kingdom. To explain
home finding from great distances at sea, one cannot reasonably invoke
the odor hypothesis as we envisage it for locating the mouth of the
river system or for stream searching, but the selection of possible guiding
beacons is very sparse indeed. In the open sea neither the bottom nor
land references are available as orienting signals. Thus, it is theorized
that the sun is one of the possible celestial cues from which salmon might
take a bearing, and indeed, both field and laboratory studies have provided unequivocal evidence that fish can use the sun for direction finding. Furthermore, it has been demonstrated that both the azimuth and
the altitude of the sun play roles, and that the fish may have the capacity
for a simple type of navigation. Field studies have thus far been less
definitive, but now that an ultrasonic transmitter has been developed
by which salmon can be equipped and tracked, new evidence may be
anticipated concerning the degree of directed orientation in a salmon's
movements in open water. Although it is felt that one of the references used in open-water orientation has been found, obviously the
surface has only been tapped. Salmon are known to travel in the sea
at night, yet there has been only limited experimentation with white
bass and as yet none with salmon on nocturnal orientation. Moreover, information is lacking on the quality and mechanisms of their orientation
on cloudy days.

Finally, after the salmon have reached the coastline within reasonable proximity to the home-river system, it is suggested that they may
detect the odors of the main rivers and thereby select the one correct
river by recognition of the odor imprinted as they left for sea. The
details of their journey through the changing salt concentrations and
variable currents of the estuarine waters to reach the mouth of the
home-river system are still vague, but it is felt that more can be learned
through ultrasonic tracking.

Having successfully entered the main river, the homing fishes move
continuously upstream in response to a positive rheotaxis. At each confluence of tributaries the salmon remain in the main river or make only
exploratory forays unless the particular tributary conveys the scent of
the natal stream. At the first perception of the home odor, the salmon
selects the tributary which carries it and continues the upstream journey,
intermittently picking up the odor track; it is presumed that the fish
does not stay within an odor cloud, for its olfactory sense would soon
become fatigued and unable to detect the guiding scent. At each new
entry of a stream the salmon must determine if it bears the unique home
odor and, if not, reject it, pressing onward until the ancestral stream
is perceived and chosen. In addition to laboratory tests of the stream

factor, fish conditioning, and odor imprinting and recall, several experiments conducted in the field have supported this hypothesis but have also indicated the need for additional definitive field tests. For example, as suggested, homing adults might be decoyed into a different stream by water diverted from the home stream; or fry could be conditioned to an odorous chemical or substitute solution which, upon their return from the sea a few years later, might be used to decoy them into a stream which is downstream from the home creek.

A plethora of new approaches and new ideas to unravel the fascinating mystery of salmon migration await future experimenters. The unsolved problems of migration are still legion, a fact which should provide an impetus toward further investigations by interested scientists. It is hoped that this account will stimulate new and original theories which can be put to test by experimentation and thus fill in the gaps in the salmon's story, which has been only partially told by present research. Moreover, it will be these future studies which will supply essential data for the wise management of salmon stocks now threatened with extinction.

## REFERENCES

Adrian, E. D., and Ludwig, C. (1938). Nervous discharges from the olfactory organs of fishes. *J. Physiol.* (*London*) **94**, 441–460.

Barnes, C. A. (1961). Personal communication.

Birman, I. B. (1958). On the occurrence and migration of Kamchatka salmon in the northwestern part of the Pacific Ocean. *Fisheries Res. Board Can. Transl. Ser.* No. 180 1–15 (mimeo).

Black, V. S. (1957). Excretion and osmoregulation. *In* "The Physiology of Fishes" (M. E. Brown, ed.), Vol. 1, pp. 169–203. Academic Press, New York.

Blair, A. A. (1956). Atlantic salmon tagged in east coast Newfoundland waters at Bonavista. *J. Fisheries Res. Board. Can.* **13**, 225–232.

Braemer, W. (1959). Versuche zu der im Richtungsfinden der Fische enthaltenen Zeitschatzung. *Zool. Anz.* (Suppl). **23**, 278–288.

Braemer, W. (1960). A critical review of the sun-azimuth hypothesis. *Cold Spring Harbor Symp. Quant. Biol.* **25**, 413–427.

Braemer, W., and Schwassmann, H. O. (1963a). Vom Rhythmus der Sonnenonentierung bei Fischen am Aquator. *Ergeb. Biol.* **26**, 181–201.

Braemer, W., and Schwassmann, H. O. (1963b). Unpublished study.

Buckland, F. (1880). "Natural History of British Fishes." Unwin Bros., S.P.C.K. (cited in Jones, 1959).

Carlin, B. (1955). Tagging of salmon smolts in the River Langan. *Rept. Inst. Freshwater Res., Drottningholm* No. 36, 57–74.

Carlin, B. (1962). Markt lax aterfangad vid Gronland. *Medd. Laxforskningsinst.* No. 8 (mimeo).

Carlin, B. (1963). Laxforskning med halkort. Sartryck ur IBM-nytt. No. 2.

Cleaver, F. C. (1964). Origins of high seas sockeye salmon. *U. S. Fish Bull.* **63**, 445–476.

Collins, G. B. (1952). Factors influencing the orientation of migrating anadromous fishes. *U. S. Fish Bull.* **52**, No. 73, 375–396.

Craigie, E. H. (1926). A preliminary experiment on the relation of the olfactory sense to the migration of the sockeye salmon (*Oncorhynchus nerka* Walbaum). *Trans. Roy. Soc. Can.* **20**, 215–224.

Creutzberg, F. (1959). Discrimination between ebb and flood tide by migrating elvers (*Anguilla vulgaris* Turt.) by means of olfactory perception. *Nature* **184**, 1961–1962.

DeLacy, A. C. (1967). Personal communication.

Donaldson, R., and Allen, G. H. (1957). Return of silver salmon, *Oncorhynchus kisutch* (Walbaum) to point of release. *Trans. Am. Fisheries Soc.* **87**, 13–22.

Fagerlund, U. H. M., McBride, J. R., Smith, M., and Tomlinson, N. (1963). Olfactory perception in migrating salmon. III. Stimulants for adult sockeye salmon (*Oncorhynchus nerka*) in home stream waters. *J. Fisheries Res. Board Can.* **20**, 1457–1463.

Fassett, N. C. (1960). Personal communication.

Griffin, D. R. (1964). "Bird Migration," Anchor Books (Sci. Study Ser.). Doubleday, New York.

Groot, C. (1965). On the orientation of young sockeye salmon (*Oncorhynchus nerka*) during their seaward migration out of lakes. *Behaviour Suppl.* **14**, 1–198.

Gross, M. G., Barnes, C. A., and Riel, G. K. (1965). Radioactivity of the Columbia River effluent. *Science* **149**, 1088–1090.

Gunning, G. E. (1959). The sensory basis for homing in the longer sunfish *Lepomis megalotis* (Rafinesque). *Invest. Indiana Lakes* **5**, 103–130.

Hara, T. J., Ueda, K., and Gorbman, A. (1965). Electrocenephalographic studies of homing salmon. *Science* **149**, 884–885.

Hartman, W. L., and Raleigh, R. F. (1964). Tributary homing of sockeye salmon at Brooks and Karluk Lakes, Alaska. *J. Fisheries Res. Board Can.* **21**, 485–504.

Hartt, A. C. (1962). Movement of salmon in the North Pacific Ocean and Bering Sea as determined by tagging 1956–58. *Intern. North Pacific Fisheries Comm.*, *Bull.* No. 6, 1–157.

Hasler, A. D. (1954). Odour perception and orientation in fishes. *J. Fisheries Res. Board. Can.* **11**, 107–129.

Hasler, A. D. (1956a). Perception of pathways by fishes in migration. *Quart. Rev. Biol.* **31**, 200–209.

Hasler, A. D. (1956b). Influence of environment reference points on learned orientation in fish (*Phoxinus*). *Z. Vergleich. Physiol.* **38**, 303–310.

Hasler, A. D. (1957). Olfactory and gustatory senses of fishes. *In* "The Physiology of Fishes" (M. E. Brown, ed.), Vol. 2, pp. 187–207. Academic Press, New York.

Hasler, A. D., and Schwassmann, H. O. (1960). Sun orientation of fish at different latitudes. *Cold Spring Harbor Symp. Quant. Biol.* **25**, 429–441.

Hasler, A. D., and Wisby, W. J. (1950). Use of fish for the olfactory assay of pollutants (phenols) in water. *Trans. Am. Fisheries Soc.* **79**, 64–70.

Hasler, A. D., and Wisby, W. J. (1951). Discrimination of stream odors by fishes and relation to parent stream behavior. *Am. Naturalist* **85**, 223–238.

Hasler, A. D., and Wisby, W. J. (1958). The return of displaced largemouth bass and green sunfish to a "home" area. *Ecology* **39**, 289–293.

Hasler, A. D., Horrall, R. M., Wisby, W. J. and Braemer, W. (1958). Sun orientation and homing in fishes. *Limnol. Oceanog.* **3**, 353–361.

Hasler, A. D., Gardella, E. S., Horrall, R. M., and Henderson, H. F. (1969). Open-water orientation of white bass *Roccus chrysops* (Rafinesque), as determined by ultrasonic tracking methods. *J. Fisheries Res. Board Can.* **26**, 2173–2192.

Heath, J. P. (1960). Penetration of fresh water through two oceanic stream bars. *Ecology* **41**, 381.

Henderson, H. F. (1963). Orientation in pelagic fishes. I. Optical problems. II. Sonic tracking. Ph.D. Thesis, University of Wisconsin [Univ. Microfilms, Ann Arbor, Michigan (64-643)].

Henderson, H. F., Hasler, A. D., and Chipman, G. G. (1966). An ultrasonic transmitter for use in studies of movements of fishes. *Trans. Am. Fisheries Soc.* **95**, 350–356.

Hoar, W. S. (1953). Control and timing of fish migration. *Biol. Rev.* **28**, 437–452.

Hoar, W. S. (1954). The behavior of juvenile Pacific salmon, with particular reference to the sockeye (*Oncorhynchus nerka*). *J. Fisheries Res. Board Can.* **11**, 69–97.

Horrall, R. M. (1961). A comparative study of two spawning populations of the White Bass, *Roccus chrysops* (Rafinesque), in Lake Mendota, Wisconsin, with special reference to homing behavior. Ph.D. Thesis, University of Wisconsin [Univ. Microfilms, Ann Arbor, Michigan (61-3116)].

Huntsman, A. G. (1942). Return of a marked salmon from a distant place. *Science* **95**, 381–382.

International North Pacific Fisheries Commission. (1963). Annual report. Vancouver, B.C., Canada.

International Pacific Salmon Fisheries Commission. (1949). Annual report. New Westminster, Canada.

Johnson, W. J., and Groot, C. (1963). Observations on the migration of young sockeye salmon (*Oncorhynchus nerka*) through a large, complex lake system. *J. Fisheries Res. Board Can.* **20**, 919–938.

Jones, J. W. (1959). "The Salmon." Harper, New York.

Kramer, G. (1950). Orientierte Zugaktivitat gikafigter Singvogel. *Naturwissenschaften* **37**, 188.

Kramer, G. (1952). Experiments on bird orientation. *Naturwissenschaften* **94**, 265–285.

Kramer, G., Pratt, J. G., and von Saint Paul, U. (1956). Directional differences in pigeon homing. *Science* **123**, 329–330.

Kyle, H. M. (1926). "The Biology of Fishes." Sidgwick & Jackson, London.

Leibowitz, H. W. (1965). "Visual Perception." Macmillan, New York.

Leibowitz, H. W. (1967). Personal communication.

Lindroth, A. (1963). Salmon conservation in Sweden. *Trans. Am. Fisheries Soc.* **92**, 286–291.

Lissmann, H. W. (1954). Direction finding in fish. *Advan. Sci.* **11**, 69–71.

McInerney, J. E. (1964). Salinity preference: an orientation mechanism in salmon migration. *J. Fisheries Res. Board Can.* **21**, 995–1018.

Manzer, J. I. (1956). Distribution and movement of young Pacific salmon during early ocean residence. *Fisheries Res. Board Can., Pacific Coast Sta. Progr. Rept.* **106**, 24–28.

Manzer, J. I. (1964). Preliminary observations on the vertical distribution of Pacific salmon (Genus *Oncorhynchus*) in the Gulf of Alaska. *J. Fisheries Res. Board Can.* **21**, 891–903.

Manzer, J. I., and Shepard, M. P. (1962). Marine survival, distribution and migration of pink salmon off the British Columbia coast. *Inst. Fish., Univ. Brit.*

Columbia, H. R. MacMillan Lectures Fish., Pink Salmon Symp., 1960 pp. 113–122.

Margolis, L., Cleaver, F. C., Fukerda, Y., and Godfrey, H. (1966). Sockeye Salmon in offshore waters. Intern. North Pacific Fisheries Comm., Bull. No. 20, Part VI, 1–70.

Neave, F. (1955). Notes on the seaward migration of pink and chum salmon fry. J. Fisheries Res. Board Can. 12, 369–374.

Neave, F. (1962). Personal communication.

Neave, F. (1964). Ocean migrations of Pacific salmon. J. Fisheries Res. Board Can. 21, 1227–1244.

Neave, F., Manzer, J. I., Godfrey, H., and LeBrasseur, R. J. (1962). High-seas salmon fishing by Canadian vessels in 1962. Fisheries Res. Board Can., Rept. No. 563 1–59 (mimeo).

Pardi, L., and Papi, F. (1952). Die Sonne als Kompass bei Talitrus saltator (Montagu) Amphipoda-Crustacea. Naturwissenschaften 39, 262–263.

Polyak, S. (1957). "The Vertebrate Visual System." Univ. of Chicago Press, Chicago, Illinois.

Powers, E. B. (1939). Chemical factors affecting the migratory movements of the Pacific salmon. Am. Assoc. Advanc. Sci., Publ. 8, 72–85.

Powers, E. B., and Clark, R. T. (1943). Further evidence on chemical factors affecting the migratory movements of fishes, especially the salmon. Ecology 24, 109–113.

Pritchard, A. L. (1943). Results of the 1942 pink salmon marking at Morrison Creek, Courtenay, B. C. Fisheries Res. Board Can., Progr. Rept. 57, 8–11.

Ricker, W. E., and Robertson, A. (1935). Observations on the behaviour of adult sockeye salmon during the spawning migration. Can. Field Naturalist 49, 132–134.

Saila, S. B., and Shappy, R. A. (1963). Random movement and orientation in salmon migration. J., Conseil Perma. Intern. Exploration Mer 28, 153–166.

Sato, R., Hiyama, Y., and Kajihara, T. (1966). The role of olfactory in return of chum salmon, Oncorhynchus keta (Walbaum), to its parent stream. Physiological basis on fish migration in the Pacific area. Proc. 11th Pacific Sci. Congr., Tokyo, 1966 Vol. 7, 20.

Schäffer, E. (1919). Der Aal auf dem Lande. Schweiz. Fischerciztg. 27, 79–80.

Scheer, B. T. (1939). Homing instinct in salmon. Quart. Rev. Biol. 14, 408–430.

Scheuring, L. (1930). Die Wanderungen der Fische. I and II. Ergeb. Biol. 1, 405–691; 7, 4–304.

Schwassmann, H. O., and Braemer, W. (1961). The effect of experimentally changed photoperiod on the sun-orientation rhythm of fish. Physiol. Zool. 34, 273–286.

Schwassmann, H. O., and Hasler, A. D. (1964). The role of the sun's altitude in sun orientation of fish. Physiol. Zool. 37, 163–178.

Snedecor, G. W. (1946). "Statistical Methods." Iowa State Coll. Press, Ames, Iowa.

Sømme, S. (1941). On the high age of smolts at migration in northern Norway. Skrifter Norske Videnskaps-Akad. Oslo, I: Mat.-Naturv. Kl. No. 16, 1–5.

Stuart, T. A. (1957). The migrations and homing behaviour of brown trout. Freshwater Salmon Fishery Res., Edinburgh, Bull. No. 18, 1–27.

Stuart, T. A. (1962). The leaping behavior of salmon and trout at falls and obstructions. Freshwater Salmon Fishery Res., Edinburgh, Bull. No. 28, 1–46.

Teichmann, H. (1957). Das Riechvermögen des Aales (Anguilla anguilla L.). Naturwissenschaften 44, 242.

von Frisch, K. (1941). Die Bedeutung des Geruchsinnes im Leben der Fische. *Naturwissenschaften* **29**, 321–333.

von Frisch, K. (1949). Die Polarisation des Himmelslichtes als orientierender Faktor bei den Tanzen der Bienen. *Experientia* **4**, 142–148.

von Frisch, K. (1950a). Die Sonne als Kompass in Leben der Bienen. *Experientia* **6**, 210–221.

von Frisch, K. (1950b). "Bees: Their Vision, Chemical Senses, and Language." Cornell Univ. Press, Ithaca, New York.

von Holst, E. (1950). Die Arbeitsweise des Statolithenapparates bei Fischen. *Z. Vergleich. Physiol.* **32**, 60–120.

Walker, T. J., and Hasler, A. D. (1949). Detection and discrimination of odors of aquatic plants by the bluntnose minnow (*Hyborhynchus notatus*, Raf.). *Physiol. Zool.* **22**, 45–63.

Walls, G. L. (1952). "The Vertebrate Eye and Its Adaptive Radiation." Cranbrooke Press, Bloomfield Hills, Michigan.

Wanemacher, J. M., Twenhofel, W. H., and Raasch, G. O. (1934). The paleozoic strata of the Baraboo area, Wisconsin. *Am. J. Sci.* **28**, 1–30.

Ward, H. B. (1921a). Some of the factors controlling the migration and spawning of the Alaska red salmon. *Ecology* **2**, 235–254.

Ward, H. B. (1921b). Some features in the migration of the sockeye salmon and their practical significance. *Trans. Am. Fisheries Soc.* **50**, 387–426.

Ward, H. B. (1939a). Psychology of salmon. *Proc. Wash. Acad. Sci.* **29**, 1–14.

Ward, H. B., (1939b). Factors controlling salmon migration. *Am. Assoc. Advance. Sci., Publ.* **8**, 60–71.

Wickett, W. P. (1958). Adult returns of pink salmon from the 1954 Fraser River planting. *Fisheries Res. Board, Can., Prog. Rept.* 111, 18–19.

Winn, H. E., Salmon, N., and Roberts, N. (1964). Sun-compass orientation by parrot fishes. *Z. Tierpsychol.* **21**, 798–812.

Wisby, W. J. (1952). Olfactory responses of fishes related to parent stream behaviour. Ph.D. Thesis, University of Wisconsin.

Wisby, W. J. (1958). Techniques for investigating the ecological aspects of the behaviour of fishes. Unpublished manuscript.

Wisby, W. J., and Hasler, A. D. (1954). The effect of olfactory occlusion on migrating silver salmon (*O. kisutch*). *J. Fisheries Res. Board Can.* **11**, 472–478.

Wright, R. H. (1964). "The Science of Smell." Basic Books, New York.

# 8

## SPECIAL TECHNIQUES

*D. J. RANDALL and W. S. HOAR*

## I. INTRODUCTION

In this chapter we have attempted to provide certain basic information regularly required by physiologists who have chosen to study the aquatic vertebrates. The coverage is by no means comprehensive. To anticipate the many questions which inevitably arise in the fish physiology laboratory would have taken us far beyond the confines of this volume. We hope, however, that this brief general account with a selected bibliography will prove useful to both the experienced worker and the beginner. Experienced workers will find tabulations of suitable anesthetics and saline solutions as well as a list of references to works on fish experimentation. The beginner will, in addition, find an outline of tried methods for the maintenance of healthy fish and some basic techniques for holding them during surgery and experimentation.

Physiologists who, for one reason or another, decide to work on fish, frequently find the technical problems of maintenance much greater than those of experimentation. Healthy fish have precise requirements of temperature, light, and dissolved solids as well as diet. Many of these requirements are specific for the particular species, and only experience and reference to the pertinent physiological literature can be recommended. Likewise, special surgical techniques such as hypophysectomy and gonadectomy vary in detail with the anatomy and physiology of the species; again it is best to start with the literature of the particular field. There are, however, the universal problems of establishing aquaria, housing and feeding fish, anesthetizing them, holding them for surgical or other procedures, and maintaining their tissues under physiological conditions. We hope that this general account will assist physiologists, particularly the beginners, with some of these more general and universal problems.

## II. MAINTENANCE OF FISH

Holding fish in a laboratory requires an adequate supply of water of the correct chemical composition and temperature; the exact requirements vary with the species being held. Fish should be kept in well-aerated water (Swift, 1963) at temperatures approaching those encountered by the fish in its natural environment. Large or rapid fluctuations in temperature should be avoided. Active fish are best kept in round rather than square aquaria. In round tanks active fish can swim continually and rapidly whereas in square tanks they tend to collide with the walls and injure themselves.

Open circulating systems are preferable to static or recirculating closed systems; an accumulation of excretory products is inevitable in closed systems and these can affect the behavior and growth of the fish (Kawamoto, 1961). In static or recirculating water systems the water should be replaced before fish show signs of stress; goldfish come to the surface in foul water. For small aquaria, recirculation of water through activated charcoal and pads of fiberglass (or glass wool) is recommended with a change of all or part of the water at weekly or more frequent intervals. The water can also be exposed to ultraviolet light to reduce the hazards of infection. Copper, galvanized iron, or asbestos cement pipes should be avoided in water systems; glass or polyvinyl tubing is preferable. Plastic lined pumps are commercially available for recirculating water. There are many article describing various fish holding facilities in the *Progressive Fish Culturist*, the *Canadian Fish Culturist*, the *Transactions of the American Fisheries Society*, and *Turtox News*. Davis (1967), Gordon (1950), Vevers (1967a,b), and Mahoney (1966) are useful sources of information on the propagation, rearing, and maintenance of fish. Spotte (1970) has produced a book on methods of fish culture; the chapters on biological, chemical, and mechanical filtration systems are particularly useful.

## A. Freshwater Systems

Supplies of freshwater are not usually a problem. Domestic tapwater frequently contain chlorine or chlorine and ammonia. Chlorine and ammonia treatment produces a series of compounds which slowly release chlorine and consequently result in prolonged bactericidal action. The compounds are difficult to remove and are extremely toxic to fish (Coventry et al., 1935), particularly at low pH, but can be removed by aeration, charcoal filtration, or by adding sodium thiosulfate to the water (Pyle, 1960). Thiosulfate, if added in large quantities to water of low pH, can result in sulfur dioxide production, which is toxic to fish (Gordon, 1950). Activated charcoal filters require periodic replacement and back-flushing to remove any holes punched in the filter bed. If water is acidic with low dissolved solids (some glacial waters), a filter bed of limestone and oyster shell is also advisable.

Domestic water supplies may exhibit pronounced variations in water pressure, $O_2$, $CO_2$, and chlorine content. In large fish holding facilities these problems are avoided by obtaining a separate water supply from a lake, or well.

Giudice (1966) and McCrimmon and Berst (1966) have described inexpensive, recirculating water systems. The McCrimmon and Berst system is able to support a biomass of 10 kg and is suitable for rearing and maintenance of most marine and freshwater species. Self-contained units for holding freshwater fish, complete with recirculation and filtration systems and temperature control are now commercially available. Höglund (1961) and Hartman (1965) have developed aquaria with simulated stream flow. Shell (1966) has evaluated the advantages of using concrete, plastic, or earthern ponds. Concrete and plastic ponds are usually less expensive to construct and easier to maintain; however, earthern ponds are preferable if some attempt is being made to simulate natural conditions. There are several published articles on the design and construction of ponds varying in capacity, cost, and intended use (Robinson and Vernesoni, 1969; Burrows and Chenoweth, 1970; Dewitt and Salo, 1960).

Handling fish causes scaling and should be avoided whenever possible. Wetting hands and gloves beforehand reduces scaling during handling. Fish can be transported in polyethylene bags (Clark, 1959) or wooden tanks (Macklin, 1959). Mortality can be reduced by cooling and supersaturating the water with oxygen and slightly anesthetizing the fish to reduce activity.

## B. Seawater Systems

Seawater supplies are often either not readily available or too expensive to install. However, a seawater substitute is now commercially available and can be obtained as a mixture of salts to which tapwater is added before use. Lockwood (1961) and Prosser and Brown (1961) give details of the composition of some artificial seawaters. These are useful for many purposes and are often mixed with small quantities of natural seawater. Recirculating seawater systems complete with filtration, aeration, and temperature control are also commercially produced. Recirculating seawater systems have been described by Chin (1959) and Parisot (1967). Large, active marine fish present special problems in the design of holding facilities. Olla *et al.* (1967) have described a large experimental aquarium system for marine pelagic fish and Magnuson (1965) has discussed the facilities developed for tuna behavior studies in Hawaii. Richards *et al.* (1968) have developed an aquarium suitable for shipboard use. Clark and Clark (1964) provide another useful source of information on seawater systems.

## C. Fish Diets

The dietary requirements vary between species of fish and with the physiological state of the animal. Gordon (1950) described methods of culture of many fish foods and listed some artificial fish diets. The nutritional requirements of fish, particularly trout, have been studied extensively (Halver, 1971; Halver and Neuhaus, 1969; Phillips 1956, 1969; Phillips and Brockway, 1956, 1957; Phillips and Balzer, 1957; Phillips and Podoliak, 1957), and diets for salmonid fish are available commercially (Locke and Linscott, 1969). Several artificial diets have been developed (Davis, 1967; Peterson *et al.*, 1967) and have been compared with natural foods (Phillips *et al.*, 1954). Antibiotics and sulfur drugs can be added to the diet to reduce the occurrence of some diseases (Snieszko, 1957; Snieszko and Bullock, 1957). The hazards of prolonged storage of dry foods as well as many problems of fish nutrition are discussed in a volume edited by Halver and Neuhaus (1969).

## D. Fish Parasites and Diseases

Davis (1967), Reichenbach-Klinke and Elkan (1965), van Duijn (1967), Petrushevskii (1957), and Halver and Neuhaus (1969) are sources of information on parasites and diseases in fish. The method of treatment is often unsophisticated; for example, most skin infections are treated by bathing freshwater fish in seawater or a dilute solution of potassium permanganate, malachite green or formalin. More sophisticated approaches include adding antibiotics or sulfonamides to the diet (Snieszko, 1957; Snieszko and Bullock, 1957; Clemens and Kermit, 1958) or injecting antibiotics into the fish.

Bacterial diseases in fish have been reviewed by Post (1965) and Snieszko (1964); Bullock (1961) has developed a schematic outline for the presumptive identification of bacterial diseases in fish.

There is an extensive Russian literature on the parasites of fish which has been translated into English (Markevich, 1951; Petrushevskii, 1957; Dogiel et al., 1958; Bauer, 1959; Bykovskaia-Pavlovskaia et al., 1964). These texts discuss the parasites of both marine and freshwater fish and consider the basis for their control. Hoffman's review (1967) of parasites of North American freshwater fish includes parasitic algae, fungi, and protozoa. The diseases of fish have been reviewed by van Duijn (1967), who presents a great deal of useful, practical information on the treatment of fish diseases. The Food and Agriculture organization of the United Nations has compiled "a provisional list of experts concerned with diseases of aquatic organisms and associated parasites" [FAO Fisheries, Biol. Tech. Paper No. 11 (1969)]. The list is of reduced value because its distribution is limited, and only names and addresses are given without details of the particular field of interest.

## E. Propagation of Fish

Culturing fish eggs requires an adequate supply of clean, aerated water of the correct temperature. Normally water flows over the eggs, and the design of the facilities used to culture the eggs is aimed at maintaining oxygen saturation, maximizing utilization of the water, and presenting an adequate substrate for the eggs. Davis (1967) described methods of propagating many fish including the Pacific salmon, grayling, pike perch, pike, muskellunge, centrarchids, channel catfish, minnows, and suckers. In general the facilities used for spawning and culture of eggs depend on the species and numbers of fish being propagated (J. M.

Shelton, 1955; Burrows and Palmer, 1955; Lindroth, 1956; Costello *et al.*, 1957; Gibor, 1958; Buss, 1959, Webster, 1962; Mason and Fessler, 1966; Vogele and Heard, 1967; Burrows and Combs, 1968; Ellis, 1969).

Sterba (1962) and Breder and Rosen (1966) present a wealth of information on care and breeding as well as a systematic description of fish. Conditions for the care and breeding of tropical fish (Innes, 1966; McInerny and Gerard, 1958) vary with the species. Tropical fish species may require a particular diet, temperature, pH and hardness of the water, or certain social condition (e.g., the presence or absence of conspecifics) before successful breeding occurs. Details for the requirements of many species can be found in the *Tropical Fish Hobbyist*. The looseleaf edition of "Exotic Tropical Fish" compiled from supplements published in the *Tropical Fish Hobbyist* is another useful source of information on tropical fish (Tropical Fish Hobbyist Publications, 245 Cornelison Avenue, Jersey City, N. J. 07302, U. S. A.).

## III. ANESTHESIA

Table I lists a few of the many anesthetics which fish physiologists have found useful. Bell (1967) has provided a more complete tabulation of the properties of these agents as well as several others worthy of consideration. Smith and Bell (1967), Klontz and Smith (1968), and McFarland and Klontz (1969) have also described their properties and, in addition, discussed techniques of application, stages of anesthesia, and hazards of use. These reviews contain comprehensive bibliographies. Several papers in the monograph by Pavlovskii (1962) and numerous articles in *Progressive Fish Culturist* during the past two decades will also be found useful. Since the response to the different anesthetics varies somewhat in different species of fish (also under different physiological conditions), the physiologist should be prepared to test more than one agent when contemplating experiments in an unfamiliar area.

Tricaine methanesulfonate (MS-222) is the most popular fish anesthetic in current use. It is readily available, easily applied, and excellent for operations, although not so highly recommended for transporting fish (Bell, 1967). Sandoz Pharmaceuticals Ltd. supplies a free bulletin of essential data (Bové, 1962). An extensive summary of information concerning MS-222 together with an annotated bibliography has been issued as a bulletin by the U. S. Department of Interior, Bureau of Sports Fisheries and Wildlife (Marking, 1967; Schoettger and Julin, 1967; Walker and Schoettger, 1967a,b; Schoettger, 1967; Schoettger *et al.*,

Table I

A Few Anesthetics for Use on Fish[a]

| Anesthetic | Suggested source | Dosage | Comments |
|---|---|---|---|
| MS-222, tricaine methanesulfonate | Sandoz Products Ltd. Sandoz House 23, Great Castle St., London W.1 | 1:10,000 to 1:45,000 | Currently most frequently used anesthetic—most expensive—reduce dose by 50% for transporting fish; toxic when in seawater exposed to light |
| Ether | Laboratory supply firms | 10–20 ml/liter | Mix well before use, can be removed from solution by aeration |
| t-Amyl alcohol Amylene hydrate | Laboratory supply firms | 5–6 ml/liter | Slow induction period, hyperactivity during recovery |
| Propoxate (R7464) 1-substituted imidazole-5-carboxylic acid ester | Janssen Pharmaceutia n.v., Beerse, Belgium | 1:1,000,000 to 1:100,000 | See Thienpont and Niemegeers (1965) |
| Quinaldine (2 methylquinoline) | Laboratory supply firms | 0.01–0.03 ml dissolved in an equal amount of acetone and added to a liter of water | Increasingly popular |
| Urethane (ethyl carbamate) | Laboratory supply firms | 5–40 mg/liter | Reliable with wide margin of safety; carcinogenic properties (see text) |

[a] The water used to make up the anesthetics should come from the same source as that in which the fish are being held.

1967). Locke (1969) has suggested that quinaldine should replace MS-222 as an anesthetic for fish, but since the fish usually retains some degree of reflex responsiveness (Schoettger and Julin, 1969) it may not be as suitable as a sedative during surgery. This problem can be overcome by using mixtures of MS-222 and quinaldine (Schoettger and Steucke, 1970). Methylpentynol is another useful anesthetic for reducing activity in salmonids (Svendsen, 1969), but much higher dose levels are required to produce an affect equivalent to MS-222 (Howland and Schoettger, 1969).

Not included in Table I or in several of the lists referred to is hypothermia. Cooling of fishes to about 4°C (depending on species and thermal history) produces a deep narcosis from which recovery is rapid on return to acclimation temperature. A gradual cooling (addition of ice to the aquarium water) is recommended since sudden chilling may produce a lethal cold shock. Although hypothermia is one of the oldest fish anesthetics, it is not generally recommended (McFarland and Klontz, 1969). However, lowered temperatures may advantageously be combined with chemical anesthetics.

The immediate response of a fish on immersion in an anesthetic varies with the agent. Excited and erratic swimming may be observed before the gradual loss of equilibrium (McFarland, 1959). With the cessation of swimming and loss of equilibrium, the opercular movements become rapid and shallow and responses to external stimulation (such as tactile stimuli or removal from the water) disappear. During deep surgical anesthesia, opercular movements may be difficult to detect (Klontz and Smith, 1968; McFarland and Klontz, 1969). Respiratory collapse is followed by cardiac collapse. For procedures requiring prolonged anesthesia, a dose level slightly less than that required to produce respiratory collapse is recommended. For brief surgical procedures such as injections or blood samplings, the fish are usually removed from the water; for prolonged operations such as hypophysectomies or abdominal surgery, the anesthetic should be circulated over the gills (flow directed into the mouth and out over the gills). Suitable arrangements for anesthesia during prolonged experimentation are described by Smith and Bell (1967) and Klontz and Smith (1968).

Up to the point of cardiac failure, recovery is usually rapid when the fish is returned to anesthetic-free and well-oxygenated water (5–30 min depending on the anesthetic). If voluntary respiratory movements are not apparent within a minute or less after the return to anesthetic-free water, the fish should be given assistance. The animal may be moved to and fro in the aquarium to force water over the gills or, better still, a current of water may be passed into the buccal cavity (Klontz and Smith, 1968).

Several of the recommended anesthetics have been frequently used in large quantities with inadequate knowledge of their long-term effects on either the users or the fish; for example, urethane (ethyl carbamate) was widely used in tagging and marking salmonids; the experimenter's hands were often exposed to the agent for long periods. It is now known that urethane has both a carcinogenic and a leucopenic action and that it is readily absorbed through the skin (Wood, 1956). A further potential risk is present in the wholesale use of such agents on young fish which will as adults be harvested for human consumption. At present, many of the fish anesthetics are classified as veterinary new drugs by the U. S. Food and Drug Administration and a veterinary new drug application is required prior to their use. Fish biologists unfamiliar with the properties of anesthetics (and hormones such as the steroids) should be warned of potential hazards.

## IV. FISH SALINES

Exposed organs and tissues studied *in vitro* remain normal and healthy only as long as they are bathed in an isotonic solution which contains a proper balance of anions and cations at an appropriate pH. Single salt solutions, even though isotonic, are toxic. This fundamental discovery was made by Sydney Ringer in 1883 and *balanced salt solutions* or *physiological salines* are often referred to as "Ringer solutions" in recognition of his classic studies (1883a,b). Lockwood (1961) has discussed the physiological bases of balanced salt solutions in modern terms. In this paper, he reviewed the properties and functions of various ions and tabulated the composition of suitable solutions for work with most groups of animals; a comprehensive bibliography was also included. In addition to Lockwood's tabulation, compositions of selected fish salines were given by Prosser and Brown (1961), Hale (1965), and Nicol (1967). Balanced salt solutions are also considered by Wolf and Quimby (Chapter 5, Volume III of this treatise).

The composition of a number of fish salines is given in Table II. As a rule, one should completely dissolve each salt in the order listed before adding the next. Bicarbonate buffered solutions evolve $CO_2$ and tend to become alkaline with time. Lockwood (1961) recommended that phosphate buffered solutions be stored in a refrigerator to slow bacterial growth or that the phosphate (and glucose if present) be added to the solution immediately before use. Distilled water should be aerated before preparing the solution.

**Table II**

Some Balanced Salt Solutions for Fishes[a]

| | NaCl | KCl | MgCl$_2$ | CaCl$_2$ | MgSO$_4$·7H$_2$O | Na$_2$SO$_4$ | NaHCO$_3$ | Na$_2$HPO$_4$ | NaH$_2$PO$_4$·2H$_2$O | KH$_2$PO$_4$ | Urea | Glucose |
|---|---|---|---|---|---|---|---|---|---|---|---|---|
| *Myxine* (Fange, 1948) | 22 | 0.4 | 0.4 | 0.5 | | | 0.3 | | | | | |
| *Petromyzon* (Young, 1933) | 5.5 | 0.14 | | 0.12 | | | | + | + | | | |
| Freshwater fish (Burnstock, 1958) | 5.9 | 0.25 | | 0.28 | 0.29 | | 2.1 | | | 1.6 | | |
| Freshwater teleosts (Wolf, 1963) | 7.25 | 0.38 | | 0.162 | 0.23 | | 1.0 | | 0.41 | | | 1.0 |
| Freshwater teleosts (Young, 1933) | 6.5 | 0.14 | | 0.12 | | | | + | + | | | |
| *Salmo clarkii* (Holmes and Stott, 1960) | 7.41 | 0.36 | | 0.17 | 0.15 | | 0.31 | 1.6 | 0.4 | | | |
| *Electrophorus* (Keynes and Martins-Ferreira, 1953) | 9.88 | 0.37 | 0.14 | 0.33 | | | | 0.17 | 0.04 | | | |
| *Electrophorus* (Altamirano and Coates, 1957) | 11.1 | 0.37 | 0.14 | 0.33 | | | | | | | | |
| *Electrophorus* (Schoffeniels, 1960) | 9.36 | 0.37 | 0.14 | 0.67 | | | | | | | | |
| Fish embryo heart (Huggel, 1959) | 6.5 | 0.2 | 0.04 | 0.1 | | | 0.2 | + | + | | | 1.0 |
| Fish heart (Jaeger, 1965) | 6.0 | 0.12 | | 0.14 | | | 0.2 | | 0.01 | | | 2.0 |

| | | | | | | | | | | |
|---|---|---|---|---|---|---|---|---|---|---|
| Cottus muscle (Hudson, 1968) | 8.32 | 0.19 | 0.1 | 0.3 | | 1.55 | | 0.38 | | 10.8 |
| Flounder (Wasserman et al., 1953) | 7.84 | 0.19 | 0.095 | 0.17 | | 3.36 | | 0.078 | | |
| Flounder (Forster and Hong, 1958) | 7.8 | 0.18 | 0.095 | 0.166 | | 0.084 | | 0.06 | | |
| Lophius (Young, 1933) | 12.0 | 0.6 | 0.35 | 0.25 | | 0.19 | | 0.068 | | |
| Uranoscopus (Young, 1933) | 13.5 | 0.6 | 0.35 | 0.25 | | | + | + | | |
| Elasmobranchs (Lutz, 1930) | 16.38 | 0.596 | | 0.555 | | 0.168 | | | 21.6 | |
| Elasmobranchs (Nichols, in Gatenby, 1937) | 16.38 | 0.89 | | 1.11 | | 0.38 | | 0.06 | 21.6 | |
| Selachians (Young, 1933) | 22 | 0.52 | 0.47 | 0.44 | | | + | + | 29 | |
| Skate (Babkin et al., 1933) | 16.38 | 0.89 | 0.38 | 1.11 | | | | 0.07 | 21.6 | |
| Rhinobatus (Salome Pereira and Sawaya, 1957) | 8.37 | 0.95 | 0.094 | 0.4 | | | | | 20.94 | 0.276 |
| Nacine (Salome Pereira and Sawaya, 1957) | 7.83 | 0.52 | 0.105 | 0.67 | | | | | 12.54 | 0.167 |
| Scylluim (Bialaszewicz and Kupfer, 1936) | 15.8 | 0.84 | 0.65 | 0.52 | 0.56 | | | | 23.6 | |
| Torpedo (Feldberg and Fessard, 1942) | 20.0 | 1.0 | 0.83 | 0.83 | | 0.17 | | | 25 | |

[a] Quantities in grams per liter of water; phosphate and glucose should be added to the solution immediately before use; aerate distilled water before preparing solutions; +, buffer with phosphate.

## V. OPERATIVE AND EXPERIMENTAL PROCEDURES

The actual technical procedure adopted during an experiment is usually specific to that study, and the reader is referred to the literature in his field of interest (Pavlovskii, 1962). Operative and experimental holding procedures may have more general application and are briefly discussed below (Klontz and Smith, 1968). Smith and Bell (1967) have described an operating table suitable for procedures involving vascular and urinary cannulations and the removal and implanation of tissues or other substances prior to experimentation. The fish is immobilized partly by anesthesia and partly by the holding facilities of the operating table. Water containing anesthetic is recycled and pumped over the gills. The fish can be operated on for periods of several hours and still recover from the anesthetic. The table described by Smith and Bell (1967) is not suitable for procedures in which the fish must be held rigidly and only lightly anesthetized. The holding clamps designed by G. Shelton (1959) for neurophysiological studies on the tench, *Tinca tinca*, are an example of a more rigid system for immobilizing fish. In this instance the fish is held firmly by an extensive system of rods and clamps surrounding the body and attached to the skull. Other investigators have used a wide variety of techniques to immobilize fish. These include nailing, tying, or screwing the fish to a large board or placing weights on the animal. These procedures are popular in studies on flatfish, skates, and rays. These techniques are difficult to apply to other fish, which have been immobilized by reducing the volume of the experimental chamber with plates or pegs until the fish is imprisoned and cannot move. Other techniques include spinalectomizing the fish just behind the head; wrapping the fish in chicken wire, nylon net, or some other suitable material, or clamping the fish in a pair of mole grips.

Fry (1957) has discussed some of the precautions to be observed when measuring oxygen consumption in fish. These precautions can be extended to many other experimental situations and are generally aimed at reducing the interaction between the fish, the investigator and the experimental surroundings, and studying animals in a normal physiological state. Technical difficulties often force a compromise between this ideal state and what is practical. Houston *et al.* (1969) were able to detect the effects of a minor operation several days after the fish had recovered from the anesthetic. It is important to assess the influence of the experimental procedure on the fish and ensure that the effect on the system under study is minimized.

REFERENCES

Altamirano, M., and Coates, C. W. (1957). Effect of potassium on *Electrophorus electricus*. *J. Cellular Comp. Physiol.* 49, 69–101.

Babkin, B. P., Bowie, D. J., and Nicholls, J. V. V. (1933). Structure and reactions of stimuli of arteries (and conus) in the elasmobranch genus *Raja*. *Contrib. Can. Biol. Fisheries* 8, No. 16 (Ser. B. Exp. No. 18), 209–225.

Bauer, O. N. (1959). The ecology of parasites of freshwater fish. *Bull. State Sci. Res. Inst. Lake River Fisheries* 49, 3–188 (translated by L. Kochva, Sivian Press, Jerusalem).

Bell, G. R. (1967). A guide to the properties, characteristics and uses of some general anaesthetics for fish. *Bull., Fisheries Res. Board Can.* 148, 1–4.

Bialaszewicz, K., and Kupfer, C. (1936). De la composition minérale des muscles des animaux marins. *Arch. Intern. Physiol.* 42, 398–404.

Bové, F. J. (1962). MS-222 Sandoz—the anaesthetic of choice for fish and other cold-blooded organisms. *Sandox News* 3, 12 p.

Breder, C. M., Jr., and Rosen, D. E. (1966). "Modes of Reproduction in Fishes." Nat. Hist. Press, Garden City, New York.

Bullock, G. L. (1961). A schematic outline for the presumptive identification of bacterial diseases of fish. *Progressive Fish Culturist* 23, 147–151.

Burnstock, G. (1958). Reversible inactivation of nervous activity in a fish gut. *J. Physiol. (London)* 141, 35–45.

Burrows, R. E., and Chenoweth, H. H. (1970). The rectangular circulating rearing pond. *Progressive Fish Culturist* 32, 67–80.

Burrows, R. E., and Combs, B. D. (1968). Controlled environments for salmon propagation. *Progressive Fish Culturist* 30, 123–136.

Burrows, R. E., and Palmer, D. D. (1955). A vertical egg and fry incubator. *Progessive Fish Culturist* 17, 147–156.

Buss, K. (1959). Jar culture of trout eggs. *Progressive Fish Culturist* 21, 26–29.

Bykovskaĭa-Pavlovskaĭa, I. E. (1964). "Key to the Parasites of Freshwater Fish of the U. S. S. R." Akad. Nauk. S. S. R. Zoologicheskii Institut. (translated from Russian by the Israel Program for Scientific Translations for the U. S. Department of the Interior and the National Science Foundation, Washington, D. C.).

Chin, E. (1959). An inexpensive recirculating seawater system. *Progressive Fish Culturist* 21, 21–93.

Clark, C. F. (1959). Experiments in the transportation of live fish in polyethylene bags. *Progressive Fish Culturist* 21, 177–182.

Clark, J. R., and Clark, R. L. (1964). "Seawater Systems for Experimental Aquariums." U. S. Department of Interior, Fish and Wildlife Service, Bureau of Sports Fisheries and Wildlife, Res. Rep. 63.

Clemens, H. P., and Kermit, E. S. (1958). The chemical control of some diseases and parasites of channel catfish. *Progressive Fish Culturist* 20, 8–15.

Costello, D. P., Davidson, M. E., Eggars, A., Fox, M. H., and Henley, C. (1957). "Methods for Obtaining and Handling Marine Eggs and Embryos." Marine Biol. Lab., Woods Hole, Massachusetts.

Coventry, F. L., Shelford, V. W., and Miller, L. F. (1935). The conditioning of a chloramine treated water supply for biological purposes. *Ecology* 16, 60–66.

Davis, H. S. (1967). "Culture and Diseases of Game Fishes." Univ. of California Press, Berkeley, California.

Dewitt, J. W., and Salo, E. O. (1960). The Humboldt State College Fish Hatchery:

An experiment with the complete recirculation of water. *Progressive Fish Culturist* **22**, 3–6.

Dogiel, V. A., Petrushevskii, G. K., and Polyanski, Yu. I. (1958). "Parasitology of Fishes." Leningrad Univ. Press, Leningrad (translated by Z. Kabata, Oliver & Boyd, Edinburgh and London).

van Duijn, C. (1967). "Diseases of Fish." Dorset House, London.

Ellis, J. N. (1969). Hydrodynamic hatching baskets. *Progressive Fish Culturist* **31**, 114–117.

Fange, R. (1948). Effect of drugs on the intestine of a vertebrate without sympathetic nervous system. *Arkiv Zool.* **40A**, No. 11, 1–9.

Feldberg, W., and Fessard, A. (1942). The cholinergic nature of the nerves to the electric organ of the torpedo (*Torpedo marmorata*). *J. Physiol. (London)* **101**, 200–216.

Forster, R. P., and Hong, S. K. (1958). *In vitro* transport of dyes by isolated renal tubules of the flounder as disclosed by direct visualisation. Intracellular accumulation and transcellular movement. *J. Cellular Comp. Physiol.* **51**, 259–272.

Fry, F. E. J. (1957). The aquatic respiration of fish. *In* "The Physiology of Fishes" (M. E. Brown, ed.), Vol. 1, pp. 1–63. Academic Press, New York.

Gatenby, J. B. (1937). "Biological Laboratory Technique; An Introduction to Research in Embryology, Cytology and Histology." Churchill, London.

Gibor, A. (1958). A simple technique used for laboratory hatching and rearing of fish. *Progressive Fish Culturist* **20**, 180–182.

Giudice, J. J. (1966). An inexpensive recirculating water system. *Progressive Fish Culturist* **28**, 28.

Gordon, M. (1950). Fishes as laboratory asimals. *In* "The Care and Breeding of Laboratory Animals" (E. J. Farris, ed.), pp. 345–559. Wiley, New York.

Hale, L. J. (1965). "Biological Laboratory Data," 2nd ed. Methuen, London.

Halver, J. E., ed. (1971). "Fish Nutrition." Academic Press, New York (in press).

Halver, J. E., and Neuhaus, O. W., eds. (1969). "Fish in Research." Academic Press, New York.

Hartman, G. F. (1965). An aquarium with simulated stream flow. *Trans. Am. Fisheries Soc.* **94**, 274–276.

Hoffman, G. L. (1967). "Parasites of North American Freshwater Fishes." Univ. of California Press, Berkeley, California.

Höglund, L. B. (1961). The reactions of fish in concentration gradients. *Rept. Inst. Freshwater Res. Drottningholm* **43**, 1–147.

Holmes, W. N., and Stott, G. H. (1960). Studies of the respiration rates of excretory tissues in the Eulthroat trout, *Salmo clarki clarki* I. Variations with body weight. *Physiol. Zool.* **33**, 9–14.

Houston, A. H., DeWilde, A. M., and Madden, J. A. (1969). Some physiological consequences of aortic catheterization in the brook trout. *J. Fisheries Res. Board Can.* **26**, 1847–1856.

Howland, R. M., and Schoettger, R. A. (1969). Efficacy of methylpentynol as an anesthetic on four salmonids. *Invest. Fish Control* No. 29, pp. 1–15. U. S. Department of the Interior, Bureau of Sport Fisheries and Wildlife.

Hudson, R. C. L. (1968). A Ringer solution for *Cottus* (teleost) fast muscle fibres. *Comp. Biochem. Physiol.* **25**, 719–725.

Huggel, H. (1959). Experimentelle Untersuchungen uber die Antomatie, ie Temperaturabhangigkeit und Arbeit des Embryonalen Fischherzen, unter besonderer

Berucksichtigung der Salmoniden und Scylliorhiniden. *Z. Vergleich. Physiol.* **42**, 63–102.

Innes, W. T. (1966). "Exotic Aquarium Fishes." Aquariums Incorporated, USA. 592p.

Jaeger, R. (1965). Aktionspotentiale des Myokardfasern des Fischherzens. *Naturwissenschaften* **52**, 482–483.

Kawamoto, N. Y. (1961). The influence of excretory substances of fishes on their own growth. *Progressive Fish Culturist* **23**, 70–75.

Keynes, R. D., and Martins-Ferreira, H. (1953). Membrane potentials in the electroplates of the electric eel. *J. Physiol. (London)* **119**, 315–351.

Klontz, G. W., and Smith, L. S. (1968). Methods of using fish as biological research subjects. *In* "Methods of Animal Experimentation" (W. I. Gay, ed.), Vol. 3, pp. 323–385. Academic Press, New York.

Lindroth, A. (1956). Salmon stripper, egg counter and incubator. *Progressive Fish Culturist* **18**, 165–170.

Locke, D. O. (1969). Quinaldine as an anesthetic for Brook Trout, Lake Trout and Atlantic Salmon. *Invest. Fish Control*, No. 24, 5p. U. S. Department of the Interior, Bureau of Sport Fisheries and Wildlife.

Locke, D. O., and Linscott, S. P. (1969). A new dry diet for landlocked Atlantic salmon and lake trout. *Progrs. Fish Culturist* **31**, 3–10.

Lockwood, A. P. M. (1961). "Ringer" solutions and some notes on the physiological basis of their ionic composition. *Comp. Biochem. Physiol.* **2**, 241–289.

Lutz, B. R. (1930). The effect of adrenalin on the auricle of elasmobranch fishes. *Am. J. Physiol.* **94**, 135–139.

McCrimmon, H. R., and Berst, A. H. (1966). A water recirculation unit for use in Fishery Laboratories. *Progressive Fish Culturist* **28**, 165–170.

McFarland, W. N. (1959). A study of the effects of anesthetics on the behaviour and physiology of fishes. *Publ. Inst. Marine Sci., Univ. Texas* **6**, 25–55.

McFarland, W. N., and Klontz, G. W. (1969). Anesthesia in fishes. *Federation Proc.* **28**, 1535–1540.

McInerny, D., and Gerard, G. (1958). "All About Tropical Fish." Harrap, London.

Macklin, R. (1959). An improved 150 gallon fish-planting tank. *Progressive Fish Culturist* **21**, 81–85.

Magnuson, J. J. (1965). Tank facilities for tuna behaviour studies. *Progressive Fish Culturist* **27**, 230–233.

Mahoney, R. (1966). "Laboratory techniques in Zoology." Butterworth, London and Washington, D. C.

Markevich, A. P. (1951). "Parasitic Fauna of Freshwater Fish of the Ukrainian S. S. R." Acad. Sci. Ukrainian, S. S. R. (translated by N. Rafael, Oldbourne, Press, London, 1963).

Marking, L. L. (1967). Toxicity of MS-222 to selected fishes. *Invest. Fish Control*, No. 12, 1–10. U. S. Department of the Interior, Bureau of Sport Fisheries and Wildlife.

Mason, J. C., and Fessler, J. L. (1966). A simple apparatus for the incubation of salmonid embryos at controlled levels of temperature, water flow and dissolved oxygen. *Progressive Fish Culturist* **28**, 171–174.

Nicol, J. A. C. (1967). "The Biology of Marine Animals," 2nd ed. Pitman, New York.

Olla, B. L., Marchioni, W. W., and Katz, H. M. (1967). A large experimental aquarium system for marine pelagic fishes. *Trans. Am. Fisheries Soc.* **96**, 143–150.

Parisot, T. J. (1967). A closed recirculated seawater system. *Progressive Fish Culturist* **29**, 133–139.

Pavlovskii, E. N. (1962). "Techniques for the Investigation of Fish Physiology." Acad. Sci. U. S. S. R. (translated from the Russian-Israel Program for Scientific Translations, Jerusalem, 1964. Available from the Office of Technical Services, U. S. Department of Commerce, Washington, D. C.).

Peterson, E. J., Robinson, R. C., and Willoughby, H. (1967). A mealgelatin diet for aquarium fish. *Progressive Fish Culturist* **29**, 170–171.

Petrushevskii, G. K. (1957). Parasites and diseases of fish. *Bull. All-Union Sci. Res. Ins. Freshwater Fisheries* **42**, 1–338 (translated by J. I. Lengy, I. Paperna, and others, S. Monson Publ., Jerusalem).

Phillips, A. M. (1956). The nutrition of trout. I. General feeding methods. *Progressive Fish Culturist* **18**, 113–119.

Phillips, A. M. (1969). Nutrition, digestion, and energy utilization. In "Fish Physiology" (W. S. Hoar and D. J. Randall, eds.), Vol. I, pp. 391–431. Academic Press, New York.

Phillips, A. M., and Balzer, G. C. (1957). The nutrition of trout. V. Ingredients for trout diets. *Progressive Fish Culturist* **19**, 158–167.

Phillips, A. M., and Brockway, D. R. (1956). The nutrition of trout. II. Protein and carbohydrate. *Progressive Fish Culturist* **18**, 159–164.

Phillips, A. M., and Brockway, D. R. (1957). The nutrition of trout. IV. Vitamin requirements. *Progressive Fish Culturist* **19**, 119–123.

Phillips, A. M., and Podoliak, H. A. (1957). The nutrition of trout. III. Fats and minerals. *Progressive Fish Culturist* **19**, 68–75.

Phillips, A. M., Nielsen, R. S., and Brockway, D. R. (1954). A comparison of hatchery diets and natural food. *Progressive Fish Culturist* **16**, 153–157.

Post, G. (1965). A review of advances in the study of diseases of fish: 1954–1964. Progressive Fish Culturist **27**, 3–12.

Prosser, C. L., and Brown, F. A. (1961). "Comparative Animal Physiology," 2nd ed. Saunders, Philadelphia, Pennsylvania.

Pyle, E. A. (1960). Neutralizing chlorine in city water for use in fish distribution tanks. *Progressive Fish Culturist* **22**, 30–33.

Reichenbach-Klinke, H., and Elkan, E. (1965). "The Principle Diseases of Lower Vertebrates." Academic Press, New York.

Richards, W. J., Palko, B. J., and Scott, E. L. (1968). A research aquarium suitable for shipboard use. *Trans. Am. Fisheries Soc.* **97**, 286–287.

Ringer, S. (1883a). A further contribution regarding the influence of the different constituents of the blood on the contraction of the heart. *J. Physiol. (London)* **4**, 29–42.

Ringer, S. (1883b). A third contribution regarding the influence of the inorganic constituents of the blood on the ventricular contraction. *J. Physiol. (London)* **4**, 222–225.                                        •

Robinson, W. R., and Vernesoni, P. (1969). Low-cost circular concrete ponds. *Progressive Fish Culturist* **31**, 180–182.

Salome Pereira, R., and Sawaya, P. (1957). Ions in the blood of elasmobranchs. *Biol. Fac. Fil. Cien. Univ. Sao Paulo Zool.* **21**, 85–92.

Schoettger, R. A. (1967). Annotated bibliography on MS-222. *Invest. Fish Control*, No. 16, 1–15. U. S. Department of the Interior, Bureau of Sport Fisheries and Wildlife.

Schoettger, R. A., and Julin, A. M. (1967). Efficacy of MS-222 as an anesthetic on four salmonids, *Invest. Fish Control*, No. 13, 1–15. U. S. Department of the Interior, Bureau of Sport Fisheries and Wildlife.

Schoettger, R. A., and Julin, A. M. (1969). Efficacy of Quinaldine as an anesthetic for seven species of Fish. *Invest. Fish Control*, No. 22, 15p. U. S. Department of the Interior, Bureau of Sport Fisheries and Wildlife.

Schoettger, R. A., and Steucke, E. W. (1970). Synergic mixtures of MS-222 and quinaldine as anesthetics for rainbow trout and northern pike. *Progressive Fish Culturist* **32**, 202–205.

Schoettger, R. A., Walker, C. R., Marking, L. L., and Julin, A. M. (1967). MS-222 as an anesthetic for channel catfish: Its toxicity, efficacy, and muscle residues. *Fish Control*, No. 17, 1–14. U. S. Department of the Interior, Bureau of Sport Fisheries and Wildlife.

Schoffeniels, E. (1960). Less bases physiques et chemiques des potentiels bio-électriques chez *Electrophorus electricus* L. *Arch. Intern. Physiol. Biochim.* **68**, 1–151.

Shell, E. W. (1966). Comparative evaluation of plastic and concrete pools and earthen ponds in fish-cultural research. *Progressive Fish Culturist* **28**, 201–205.

Shelton, G. (1959). The respiratory centre in the tench. I. The effects of brain transection on respiration. *J. Exptl. Biol.* **36**, 191–202.

Shelton, J. M. (1955). The hatching of chinook salmon eggs under simulated stream conditions. *Progressive Fish Culturist* **17**, 20–35.

Smith, L. S., and Bell, G. R. (1967). Anesthetic and surgical techniques for Pacific salmon. *J. Fisheries Res. Board Can.* **24**, 1579–1588.

Snieszko, S. F. (1957). Use of antibiotics in the diet of salmonid fishes. *Progressive Fish Culturist* **19**, 81–84.

Snieszko, S. F. (1964). Selected topics on bacterial fish diseases. *Can. Fish Culturist* **32**, 19–24.

Snieszko, S. F., and Bullock, G. L. (1957). Determination of the susceptibility of *Aeromonas salmonicida* to sulfonamides and antibiotics, with a summary report on the treatment and prevention of Fununculosis. *Progressive Fish Culturist* **19**, 99–107.

Spotte, S. H. (1970). "Fish and invertebrate culture. Water management in closed systems." Wiley (Interscience), New York.

Sterba, G. (1962). "Freshwater Fishes of the World." Studio Vista, London.

Svendsen, G. E. (1969). Annotated bibliography on Methylpentynol. *Invest. Fish Control*, No. 31, 1–7. U. S. Department of the Interior, Bureau of Sport Fisheries and Wildlife.

Swift, D. R. (1963). Influence of oxygen concentration on growth of Brown Trout *Salmo trutta*. *Trans. Am. Fisheries Soc.* **92**, 300–301.

Thienpont, D., and Niemegeers, C. J. E. (1965). Porpoxate (R7464): A new potent anaesthetic agent in Cold-blooded vertebrates. *Nature* **205**, 1018–1019.

Vevers, H. G. (1967a). Freshwater fish. *In* "The UFAW Handbook on the Care and Management of Laboratory Animals," 3rd ed., pp. 893–897. Livingstone, Edinburgh and London.

Vevers, H. G. (1967b). Marine aquaria. *In* "The UFAW Handbook on the Care and Management of Laboratory Animals" 3rd ed., pp. 898–905. Livingstone, Edinburgh and London.

Vogele, L. E., and Heard, W. R. (1967). An experimental hatching and rearing facility for larval reservoir fishes. *Progressive Fish Culturist* **29**, 177–179.

Walker, C. R., and Schoettger, R. A. (1967a). Methods of determining MS-222 residues in fish. *Invest. Fish Control*, No. 14, 1–10. U. S. Department of the Interior, Bureau of Sport Fisheries and Wildlife.

Walker, C. R., and Schoettger, R. A. (1967b). Residues of MS-222 in four salmonids following anesthesia. *Invest. Fish Control*, No. 15, 1–11. U. S. Department of the Interior, Bureau of Sport Fisheries and Wildlife.

Wasserman, K., Becker, E. L., and Fishman, A. P. (1953). Transport of phenol red in the flounder renal tubule. *J. Cellular Comp. Physiol.* **42**, 385–393.

Webster, D. A. (1962). Artificial spawning facilities for brook trout, *Salvelinus fontinalis*. *Trans. Am. Fisheries Soc.* **91**, 168–174.

Wolf, K. (1963). Physiological salines for freshwater teleosts. *Progressive Fish Culturist* **25**, 135–140.

Wood, E. M. (1956). Urethane as a carcinogen. *Progressive Fish Culturist* **18**, 135–136.

Young, J. Z. (1933). The preparation of isotonic solutions for use in experiments with fish. *Publ. Staz. Zool. Napoli* **12**, 425–431.

# AUTHOR INDEX

Numbers in italics refer to the pages on which the complete references are listed.

## A

Abu Gideiri, Y. B., 326, *355*
Adams, C. K., 213, *275*
Adler, N., 212, 264, *269*
Adrian, E. D., 298, *355*, 436, *506*
Agranoff, B. W., 243, 244, 245, 246, 262, *269, 270, 271*
Ahsan, S. N., 402, *416*
Alabaster, J. S., 81, *87*
Albrecht, H., 286, *355*
Alderdice, D. F., 27, 47, 48, 87, 89, 94, 169, *188*
Alexander, A. E., 184, *187*
Allanson, B. R., 29, *87*
Allard, H. A., 378, 397, *420*
Allen, G. H., 195, *271*, 452, *507*
Allen, K. O., 21, *88*
Allison, L. N., 401, *416*
Altamirano, M., 520, *523*
Amdur, B. H., *151*, 162, 165, 166, 169, 171, 172, 173, *188*
Ames, L., 216, *269*
Anderson, J. M., 10, 12, 50, 80, 82, 84, 85, *93, 95*
Andreasson, S., 383, *416*
Andrewartha, H. G., 42, *88*
Andrews, C. W., 160, *187*
Andriashev, A. P., 166, *187*
Apfelbach, R., 286, *355*
Arai, M. N., 23, 24, *88*
Arendsen de Wolf-Exalto, E., 286, *355*
Arey, L. B., 391, *416*
Arnold, H., 147, *148*
Aronson, L. R., 203, 204, 238, 256, *269*, 286, 354, *355, 358, 362*, 408, 411, *416*
Arora, H. L., 268, *269*
Arrhenius, S., *88*
Aschoff, J., 374, 375, 376, 378, 380, 387, 388, 389, 409, 415, *417*

## B

Assaf, S. A., 136, 137, 138, 148, *149*
Atkinson, D. E., 100, 116, 118, 120, *149*
Atz, J. W., 397, 407, *417*

## B

Babkin, B. P., 521, *523*
Baenninger, R., 323, *355*
Baerends, G. P., 264, 268, *269*, 284, 285, 286, 289, 290, 292, 293, 294, 295, 307, 314, 316, 317, 319, 320, 321, 323, 324, 328, 333, 334, 335, 339, 341, 345, 348, 350, 351, 352, 354, *355, 356, 360, 364*
Baerends-van Roon, J. M., 268, *269*, 286, 294, 307, 314, 316, 328, 335, 351, *356*
Baggerman, B., 267, *269*, 354, ?55, *356*, 398, 400, 403, *417*
Baldwin, E., 145, *149*
Baldwin, J., 101, 103, 106, 126, 127, 135, 137, *149*
Baldwin, S., 102, 114, *149*
Balinsky, J. B., 145, *149*
Ball, E. Q., 143, *149*
Ballintijn, C. M., 288, *356, 362*
Ballintijn-de Vries, G., 285, 354, *364*
Balls, R., 381, *417*
Balzer, G. C., 514, *526*
Bardoch, J. E., 37, 78, *88*, 203, 204, *270*, 372, 383, *419, 423*
Barlow, G. W., 7, *88*, 285, 286, 327, 350, 352, *356, 357, 365, 369*, 381, *417*
Barnes, C. A., 454, *506, 507*
Barnum, C. P., 374, *420*
Barr, L. M., 54, 55, 58, *96*
Barrett, I., 75, *88*
Barrington, E. J. W., 400, *417*
Basford, R. E., 147, *154*
Baslow, M. H., 75, *88*, 126, *149*
Bastock, M., 294, 336, 343, *357*

**539**

# SYSTEMATIC INDEX

*Note:* Names listed are those used by the authors of the various chapters. No attempt has been made to provide the current nomenclature where taxonomic changes have occurred.

## A

*Aequidens*
  *A. latifrons*, 316, 326
  *A. maroni*, 349
  *A. portalegrensis*, 285, 316, 493
African mouthbreeders 210, 214, 222, 229, 230, 232, 239
*Ameiurus*, 391
  *A. nebulosus*, 25, 32–33, 61–62, 400
*Amphiprion*, 283
Anabantid, 284, 286, 340, 352
*Anableps microlepis*, 412–413
Antennariids, 349
*Antennarius*, 349
*Apeltes*, 399
*Apistogramma*
  *A. borelli*, 316, 328
  *A. reitzigi*, 316, 328
*Arapaima gigas*, 408
*Aspidontus tueniatus*, 205, 282, *see also* Blenny
*Aspinodontus*, 349
*Astatotilapia strigigena*, 318, 326, 328
*Astronotus ocellatus*, 268, 350. 408
*Astyanax*, 407
  *A. mexicanus*, 400
*Atherina laticeps*, 206

## B

*Badis badis*, 286
*Balistes*, 282
*Barbus stoliczkanus*, 349
Barracuda, 205
Bass *see also Micropterus, Morone*
  large mouth, 56, 61, 382–383
  rock, 381
  small mouth, *see Micropterus dolomieui*

white, 382, 452, 466–470, 488, 497, 499, 505, *see also Morone chrysops Raf.*
*Bathygobius*, 310
*B. soporator*, 203, 411
*Bathystoma rimator*, 381
*Betta splendens*, 241, 312, 318, 323, 326, 340, *see also* Siamese fighting fish
Bitterling *see also Rhodeus*, 307, 309–310, 336, 344
*Blennius*
  *B. fluvialtilis*, 286 *see also* Blenny
  *B. pholis* 411
Blenny, 411, *see also Aspidontus, Blennius*
Bluegill, 383, 391, *see also Lepomis*
*Boreogadus saida*, 160
Brachydanio, 327
Bullhead, *see also Ameiurus*
  black, 382
  brown, *see A. nebulosus*
Burbot, 383

## C

*Carassius*
  *C. auratus*, 4–5, 61, 254, *see also* Goldfish
  *C. carassius*, 158, *see also* Carp
*Carcinus maecnas*, 410, 414–415
Carp, 24, 61, 126, 141, 206, 382, 402
  Crucian, *see carassius*
  European, 141
Catfish, *see also Ictalurus punctatus*
  channel, 515
Centrachids, 284, 286, 393, 403, 407, 409, 488, 490, 515, *see also Enneacanthus*
Chaenicthyidae, 169

# SUBJECT INDEX

## A

Acclimation, 14–15, 102, *see also* Temperature, acclimation
Acclimatization, 14–15, 102
Acetoxy cycloheximide, 244, 246
Acetylcholine, 106
Acetylcholinesterase, 103, 106, 126, 129–137
Acetyl CoA, 120
N-Acetyl galactosamine, 176, 181
Acquisition, 221–223, 226–227, 241, 244, 252, 254–256, *see also* Learning
Actinomycin, 244, 246
Activation, 295, 303–304, 306–307, 311, 321–322, 334–335, 345–347
  energy of, 135–139
  entropy, 136–137
Adaptation, 206, 267
  cold, 157
  evolutionary, 102, 135–140, *see also* Enzymes
Adaption, 314
Adenosine diphosphate (ADP), 118, 120–121
Adenosine monophosphate (AMP), 103–104, 115–118
Adenosine triphosphate (ATP), 104, 116, 118, 120, 123, 147
Adenylate, 116, 120, 145
Aggression
  activities, 323, 336
  behavior, 292, 294, 309, 312, 322–324, 326, 336–337, 340–341, 344, 354–355, *see also* Behavior, aggressive
  motivation, 310
  responses, 285, 322–324
  system, 351
  tendency, 310–311, 321, 337, 346–347
Aggressive behavior, *see* Behavior, aggressive
Agnostic behavior, *see* Behavior, agnostic
Alanine, 176

Aldolase, 104
Alevin, 431
Amines, 174
Amino acids, 133, 144, 173–175
Amino oxidases, 136
Ammonia, 38, 145, 513
Amnesia, 243–246, 248
  amnesic agents, 246
Anabolism, 2
Anaerobiosis, 141, 159
Analyzers, 220–222, 233–234, 259
Anchor ice, 167–168
Anesthesia, 255, 516–519, 522
  table, 517
  in transport of fish, 514, 516
Appetitive behavior, *see* Behavior, appetitive
Areola, 295
Arginine, 147
Aromatic substances, 436, 444
Arrhenius plots, 106, 118–119, 135–136
Aschoff's rule, 375, 387
Aspartate transcarbamylase, 147
Assimilation, 6
Avoidance, *see also* Freezing
  learning, 205, 206, 216, 243, 247, 250, 256–260, 262–263, 283

## B

Behavior, 76
  aggressive, 292, 294, 309, 312, 322–324, 326, 336–337, 340–341, 344, 354–355, *see also* Aggression
  attacking in, 214, 282, 306, 310, 312, 314, 317–318, 321, 336, 342
  biting, 341–342, 346
  fighting in, 284–285, 307, 310–312, 324, 328, 337, 340, 349
  fleeing, *see* Fleeing
  head batting, 288
  threat, 282, 310

Forgetting, 223–224, 245, 251–255, *see also* Memory
Freezing, 157–187
Fructose diphosphatase (FDPase), 103, 105, 115–118, 137, 145
Fructose diphosphate, 118–120, 145
Fructose diphosphate aldolases, 104, 126, 137
Fructose-6-phosphate, 104
Fry, 194–196, 452, 501, 503–504
  definition, 431–432
Fuscin, 391

### G

Galactose, 175–185
Gametogenesis, 134, 401
Genital papilla, 350
Geotaxis, 322
Gills, 37–38, 65, 73, 518, 522
  cavity, 301
  covers, 264, 307, 340
Gluconeogenesis, 114–115, 118, 144–145
Glucose, 173, 175
Glucose-6-phosphate dehydrogenase, 108, 123, 147
Glutamate dehydrogenase, 145
Glutamine, 145
Glutamine synthetase, 145
Glyceraldehyde-3-phosphate dehydrogenase, 137, 148
Glycerol phosphate dehydrogenase, 104
Glycogen, 122–123, 130
Glycogen phosphorylase, 123, 137
Glycogen synthetase, 123
Glycolysis, 115, 118, 121, 123, 130
Glycoproteins, 175–185
Gonadal cycle, 285
Gonadal development, 396–397, 399
Gonadal maturation, 400–405, 408–409, 412
Gonadal morphological changes, 398
Gonadal phase, 406
Gonadal regression, 408
Gonadal tissue restitution, 401
Gonadal weight increase, 400
Grafting, 225
Growth, 134, 157
  processes, 375
Gymnotid behavior, 384–385

### H

Habit reversal, 217–226, *see also* Learning
Habituation, 212, 324
Hemoglobin, 140
Heterogeneous summation rule, 318, 320–321
Hexokinase, 147
Homeostasis, 82–83, 102
  homeostatic drives, 264
Homing, 197, 200, 202–203, *see also* Orientation, Migration
  salmon, 430–432
Homiotherms, 319, 377
Hormones
  FSH, 354
  gonadal, 354
  gonadotropic, 354
  hormonal control, 354–355
  LH, 354
  pituitary, 407
  prolactin, 354–355
Hunger, 286, 339, 345, 353, 440
Hybrid, 197, 325
Hyperactivity, 203–204
Hypoaggressive, 324
Hypothalamus, 257, 261, 265
Hypothermia, 518

### I

Imprinting, 196–200, 208, 328–329, 444, 451, 454
  artificial, 451
  natural, 452–453
Incompatibility, 342, 345
Incubation, 295, 319
Information
  feedback, 295, 336, 345
  processing, 311–320, 332
Inhibition
  in enzyme regulation, 115–121
  feedback, 120–121
  metabolic, 248
  mutual, 330, 344, 346
  proactive (PI), 223–224, 251
  retroactive, 251
Instinct, 300, 334–335, 429
Interocular transfer, 257–259